Ants of Africa and Madagascar

Ants of Africa and Madagascar

A Guide to the Genera

BRIAN L. FISHER and BARRY BOLTON

Illustrated by Jessica Huppi

UNIVERSITY OF CALIFORNIA PRESS

University of California Press, one of the most distinguished university presses in the United States, enriches lives around the world by advancing scholarship in the humanities, social sciences, and natural sciences. Its activities are supported by the UC Press Foundation and by philanthropic contributions from individuals and institutions. For more information, visit www.ucpress.edu.

University of California Press
Oakland, California

Cataloging-in-Publication Data on file at the Library of Congress.

ISBN 978-0-520-27866-0 (cloth)
ISBN 978-0-520-29089-1 (paper)
ISBN 978-0-520-96299-6 (ebook)

Manufactured in the United States of America

22 21 20 19 18 17 16
10 9 8 7 6 5 4 3 2 1

We dedicate this book to those who have chosen to explore the rich ant fauna of Africa and Madagascar, such as Roy Snelling, who spent his last few years working on the Kenya fauna. Although the region offers unique challenges for ant researchers, it presents the joys of exploration and discovery at the same time.

CONTENTS

Color plates follow page 58.

ACKNOWLEDGMENTS

One day ants will be on equal footing with birds in terms of their appreciation by the public and our understanding of their biology and their role in conservation. We hope our illustrated generic key will stimulate interest in African and Malagasy ants and open the door for ant studies across the region.

This key arrives at a particularly important time in our understanding of the regional fauna. Recent efforts to organize ant classification to reflect phylogenetic history have shuffled ant classification. For many, this key will be their first introduction to these changes. At the same time, large-scale inventories are providing improved distribution ranges for species and genera, and morphological studies have provided a greater understanding of diagnostic characters for genera. In Bolton's 1994 generic key, for example, 89 and 46 genera were recognized for Africa and Madagascar, while in this guide we recognize 104 and 72, respectively. As one gauge of the taxonomic effort in the region, since 1994, over 650 species have been described from the African and Malagasy regions.

We could not have accomplished this work without the efforts of many, including a great number of collectors and institutions like the Natural History Museum, London and the California Academy of Sciences (CAS), which continue to support and care for their growing ant collections. Our special gratitude goes to Peter Hawkes for making his collections from southern Africa available for study and for his thoughtful comments on earlier drafts of the keys. In addition, our thanks to the authors of recent publications who kept us up-to-date with their findings and novelties, including Flavia Esteves, Georg Fischer, Francisco Hita Garcia, John S. LaPolla, Jean Claude Rakotonirina, and Masashi Yoshimura, with special thanks to Marek Borowiec for permission to include findings from his ongoing research on Dorylinae.

We are very grateful to a number of people who, during the construction of these keys and definitions, took the time to help us by checking and improving various aspects of their structure, including Gary D. Alpert and Philip S. Ward. We would like to thank the instructors and students of previous African and Malagasy ant courses for their insightful criticisms of earlier versions of these keys. In addition, we thank Jessica Huppi for working over three years to complete the line illustrations, with additional work by Ginny Kirsch. Michele Esposito provided help in databasing, specimen preparation, and creating the color plates. We also thank the team of AntWeb imagers for their effort in imaging African species: Cerise Chen, Will Ericson, Michele Esposito, Shannon Hartman, Zach Lieberman, April Nobile, Estella Ortega, Ryan Perry, Erin Prado, Jean Claude Rakotonirina, and Alexandra Westrich. The virtual collection of African ants available on AntWeb helped in all stages in preparing this book.

B. Fisher would like to thank those who helped explore Africa and Madagascar, including Marius Burger, Flavia Esteves, Steve Goodman, and Simon Van Noort as well as those at the Madagascar Biodiversity Center: Balsama Rajemison, Jean-Claude Rakotonirina, Jean-Jacque Rafanomezantsoa, Chrislain Ranaivo, Hanitriniana Rasoazanamavo, Nicole Rasoamanana, Clavier Randrianandrasana, Dimby Raharinjanahary, Njaka Ravelomanana, and Manoa Ramamonjisoa. We would also like to thank the team at the University of California Press, including Merrik Bush-Pirkle, Kate Hoffman, and Claudia Smelser.

INTRODUCTION TO THE ANT GENERA

It has been more than 20 years since the last keys to the ant genera of the Afrotropical and Malagasy regions were published (Bolton, 1994). Taxonomy has advanced at a startling rate since then, much of the advancement fueled by the development of DNA analysis, which has revealed numerous relationships that were not apparent from the study of morphology alone. In recent years many researchers have become aware that the phenomena of convergence of characters and parallel evolution, especially in the huge subfamily Myrmicinae, are extensive. But progress toward untangling the mass of suppositions has been hampered by a lack of knowledge concerning which morphological characters were trustworthy enough to produce monophyletic groups, and which were the products of convergence and parallelism. DNA analysis has indicated the existence of numerous monophyletic groups that were previously unsuspected, and this in turn has allowed a reexamination of morphological features and a re-sorting of characters thus isolated.

The purpose of this volume is to reflect changes in, and additions to, the genus-rank taxonomy in the Afrotropical and Malagasy regions that have accrued through the intervening years and to present up-to-date keys and definitions that indicate the present state of the taxonomy. For the purposes of this book the Afrotropical region consists of sub-Saharan Africa and the islands in the Gulf of Guinea; the Malagasy region consists of Madagascar and the Indian Ocean islands of Aldabra, the Chagos Archipelago, Comoros, Europa, Farquhar, Mauritius, Mayotte, Réunion, Rodrigues, and Seychelles. In these 2 regions we currently recognize a total of 122 genera, distributed through 11 subfamilies. Many of these genera are common to both regions, but some are restricted to one or the other; some are represented by introductions from other zoogeographical regions, and 4 known genera await descriptions of their newly discovered regional species.

Among the endemic genera listed for the Afrotropical and Malagasy regions, 23 are currently monotypic. Of these genera, 21 contain only a single named taxon, of species rank, but in 2 genera (*Megaponera* and *Paltothyreus*) there are also formally described subspecies whose status has not been tested by modern techniques. In addition, there is *Oecophylla*, only 1 species of which is Afrotropical; but again, this species possesses 7 described African subspecies that have never been properly scrutinized.

Of the remaining genera, 60 have had their species-rank taxonomy revised since 1960, for one or both regions, so that relatively modern keys are available for the identification of the species in these genera. In some genera there are keys that were produced much earlier than 1960, but these are generally overloaded with infraspecific names and now

TABLE 1   Endemicity of Regional Genera

+ = present; +i = present as an introduction, either certain or suspected; +X = genus present but species undescribed or indeterminate; 0 = absent; ? = taxonomy dubious.

Subfamilies and genera	Occurrence of genera	
	Afrotropical	Malagasy
AGROECOMYRMECINAE		
Ankylomyrma	+	0
Agroecomyrmecinae total genera	1	0
AMBLYOPONINAE		
Adetomyrma	0	+
Concoctio	+	0
Mystrium	+	+
Prionopelta	+	+
Stigmatomma	+	+
Xymmer	+	+X
Amblyoponinae total genera	5	5
APOMYRMINAE		
Apomyrma	+	0
Apomyrminae total genera	1	0
DOLICHODERINAE		
Aptinoma	0	+
Axinidris	+	0
Ecphorella	+	0
Linepithema	+i	0
Ochetellus	0	+i
Ravavy	?	+
Tapinoma	+	+
Technomyrmex	+	+
Dolichoderinae total genera	5	5
DORYLINAE		
Aenictogiton	+	0
Aenictus	+	0
Chrysapace	0	+X
Dorylus	+	0
Eburopone	+	+
Lioponera	+	+
Lividopone	0	+
Ooceraea	0	+i
Parasyscia	+	+
Simopone	+	+
Tanipone	0	+
Vicinopone	+	0
Zasphinctus	+	0
Dorylinae total genera	9	8

(continued)

TABLE 1 (continued)

+ = present; +i = present as an introduction, either certain or suspected; +X = genus present but species undescribed or indeterminate; 0 = absent; ? = taxonomy dubious.

Subfamilies and genera	Occurrence of genera	
	Afrotropical	*Malagasy*
FORMICINAE		
Acropyga	+	0
Agraulomyrmex	+	0
Anoplolepis	+	+i
Aphomomyrmex	+	0
Brachymyrmex	?	+i
Camponotus	+	+
Cataglyphis	+	0
Lepisiota	+	+X
Nylanderia	+	+
Oecophylla	+	0
Paraparatrechina	+	+
Paratrechina	+	+
Petalomyrmex	+	0
Phasmomyrmex	+	0
Plagiolepis	+	+
Polyrhachis	+	0
Santschiella	+	0
Tapinolepis	+	+X
Formicinae total genera	**18**	**9**
LEPTANILLINAE		
Leptanilla	+	0
Leptanillinae total genera	**1**	**0**
MYRMICINAE		
Adelomyrmex	0	+X
Anillomyrma	+X	0
Aphaenogaster	0	+
Atopomyrmex	+	0
Baracidris	+	0
Bondroitia	+	0
Calyptomyrmex	+	+X
Cardiocondyla	+	+
Carebara	+	+
Cataulacus	+	+
Crematogaster	+	+
Cyphoidris	+	0
Cyphomyrmex	0	+i
Dicroaspis	+	+X
Diplomorium	+	0
Erromyrma	+i	+i

+ = present; +i = present as an introduction, either certain or suspected; +X = genus present but species undescribed or indeterminate; 0 = absent; ? = taxonomy dubious.

Subfamilies and genera	Occurrence of genera	
	Afrotropical	*Malagasy*
Eurhopalothrix	0	+X
Eutetramorium	0	+
Malagidris	0	+
Melissotarsus	+	+
Meranoplus	+	+
Messor	+	0
Metapone	+X	+
Microdaceton	+	0
Monomorium	+	+
Myrmicaria	+	0
Nesomyrmex	+	+
Ocymyrmex	+	0
Pheidole	+	+
Pilotrochus	0	+
Pristomyrmex	+	+
Royidris	0	+
Solenopsis	+	+
Strumigenys	+	+
Syllophopsis	+	+
Temnothorax	+	0
Terataner	+	+
Tetramorium	+	+
Trichomyrmex	+	+
Vitsika	0	+
Vollenhovia	0	+i
Wasmannia	+i	0
Myrmicinae total genera	**32**	**30**
PONERINAE		
Anochetus	+	+
Asphinctopone	+	0
Boloponera	+	0
Bothroponera	+	+
Brachyponera	+	+i
Centromyrmex	+	0
Cryptopone	+	0
Dolioponera	+	0
Euponera	+	+
Feroponera	+	0
Fisheropone	+	0
Hagensia	+	0
Hypoponera	+	+
Leptogenys	+	+

(continued)

TABLE 1 (continued)

+ = present; +i = present as an introduction, either certain or suspected; +X = genus present but species undescribed or indeterminate; 0 = absent; ? = taxonomy dubious.

Subfamilies and genera	Occurrence of genera	
	Afrotropical	*Malagasy*
Loboponera	+	0
Megaponera	+	0
Mesoponera	+	+
Odontomachus	+	+
Ophthalmopone	+	0
Paltothyreus	+	0
Parvaponera	+	+
Phrynoponera	+	0
Platythyrea	+	+
Plectroctena	+	0
Ponera	+i	+i
Promyopias	+	0
Psalidomyrmex	+	0
Streblognathus	+	0
Ponerinae total genera	28	11
PROCERATIINAE		
Discothyrea	+	+
Probolomyrmex	+	+
Proceratium	+	+
Proceratiinae total genera	3	3
PSEUDOMYRMECINAE		
Tetraponera	+	+
Pseudomyrmecinae total genera	1	1
Total genera per region	104	72
Total genera, both regions	122	

unavailable infrasubspecific names. These early keys were often produced only by reference to preexisting descriptions; the actual type specimens, the material upon which the names were based, were usually not consulted. As a result, many of the pre-1960 keys were largely guesswork and consequently inaccurate, difficult to use, or both. Recent keys for the identification of species are noted following the descriptions of the individual genera.

Large Afrotropical genera that have a history of contributions by multiple authors over a long period of time usually show, just before the commencement of a full revision, a considerable number of species-rank names, surrounded by a cloud of infraspecific names, together with a number of infrasubspecific (unavailable) names. For instance, B. Bolton's (1980) study of Afrotropical *Tetramorium* commenced with about 104 previously described names of species rank, 105 names of infraspecific rank, and 19 unavailable names. A number of the species were obviously valid, but several had been described twice or more, by different authors, because of the inadequacies of the original descriptions. At the same time, and for the same reason, a good number of infraspecific names had been attached to species to which they were not truly related. After revision this mass resolved into 175 valid species-rank taxa, an increase in the number of regional species of about 68 percent.

Interestingly, a similar analysis of Bolton's (1987) study of Afrotropical *Monomorium* shows an increase after revision of about 67 percent in terms of number of species. The percentage increase in numbers of *Strumigenys*, however, does not follow this pattern. Bolton (1983) shows a large 168 percent increase in the number of Afrotropical *Strumigenys* species, and B. L. Fisher (2000) an incredible 1775 percent in the same genus in the Malagasy region. The reason for these huge increases is not hard to understand. *Strumigenys* is predominantly a genus of small to minute species, of retiring or cryptic habits, that mostly inhabits leaf litter and topsoil, and so it is hardly ever collected by hand. The vast majority of its species therefore remained unknown to the early authors, and its real numbers did not become apparent until the advent of collections by Winkler bag technique (Fisher, 1998).

But for large genera whose species are generally collectable by hand, and which have been largely described by pre-1960 authors, let us casually assume that the increase in number of species following a formal revision will average about 65 percent. Applying this increase to some other unrevised genera would very roughly indicate a regional fauna of 279 Afrotropical and 83 Malagasy species of *Camponotus*, 223 Afrotropical and 36 Malagasy species of *Crematogaster* (the actual number of Malagasy *Crematogaster* species is currently 37, the figure culled from the various recent revisionary works of B. B. Blaimer; see references). A crude application of the 65 percent guess across the entire fauna yields a very rough total of about 3,000 species in the Afrotropical region and about 1,000 in the Malagasy region. This estimate, however, does not take into account the additional increase likely to result from more intensive sampling across ecoregions of the African mainland, and so likely underestimates the total species for the Afrotropical region. The total number of species in the Afrotropical and Malagasy regions may be as high as 5,000 species.

Of the 122 genera listed in Table 1, some have very restricted distributions, while others are considerably more widespread. The relative distribution of the genera found in the Afrotropical and Malagasy regions, on a worldwide scale, is summarized in Table 2.

During the long history of ant taxonomy, from 1758 to the present day, many names in the family-group (names applied to families, subfamilies, and tribes) and names in the genus-group (names applied to genera and subgenera) have been proposed. A large proportion of these have survived unchanged to the present, but a number were proposed for supposed groups that were later found to be synonyms of earlier names or were inadmissible because the name was a junior homonym—one that had already been used elsewhere, and earlier, for a different group of insects. Another category of discarded names includes those that were the results of misidentifications, where an author had placed a name in one group, only for it to be discovered later that the grouping was incorrect. This book utilizes only the

TABLE 2 Relative Distributions of the Genera

Genera restricted to the Afrotropical and Malagasy regions
(genera that are present in both regions but are absent from all other regions)

Amblyoponinae: *Xymmer*
Dorylinae: *Eburopone*
Myrmicinae: *Dicroaspis, Melissotarsus, Terataner*

Number of genera: 5

Genera with uniquely Afrotropical distribution
(genera that are restricted to the Afrotropical region and occur nowhere else)

Agroecomyrmecinae: *Ankylomyrma*
Amblyoponinae: *Concoctio*
Apomyrminae: *Apomyrma*
Dolichoderinae: *Axinidris, Ecphorella*
Dorylinae: *Aenictogiton, Vicinopone*
Formicinae: *Agraulomyrmex, Aphomomyrmex, Petalomyrmex, Phasmomyrmex, Santschiella*
Myrmicinae: *Atopomyrmex, Baracidris, Bondroitia, Cyphoidris, Diplomorium, Microdaceton, Ocymyrmex*
Ponerinae: *Asphinctopone, Boloponera, Dolioponera, Feroponera, Fisheropone, Hagensia, Loboponera, Megaponera, Ophthalmopone, Paltothyreus, Phrynoponera, Plectroctena, Promyopias, Psalidomyrmex, Streblognathus*

Number of genera: 34

Genera with uniquely Malagasy distribution
(genera that are restricted to the Malagasy region and occur nowhere else)

Amblyoponinae: *Adetomyrma*
Dolichoderinae: *Aptinoma, Ravavy*
Dorylinae: *Lividopone, Tanipone*
Myrmicinae: *Eutetramorium, Malagidris, Pilotrochus, Royidris, Vitsika*

Number of genera: 10

Genera in the Afrotropical and other region(s) but absent from the Malagasy region
(genera that occur in the Afrotropical region plus one or more other regions but are absent from the Malagasy region)

Dorylinae: *Aenictus, Dorylus, Zasphinctus*
Formicinae: *Acropyga, Cataglyphis, Oecophylla, Polyrhachis*
Myrmicinae: *Anillomyrma, Messor, Myrmicaria, Temnothorax, Wasmannia*
Ponerinae: *Centromyrmex, Cryptopone*

Number of genera: 14

Genera in the Malagasy and other region(s) but absent from the Afrotropical region
(genera that occur in the Malagasy region plus one or more other regions but are absent from the Afrotropical region)

Dolichoderinae: *Ochetellus*
Dorylinae: *Chrysapace, Ooceraea*
Myrmicinae: *Adelomyrmex, Aphaenogaster, Cyphomyrmex, Eurhopalothrix, Vollenhovia*
Number of genera: 8

Genera common to the Afrotropical and Malagasy and other regions
(genera that occur in both the Afrotropical region and the Malagasy region and are also present in one or more other regions)

Amblyoponinae: *Mystrium, Prionopelta, Stigmatomma*
Dolichoderinae: *Tapinoma, Technomyrmex*
Dorylinae: *Lioponera, Parasyscia, Simopone*
Formicinae: *Anoplolepis, Brachymyrmex, Camponotus, Lepisiota, Nylanderia, Paraparatrechina, Paratrechina, Plagiolepis, Tapinolepis*
Myrmicinae: *Calyptomyrmex, Cardiocondyla, Carebara, Cataulacus, Crematogaster, Erromyrma, Meranoplus, Metapone, Monomorium, Nesomyrmex, Pheidole, Pristomyrmex, Solenopsis, Strumigenys, Syllophopsis, Tetramorium, Trixchomyrmex*
Ponerinae: *Anochetus, Bothroponera, Brachyponera, Euponera, Hypoponera, Leptogenys, Mesoponera, Odontomachus, Parvaponera, Platythyrea, Ponera*
Proceratiinae: *Discothyrea, Probolomyrmex, Proceratium*
Pseudomyrmecinae: *Tetraponera*

Number of genera: 49

Genera that occur in both the Afrotropical region and the Malagasy region
(genera present in both regions, regardless of their occurrence in other regions)

Amblyoponinae: *Mystrium, Prionopelta, Stigmatomma, Xymmer*
Dolichoderinae: *Tapinoma, Technomyrmex*
Dorylinae: *Eburopone, Lioponera, Parasyscia, Simopone*
Formicinae: *Anoplolepis, Brachymyrmex, Camponotus, Lepisiota, Nylanderia, Paraparatrechina, Paratrechina, Plagiolepis, Tapinolepis*
Myrmicinae: *Calyptomyrmex, Cardiocondyla, Carebara, Cataulacus, Crematogaster, Dicroaspis, Erromyrma, Melissotarsus, Meranoplus, Metapone, Monomorium, Nesomyrmex, Pheidole, Pristomyrmex, Solenopsis, Strumigenys, Syllophopsis, Terataner, Tetramorium, Trichomyrmex*
Ponerinae: *Anochetus, Bothroponera, Brachyponera, Euponera, Hypoponera, Leptogenys, Mesoponera, Odontomachus, Parvaponera, Platythyrea, Ponera*
Proceratiinae: *Discothyrea, Probolomyrmex, Proceratium*
Pseudomyrmecinae: *Tetraponera*

Number of genera: 54

most recent applications of the various names, but older literature will often show these now discarded names, whose fates can be tracked in Box 1.

The genera recognized here vary enormously in terms of the number of species that each contains, but the figures given for numbers of species are at best only an approximation of the true numbers of species represented in the wild. Collections in natural history museums, and in other collections of ants in the world, contain large numbers of species that are known to be undescribed. The task of identifying and describing these species is far from complete. Furthermore, the species-rank taxonomy of some of the largest and most important genera remains unstudied in detail and consequently rather confused. For the Afrotropical and Malagasy regions, the numbers of species currently recognized in the 110 native genera are summarized in Table 3.

BOX 1 Current Synonyms of Family-Group and Genus-Group Names

A number of subfamilies and genera that occur in the Afrotropical and Malagasy regions are senior synonyms of other names in the family-group (families, subfamilies, tribes), and names in the genus-group (genera, subgenera), that were originally proposed in the regions. These synonymized names may be encountered in the older literature. This alphabetically arranged list indicates these synonyms, with the valid senior synonym in **bold**; names that lack junior synonyms, and have never been subject to changes, are omitted. In addition, some species have been referred incorrectly to genera that do not occur in the regions under consideration; the genera in question are also listed here.

Acantholepis: homonymous name replaced by *Lepisiota*
Acidomyrmex: junior synonym of *Tetramorium*
Acrocoelia: junior synonym of *Crematogaster (Crematogaster)*
Acropyga Roger, 1862 = *Malacomyrma* Emery, 1922
Aenictinae: junior synonym of Dorylinae
Aenictogitoninae: junior synonym of Dorylinae
Aeromyrma: junior synonym of *Carebara*
Aethiopopone: junior synonym of *Zasphinctus*
Afroxyidris: junior synonym of *Carebara*
Anacantholepis: junior synonym of *Plagiolepis*
Aneleus: junior synonym of *Carebara*
Anergatides: junior synonym of *Pheidole*
Anoplolepis Santschi, 1914 = *Zealleyella* Arnold, 1922
Aphaenogaster Mayr, 1853 = *Deromyrma* Forel, 1913
Asphinctopone Santschi, 1914 = *Lepidopone* Bernard, 1953
Atopogyne: junior synonym of *Crematogaster (Crematogaster)*
Atopula: junior synonym of *Tetramorium*
Axinidrini: junior synonym of Dolichoderinae
Brunella: homonymous name replaced by *Malagidris*
Cacopone: junior synonym of *Plectroctena*
Cardiocondyla Emery, 1869 = *Emeryia* Forel, 1890, = *Dyclona* Santschi, 1930, = *Loncyda* Santschi, 1930
Carebara Westwood, 1840 = *Pheidologeton* Mayr, 1862, = *Oligomyrmex* Mayr, 1867, = *Aeromyrma* Forel, 1891, = *Aneleus* Emery, 1900, = *Paedalgus* Forel, 1911, = *Crateropsis* Patrizi, 1948, = *Sporocleptes* Arnold, 1948, = *Nimbamyrma* Bernard, 1953, = *Afroxyidris* Belshaw and Bolton, 1994
Cataglyphis Foerster, 1850 = *Monocombus* Mayr, 1855
Cataulacus Smith, F., 1853 = *Otomyrmex* Forel, 1891
Centromyrmecini: junior synonym of Ponerini
Centromyrmex Mayr, 1866 = *Glyphopone* Forel, 1913, = *Leptopone* Arnold, 1916
Cephaloxys: homonymous name replaced by *Smithistruma* (itself now a junior synonym of *Strumigenys*)
Cerapachyinae: junior synonym of Dorylinae
Cerapachys Smith, F., 1857. Species of this genus are absent from the Afrotropical and Malagasy regions.
Champsomyrmex: junior synonym of *Odontomachus*
Cladarogenys: junior synonym of *Strumigenys*
Crateropsis: junior synonym of *Carebara*
Cratomyrmex: junior synonym of *Messor*

Crematogaster Lund, 1831. Two subgenera are currently retained, from an earlier 8 that were recognized in the regions. Subgenus *C. (Crematogaster) = C. (Acrocoelia)* Mayr, 1853, = *C. (Oxygyne)* Forel, 1901, = *C. (Decacrema)* Forel, 1910, = *C. (Atopogyne)* Forel, 1911, = *C. (Nematocrema)* Santschi, 1918, = *C. (Sphaerocrema)* Santschi, 1918. Subgenus *C. (Orthocrema)* Santschi, 1918 = *C. (Eucrema)* Santschi, 1918

Cysias: junior synonym of *Ooceraea*

Decacrema: junior synonym of *Crematogaster (Crematogaster)*

Decamorium: junior synonym of *Tetramorium*

Deromyrma: junior synonym of *Aphaenogaster*

Diplorhoptrum: junior synonym of *Solenopsis*

Discothyrea Roger, 1863 = *Pseudosysphincta* Arnold, 1916

Dodous: junior synonym of *Pristomyrmex*

Dolichoderinae Forel, 1878 = Axinidrini Weber, 1941

Dorylinae Leach, 1815 = Cerapachyinae Forel, 1893, = Aenictinae Emery, 1901, = Aenictogitoninae Ashmead, 1905

Dyclona: junior synonym of *Cardiocondyla*

Ectomomyrmex Mayr, 1867. Species of this genus are absent from the Afrotropical and Malagasy regions.

Emeryia: junior synonym of *Cardiocondyla*

Engramma: junior synonym of *Technomyrmex*

Epitritus: junior synonym of *Strumigenys*

Epixenus: junior synonym of *Monomorium*

Equestrimessor: junior synonym of *Trichomyrmex*

Escherichia: junior synonym of *Probolomyrmex*

Eucrema: junior synonym of *Crematogaster (Orthocrema)*

Euponerinae: junior synonym of Ponerini

Glamyromyrmex: junior synonym of *Strumigenys*

Glyphopone: junior synonym of *Centromyrmex*

Goniothorax: homonyous name replaced by *Nesomyrmex*

Heptacondylus: junior synonym of *Myrmicaria*

Holcomyrmex: junior synonym of *Trichomyrmex*

Hoplomyrmus: junior synonym of *Polyrhachis*

Hylidris: junior stynonym of *Pristomyrmex*

Icothorax: junior synonym of *Temnothorax*

Ireneopone: junior synonym of *Nesomyrmex*

Iridomyrmex Mayr, 1862. Species of this genus are absent from the Afrotropical and Malagasy regions.

Isolcomyrmex: junior synonym of *Trichomyrmex*

Lampromyrmex: junior synonym of *Monomorium*

Lepidopone: junior synonym of *Asphinctopone*

Lepisiota Santschi, 1926 = *Acantholepis* Mayr, 1861 (homonym), = *Pseudacantholepis* Bernard, 1953 (unavailable name)

Leptogenyini: junior synonym of Ponerini

Leptogenys Roger, 1861 = *Lobopelta* Mayr, 1862, = *Machaerogenys* Emery, 1911, = *Microbolbos* Donisthorpe, 1948

Leptopone: junior synonym of *Centromyrmex*

Limnomyrmex: junior synonym of *Nesomyrmex*

Lioponera Mayr, 1879 = *Phyracaces* Emery, 1902

Lobopelta: junior synonym of *Leptogenys*

Loncyda: junior synonym of *Cardiocondyla*

(continued)

Machaerogenys: junior synonym of *Leptogenys*

Macromischoides: junior synonym of *Tetramorium*

Malacomyrma Emery, 1922: junior synonym of *Acropyga*

Malagidris Bolton and Fisher, 2014 = *Brunella* Forel, 1917 (homonym)

Mesanoplolepis: junior synonym of *Tapinolepis*

Mesoponera Emery, 1900 = *Xiphopelta* Forel, 1913

Messor Forel, 1890 = *Cratomyrmex* Emery, 1892, = *Sphaeromessor* Bernard, 1985 (unavailable name)

Miccostruma: junior synonym of *Strumigenys*

Microbolbos: junior synonym of *Leptogenys*

Monocombus: junior synonym of *Cataglyphis*

Monomorium Mayr, 1855 = *Lampromyrmex* Mayr, 1868, = *Epixenus* Emery, 1908, = *Xeromyrmex* Emery, 1915, = *Paraphacota* Santschi, 1919, = *Pharaophanes* Bernard, 1967

Myopias Roger, 1861. Species of this genus are absent from the Afrotropical and Malagasy regions.

Myrmicaria Saunders, W.W., 1842 = *Heptacondylus* Smith, F., 1857, = *Physatta* Smith, F., 1857

Myrmisaraka: junior synonym of *Vitsika*

Nematocrema: junior synonym of *Crematogaster (Crematogaster)*

Nesomyrmex Wheeler, W.M., 1910 = *Goniothorax* Emery, 1896 (homonym), = *Tetramyrma* Forel, 1912, = *Limnomyrmex* Arnold, 1948, = *Ireneopone* Donisthorpe, 1946

Nimbamyrma: junior synonym of *Carebara*

Odontomachidae: junior synonym of Ponerini

Odontomachus Latreille, 1804 = *Champsomyrmex* Emery, 1892

Oligomyrmex: junior synonym of *Carebara*

Ooceraea Roger, 1862 = *Cysias* Emery, 1902

Otomyrmex: junior synonym of *Cataulacus*

Oxygyne: junior synonym of *Crematogaster (Crematogaster)*

Paedalgus: junior synonym of *Carebara*

Paraphacota: junior synonym of *Monomorium*

Parapheidole: junior synonym of *Pheidole*

Parholcomyrmex: junior synonym of *Trichomyrmex*

Pharaophanes: junior synonym of *Monomorium*

Pheidole Westwood, 1839 = *Parapheidole* Emery, 1915, = *Anergatides* Wasmann, 1915

Pheidologeton: junior synonym of *Carebara*

Phyracaces: junior synonym of *Lioponera*

Physatta: junior synonym of *Myrmicaria*

Plagiolepis Mayr, 1861 = *Anacantholepis* Santschi, 1914

Plectroctena Smith, F., 1858 = *Cacopone* Santschi, 1914

Plectroctenini: junior synonym of Ponerini

Polyrhachis Smith, F., 1857 = *Hoplomyrmus* Gerstäcker, 1859, = *Pseudocyrtomyrma* Emery, 1921

Ponerini Lepeletier de Saint-Fargeau, 1835 = Odontomachidae Mayr, 1862, = Leptogenyini Forel, 1893, = Euponerinae Emery, 1909, = Centromyrmecini Emery, 1911, = Plectroctenini Emery, 1911

Prenolepis Mayr, 1861. Species of this genus are absent from the Afrotropical and Malagasy regions.

Pristomyrmex Mayr, 1866 = *Hylidris* Weber, 1941, = *Dodous* Donisthorpe, 1946

Probolomyrmex Mayr, 1901 = *Escherichia* Forel, 1910

Proceratiinae Emery, 1895 = Discothyrinae Clark, 1951

Proceratium Roger, 1863 = *Sysphingta* Roger, 1863

Proscopomyrmex: junior synonym of *Strumigenys*

Pseudacantholepis: unavailable name, the material of which is referable to *Lepisiota*

Pseudocyrtomyrma: junior synonym of *Polyrhachis*

Pseudolasius Emery, 1887. Species of this genus are absent from the Afrotropical and Malagasy regions.

Pseudoponera Emery, 1900. Species of this genus are absent from the Afrotropical and Malagasy regions.

Pseudosysphincta: junior synonym of *Discothyrea*

Quadristruma: junior synonym of *Strumigenys*

Pyramica: junior synonym of *Strumigenys*

Rhoptromyrmex: junior synonym of *Tetramorium*

Semonius: junior synonym of *Tapinoma*

Serrastruma: junior synonym of *Strumigenys*

Smithistruma: junior synonym of *Strumigenys*

Solenopsis Westwood, 1840 = *Diplorhoptrum* Mayr, 1855

Sphaerocrema: junior synonym of *Crematogaster (Crematogaster)*

Sphaeromessor: unavailable name, the material of which is referable to *Messor*

Sphinctomyrmex Mayr, 1866. Species of this genus are absent from the Afrotropical and Malagasy regions.

Sporocleptes: junior synonym of *Carebara*

Strumigenys Smith, F., 1860 = *Pyramica* Roger, 1862, = *Cephaloxys* Smith, F., 1865 (homonym), = *Epitritus* Emery, 1869, = *Trichoscapa* Emery, 1869, = *Glamyromyrmex* Wheeler, W.M., 1915, = *Proscopomyrmex* Patrizi, 1946, = *Smithistruma* Brown, 1948, = *Serrastruma* Brown, 1948, = *Miccostruma* Brown, 1948, = *Quadristruma* Brown, 1949, = *Cladarogenys* Brown, 1976

Sysphingta: junior synonym of *Proceratium*

Tapinolepis Emery, 1925 = *Mesanoplolepis* Santschi, 1926

Tapinoma Foerster, 1850 = *Semonius* Forel, 1910

Tapinoptera: junior synonym of *Technomyrmex*

Technomyrmex Mayr, 1872 = *Engramma* Forel, 1905 = *Tapinoptera* Santschi, 1925

Temnothorax Mayr, 1861 = *Icothorax* Hamann and Klemm, 1967

Terataner Emery, 1912 = *Tranetera* Arnold, 1952

Tetramorium Mayr, 1855 = *Xiphomyrmex* Forel, 1887, = *Triglyphothrix* Forel, 1890, = *Rhoptromyrmex* Mayr, 1901, = *Atopula* Emery, 1912, = *Decamorium* Forel, 1913, = *Acidomyrmex* Emery, 1915, = *Macromischoides* Wheeler, W.M., 1920

Tetramyrma: junior synonym of *Nesomyrmex*

Tetraponera Smith, F., 1852 = *Sima* Roger, 1863, = *Pachysima* Emery, 1912, = *Viticicola* Wheeler, W.M., 1919

Trachymesopus Emery, 1911. This genus is a junior synonym of *Pseudoponera* Emery, 1900, of which no species are Afrotropical; some African species were wrongly referred to as *Trachymesopus*.

Tranetera: junior synonym of *Terataner*

Trichomyrmex Mayr, 1865 = *Holcomyrmex* Mayr, 1879, = *Parholcomyrmex* Emery, 1915, = *Isolcomyrmex* Santschi, 1917, = *Equestrimessor* Santschi, 1919

Trichoscapa: junior synonym of *Strumigenys*

Triglyphothrix: junior synonym of *Tetramorium*

Vitsika Bolton and Fisher, 2014 = *Myrmisaraka* Bolton and Fisher, 2014

Xeromyrmex: junior synonym of *Monomorium*

Xiphomyrmex: junior synonym of *Tetramorium*

Xiphopelta: junior synonym of *Mesoponera*

Zasphinctus Wheeler, W.M., 1918 = *Aethiopopone* Santschi, 1930 (Afrotropical species were originally described in *Sphinctomyrmex*)

Zealleyella: junior synonym of *Anoplolepis*

TABLE 3 Number of Described Species in Endemic Regional Genera

Subfamilies and genera	Number of species	
	Afrotropical	Malagasy
AGROECOMYRMECINAE		
Ankylomyrma	1	0
Agroecomyrmecinae total species	1	0
AMBLYOPONINAE		
Adetomyrma	0	9
Concoctio	1	0
Mystrium	1	10
Prionopelta	3	6
Stigmatomma	2	1
Xymmer	1	0
Amblyoponinae total species	8	26
APOMYRMINAE		
Apomyrma	1	0
Apomyrminae total species	1	0
DOLICHODERINAE		
Aptinoma	0	2
Axinidris	21	0
Ecphorella	1	0
Ravavy	0	1
Tapinoma	14 [6]	5
Technomyrmex	27	12
Dolichoderinae total species	63 [6]	20
DORYLINAE		
Aenictogiton	7	0
Aenictus	34 [17]	0
Dorylus	55 [63]	0
Eburopone	1	1
Lioponera	10	2
Lividopone	0	1
Ooceraea	0	1
Parasyscia	14	1
Simopone	18	16
Tanipone	0	10
Vicinopone	1	0
Zasphinctus	2	0
Dorylinae total species	142 [80]	32
FORMICINAE		
Acropyga	3	0
Agraulomyrmex	2	0
Anoplolepis	9 [4]	0 [1]
Aphomomyrmex	1	0

Subfamilies and genera	Number of species	
	Afrotropical	*Malagasy*
FORMICINAE *(continued)*		
Camponotus	169 [116]	50 [33]
Cataglyphis	4 [4]	0
Lepisiota	47 [29]	0
Nylanderia	17	10 [6]
Oecophylla	1 [7]	0
Paraparatrechina	11	5
Paratrechina	3	3
Petalomyrmex	1	0
Phasmomyrmex	4 [2]	0
Plagiolepis	20 [7]	2
Polyrhachis	48	0
Santschiella	1	0
Tapinolepis	11 [3]	0
Formicinae total species	**352** [172]	**70** [40]
LEPTANILLINAE		
Leptanilla	3	0
Leptanillinae total species	**3**	**0**
MYRMICINAE		
Aphaenogaster	0	3 [3]
Atopomyrmex	3	0
Baracidris	3	0
Bondroitia	2	0
Calyptomyrmex	16	0
Cardiocondyla	14	5 [1]
Carebara	62 [11]	3
Cataulacus	39	8
Crematogaster	135 [173]	37
Cyphoidris	4	0
Dicroaspis	2	0
Diplomorium	1	0
Eutetramorium	0	3
Malagidris	0	6
Melissotarsus	3	1
Meranoplus	8	4
Messor	15 [1]	0
Metapone	0	3
Microdaceton	4	0
Monomorium	132	19
Myrmicaria	22 [27]	0
Nesomyrmex	25	4
Ocymyrmex	34	0
Pheidole	72 [69]	25 [6]
Pilotrochus	0	1

(continued)

TABLE 3 (continued)

Subfamilies and genera	Number of species	
	Afrotropical	*Malagasy*
MYRMICINAE *(continued)*		
Pristomyrmex	5	3
Royidris	0	15
Solenopsis	12 [9]	2
Strumigenys	135	90
Syllophopsis	7	10
Temnothorax	6	0
Terataner	6	6
Tetramorium	235	108
Trichomyrmex	7	2
Vitsika	0	16
Myrmicinae total species	**1,009** [290]	**374** [10]
PONERINAE		
Anochetus	19	5
Asphinctopone	3	0
Boloponera	1	0
Bothroponera	22 [14]	8
Brachyponera	1 [2]	0
Centromyrmex	10	0
Cryptopone	1	0
Dolioponera	1	0
Euponera	5 [1]	14
Feroponera	1	0
Fisheropone	1	0
Hagensia	2 [4]	0
Hypoponera	54	7 [4]
Leptogenys	56	60
Loboponera	9	0
Megaponera	1 [5]	0
Mesoponera	15 [5]	1 [1]
Odontomachus	2	3
Ophthalmopone	5 [1]	0
Paltothyreus	1 [6]	0
Parvaponera	2 [1]	1 [1]
Phrynoponera	5	0
Platythyrea	14	4
Plectroctena	16	0
Promyopias	1	0
Psalidomyrmex	6	0
Streblognathus	2	0
Ponerinae total species	**256** [39]	**103** [6]

Subfamilies and genera	Number of species	
	Afrotropical	*Malagasy*
PROCERATIINAE		
Discothyrea	7	1
Probolomyrmex	3	3
Proceratium	9	3
Proceratiinae total species	**19**	**7**
PSEUDOMYRMECINAE		
Tetraponera	**30** [14]	**21** [5]
Pseudomyrmecinae total species	**30** [14]	**21** [5]
Total species	1,884	644
Total subspecies	[601]	[61]
Total species + subspecies	2,485	705

The entries represent the current number of validly described species for each native genus in each region; introduced genera (known or suspected), and known but undescribed species, are ignored. Numbers of unresolved infraspecific taxa (subspecies) in taxonomically unrevised genera are indicated by [*n*].

Table 3 includes only genera that occur naturally in the Afrotropical and Malagasy regions. Deliberately omitted are the few species that represent known or suspected introductions from other zoogeographical regions, which belong to the Neotropical genera *Brachymyrmex*, *Cyphomyrmex*, *Linepithema*, and *Wasmannia*, and the Oriental-Malesian genera *Erromyrma*, *Ochetellus*, *Ponera*, and *Vollenhovia*. When those are taken into account, the genera fall into the size categories listed in Box 2. It is interesting to note that the sum of species in just the 7 largest genera exceeds the sum of all of the other 111 genera combined, and that the most species-rich genus, hyperdiverse *Tetramorium*, has 115 more species than the second largest genus, *Strumigenys*. In other words, *Tetramorium* is so successful in the Afrotropical and Malagasy regions that it contains more species than the combined total of the first 82 genera listed in Box 2.

A SHORT HISTORY OF ANT TAXONOMY IN THE REGIONS

In species-rank taxonomy, our understanding of the Malagasy region's ants began with what should have been a great advantage: the early production of a couple of authoritative volumes by A. Forel (1891, 1892), which summarized all the small taxonomic contributions to date and added a large number of new taxa, all in a unified system. Unfortunately, this excellent beginning was not developed further, and for the next hundred years only minor contributions were added. Most of these took the form of small papers that described a few new taxa collected by a single individual on Madagascar itself as well as additions to the restricted faunas of the Indian Ocean islands that constitute part of the region.

Real comprehension of the entire region's extensive ant fauna began only with the publications of B. L. Fisher and his associates (see references), which focused on the revisionary

BOX 2 Summary of Genera by Number of Species

This list provides a simple estimate of the relative sizes of the 118 genera recorded from the Afrotropical and Malagasy regions, in terms of number of described species per genus. Described introductions are included, but known species that remain undescribed are excluded. Species common to both regions are counted only once. Omitted are 4 genera (*Adelomyrmex, Anillomyrma, Chrysapace, Eurhopalothrix*) because although the genera have been collected in the regions, their species remain undescribed.

GENERA WITH ONLY 1 SPECIES
Ankylomyrma, Aphomomyrmex, Apomyrma, Boloponera, Brachymyrmex, Concoctio, Cryptopone, Cyphomyrmex, Diplomorium, Dolioponera, Eburopone, Ecphorella, Erromyrma, Feroponera, Fisheropone, Linepithema, Lividopone, Megaponera, Ochetellus, Oecophylla, Ooceraea, Paltothyreus, Petalomyrmex, Pilotrochus, Promyopias, Ravavy, Santschiella, Vicinopone, Wasmannia, Xymmer

GENERA WITH 2-3 SPECIES
Agraulomyrmex (2), *Aptinoma* (2), *Bondroitia* (2), *Brachyponera* (2), *Dicroaspis* (2), *Hagensia* (2), *Parvaponera* (2), *Streblognathus* (2), *Vollenhovia* (2), *Zasphinctus* (2), *Acropyga* (3), *Aphaenogaster* (3), *Asphinctopone* (3), *Atopomyrmex* (3), *Baracidris* (3), *Eutetramorium* (3), *Leptanilla* (3), *Metapone* (3), *Ponera* (3), *Stigmatomma* (3)

GENERA WITH 4-6 SPECIES
Cataglyphis (4), *Cyphoidris* (4), *Melissotarsus* (4), *Microdaceton* (4), *Odontomachus* (4), *Phasmomyrmex* (4), *Ophthalmopone* (5), *Paratrechina* (5), *Phrynoponera* (5), *Malagidris* (6), *Probolomyrmex* (6), *Psalidomyrmex* (6), *Temnothorax* (6)

GENERA WITH 7-10 SPECIES
Aenictogiton (7), *Trichomyrmex* (7), *Discothyrea* (8), *Pristomyrmex* (8), *Adetomyrma* (9), *Anoplolepis* (9), *Loboponera* (9), *Prionopelta* (9), *Centromyrmex* (10), *Tanipone* (10)

GENERA WITH 11-15 SPECIES
Mystrium (11), *Tapinolepis* (11), *Lioponera* (12), *Meranoplus* (12), *Proceratium* (12), *Terataner* (12), *Solenopsis* (14), *Messor* (15), *Parasyscia* (15), *Royidris* (15), *Syllophopsis* (15)

GENERA WITH 16-20 SPECIES
Calyptomyrmex (16), *Cardiocondyla* (16), *Mesoponera* (16), *Paraparatrechina* (16), *Plectroctena* (16), *Vitsika* (16), *Platythyrea* (18), *Tapinoma* (18), *Euponera* (20)

GENERA WITH 21-30 SPECIES
Axinidris (21), *Plagiolepis* (21), *Myrmicaria* (22), *Anochetus* (24), *Nylanderia* (25), *Nesomyrmex* (29), *Bothroponera* (30)

GENERA WITH 31-40 SPECIES
Aenictus (34), *Ocymyrmex* (34), *Simopone* (34), *Technomyrmex* (36)

GENERA WITH 41-50 SPECIES
Cataulacus (46), *Lepisiota* (47), *Polyrhachis* (48)

GENERA WITH 51-60 SPECIES
Tetraponera (51), *Dorylus* (55), *Hypoponera* (60)

GENUS WITH 61-70 SPECIES
Carebara (65)

GENUS WITH 91-100 SPECIES
Pheidole (95)

GENUS WITH 101-120 SPECIES
Leptogenys (112)

GENUS WITH 141-150 SPECIES
Monomorium (144)

GENUS WITH 161-170 SPECIES
Crematogaster (169)

GENUS WITH 211-220 SPECIES
Camponotus (216)

GENUS WITH > 220 SPECIES
Strumigenys (221)

GENUS WITH > 300 SPECIES
Tetramorium (336)

taxonomy of whole genera, or groups of genera, from the entire region. These were based on exhaustive collecting conducted over many years by Fisher himself or by his students and colleagues. The results of these endeavors have so far covered 32 genera as represented in the region, but perhaps the most spectacular result was Fisher's (2000) revision of the Malagasy species of the genus *Strumigenys*. In this genus of small, cryptic ants only 6 species had been recorded in the entire Malagasy region up to that date. Fisher's work, coupled with a very minor contribution by Bolton (2000), raised the number to 90 well-defined, valid species.

The species-rank taxonomic situation in the Afrotropical region had no initial unified system such as was available for the Malagasy. From the earliest times to about 1950, taxonomic input for the region consisted almost entirely of scattered descriptions of whatever taxa occurred in a particular area. Frequently, these were reports on collections made in a very small area, over a very limited period, by a single entomologist. Dozens of papers appeared, year after year, and each of them merely added to the confused mass of names that had already been published. Over the years, the descriptions became more and more superficial, and the real identities and affinities of the nominal species, and their infraspecific taxa, became more obscure. It was almost as if the main taxonomists of those earlier times were in a race to see who could produce the most names, regardless of their uniqueness, accuracy, or validity. Very occasionally, an author would produce a revision, or a monograph of a particular genus, but such an offering often became just another production line for dubious names.

There were, of course, examples of authors trying to break this monotonous cycle. Outstanding among these was the production by G. Arnold (1915, 1916, 1917, 1920, 1922, 1924, 1926) of a multivolume study of the entire South African ant fauna. This survey presented keys and descriptions for all the named ant taxa of the country in a systematic order and also successfully added many new taxa to the total. Although now out of date, the work still strikes a modern reader as refreshingly different from the usual scattering of minimalist descriptions that then prevailed. Another landmark was W. M. Wheeler's (1922) production of the monumental faunal study, "The Ants of the Belgian Congo." Not only did this work treat whole genera, but it also included keys to the genera themselves, biological notes, a detailed catalogue, and much more.

By the 1950s it was apparent that the species-rank taxonomy was grossly inflated, if not almost impenetrable, and that a shift away from small-area faunistics and one-by-one descriptions and toward revisionary studies of species groups or whole genera was needed, to pin down which names were truly valid and which were synonyms or even invalid. The impetus for this was provided initially by W. L. Brown, who in the early 1950s began work on the genera of dacetine ants. The task of constructing taxonomic revisions of particular genera, as they occurred in the entire Afrotropical region, was later taken up by Bolton, his colleagues and associates, and other taxonomists between 1973 and the present (see references), so that today a good proportion of the genera (64) have received some relatively recent taxonomic attention. The task is by no means complete, as there is easily more than a lifetime's accumulation of work remaining, but a scan through the genera included in this volume will show interested taxonomists which genera are still in need of a modern synthesis of their species.

Among higher taxa, such as subfamilies and the genera themselves, there was generally more certainty and stability than at species level. This was because, from very early times, a number of authors had striven to define the groups as accurately as was possible (for example, G. Mayr, 1865). The most influential of these was C. Emery's (1910, 1911, 1913, 1921, 1922, 1924, 1925) masterpiece in the *Genera Insectorum* series. These volumes provided diagnoses and keys to the genera and higher taxa as well as a full catalogue of all named forms. It was extremely influential and was reinforced by W. M. Wheeler's (1922) inclusion of keys to subfamilies and genera in "The Ants of the Belgian Congo." The two works were very interdependent and together formed the Emery-Wheeler classification of ants, some of which still survives today. But in the years after 1925, many changes and additions were made to the Emery-Wheeler system, which gradually lost its uniformity and became partially decrepit. An attempt to update the classification and rectify the many introduced errors was made by Bolton (1994), who presented a unified set of keys to the genera of the world, treating the Afrotropical and Malagasy regions together as a single unit. The most recent printed synopsis of higher ant taxa is that of Bolton (2003), but a considerable amount of work that has improved on this study has been published in the intervening years. These contributions are noted in the text under the entries of the various subfamilies and genera.

Taxonomic catalogues are useful as they show the condition of the classification at a given time. Not only do they list described taxa in the species-group as they stood at the time of the particular catalogue's production, but they also indicate the genera and subfamilies to which those taxa were assigned, which provides a good overview of which higher taxa were considered valid at the time. Early catalogues were published by J. Roger (1863), Mayr (1863), and C.G. de Dalla Torre (1893). In the intermediate period were the works of Emery, in the *Genera Insectorum* series mentioned earlier, and Wheeler's (1922) catalogues of Afrotropical and Malagasy taxa. After a long hiatus, Bolton (1995) produced his world ant catalogue, which is now kept up-to-date online. In addition, reputable revisions of genera or higher taxa may also provide lists of included species, such as in the ponerine revision of C.A. Schmidt and S.O. Shattuck (2014).

The system of nomenclature developed for ants, from very early in its history, was blighted by an overinflated set of subdivisions of names: the weird and unnecessary pentanomial system. Under this system any taxon could have up to five names: 1. Genus; 2. Subgenus; 3. Species; 4. Subspecies (or Race, or Stirps, names of apparently equivalent, or near-equivalent, rank); and 5. Variety. Complicating matters further, a varietal name could be attached directly to a species-rank name as well as to one of subspecies/race/stirps rank. No two authors seemed able to agree on a consistent status for any one name, so that one author would call a taxon a species or a subspecies, while another would call it a subspecies, or a variety of a species, or a variety of a subspecies. For instance, the tortuously long *Camponotus (Myrmoturba) maculatus* st. *melanocnemis* var. *lohieri* Santschi, 1913, was referred to just a couple of years later as *Camponotus (Myrmoturba) maculatus* var. *lohieri* Emery, 1915; it is currently regarded as a straight synonym of *C. maculatus*. This complexity was complicated further by the fact that a single author often did not show any consistency, referring a name to one grade in one paper and a different grade in another. The International Code of Zoological Nomenclature (fourth edition, 1999) now regulates these excesses. Readers of older taxonomic papers should be aware of its provisions and bear them in mind when trying to interpret the status of the published names.

TAXONOMIC NOVELTIES

A number of modifications to the preexisting taxonomy are initiated in this volume. They are discussed at the appropriate places in the text:

Subfamily Apomyrminae is revived from synonymy and reinstated.

Two genera are newly described: *Erromyrma* (Myrmicinae) and *Lividopone* (Dorylinae).

One new genus-rank synonym is proposed: *Vitsika* = *Myrmisaraka* (Myrmicinae).

One species is transferred between genera: *Euponera suspecta* Santschi, 1914, is newly combined as *Parvaponera suspecta* (Santschi, 1914).

2 new synonyms are proposed of names in the species-group: *Messor galla* = *M. galla obscurus* (Myrmicinae); *Bothroponera cambouei* = *Pachycondyla kipyatkovi* (Ponerinae).

FAMILY FORMICIDAE

The Ants

All ants are contained within the single family Formicidae within the Aculeata, which is a monophyletic group of families within the monophyletic suborder Apocrita, of the order Hymenoptera (Gauld and Bolton, 1988; Goulet and Huber, 1993). The family is defined by a unique combination of biological and morphological traits, which is given here. In recent years a number of molecular phylogenies that span the whole family have been produced. Notable among these are Moreau, et al. (2006); Brady, et al. (2006); Ouellette, Fisher, and Girman (2006); Moreau (2009); and Moreau and Bell (2013). Phylogenetic studies that deal in detail with single subfamilies are mentioned in the introductions to the subfamilies concerned. A useful overview of ant phylogeny and classification has been published by Ward (2007), and Bolton (2003) provided the most recent morphological analysis of all the subfamilies.

Family FORMICIDAE Latreille, 1809

BIOLOGY: The vast majority of ants can easily be recognized because they are eusocial hymenopterous insects that have a wingless worker caste, and their colonies persist for more than one season, usually producing a new generation of reproductives (queens and males) annually. Alate sexual forms engage in mass nuptial flights, and the wings of alate queens are deciduous and shed after mating.

MORPHOLOGY: Hymenoptera Aculeata with the following combination of characters:

The head is prognathous.

An infrabuccal sac is present between the labium and the hypopharynx.

The antenna is geniculate between the elongated first segment (the scape) and the remaining segments together (the funiculus).

The antenna has 4–12 segments in female castes, and 9–13 in males.

A metapleural gland is usually present in the female castes.

Abdominal segment 2 forms a differentiated petiole; it is always isolated from the mesosoma by a constriction and an articulation, and there is almost always a constriction between it and the following segment (A3).

In the forewing of alate sexual forms, the cross-veins 3rs-m and 2m-cu are always absent.

SEXES AND CASTES: All ant species have 2 sexes, female and male—the former of which has a diploid chromosome number, the latter a haploid chromosome number, as in all Hymenoptera. In the overwhelming majority of ant species, the female sex is divided into 2 castes: a wingless worker (ergates) caste, with a relatively simple mesosoma, and an alate (winged) queen (gyne) caste in which the mesosoma, as well as bearing wings when virgin, has a greater number of mesosomal sclerites. Queens that have mated and shed their wings may continue to be referred to as alate, reflecting their original appearance, or be called dealate.

There are, however, numerous exceptions to this general rule. The worker caste is usually monomorphic, of a single form where all the workers of a colony are morphologically alike, but in some genera the workers may be dimorphic (of 2 different forms) or polymorphic (of 3 or more different forms). In polymorphic forms the workers of each morph may be distinct, but usually their morphologies form a continuous sequence of structural modification—from smallest (usually the simplest) to largest (usually the most specialized). In a relatively few genera, workers take over the role of reproductives in the colony, where they are called gamergates. In these societies no queen is present, the caste having been abandoned. In advanced socially parasitic species, the worker caste may be entirely lost.

The queen caste is usually easily distinguished from the worker, but in some genera the queen never develops wings, has the overall appearance of a worker, though often larger, and is termed an ergatoid (worker-like) queen. Most species have either one or the other form of queen, but in some species a colony may contain both alate and ergatoid queens. Even less commonly, some species have both alates and ergatoids, as well as one or more intercastes—forms that are morphologically intermediate between the alate and the ergatoid forms. Conspicuous among the nomadic, mass predatory taxa of the Dorylinae is a monstrous ergatoid queen with very modified morphology, termed a dichthadiigyne.

The male is usually alate, and this sex is only rarely subdivided into separate castes, but in some species wingless worker-like males are produced, termed ergatoid males. These ergatoid males may replace the usual alate males or may exist together with them in a single species. Extremely rarely, the ergatoid male caste is present as 2 discrete subcastes.

In the part of the book that details the individual genera, comments are included for those with peculiar reproductive forms, but the majority (those with a monomorphic worker, alate queen, and alate male) are not specially mentioned. The definitions of subfamilies and genera in the text are based on the worker caste. This is because it is the most numerous and commonly encountered caste, present year-round both within and outside of colonies. Queens, on the other hand, are usually restricted to a single individual, always concealed within a colony, and males are transient, only produced once each season and not leaving the nest until the time of the nuptial flight, after which they soon die. There are a number of genera in which the males remain unknown, and many in which the number of recorded queens falls far short of the actual number of known species. As a consequence, almost all morphology-based ant taxonomy depends on the workers. In recent years some effort has been made to increase the numbers of known males, but this endeavor is far from complete.

AFROTROPICAL AND MALAGASY SUBFAMILIES

The family Formicidae is currently divided into 17 extant subfamilies and 4 that are wholly extinct. In the regions under consideration here, 11 extant subfamilies occur. Present in both the Afrotropical and the Malagasy regions are 8 subfamilies, including the world's 4 largest subfamilies Formicinae, Dolichoderinae, Myrmicinae, and Ponerinae together with the 4 smaller subfamilies Amblyoponinae, Dorylinae, Proceratiinae, and Pseudomyrmecinae. In addition, the Afrotropical region has representatives of 3 very small subfamilies (Agroecomyrmecinae, Apomyrminae, and Leptanillinae) that do not occur in the Malagasy region. Of the 6 subfamilies that are entirely absent from the Afrotropical and the Malagasy regions, Martialinae and Paraponerinae are both monotypic and restricted to the Neotropical region. Aneuretinae, another monotypic group, is found only in Sri Lanka, although it has an extensive fossil record, and Myrmeciinae is almost entirely restricted to Australia. The final 2 subfamilies both exhibit a mainly Gondwanic distribution. Heteroponerinae occurs in the Neotropical and Austral regions, with a strange isolated genus in the Palaearctic. Ectatomminae has mainly the same distribution but also extends into the Malesian (Indo-Australian) and Oriental regions, with a few species penetrating the southern Nearctic.

The subfamilies of Formicidae that occur in the Afrotropical and Malagasy regions together are identified below in a single dichotomous key. This is followed by comments and morphological definitions concerning each subfamily, in alphabetical order. Definitions of specialized anatomical terms encountered in the key are given in the morphological glossary at the end of the book.

Key to Afrotropical and Malagasy Subfamilies (Workers)

1a Body with a single isolated or reduced segment (A2 = petiole) between the mesosoma and gaster **(A)**; A3 is either entirely confluent with A4 **(B)** or is separated from it by a girdling constriction **(C)**; if the latter, then A3 is not markedly reduced in size with respect to A4 2

1b Body with 2 isolated or distinctly reduced segments (petiole and postpetiole = A2 and A3) between the mesosoma and gaster **(AA)**; either both segments are much reduced or the second is somewhat larger than the first, but if the latter, then A3 is distinctly very much smaller than A4 **(DD)** 8

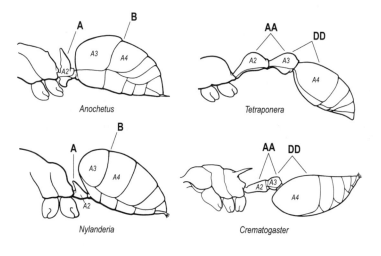

Anochetus   *Tetraponera*

Nylanderia   *Crematogaster*

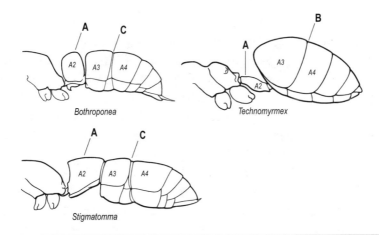

Bothroponea

Technomyrmex

Stigmatomma

2a (1)   Apex of abdomen with a semicircular to circular orifice, the acidopore, formed from the hypopygium **(A)**. This structure often projects as a nozzle that is fringed with setae, but sometimes there is no nozzle or setae, and the acidopore is overlapped by the pygidium when not in use (becomes visible if the sclerites are separated); in this latter condition, the antennal sockets are located well behind the posterior clypeal margin. Sting absent . . . . **Formicinae**

2b   Apex of abdomen with hypopygium lacking a semicircular to circular acidopore **(AA)**. Sting present or absent; when present, it is usually easily visible, but when reduced or vestigial (or sometimes if present but completely retracted), the hypopygium has a smooth posterior margin and the antennal sockets abut the posterior margin of the clypeus. 3

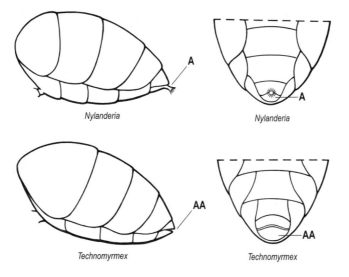

Nylanderia

Nylanderia

Technomyrmex

Technomyrmex

3a (2)   With head in ventral view and mouthparts closed, the prementum is not visible; it is entirely concealed by the labrum anteriorly and the maxillary stipites
on each side; the stipites meet along the midline **(A)**. Metapleural gland orifice overhung and concealed from above by a cuticular lip or flange. Spiracles
on the tergites of A5–A7 exposed, not overlapped and concealed by the ter-
gites of the preceding segments **(B)**. . . . . **Dorylinae** (part)

3b   With head in ventral view and mouthparts closed, the prementum is clearly
visible; it is bounded by the labrum anteriorly and the maxillary stipites on
each side of it **(AA)**. Metapleural gland orifice not overhung and concealed
from above by a cuticular lip or flange. Spiracles on the tergites of A5–A7
overlapped and concealed by the tergites of the preceding segments **(BB)** **4**

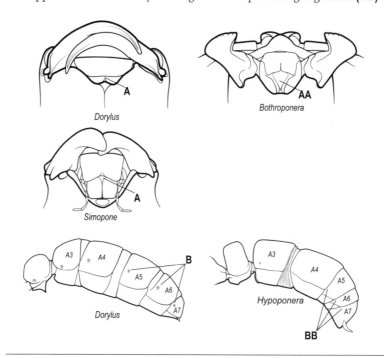

Dorylus

Bothroponera

Simopone

Dorylus

Hypoponera

4a (3)   Tergite of the helcium with an extensive U-shaped or V-shaped emargination
dorsally in its anterior margin, easily visible if the gaster (= A3–A7) is slightly
depressed from the petiole (A2) **(A)**. Sting vestigial or absent, in any case the
vestiges not detectable without dissection . . . . **Dolichoderinae**

4b   Tergite of the helcium entire, its anterior margin dorsally without a U-shaped
or V-shaped emargination **(AA)**. Sting present and functional, often projecting in dead specimens; in many species the sting shaft is visible through the
cuticle of the abdominal ventral apex, even when it is fully retracted **5**

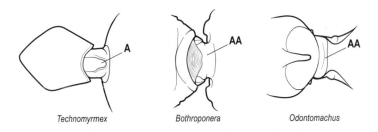

Technomyrmex   Bothroponera   Odontomachus

5a (4)   Promesonotal suture absent from the dorsum of the mesosoma **(A)**
. . . . **Proceratiinae**

5b   Promesonotal suture fully developed across the dorsum of the mesosoma, articulated, the suture flexible in fresh specimens **(AA)**. 6

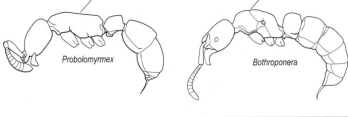

Probolomyrmex   Bothroponera

6a (5)   Antennal sockets entirely exposed in full-face view **(A)**. Labrum with numerous peg-like dentiform setae arranged in transverse rows **(B)**. A transverse sulcus present across the sternite of A3 posterior to the helcium **(C)**
. . . . **Apomyrminae**

6b   Antennal sockets partially to entirely covered by the frontal lobes in full-face view **(AA)**. Labrum without transverse rows of peg-like dentiform setae **(BB)**. Sternite of A3, posterior to the helcium, without a transverse sulcus **(CC)**
. 7

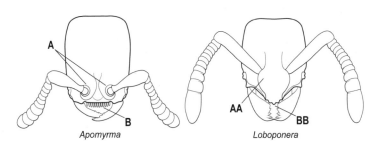

Apomyrma   Loboponera

(Figure continues on next page)

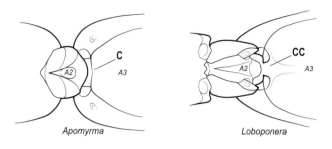

Apomyrma | Loboponera

7a (6) Petiole (A2) very broadly attached to A3, so that in profile the petiole does not have a free, descending posterior face **(A)**. Helcium in profile with its dorsal surface projecting from very high on the anterior face of A3, so that above the helcium A3 has no free anterior face **(B)**. Orifice of the metapleural gland is directed prodominantly dorsally and posteriorly, on a curved surface mesad of a posterolateral swelling or plate . . . . **Amblyoponinae**

7b Petiole (A2) narrowly attached to A3, so that in profile the petiole has a distinct free, descending posterior face **(AA)**. Helcium in profile with its dorsal surface projecting from the midheight or lower on the anterior face of A3, so that above the helcium A3 has a free anterior face **(BB)**. Orifice of the metapleural gland a simple hole, directed laterally or posteriorly, not on a curved surface mesad of a posterolateral swelling or plate . . . . **Ponerinae**

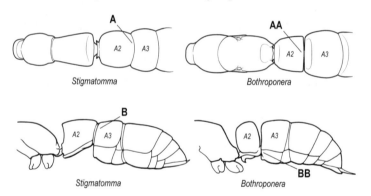

Stigmatomma | Bothroponera

Stigmatomma | Bothroponera

8a (1)   Eyes present and usually conspicuous, composed of 1 to many ommatidia **(A)** . . . 9

8b   Eyes entirely absent **(AA)** 11

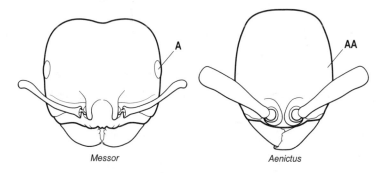

Messor   Aenictus

9a (8)   Eyes located at the extreme posterolateral corners of the head, at the posterior apex of a broad, shallow scrobe **(A)**. First gastral segment (A4) with tergosternal fusion, the tergite massively hypertrophied and ball-like, the sternite extremely reduced **(B)**. Segments A5–A7 reduced and telescoped within A4, together directed anteriorly. Metapleural gland orifice opens laterally

. . . . **Agroecomyrmecinae** (genus *Ankylomyrma* only)

9b   Eyes not located at the extreme posterolateral corners of the head, not at the posterior apex of a broad shallow scrobe **(AA)**. First gastral segment (A4) with the tergite and sternite not fused, the tergite not ball-like, the sternite not reduced **(BB)**. Segments A5–A7 not reduced, not telescoped within A4, not directed anteriorly. Metapleural gland orifice opens directly laterally or is a narrow slit on the side of the sclerite. 10

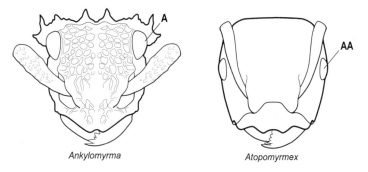

Ankylomyrma   Atopomyrmex

(Figure continues on next page)

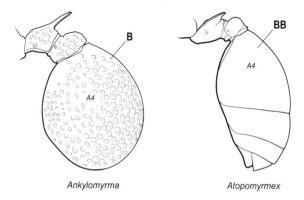

Ankylomyrma   *Atopomyrmex*

10a (9) Promesonotal suture conspicuously present and fully articulated across the dorsum of the mesosoma **(A)**. Metapleural gland orifice simple, opening directly laterally, the orifice conspicuous and located immediately above the lower margin of the metapleuron. Metatibia with 2 spurs (anterior may be very small) **(B)**. Pretarsal claw of hind leg with a preapical tooth present on the inner curvature (may be difficult to see in small species) **(C)**
. . . .**Pseudomyrmecinae** (genus *Tetraponera* only)

10b Promesonotal suture usually absent across the dorsum of the mesosoma, but sometimes a weak impression or a narrow, fully fused transverse line can be seen **(AA)**. Metapleural gland orifice a longitudinal slit or narrow crescent that opens dorsally to posterodorsally and is located on the side, well above the lower margin of the metapleuron (slit may be very narrow and difficult to see). Metatibia with 0–1 spur **(BB)**. Pretarsal claw of hind leg simple, without a preapical tooth on the inner curvature **(CC)** . . . . **Myrmicinae** (part)

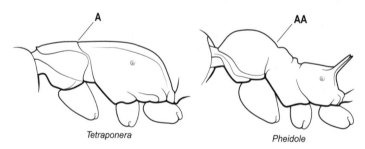

Tetraponera   *Pheidole*

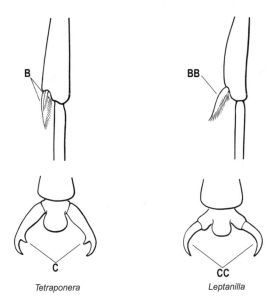

Tetraponera

Leptanilla

11a (8) Promesonotal suture conspicuously present across the dorsum of the mesosoma, deeply impressed and fully articulated **(A)**. Propodeal lobes absent **(B)**.
Pygidium large and conspicuous. **(C)** . . . . **Leptanillinae** (genus *Leptanilla* only)

11b Promesonotal suture absent across the dorsum of the mesosoma **(AA)**. Propodeal lobes present **(BB)**. Pygidium small and inconspicuous **(CC)**. **12**

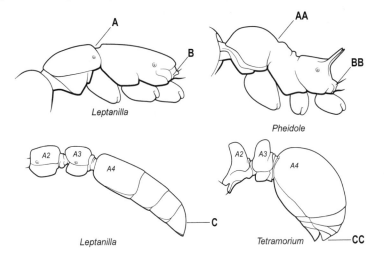

Leptanilla

Pheidole

Leptanilla

Tetramorium

12a (11) Metatibial gland present: apical half of ventral surface of metatibia with an elongate impression or area of extremely thin translucent cuticle **(A)**. Metapleural gland orifice opens laterally but is overhung from above by a flap of cuticle. Antennal sockets very close to the anterior margin of the head, always fully exposed; frontal lobes entirely absent **(B)**. Parafrontal ridges present **(C)**. Antenna with 10 segments . . . . **Dorylinae** (part; genus *Aenictus*)

12b Metatibial gland absent: apical half of ventral surface of the metatibia without an elongate impression or an area of extremely thin translucent cuticle. Metapleural gland orifice a longitudinal slit or narrow crescent that opens dorsally to posterodorsally (may be difficult to see). Antennal sockets well behind the anterior margin of the head; frontal lobes present that partially to entirely conceal the antennal sockets in full-face view **(BB)**. Parafrontal ridges absent **(CC)**. Antenna with 4–12 segments . . . . **Myrmicinae** (part)

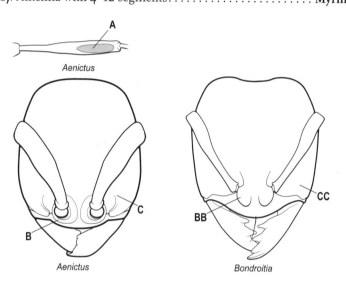

Aenictus

Aenictus

Bondroitia

Subfamily Accounts

AGROECOMYRMECINAE Carpenter, 1930

INTRODUCTION: Agroecomyrmecinae is a very small subfamily that until now has contained only the monotypic extant Neotropical genus *Tatuidris* Brown and Kempf together with 2 fossil genera, *Agroecomyrmex* Wheeler, W.M., from the Eocene Baltic Amber, and *Eulithomyrmex* Carpenter from the Oligocene of the USA. To these, the Afrotropical genus *Ankylomyrma* was recently transferred here from Myrmicinae based on molecular phylogenetic analysis (Ward, et al. 2015).

Following a full dissection of the worker of *Ankylomyrma* by one of us (Bolton), it became clear that the genus certainly did not belong in Myrmicinae, as it differs from the uniform morphology of all members of that very large subfamily in terms of structure of the torulus, form of the metapleural gland orifice, very different structure of the helcium, and complete tergosternal fusion of A4. The establishment of a separate subfamily to contain this genus, based on the tribe name Ankylomyrmini proposed by Bolton (2003), was considered, but the publication by Ward, et al. (2015) of DNA analyses of myrmicine genera indicated the close affinity of *Ankylomyrma* with *Tatuidris*. The new definition that follows applies to both extant representatives of the subfamily, and the characters of both are compared for clarity; some other characters are noted under the definition of *Ankylomyrma*. Although the differences seem great, their range is in fact no greater than is exhibited by the various genera of Myrmicinae. However, it is probably best to regard the subfamily as having 2 tribes, based on the differences noted here: Ankylomyrmini Bolton (2003), containing only the arboreal Afrotropical *Ankylomyrma coronacantha*, and Agroecomyrmecini Carpenter (1930), which contains *Tatuidris* and the fossil genera mentioned above.

DIAGNOSTIC REMARKS: *Ankylomyrma* has a bizarre appearance and can immediately be identified by the following combination of characters: eyes present, at extreme posterior apices of antennal scrobes; pronotum strongly convex and with 4 pairs of tubercles; metapleural gland orifice a simple round to elliptical hole that opens laterally in the lower posterior corner of the metapleuron; waist of 2 segments; propodeum and petiole (A2) each bispinose; abdominal segment 4 (first gastral) hypertrophied and ball-like, consisting almost entirely of the tergite.

GENUS INCLUDED IN THIS STUDY: *Ankylomyrma*.

DEFINITION OF THE SUBFAMILY (WORKER): Formicidae with the following characters in combination:

Prementum exposed when mouthparts closed (ventral view of head); prementum bounded by the labrum anteriorly and the maxillary stipites on each side. [In *Tatuidris* the hypertrophied labrum covers the entire buccal cavity, but the stipites do not overlap the prementum.]

Eyes present, located at extreme posterior apex of the antennal scrobe; ocelli absent.

Clypeus large, triangular, broadly inserted between the frontal lobes, conspicuously present in front of the antennal sockets; anterior margins of the sockets located behind the anterior margin of the head.

Frontal carinae and frontal lobes present.

Torulus a simple, near-vertical annulus; its upper arc does not form a laterally projecting flange or lobe below the frontal carina. [In *Tatuidris* the torulus has a ventrally directed lobe on its upper arc below the frontal carina.]

Antennal sockets in full-face view entirely concealed by the frontal lobes.

Antenna with 12 segments, gradually incrassate apically. [In *Tatuidris* the antenna with 7 segments, with a strong apical club of 2 segments.]

Antennal scrobes present, the eyes located at their posterior apices.

Promesonotal suture absent on the dorsum of the mesosoma.

Epimeral sclerite absent.

Metapleural gland orifice a simple round to elliptical hole that opens laterally in the lower posterior corner of the metapleuron.

Metacoxal cavities fully closed, with an uninterrupted band of cuticle between the propodeal foramen and the metacoxal cavity.

Propodeal lobes present.

Tibial spurs: mesotibia 1; metatibia 1.

Metatibial gland absent.

Pretarsal claws simple.

Waist of 2 segments (A2 and A3 = petiole and postpetiole).

Abdominal segment 2 (petiole) sessile or stoutly short-pedunculate.

Abdominal segment 2 with the tergite and sternite fused.

Helcium sternite visible in profile. With the helcium detached and in anterior view, the tergite overlaps the sternite on each side; the sternite is convex and weakly bulges ventrally to slightly below the level of the tergal apices.

Abdominal segment 3 (postpetiole) with the tergite and sternite fused. [In *Tatuidris* unfused.]

Abdominal segment 4 (first gastral segment) with narrow presclerites.

Abdominal segment 4 with the tergite and sternite fused. The tergite is either enlarged and vaulted (*Tatuidris*) or is enormously hypertrophied and ball-like (*Ankylomyrma*); in the latter case, the sternite is reduced to a narrow flange, fused to the anteroventral margin of the tergite so that together they form an anteriorly directed circular orifice within which the very reduced remaining segments (A5–A7) are telescoped.

Abdominal segments 5–7 without presclerites.

Abdominal tergites of 5–7 with spiracles not visible without dissection or distension.

Stridulitrum present on the pretergite of A4.

Sting present and functional.

AMBLYOPONINAE Forel, 1893

INTRODUCTION: Amblyoponinae is a small subfamily of 10 genera, with an almost worldwide distribution, 6 genera of which occur in the regions under consideration here. For most of its taxonomic history it has been considered as a tribe of Ponerinae, as in an earlier review of its genera by Brown (1960). The group was established at subfamily rank by Bolton (2003), and a recent review of its component genera has been published by Yoshimura and Fisher (2014). The Afrotropical and Malagasy regions together currently have only 29 described species, but many more await description in museum collections. All amblyoponine species are hypogaeic or cryptic. They are most often found in leaf litter and the soil, sometimes very deep down, but some species nest and forage in rotten wood.

DIAGNOSTIC REMARKS: Amblyoponines are mostly unlikely to be confused with other subfamilies as their habitus is generally obvious. Eyes may be absent, but when present they are usually located behind the midlength of the side of the head; the anterior clypeal margin usually has tooth-like stout setae present; the promesonotal suture is present and articulated; the metapleural gland orifice is directed dorsally and posteriorly; the waist is of 1 segment, the petiole (A2), which is broadly attached to A3 and which lacks a differentiated posterior surface; the helcium projects from very high on the anterior surface of A3; a sting is always present. Amblyoponines are only likely to be confused with some Ponerinae, but in the latter the petiole is always narrowly attached to the gaster, and the helcium arises at the midheight or more usually very low down, on the anterior face of A3. In addition, the metapleural gland orifice in ponerines is simple, usually in the lower posterior corner of the metapleuron, and opens laterally or posteriorly; it is never located on a curved surface mesad of a posterolateral swelling or plate.

GENERA INCLUDED IN THIS STUDY: *Adetomyrma, Concoctio, Mystrium, Prionopelta, Stigmatomma, Xymmer.*

DEFINITION OF THE SUBFAMILY (WORKER): Formicidae with the following characters in combination:

Prementum exposed when mouthparts closed (ventral view of head); prementum bounded by the labrum anteriorly and the maxillary stipites on each side.

Eyes present or absent; when present, usually located behind the midlength of the head; ocelli absent.

Labrum without rows of peg-like dentiform setae that project anteriorly.

Anterior margin of the clypeus with dentiform setae usually present that generally appear as teeth (absent in *Xymmer*).

Clypeus varies from broad to very narrow in front of the antennal sockets; anterior margins of the sockets are usually located well behind the anterior margin of the head but are close to the anterior margin in some.

Frontal carinae present, expanded anteriorly into frontal lobes.

Torulus with a dorsal flange that is visible immediately below the frontal carina or is fused to the frontal carina.

Antennal sockets partially to entirely concealed by the frontal lobes.

Antenna with 8–12 segments.

Antennal scrobes absent.

Promesonotal suture present on the dorsum of the mesosoma, articulated and flexible in fresh specimens.

Epimeral sclerite absent.

Metapleural gland orifice directed dorsally and posteriorly, on a curved surface mesad of a posterolateral swelling or plate.

Metacoxal cavities fully closed or with a distinct flexible suture between the metacoxal cavity and the propodeal foramen.

Propodeal lobes present.

Metatibial gland absent.

Pretarsal claws small and simple.

Waist of 1 segment (A2 = petiole).

Abdominal segment 2 (petiole) sessile; in profile without a differentiated posterior surface.

Petiole (A2) with the tergite and sternite fused anteriorly (sometimes with no trace of suture) but unfused and free posteriorly.

Helcium projects from very high on the anterior surface of A3.

Helcium sternite not visible in profile. With the helcium detached and in anterior view, the tergite overlaps the sternite on each side; the sternite is transverse and does not bulge ventrally below the level of the tergal apices.

Abdominal segment 3 (first gastral segment) with the tergite and sternite usually fully fused (unfused in *Adetomyrma*).

Abdominal segment 4 (second gastral segment) with the tergite and sternite usually fused (unfused in *Adetomyrma*).

Abdominal segment 4 usually with presclerites present (absent in *Adetomyrma*).

Abdominal segments 5–7 without presclerites.

Spiracles on the tergites of A5–A7 concealed under the posterior margin of the preceding tergite.

Stridulitrum absent from the tergite of A4.

Pygidium distinct, unarmed (hypertrophied in *Adetomyrma*).

Sting present and functional, usually large and conspicuous.

APOMYRMINAE Dlussky and Fedoseeva, 1988 stat. rev.

INTRODUCTION: This small subfamily contains only the monotypic Afrotropical genus *Apomyrma*. Apomyrmini was initially proposed as a tribe of Ponerinae by Dlussky and Fedoseeva (1988) and soon after was elevated to subfamily rank by Baroni Urbani, Bolton, and Ward (1992). Some years later, Saux, Fisher, and Spicer (2004) placed *Apomyrma* in Amblyoponinae, despite its extremely different and bizarre morphology. However, recent advanced DNA analysis by one of us (Fisher, unpublished) indicates that *Apomyrma* belongs to a clade that is sister to Amblyoponinae, not part of it, and therefore we reinstate Apomyrminae at subfamily rank here. The single Afrotropical species is hypogeic and has been recovered from leaf litter and soil samples.

DIAGNOSTIC REMARKS: The unique species of Apomyrminae is immediately diagnosed by its double row of anteriorly directed peg-like setae on the labrum (and their absence from the clypeus), lack of eyes and frontal carinae, fully exposed antennal sockets that are very close to the anterior margin of the head, very deeply impressed and articulated promesonotal suture, 1-segmented nodiform waist with the petiole (A2) short-pedunculate anteriorly and with a well-defined posterior surface, and the helcium attached low down on the anterior face of the the first gastral segment (A3).

GENUS INCLUDED IN THIS STUDY: *Apomyrma*.

DEFINITION OF THE SUBFAMILY (WORKER): Formicidae with the following characters in combination:

Prementum exposed when mouthparts closed (ventral view of head); prementum
   bounded by the labrum anteriorly and the maxillary stipites on each side.

Labrum with a transverse double row of peg-like dentiform setae that project anteriorly;
   anterior margin of the clypeus without dentiform setae.

Eyes absent; ocelli absent.

Clypeus very narrow in front of the antennal sockets, anterior margins of the sockets
   close to the anterior margin of the head.

Frontal carinae entirely absent; a short and low median tumulus or ridge present
   between the antennal sockets.

Antennal sockets entirely exposed in full-face view.

Torulus a simple annulus.

Antenna with 12 segments.

Antennal scrobes absent.

Promesonotal suture present on the dorsum of the mesosoma, articulated and very
   deeply impressed; flexible in fresh specimens; metanotal groove absent.

Metapleural gland orifice in the lower posterior corner of the metapleuron, opening
   lateroventrally.

Metacoxal cavities fully closed, with an unbroken annulus around each foramen.

Propodeal spiracle posterior and low down on the sclerite.

Propodeal lobes present but reduced.

Tibial spurs: mesotibia 2, metatibia 2; on each tibia the anterior spur small and simple, the posterior spur larger and pectinate.

Pretarsal claws small and simple.

Waist of 1 segment (A2 = petiole).

Abdominal segment 2 (petiole) short-pedunculate anteriorly, nodiform, with well-defined posterior surface.

Petiole (A2) with the tergite and sternite fused for most of the length, but an unfused minute medioventral sternal sclerite remains present posteriorly.

Helcium attached low on the anterior face of the first gastral segment (A3), so that in profile A3 above the helcium has a free anterior face.

Helcium sternite not visible in profile. With the helcium detached and in anterior view, the tergite overlaps the sternite on each side; the sternite is transverse and does not bulge ventrally below the level of the tergal apices.

Abdominal segment 3 (first gastral segment) with the tergite and sternite fused. Sternite of A3 with a transverse sulcus, posterior to the helcium.

Abdominal segment 4 (second gastral segment) with the tergite and sternite unfused.

Abdominal segment 4 without presclerites.

Spiracles on the tergites of A5–A7 concealed under the posterior margin of the preceding tergite.

Stridulitrum absent from the tergite of A4.

Pygidium hypertrophied, unarmed, convex across and downcurved posteriorly.

Sting present and functional, large and conspicuous.

DOLICHODERINAE Forel, 1878

INTRODUCTION: One of the 4 largest subfamilies, Dolichoderinae has a worldwide distribution, with 28 extant and 18 fossil genera. The regions under consideration here are depauperate in dolichoderine genera, with only 8 present. Among these one is known to be, and another is suspected of being, introduced from other regions, leaving just 6 genera with definitely endemic species. The most recent comprehensive review of extant dolichoderine genera of the world is by Shattuck (1992), and the phylogeny of the subfamily has been covered by Ward, et al. (2010). The Afrotropical and Malagasy regions together currently have 89 species and subspecies, some of which are hypogeic, both nesting and foraging in the leaf litter. A small number are epigeic, but most are associated with plants, either nesting and foraging entirely on them or nesting terrestrially but foraging predominantly arboreally.

DIAGNOSTIC REMARKS: Superficially similar to Formicinae but always lacking an acidopore; instead, the pygidium and hypopygium of dolichoderines meet in a transverse slit. In both subfamilies the dorsal surface of the helcium usually has a deep, U-shaped notch or impression in its anterior border. This notch is present in all dolichoderines and in all formicines except *Polyrhachis*, where it has been secondarily reduced or lost. The notch is clearly visible if the gaster (A3–A7) is slightly depressed relative to A2 (petiole). In addition, dolichoderine mandibles have serrations or denticles on the basal margin; eyes are always present; the promesonotal suture is present and articulated; the waist is of one small segment, the petiole (A2), which in many common genera is extremely reduced and overhung from behind by the first gastral segment (A3); the sting is vestigial and nonfunctional or more usually absent.

GENERA INCLUDED IN THIS STUDY: *Aptinoma, Axinidris, Ecphorella, Linepithema, Ochetellus, Ravavy, Tapinoma, Technomyrmex.*

DEFINITION OF THE SUBFAMILY (WORKER): Formicidae with the following characters in combination:

Prementum exposed when mouthparts closed (ventral view of head); prementum bounded by the labrum anteriorly and the maxillary stipites on each side.
Mandible multidentate; basal margin of the mandible armed with serrations or denticles (unknown in *Ecphorella*).
Eyes present; ocelli absent.
Clypeus not reduced, conspicuously present in front of the antennal sockets; anterior margins of the sockets located well behind the anterior margin of the head.
Frontal carinae present to absent; when present, not expanded into the frontal lobes.
Torulus simple, usually a roughly circular sclerite; upper arc of the torulus projects laterally beyond the frontal carina in full-face view; entire torulus visible in those taxa that lack frontal carinae.
Antennal sockets partially to mostly exposed in full-face view.
Antenna with 11–12 segments.
Antennal scrobes absent.
Promesonotal suture present on the dorsum of the mesosoma, articulated and flexible in fresh specimens.
Metathoracic spiracles exposed.
Metapleural gland orifice present, opening laterally or posterolaterally near the lower posterior corner of the metapleuron.
Metacoxal cavities fully closed by a slender bar of cuticle that separates the metacoxal cavity from the propodeal foramen.
Propodeal lobes absent.
Metatibial gland absent.
Pretarsal claws simple.
Waist of 1 segment (A2 = petiole); segment extremely reduced and overhung by base of the gaster in some genera.
Abdominal segment 2 (petiole) with the tergite and sternite fused.
Petiole (A2) sessile to weakly pedunculate.
Helcium projects from A3 (first gastral segment) very low down on its anterior face.
Helcium sternite not visible in profile. With the helcium detached and in anterior view,

the tergite overlaps the sternite on each side; the sternite is transverse and retracted, and does not bulge ventrally below the level of the tergal apices.

Helcium tergite dorsally with a U-shaped emargination in its anterior margin, visible if A3–A7 (gaster) is slightly depressed relative to A2 (petiole).

Abdominal segment 3 (first gastral segment) with the tergite and sternite usually not fused, but in some genera fused basally on each side of the helcium.

Abdominal segment 4 (second gastral segment) with the tergite and sternite not fused.

Abdominal segment 4–7 without presclerites.

Abdominal segments 5–7 with spiracles often concealed under the posterior margin of preceding tergite, but sometimes these spiracles are visible, located just posterior to the limit of overlap of the preceding tergite.

Stridulitrum absent from the tergite of A4 (second gastral segment).

Pygidium small or very small, simple; in several genera reflexed so that it is on the ventral surface of the abdomen.

Sting vestigial to absent, nonfunctional and never visible (sting remnants not detectable without dissection); the hypopygium without an acidopore; the pygidium and hypopygium meet in a transverse slit.

DORYLINAE Leach, 1815

INTRODUCTION: A recent detailed phylogeny by Brady, et al. (2014) has redefined this subfamily by amalgamating the original 4 small subfamilies that occur in the Afrotropical and Malagasy regions (Aenictinae, Aenictogitoninae, Cerapachyinae, Dorylinae), plus 2 extralimital family-group taxa (Ecitoninae, Leptanilloidinae), under the single name Dorylinae. This survey refined and greatly expanded upon several earlier studies by Bolton (1990b), Brady (2003), and Brady and Ward (2005).

The same study informally resurrected a number of names in the genus-group that occur in the regions under consideration here: *Chrysapace, Lioponera, Ooceraea, Parasyscia,* and *Zasphinctus*. Formal taxonomic treatment of these genera, together with further modifications, will be presented by Marek Borowiec in a forthcoming study, currently in preparation. These changes have resulted in a subfamily that contains 28 extant genera and 1 fossil genus, of which 13 occur in the regions under consideration here. The subfamily has a worldwide distribution but is most species-rich in the tropics. The Afrotropical and Malagasy regions together have more than 250 doryline species and subspecies. Most of these are hypogeic or cryptic, nesting and foraging in leaf litter, soil, or rotten wood, and often prey on other social insects. Some common species are epigeic surface raiders, including the often encountered driver ants, a conspicuous group of the genus *Dorylus*. In contrast to the majority of dorylines, the species of a few genera are mostly or entirely arboreal, have large conspicuous eyes, and appear to prey on the brood of other ants.

DIAGNOSTIC REMARKS: Doryline ants possess unique characters that isolate them from all others. The prementum of the mouthparts is usually entirely concealed (described below); the clypeus is very narrow in front of the antennal sockets, and the latter are usually fully exposed; the promesonotal suture is often vestigial or absent but even when present and distinct is entirely fused and inflexible; the metapleural gland orifice is concealed below and behind a ventrally directed cuticular flange; the epimeral sclerite is always absent; a metatibial gland is usually present; the waist may be of 1 or 2 segments; the sternite of the helcium is visible in profile; the spiracles on abdominal tergites 5–7 are shifted posteriorly and exposed, not concealed under the posterior margin of the preceding tergite; the pygidium is always modified, being large and armed with teeth or denticles in all genera except in *Aenictus*, where it is very reduced.

In addition, a study in progress on the worker mesosomal endoskeleton suggests that in Dorylinae the endoskeletal arrangement is different from all other Formicidae. Dissections of *Dorylus* and *Simopone* show that the Y-shaped struts of the ascending process of the mesendoskeleton slope posteriorly; the dorsal apodemes of the mesendoskeleton extend directly laterally and terminate in a pleural endophragmal pit on each side (frequently visible externally); and the dorsal apodemes of the metendoskeleton are fused to the posterior surface of the mesendoskeletal dorsal apodemes. In effect, all the struts of the endoskeleton are fused into a rigid structure. Elsewhere in Formicidae (*Camponotus, Messor, Myrmecia, Pachycondyla, Paltothyreus*), the Y-shaped ascending process of the mesendoskeleton slopes anteriorly; the dorsal apodemes of the mesendoskeleton are divergent and directed anterolaterally; and the dorsal apodemes of the metendoskeleton are long and divergent, extend anteriorly dorsal to those of the mesendoskeleton, and are not fused with them.

GENERA INCLUDED IN THIS STUDY: *Aenictogiton, Aenictus, Chrysapace, Dorylus, Eburopone, Lioponera, Lividopone, Ooceraea, Parasyscia, Simopone, Tanipone, Vicinopone, Zasphinctus*.

DEFINITION OF THE SUBFAMILY (WORKER): Formicidae with the following characters in combination:

Prementum usually not visible when mouthparts closed (ventral view of head); the prementum is concealed by the labrum anteriorly and the maxillary stipites on each side, which overlap the prementum and meet along the ventral midline (prementum partially visible in *Aenictus*).

Eyes absent or present; ocelli absent or present.

Clypeus reduced, narrow from front to back, especially in front of the antennal sockets; anterior margins of the sockets extremely close to, or at, the anterior margin of the head.

Frontal carinae present as a pair of raised ridges that may be very closely approximated (sometimes almost fused into a single median crest), in *Simopone* expanded into narrow, horizontal frontal lobes.

Torulus without a laterally projecting flange or lobe above the socket but below the frontal carina.

Antennal sockets usually fully exposed in full-face view (partially concealed in *Simopone*).

Antenna with 7–12 segments.

Antennal scrobes usually absent (present only in 1 species of *Simopone*).

Promesonotal suture absent to strongly present on the dorsum of the mesosoma, but when present always completely fused and inflexible.

Epimeral sclerite absent.

Metapleural gland orifice in the lower posterior corner of the metapleuron, opening laterally, the orifice concealed below and behind a ventrally directed cuticular flange.

Metacoxal cavities fully closed by a continuous broad cuticular annulus that separates the propodeal foramen from the metacoxal cavity.

Propodeal lobes present or absent.

Endoskeletal arrangement of the mesosoma apparently unique in Formicidae (discussed above).

Metatibial gland usually present, located distally on the ventral surface (undetectable or absent in some).

Pretarsal claws usually simple, sometimes with a preapical tooth.

Waist of 1 or 2 segments (usually of 1 segment [petiole = A2] but in *Aenictus* and *Ooceraea* a distinctly reduced postpetiole [A3] is also present).

Abdominal segment 2 (petiole) with the tergite and sternite not fused.

Petiole (A2) usually sessile, uncommonly subsessile.

Helcium usually projects from about the midheight on the anterior surface of A3 (in *Lividopone* it is well above the midheight, close to the dorsum).

Helcium sternite visible in profile. With the helcium detached and in anterior view, the tergite overlaps the sternite on each side but the sternite is convex and bulges ventrally to below the level of the tergal apices.

Abdominal segment 3 with the tergite and sternite fused.

Abdominal segment 4 with the tergite and sternite unfused.

Abdominal segment 4 with strongly differentiated presclerites (presclerites also present on segments A5 and A6 in *Aenictogiton, Dorylus, Zasphinctus*).

Abdominal segments 5–7 with spiracles shifted posteriorly on the tergites and exposed, not concealed under posterior margin of the preceding tergite.

Stridulitrum absent from the pretergite of A4.

Pygidium (tergite of A7) modified: usually large and distinct, flattened to concave dorsally and usually armed with a tooth or row of denticles or peg-like setae on each side. In *Aenictus* the pygidium is reduced to a very narrow U-shaped sclerite.

Sting present (usually conspicuous but reduced and nonfunctional in *Dorylus*).

FORMICINAE Latreille, 1809

INTRODUCTION: In terms of numbers of species, Formicinae is second only in size to Myrmicinae. The subfamily has a worldwide distribution and contains 51 extant genera, together with about 18 extinct genera. Of the extant genera, 18 occur in the Afrotropical and Malagasy regions. There is no recent review of the component genera of Formicinae. The most recent overview of the current structure of the subfamily is provided by Bolton (2003), but this is admitted to contain a number of only weakly supported hypotheses concerning the relationships of some groups. The Afrotropical and Malagasy regions together contain over 600 formicine species and subspecies that occupy an enormous diversity of ecologies. Many smaller species are abundant in the leaf litter and the soil, and in rotten wood, but numerous taxa forage on the ground surface or on low vegetation, and many are entirely arboreal. Some arboreal forms nest in hollow stems, or in cavities in rotten wood, or in burrows formed by other insects, but others tunnel their own nests in living timber, and one species makes its nests by using larval silk to stitch leaves together.

DIAGNOSTIC REMARKS: The presence of an acidopore (descibed below) at the apex of the gaster is immediately diagnostic of formicine ants; there is no trace of a sting, and there is no structure similar to an acidopore in any other subfamily. In the field, formicines are only likely to be confused with dolichoderines, as the habitus of some may be similar. However, apart from the absence of an acidopore, dolichoderines have serrations or denticles on the basal margin of the mandible; the petiole (A2) is small and in many common genera is extremely reduced and overhung from behind by the first gastral segment (A3).

GENERA INCLUDED IN THIS STUDY: *Acropyga, Agraulomyrmex, Anoplolepis, Aphomomyrmex, Brachymyrmex, Camponotus, Cataglyphis, Lepisiota, Nylanderia, Oecophylla, Paraparatrechina, Paratrechina, Petalomyrmex, Phasmomyrmex, Plagiolepis, Polyrhachis, Santschiella, Tapinolepis.*

DEFINITION OF THE SUBFAMILY (WORKER): Formicidae with the following characters in combination:

Prementum exposed when mouthparts closed (ventral view of head); prementum
    bounded by the labrum anteriorly and the maxillary stipites on each side.
Eyes usually present, only very rarely absent; ocelli present or absent.
Clypeus not reduced, conspicuously present in front of the antennal sockets; anterior
    margins of the sockets located well behind the anterior margin of the head.
Frontal carinae absent to strongly present; when present, sometimes expanded into
    frontal lobes.
Torulus simple, usually a roughly circular sclerite; upper arc of the torulus usually proj-
    ects laterally beyond the frontal carina in full-face view (but not in those species with
    broad frontal lobes); entire torulus visible in those taxa that lack strong frontal carinae.
Antennal sockets partially to mostly exposed in full-face view.
Antenna with 9–12 segments.
Antennal scrobes absent.
Promesonotal suture usually present on the dorsum of the mesosoma, articulated and
    flexible in fresh specimens (suture is fused, very reduced or absent only in a few
    species of *Polyrhachis*).
Metathoracic spiracle exposed.
Metapleural gland orifice absent or present; when present, it opens laterally or posteri-
    orly near the lower posterior corner of the metapleuron (metapleural gland absent
    in *Oecophylla, Phasmomyrmex,* and *Polyrhachis*; usually absent in *Camponotus* but
    present in some Madagascan species).

Metacoxal cavities fully closed by a slender bar of cuticle that separates the metacoxal cavity from the propodeal foramen.

Propodeal lobes usually absent (present only in *Oecophylla*).

Metatibial gland absent.

Pretarsal claws simple.

Waist of 1 segment (A2 = petiole).

Abdominal segment 2 (petiole) with the tergite and sternite fused.

Petiole (A2) sessile to pedunculate.

Helcium projects from anterior surface of A3 very low down on its anterior face (at about the midheight only in *Oecophylla*).

Helcium sternite not visible in profile. With the helcium detached and in anterior view, the tergite overlaps the sternite on each side; the sternite is transverse and retracted, and does not bulge ventrally below the level of the tergal apices.

Helcium tergite dorsally usually with a U-shaped emargination in its anterior margin, visible if A3–A7 (gaster) is slightly depressed relative to A2 (petiole); emargination is usually extensive but is very shallow to absent in many *Polyrhachis* species.

Abdominal segment 3 (first gastral segment) with the tergite and sternite never entirely fused; either entirely unfused or fused basally for some distance on each side of the helcium.

Abdominal segment 4 (second gastral segment) with the tergite and sternite not fused.

Abdominal segments 4–7 without presclerites.

Abdominal segments 5–7 with spiracles often concealed under the posterior margin of the preceding tergite, but sometimes these spiracles are visible, located just posterior to the limit of overlap of the preceding tergite.

Stridulitrum absent from the tergite of A4 (second gastral tergite).

Pygidium (tergite of A7) usually large, unarmed.

Acidopore present, sting absent. The acidopore is a U-shaped to almost circular excavation of the posteromedian margin of the hypopygium; sometimes it is at the apex of a prominent nozzle that is fringed with setae, but sometimes both nozzle and setae are absent (*Camponotus, Polyrhachis*), in which case the acidopore may be concealed by the apex of the pygidium.

LEPTANILLINAE Emery, 1910

INTRODUCTION: A small subfamily of mostly minute, cryptic ants that is divided into 2 tribes, each tribe containing 3 genera. The genera of Leptanillinae were most recently reviewed by Bolton (1990a), but there have been some modifications since that publication. Recent molecular phylogenetic analyses suggest that Leptanillinae is the sister group to all other ants (Brady, et al. 2006; Moreau, 2009; Moreau and Bell, 2013). Distribution of the subfamily is throughout the Old World tropics and subtropics, in places with species present in the temperate zones. Only a single genus is present in the Afrotropical region, represented by a mere 3 described species; the subfamily is absent from, or at least has not yet been discovered in, the Malagasy region.

DIAGNOSTIC REMARKS: These minute and uncommon ants are unlikely to be confused with any others. Eyes and frontal carinae are absent; the clypeus is very reduced so that the antennal sockets, which are extremely close to the anterior margin of the head, are fully exposed; antennae are always of 12 segments; the mesosoma is slender, with a strongly developed promesonotal suture that is fully articulated; the waist is of 2 segments (A2 and A3), of which the petiole is sessile; the pygidium is large and simple, and a sting is present.

GENUS INCLUDED IN THIS STUDY: *Leptanilla.*

DEFINITION OF THE SUBFAMILY (WORKER): Formicidae with the following characters in combination:

Prementum exposed when mouthparts closed (ventral view of head); prementum bounded by the labrum anteriorly and the maxillary stipites on each side.

Eyes absent; ocelli absent.

Clypeus reduced, narrow from front to back especially in front of the antennal sockets; anterior margins of the sockets extremely close to, or at, the anterior margin of the head.

Frontal carinae usually absent; sometimes a pair of extremely closely approximated minute ridges, or a low median ridge, present.

Torulus a simple annulus without a laterally projecting flange or lobe above the socket but below the frontal carina.

Antennal sockets fully exposed in full-face view.

Antenna with 12 segments.

Antennal scrobes absent.

Promesonotal suture present on the dorsum of the mesosoma, deeply impressed, fully articulated and flexible in fresh specimens.

Epimeral sclerite absent.

Metapleural gland orifice in the lower posterior corner of the metapleuron, the orifice situated at the base of a longitudinal groove; both orifice and groove are overhung by a cuticular rim.

Metacoxal cavities small, fully closed by a band of cuticle that separates the propodeal foramen from the metacoxal cavity.

Propodeal lobes absent.

Metatibial gland absent.

Pretarsal claws simple.

Waist of 2 segments (A2 and A3 = petiole and postpetiole).

Abdominal segment 2 (petiole) with the tergite and sternite fused, without trace of a suture.

Petiole (A2) sessile.

Helcium sternite not visible in profile. With the helcium detached and in anterior view, the tergite overlaps the sternite on each side; the sternite is transverse and does not bulge ventrally below the level of the tergal apices.

Abdominal segment 3 (postpetiole) with the tergite and sternite fused.

Abdominal segment 4 (first gastral) with the tergite and sternite unfused.

Abdominal segment 4 with presclerites.

Abdominal segments 5–7 without presclerites.

Abdominal tergites 5–7 with spiracles concealed under the posterior margin of the preceding tergite.

Stridulitrum absent from the pretergite of A4.

Pygidium (tergite of A7) large and simple, unarmed, convex across and downcurved posteriorly.

Sting present and fully functional.

MYRMICINAE Lepeletier de Saint-Fargeau, 1835

INTRODUCTION: Easily the largest subfamily, both in terms of number of genera and number of species, and with a worldwide distribution. The subfamily contains approximately 145 extant genera and about 23 that are extinct. Represented in the regions under consideration here are 42 extant genera, 4 of which are known or suspected to be introductions from other regions. There is no recent review of the component genera of Myrmicinae, but Ward, et al. (2015) have published a DNA-based phylogeny that encompasses most of the genera. The Afrotropical and Malagasy regions together contain over 1,600 myrmicine species and subspecies, far more than any other subfamily; more in fact than all the other subfamilies combined. They are morphologically very diverse, numerous to abundant in all ecosystems, and nest and forage in all available niches, from deep underground to the tips of the tallest trees.

DIAGNOSTIC REMARKS: Unfortunately, most characters critical to the diagnosis of members of this vast and morphologically diverse subfamily require microscopic study and some degree of disarticulation or dissection. In the field, the vast majority of myrmicines may be recognized through their possession of a 2-segmented waist, combined with a broad clypeus, usual presence of frontal lobes that partially to entirely conceal the antennal sockets, and the absence of the promesonotal suture. In the laboratory these characters, combined with the form of the torulus, the structure of the metapleural gland orifice, and the unique morphology of the petiole and helcium (all described below), are diagnostic.

GENERA INCLUDED IN THIS STUDY: *Adelomyrmex, Anillomyrma, Aphaenogaster, Atopomyrmex, Baracidris, Bondroitia, Calyptomyrmex, Cardiocondyla, Carebara, Cataulacus, Crematogaster, Cyphoidris, Cyphomyrmex, Dicroaspis, Diplomorium, Erromyrma, Eurhopalothrix, Eutetramorium, Malagidris, Melissotarsus, Meranoplus, Messor, Metapone, Microdaceton, Monomorium, Myrmicaria, Nesomyrmex, Ocymyrmex, Pheidole, Pilotrochus, Pristomyrmex, Royidris, Solenopsis, Strumigenys, Syllophopsis, Temnothorax, Terataner, Tetramorium, Trichomyrmex, Vitsika, Vollenhovia, Wasmannia.*

DEFINITION OF THE SUBFAMILY (WORKER): Formicidae with the following characters in combination:

Prementum exposed when mouthparts closed (ventral view of head); prementum bounded by the labrum anteriorly and the maxillary stipites on each side. [In a couple of extralimital genera the hypertrophied labrum conceals the entire buccal cavity.]

Eyes present or absent; ocelli usually absent; very rarely the median ocellus may be present in genera where ocelli are normally absent; only *Metapone* often with ocelli.

Clypeus not reduced, present in front of the antennal sockets, usually conspicuously so; anterior margins of the sockets located behind the anterior margin of the head. Median portion of the clypeus projects back between the frontal carinae (except in *Melissotarsus*).

Frontal carinae present; usually expanded anteriorly into frontal lobes that may be very broad (frontal lobes vestigial to absent in *Pristomyrmex*).

Torulus with a laterally projecting dorsal flange or lobe immediately below the frontal carina; in full-face view this torulus flange or lobe is visible in genera with narrow frontal lobes, where it projects laterally or posterolaterally. In genera with broad frontal lobes the torulus flange can usually be seen in profile, below the margin of the lobe.

Antennal sockets in full-face view usually mostly to entirely concealed by the frontal lobes, only rarely mostly exposed.

Antenna with 4–12 segments.

Antennal scrobes present or absent; when present, may be above or below the eyes.

Promesonotal suture usually entirely absent on the dorsum of the mesosoma, but in some genera represented by a fused remnant that appears as a line or feeble indentation.

Epimeral sclerite absent.

Metapleural gland orifice a longitudinal slit or narrow crescent that opens dorsally to posterodorsally some distance above the metapleural ventral margin.

Metacoxal cavities fully closed; with an uninterrupted band of cuticle that separates the propodeal foramen from the metacoxal cavity.

Propodeal lobes usually present (minute to absent only in *Melissotarsus* and *Crematogaster*).

Metatibial gland absent.

Pretarsal claws simple.

Waist of 2 segments (A2 and A3 = petiole and postpetiole).

Abdominal segment 2 (petiole) usually distinctly pedunculate (sessile in *Cataulacus* and *Meranoplus*).

Petiole (A2) with the tergite and sternite fully fused.

Petiole (A2), when disarticulated from postpetiole (A3) and in posterior view, with the fused tergite and sternite equally convex, their inner margins forming a circle; the tergite and sternite meet edge to edge and the former does not overlap the latter.

Helcium sternite visible in profile. With the helcium detached and in anterior view, the tergite does not overlap the sternite; instead, the 2 sclerites meet edge to edge and form a circle.

Abdominal segment 3 (postpetiole) with the tergite and sternite usually not fused (fused in *Cataulacus* and apparently also in *Myrmicaria*).

Abdominal segment 4 (first gastral) with the tergite and sternite not fused.

Abdominal segment 4 with presclerites (except in *Melissotarsus*); the postsclerites of A4 not constricted to a narrow neck immediately behind the presclerites except in some species of *Ocymyrmex*.

Abdominal segments 5–7 without presclerites.

Abdominal tergites 5–7 with spiracles concealed under the posterior margin of the preceding tergite.

Stridulitrum present or absent on the pretergite of A4.

Pygidium (tergite of A7) small to distinct, biconvex and unarmed.

Sting usually present and functional but sometimes spatulate or thread-like, sometimes very reduced.

PONERINAE Lepeletier de Saint-Fargeau, 1835

INTRODUCTION: One of the 4 largest subfamilies, and with a worldwide distribution, Ponerinae includes 47 extant and about 8 extinct genera. Of the extant genera, 28 occur in the Afrotropical and Malagasy regions; only 1 genus is represented by introduced species. A new analysis of ponerine genera has recently been published by Schmidt and Shattuck (2014), following Schmidt's (2013) molecular phylogeny, and their revised system of genera is applied in this study. In the Afrotropical and Malagasy regions together, the ponerines number just over 400 species and subspecies. They are characteristically ants of the topsoil, leaf litter, ground surface, and rotten wood, but a few are entirely arboreal and others may inhabit termitaries or other specialized hunting grounds.

DIAGNOSTIC REMARKS: Ponerine ants generally have a thick and armor-like integument. The clypeus is not reduced, so that the antennal sockets are well behind the anterior margin of the head. The frontal carinae are restricted to frontal lobes whose outlines are semicircular or bluntly triangular in full-face view, have a pinched-in appearance posteriorly, and partially to entirely conceal the antennal sockets. The torulus medially is fully fused to the frontal lobe. The promesonotal suture is complete, conspicuous, and fully articulated. The waist is of a single segment (petiole) that is a node or scale and is sessile anteriorly. The helcium usually projects from very low down on the anterior surface of A3 but is placed somewhat higher in a few genera (up to about the midheight of A3). A sting is always present, usually large and conspicuous.

GENERA INCLUDED IN THIS STUDY: *Anochetus, Asphinctopone, Boloponera, Bothroponera, Brachyponera, Centromyrmex, Cryptopone, Dolioponera, Euponera, Feroponera, Fisheropone, Hagensia, Hypoponera, Leptogenys, Loboponera, Megaponera, Mesoponera, Odontomachus, Ophthalmopone, Paltothyreus, Parvaponera, Phrynoponera, Platythyrea, Plectroctena, Ponera, Promyopias, Psalidomyrmex, Streblognathus.*

DEFINITION OF THE SUBFAMILY (WORKER): Formicidae with the following characters in combination:

Prementum exposed when mouthparts closed (ventral view of head); prementum bounded by the labrum anteriorly and the maxillary stipites on each side.

Eyes present or absent; ocelli absent.

Clypeus not reduced, present in front of the antennal sockets; anterior margins of the sockets located behind the anterior margin of the head.

Clypeus with median portion narrowed posteriorly, narrowly inserted between the frontal lobes as a slender triangle or linear strip (broad in *Platythyrea* and *Megaponera*).

Frontal carinae present only as expanded frontal lobes that may be small to very broad. Inner borders of the frontal lobes very closely approximated mediodorsally or confluent for most of their length (not in *Megaponera* or *Platythyrea*).

Frontal lobes with lateral margins short semicircular or bluntly triangular in full-face view, with a pinched-in appearance posteriorly.

Torulus medially fully fused with the frontal lobe; in full-face view the torulus is sometimes discernible as an outer marginal strip on the frontal lobe.

Antennal sockets partially to entirely concealed by the frontal lobes.

Antenna with 12 segments; funiculus usually gradually incrassate apically, uncommonly with a club.

Antennal scrobes absent.

Promesonotal suture present on the dorsum of the mesosoma, fully articulated, flexible in fresh specimens.

Epimeral sclerite usually present, subcircular and discrete (absent in *Cryptopone, Dolioponera, Hypoponera*, some *Loboponera*).

Metapleural gland orifice usually in the lower posterior corner of the metapleuron (higher and more anterior in *Centromyrmex*); orifice opens laterally or posteriorly, not concealed by cuticular flanges either dorsally or ventrally.

Metacoxal cavities usually with a suture present across the cuticular annulus between the propodeal foramen and metacoxal cavity; uncommonly open so that the propodeal foramen and metacoxal cavity are confluent (some *Platythyrea*) or fully closed, with an unbroken band of cuticle separating the 2 foramina (e.g., some *Hypoponera*).

Mesotibiae, metatibiae, mesobasitarsi, and metabasitarsi usually without spines and without thick, enlarged spiniform traction setae (present in *Centromyrmex, Feroponera, Promyopias*; present only on mesotibia in *Cryptopone* and 1 species of *Psalidomyrmex*).

Pretarsal claws simple, or with 1 or more preapical teeth, or pectinate.

Propodeal lobes present.

Metatibial gland usually absent; when present, located distally on the posterior (not ventral) surface, close to the base of the spur.

Waist of 1 segment (A2 = petiole).

Abdominal segment 2 (petiole) with the tergite and sternite not fused.

Petiole (A2) sessile.

Petiole sternite usually simple, not bifurcated into an internal and an external plate posteriorly (bifurcated in *Asphinctopone, Brachyponera, Phrynoponera*, and some *Platythyrea*).

Petiole sternite anteroventrally, immediately behind its articulation with the propodeum, with an arched or broadly horseshoe-shaped strip of cuticle. Immediately behind this is a transverse impression or depressed area, usually followed by a zone with proprioceptor hairs (not in *Loboponera, Plectroctena, Psalidomyrmex*).

Helcium usually projects from the anterior surface of A3 (first gastral segment) very low down on its anterior face; less commonly the helcium is about at the midheight of A3 (*Centromyrmex, Feroponera, Platythyrea, Promyopias*); uncommonly it arises between these positions (*Cryptopone*).

Helcium sternite not visible in profile. With the helcium detached and in anterior view, the tergite overlaps the sternite on each side; the sternite is transverse and retracted, and does not bulge ventrally below the level of the tergal apices.

Abdominal segment 3 (first gastral segment) with the tergite and sternite fused.

Abdominal segment 4 (second gastral segment) with the tergite and sternite fused.

Abdominal segment 4 usually with distinct presclerites, but differentiated presclerites are absent in some genera (*Asphinctopone, Streblognathus*, some *Leptogenys*).

Abdominal segments 5–7 without presclerites.

Abdominal tergites 5–7 with spiracles concealed under the posterior margin of the preceding tergite.

Stridulitrum present or absent on the pretergite of A4.

Pygidium (tergite of A7) distinct, biconvex and unarmed.

Hypopygium (sternite of A7) usually unarmed (armed with spines or spiniform setae in *Ophthalmopone* and *Paltothyreus*).

Sting present and functional, usually large and conspicuous.

PROCERATIINAE Emery, 1895

INTRODUCTION: This small subfamily is present throughout the world's tropical zones and contains only 3 extant and 1 extinct genus. All 3 extant genera occur in both the regions studied here. Proceratiinae was established as a subfamily by Bolton (2003). Prior to this, its main genera (*Discothyrea, Proceratium*) were usually regarded as components of Ectatommini (e.g., Brown, 1958), with *Probolomyrmex* treated as a ponerine (e.g., Brown, 1975). The Afrotropical and Malagasy regions together currently have 26 described proceratiine species, the vast majority of which are entirely hypogeic or cryptic, but some will ascend shrubs and trees to forage, and 1 nests in rotten branches that are still attached to the tree.

DIAGNOSTIC REMARKS: The proceratiines are characterized by a narrow clypeus and antennal sockets that are close to the anterior margin of the head; the antennal sockets are mostly to entirely exposed; the promesonotal suture is entirely absent from the dorsum of the mesosoma, and the metanotal groove is usually also absent; the metapleural gland orifice is a simple hole; the waist is usually of 1 segment (petiole), but in some species the following segment (A3) is also somewhat reduced; the helcium is located at, or just above, the midheight of the anterior surface of A3; a fully functional sting is present.

GENERA INCLUDED IN THIS STUDY: *Discothyrea, Probolomyrmex, Proceratium.*

DEFINITION OF THE SUBFAMILY (WORKER): Formicidae with the following characters in combination:

Prementum exposed when mouthparts closed (ventral view of head); prementum bounded by the labrum anteriorly and the maxillary stipites on each side.

Eyes present or absent; ocelli absent (median ocellus extremely rarely present in *Proceratium*).

Clypeus reduced, narrow from front to back especially in front of the antennal sockets; anterior margins of antennal sockets close to or at the anterior margin of the head.

Frontal carinae present or fused into a single median longitudinal crest; when carinae present, sometimes expanded into narrow frontal lobes.

Torulus not fused to frontal carina, without a laterally projecting flange or lobe above the socket but below the frontal carina.

Antennal sockets fully exposed in full-face view, or at most with the inner margins of the sockets concealed.

Antenna with 6–12 segments.

Antennal scrobes usually absent (broadly, shallowly present in some *Discothyrea*).

Promesonotal suture absent on the dorsum of the mesosoma.

Metanotal groove usually absent, very rarely vestigially present.

Epimeral sclerite absent.

Metapleural gland orifice lateral, simple, without cuticular lobes or flanges.

Metacoxal cavities either fully closed or with a narrow suture across the cuticular annulus between the metacoxal cavity and the propodeal foramen.

Propodeal lobes present.

Propodeal spiracle with orifice circular.

Metatibial gland absent.

Pretarsal claws simple.

Waist usually of 1 segment (A2 = petiole); in some the third abdominal segment reduced and subpostpetiolate.

Abdominal segment 2 (petiole) with the tergite and sternite unfused (*Probolomyrmex*), or fused (*Discothyrea, Proceratium*).

Petiole (A2) sessile to subsessile.

Helcium projects from the anterior surface of A3 (first gastral segment) at about its midheight or somewhat above its midheight.

Abdominal segment 3 (first gastral segment) with the tergite and sternite fused.

Abdominal segment 4 (second gastral segment) with the tergite and sternite fused.

Abdominal segment 4 with presclerites.

Abdominal segments 5–7 without presclerites.

Abdominal tergites 5–7 with spiracles concealed under the posterior margin of the preceding tergite.

Stridulitrum absent from the pretergite of A4.

Pygidium (tergite of A7) distinct but small, unarmed.

Sting present and functional, usually conspicuous.

PSEUDOMYRMECINAE Smith, M.R., 1952

INTRODUCTION: This small subfamily, all species of which are arboreal, contains only 3 genera. Of these, 2 are restricted to the New World; the third occurs throughout the Old World tropics and is represented by 70 species and subspecies in the Afrotropical and Malagasy regions combined. The genera that constitute Pseudomyrmecinae were most recently revised by Ward (1990) and the phylogeny investigated by Ward and Downie (2005).

DIAGNOSTIC REMARKS: The pseudomyrmecines can only be mistaken for myrmicines, but in the former there is always a fully developed promesonotal suture that is fully articulated and flexible in fresh specimens—a character that is always absent from myrmicines. In addition, in pseudomyrmecines eyes are always present, as are slender frontal carinae; the metapleural gland orifice is a simple hole; a strong sting is always apparent. With disarticulation, the tergite of the helcium, in anterior view, overlaps the sternite on each side. This contrasts with the structure seen in myrmicines, where the tergite and sternite of the helcium form a circle, without an overlap of sclerites.

GENUS INCLUDED IN THIS STUDY: *Tetraponera*.

DEFINITION OF THE SUBFAMILY (WORKER): Formicidae with the following characters in combination:

Prementum exposed when mouthparts closed (ventral view of head); prementum bounded by the labrum anteriorly and the maxillary stipites on each side.

Eyes present, frequently large; ocelli present or absent.

Clypeus not reduced, present in front of the antennal sockets; the anterior margins of the sockets located well behind the anterior margin of the head. Median portion of the clypeus not extended back between the frontal carinae.

Frontal carinae present, frontal lobes insignificant to small.

Torulus with a laterally projecting dorsal flange or lobe immediately below the frontal carina; in full-face view this torulus flange or lobe is usually visible, projecting laterally or posterolaterally beyond the frontal carina.

Antennal sockets partially to mostly concealed in full-face view.

Antenna with 12 segments.

Antennal scrobes absent.

Promesonotal suture present on the dorsum of the mesosoma, articulated and flexible in fresh specimens.

Epimeral sclerite present or absent.

Metapleural gland orifice a simple hole that opens laterally in the lower posterior corner of the metapleuron, very close to the metapleural ventral margin.

Metacoxal cavities fully closed by an uninterrupted band of cuticle between the propodeal foramen and the metacoxal cavity.

Propodeal spiracle located high on the side, at or in front of the propodeal midlength.

Propodeal lobes present.

Metatibial gland absent.

Pretarsal claws each with a single tooth on its inner curvature (inconspicuous in some small species).

Waist of 2 segments (A2 and A3 = petiole and postpetiole).

Tergite and sternite of A2 (petiole) not fused.

Abdominal segment 2 (petiole) subsessile to distinctly pedunculate.

Helcium sternite not visible in profile. With the helcium detached and in anterior view, the tergite overlaps the sternite on each side; the sternite is transverse and does not bulge ventrally below the level of the tergal apices.

Tergite and sternite of abdominal segment 3 (postpetiole) not fused.

Tergite and sternite of abdominal segment 4 (first gastral segment) not fused.

Abdominal segment 4 with presclerites.

Abdominal segments 5–7 without presclerites.

Spiracles on tergites of A5–A7 concealed under the posterior margin of the preceding tergite.

Stridulitrum present on the pretergite of A4.

Pygidium distinct, biconvex and unarmed.

Sting present and functional, usually large and conspicuous.

PLATE 1

Many undersampled regions of Africa and Madagascar present exciting logistical challenges to visit but offer a chance to discover new insights into the region's diversity and biogeography. A recent expedition to the Makay Massif in southwestern Madagascar revealed many endemic species. Photo by Brian Fisher.

The expedition to Ambatovaky in eastern Madagascar revealed one of the most diverse lowland forests in Madagascar but also documented threats from illegal logging. The long-term survival of this forest block, and all the newly discovered species, are at risk. Photo by Brian Fisher.

PLATE 2

While the lowland forest in Seychelles is home to mostly introduced ant species, the mountains above 600 meters harbor such endemics as *Terataner scotti*. Photo by Brian Fisher.

The Central African rainforest is the least explored but most species-rich ecoregion of Africa. Photo by Brian Fisher.

PLATE 3

Biological inventories and resulting reference collections are essential first steps in the exploration and documentation of ant diversity. Expeditions, such as the one to the deep reaches of canyons in the Makay Massif, require cooperation of scientists, local guides and porters, and in this case, a hot air balloon pilot. Photo by Brian Fisher.

Sifting leaf litter combined with mini-winkler extractions of samples along a transect offers a standardized approach to comparing leaf litter ant communities across localities. Photo by Brian Fisher.

PLATE 4

Digging is slow and hard work but offers a chance to find some of the rarest subterranean ants, such as *Leptanilla*. Photo by Alex Wild.

Methods for collecting arboreal ants include beating low vegetation. After whipping the leaves with a stick, ants are collected off a white sheet. Photo by Brian Fisher.

PLATE 5

Field expeditions offer the rare chance to study the behavior of such ants as (top) *Tetraponera manangotra* (Ward, 2009) and (bottom) *Syllophopsis hildebrandti* (Forel, 1892) from the Col de Tanatana in Andohahela National Park, northwest of Tolagnaro in Madagascar. Photos by Brian Fisher.

PLATE 6

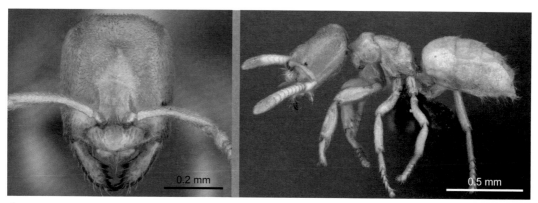

Acropyga

Acropyga silvestrii CASENT0235344, Tanzania

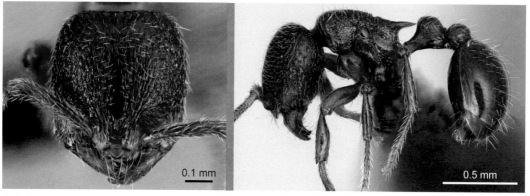

Adelomyrmex

Adelomyrmex SC02 CASENT0159515, Seychelles

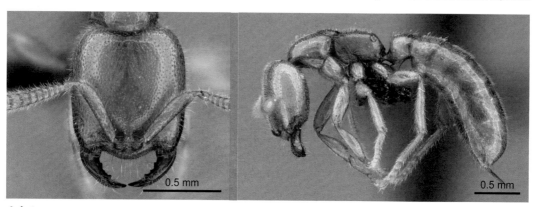

Adetomyrma

Adetomyrma bressleri CASENT0205995, Madagascar

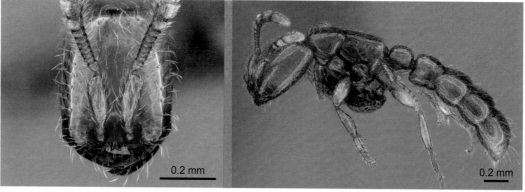

Aenictogiton

Aenictogiton UG01 CASENT0906052, Uganda

PLATE 7

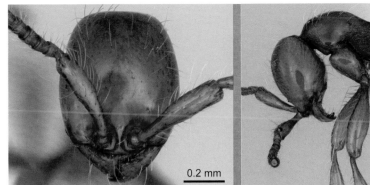

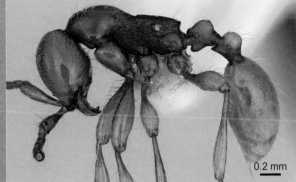

Aenictus

Aenictus GA03 CASENT0005956, Gabon

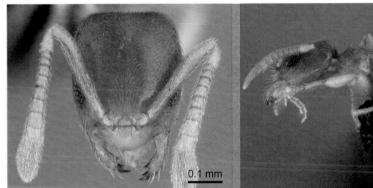

Agraulomyrmex

Agraulomyrmex ZA02 CASENT0217131, South Africa

Anillomyrma

Anillomyrma AFRC-LIM-01 CASENT0249583, South Africa

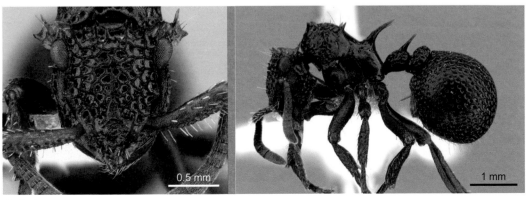

Ankylomyrma

Ankylomyrma coronacantha CASENT0902014, Cameroon

PLATE 8

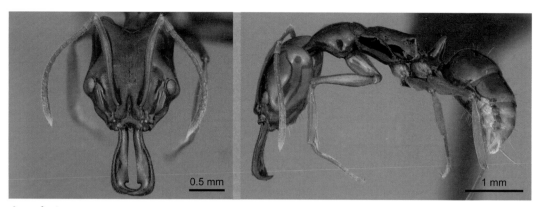

Anochetus

Anochetus madagascarensis CASENT0447670, Madagascar

Anoplolepis

Anoplolepis custodiens CASENT0249925, South Africa, head, CASENT0170536, South Africa, profile

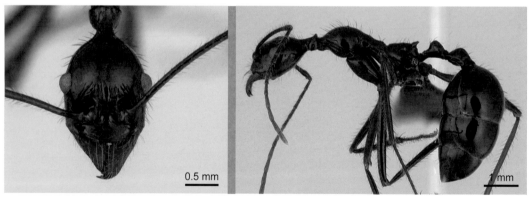

Aphaenogaster

Aphaenogaster swammerdami CASENT0002607, Madagascar

Aphomomyrmex

Aphomomyrmex afer CASENT0900319, Cameroon

PLATE 9

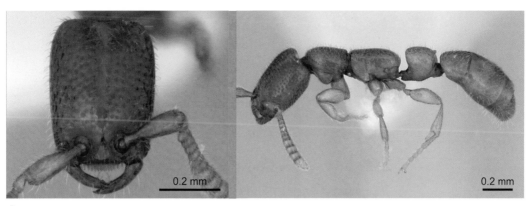

Apomyrma

Apomyrma stygia CASENT0000077, Cameroon

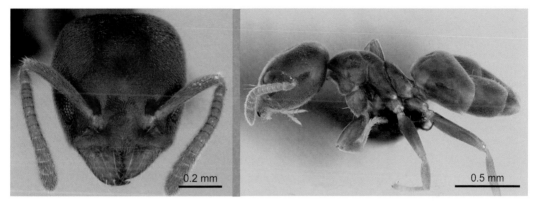

Aptinoma

Aptinoma antongil CASENT0489161, Madagascar

Asphinctopone

Asphinctopone silvestrii CASENT0406788, Central African Republic

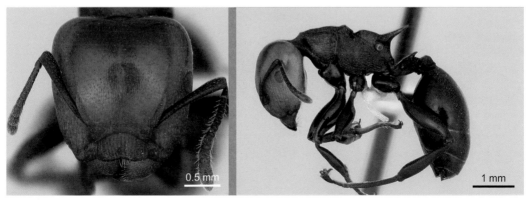

Atopomyrmex

Atopomyrmex mocquerysi CASENT0415562, Central African Republic

PLATE 10

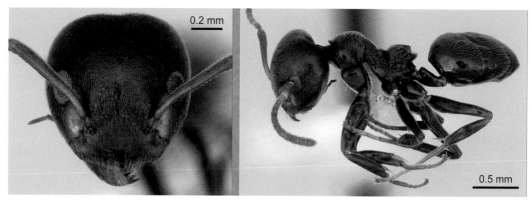

Axinidris

Axinidris luhya CASENT0172612, Kenya

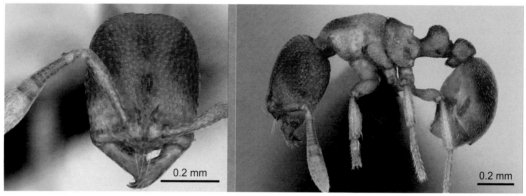

Baracidris

Baracidris sitra CASENT0006829, Gabon

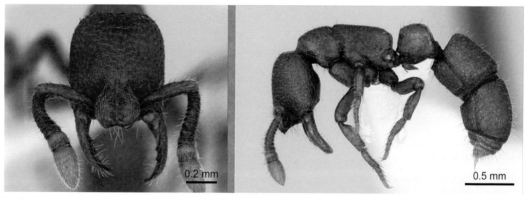

Boloponera

Boloponera vicans CASENT0401737, Central African Republic

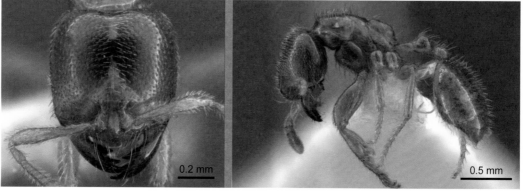

Bondroitia

Bondroitia lujae CASENT0900309, Angola

PLATE 11

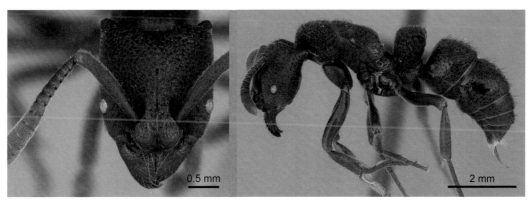

Bothroponera

Bothroponera cambouei CASENT0034270, Madagascar

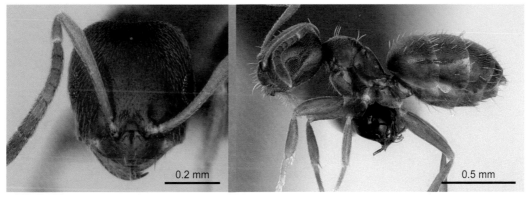

Brachymyrmex

Brachymyrmex cordemoyi CASENT0103230, Mauritius, head, CASENT0109850, Mauritius, profile

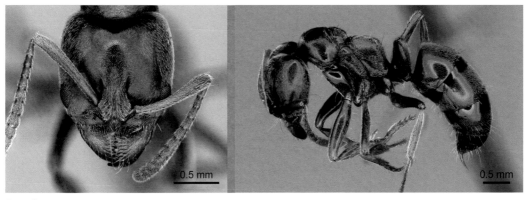

Brachyponera

Brachyponera sennaarensis CASENT0264290, Saudi Arabia

Calyptomyrmex

Calyptomyrmex KM01 CASENT0136410, Comoros

PLATE 12

Camponotus

Camponotus imitator CASENT0452863, Madagascar

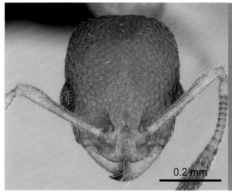

Cardiocondyla

Cardiocondyla wroughtonii CASENT0060052, Mauritius

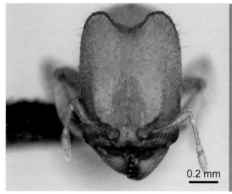

Carebara (major)

Carebara FAS12 CASENT0493154, Madagascar

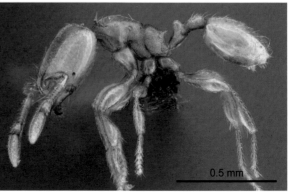

Carebara (minor)

Carebara AFRC-GAU-02 CASENT0235950, South Africa

PLATE 13

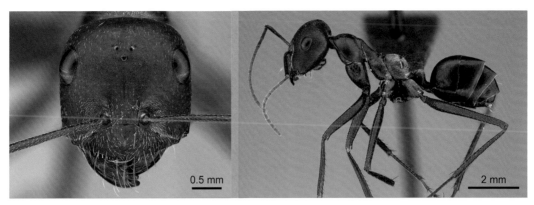

Cataglyphis

Cataglyphis SD01 CASENT0073619, Sudan

Cataulacus

Cataulacus wasmanni CASENT0498558, Madagascar

Centromyrmex

Centromyrmex raptor CASENT0066716, Zambia

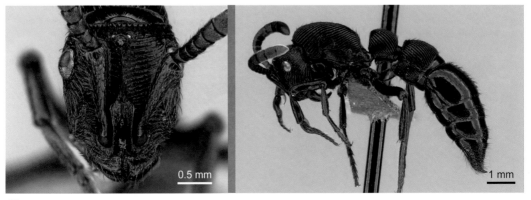

Chrysapace

Chrysapace MG01 CASENT0906558, Madagascar

PLATE 14

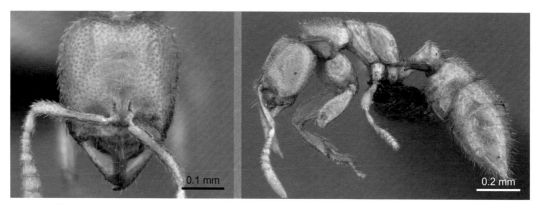

Concoctio

Concoctio AFRC-GAU-01 CASENT0235871, South Africa

Crematogaster

Crematogaster tricolor CASENT0178194, Mayotte

Cryptopone

Cryptopone hartwigi CASENT0235879, South Africa

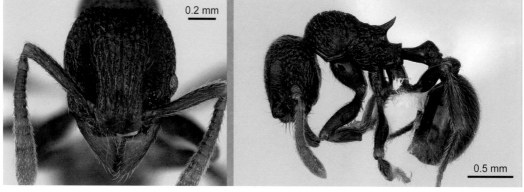

Cyphoidris

Cyphoidris exalta CASENT0405993, Central African Republic

PLATE 15

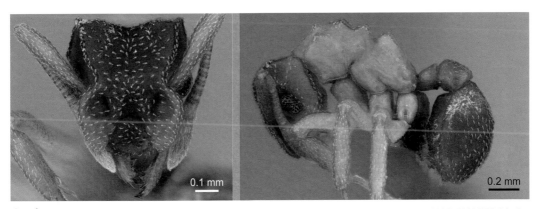

Cyphomyrmex

Cyphomyrmex minutus CASENT0264753, Réunion

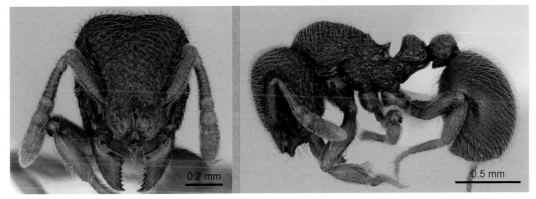

Dicroaspis

Dicroaspis cryptocera CASENT0172774, Gabon

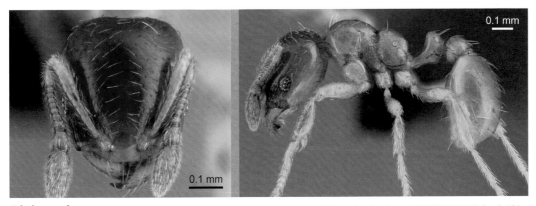

Diplomorium

Diplomorium longipenne CASENT0217046, South Africa

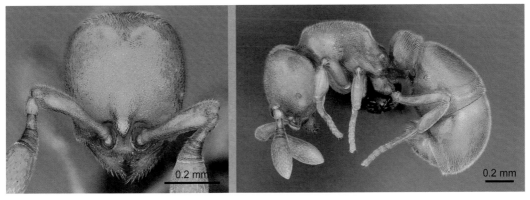

Discothyrea

Discothyrea berlita CASENT0007016, Mauritius

PLATE 16

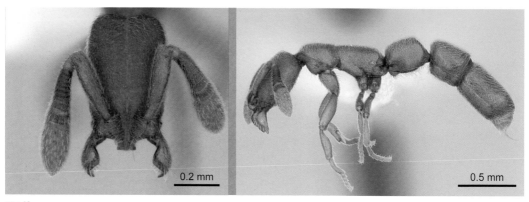

Dolioponera

Dolioponera fustigera CASENT0412032, Central African Republic

Dorylus

Dorylus gribodoi CASENT0172626, Côte d'Ivoire

Eburopone

Eburopone MG14 CASENT0494929, Madagascar

Ecphorella

Ecphorella wellmani CASENT0102456, Angola

PLATE 17

Erromyrma

Erromyrma latinode CASENT0107541, Madagascar

Euponera

Euponera sjostedti CASENT0296396, Uganda

Eurhopalothrix

Eurhopalothrix KM01 CASENT0147491, Comoros

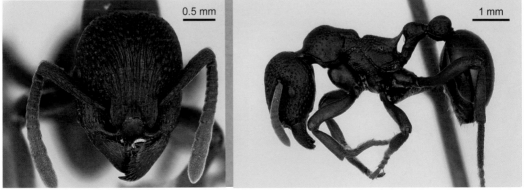

Eutetramorium

Eutetramorium mocquerysi CASENT0077435, Madagascar

PLATE 18

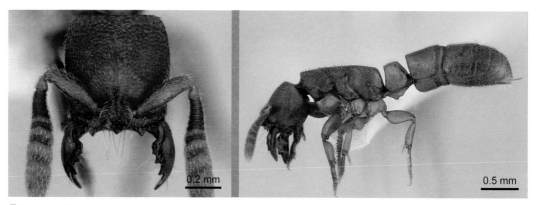

Feroponera

Feroponera ferox CASENT0102994, Cameroon

Fisheropone

Fisheropone ambigua CASENT0906216, Gabon

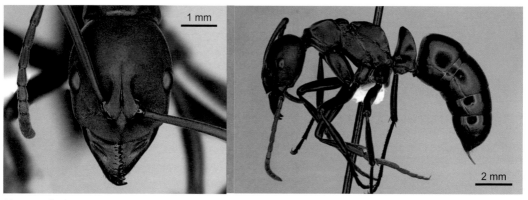

Hagensia

Hagensia havilandi CASENT0249202, South Africa

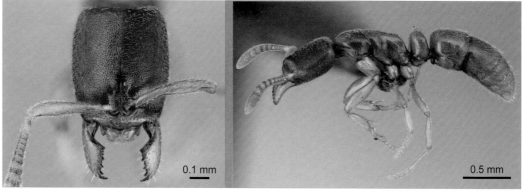

Hypoponera

Hypoponera jeanneli CASENT0217336, Tanzania

PLATE 19

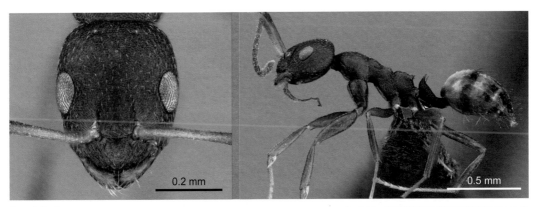

Lepisiota

Lepisiota AFRC-TZ-11 CASENT0235502, Tanzania

Leptanilla

Leptanilla UG01 CASENT0350575, Uganda

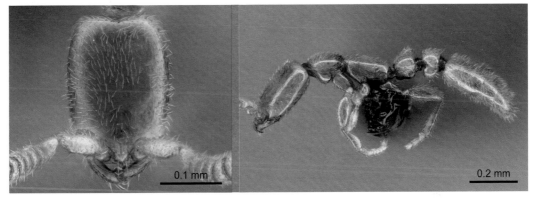

Leptogenys

Leptogenys maxillosa CASENT0137950, Madagascar

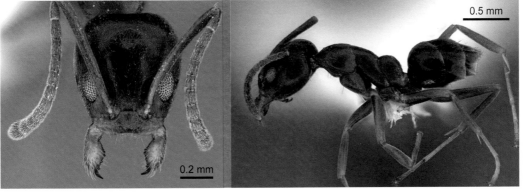

Linepithema

Linepithema humile CASENT0281314, South Africa

PLATE 20

Lioponera

Lioponera MG02 CASENT0170756, Madagascar

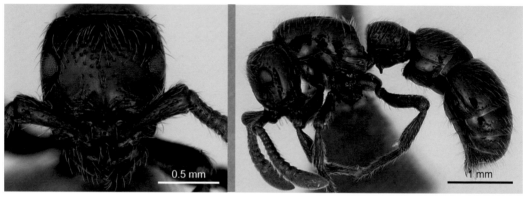

Lividopone

Lividopone MG12 CASENT0120028, Madagascar

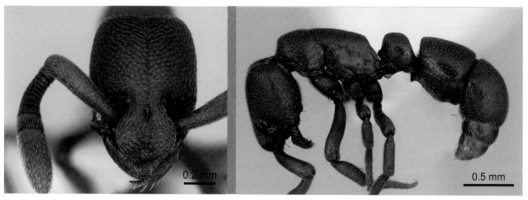

Loboponera

Loboponera trica CASENT0003113, Democratic Republic of the Congo

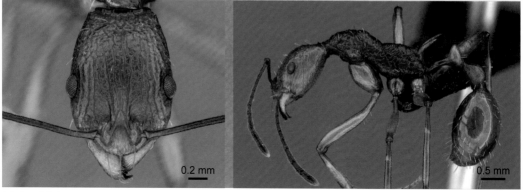

Malagidris

Malagidris jugum CASENT0054119, Madagascar

PLATE 21

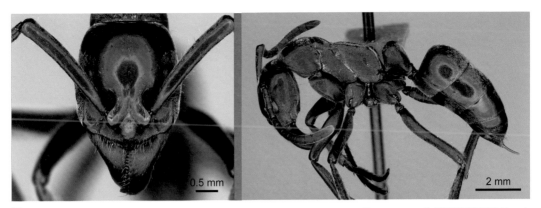

Megaponera

Megaponera analis CASENT0352477, Uganda

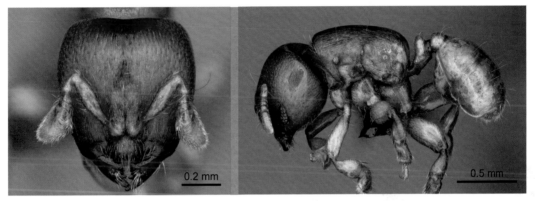

Melissotarsus

Melissotarsus UG01 CASENT0362244, Uganda

Meranoplus

Meranoplus magrettii CASENT0235913, South Africa

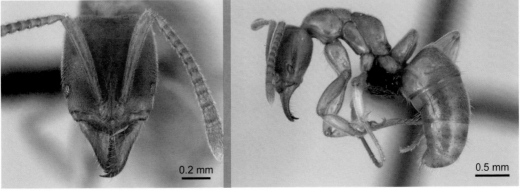

Mesoponera

Mesoponera ambigua CASENT0132515, Mayotte

PLATE 22

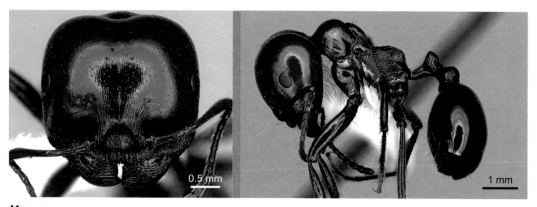

Messor

Messor angularis CASENT0217539, Kenya

Metapone

Metapone madagascarica CASENT0004528, Madagascar

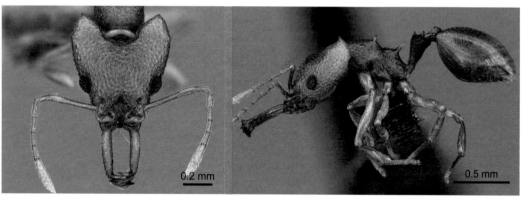

Microdaceton

Microdaceton exornatum CASENT0235521, Tanzania

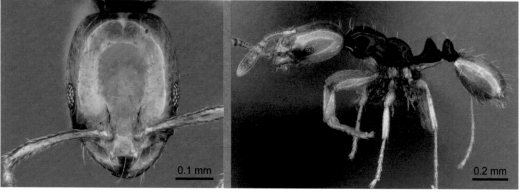

Monomorium

Monomorium mirandum CASENT0235525, Tanzania

PLATE 23

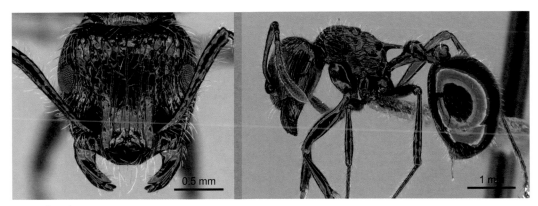

Myrmicaria

Myrmicaria WC01 CASENT0264037, South Africa

Mystrium

Mystrium eques CASENT0317399, Madagascar

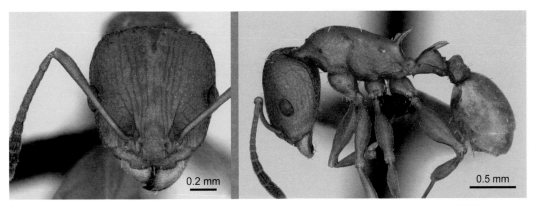

Nesomyrmex

Nesomyrmex hafahafa CASENT0460666, Madagascar

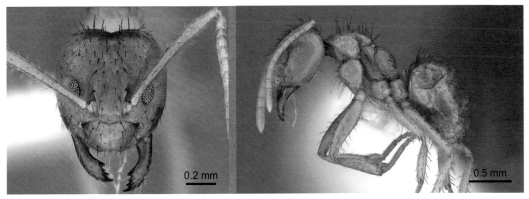

Nylanderia

Nylanderia amblyops CASENT0281132, Madagascar

PLATE 24

Ochetellus

Ochetellus glaber CASENT0173321, Réunion

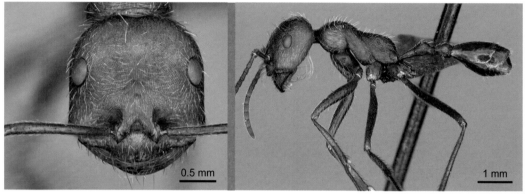

Ocymyrmex

Ocymyrmex foreli CASENT0235947, South Africa

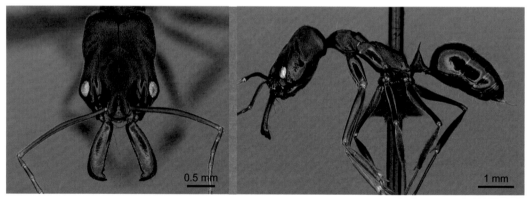

Odontomachus

Odontomachus troglodytes CASENT0235559, Tanzania

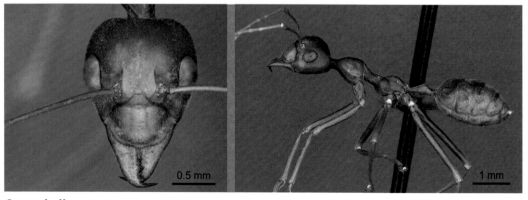

Oecophylla

Oecophylla longinoda CASENT0235557, Tanzania

PLATE 25

Ooceraea

Ooceraea biroi CASENT0055090, Madagascar

Ophthalmopone

Ophthalmopone hottentota CASENT0914340, South Africa

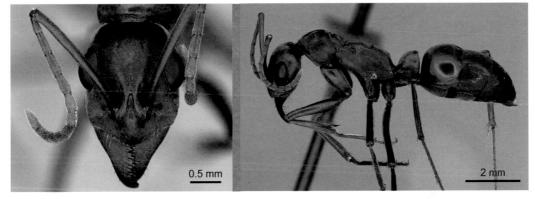

Paltothyreus

Paltothyreus tarsatus CASENT0249209, Kenya

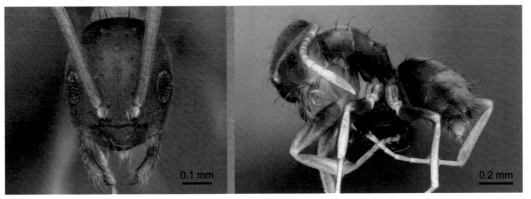

Paraparatrechina

Paraparatrechina luminella CASENT0159944, Seychelles, head, CASENT0159693, Seychelles, profile

PLATE 26

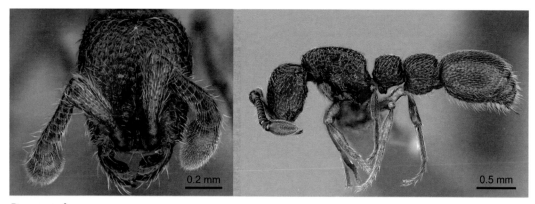

Parasyscia

Parasyscia MG03 CASENT0906552, Madagascar, head, CASENT0299970, Madagascar, profile

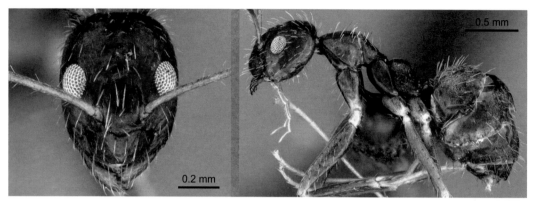

Paratrechina

Paratrechina ankarana CASENT0906652, Madagascar

Parvaponera

Parvaponera darwinii CASENT0906556, Madagascar

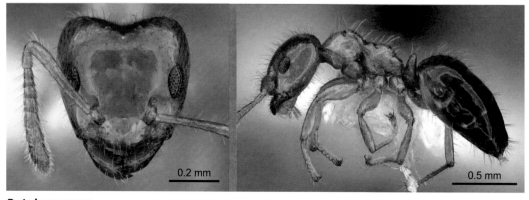

Petalomyrmex

Petalomyrmex phylax CASENT0281158, Cameroon

PLATE 27

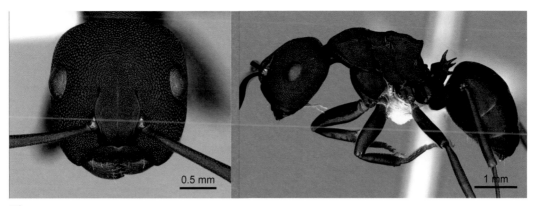

Phasmomyrmex

Phasmomyrmex aberrans CASENT0906849, Cameroon

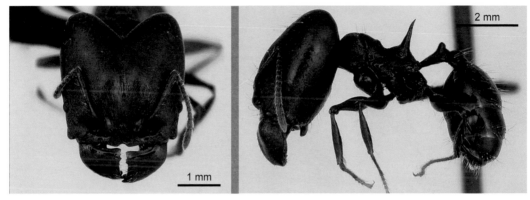

Pheidole (major)

Pheidole longispinosa CASENT0494701, Madagascar

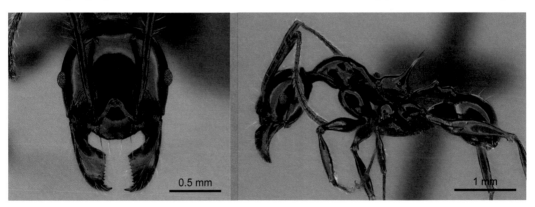

Pheidole (minor)

Pheidole longispinosa CASENT0148996, Madagascar

Phrynoponera

Phrynoponera gabonensis CASENT0411299, Central African Republic

PLATE 28

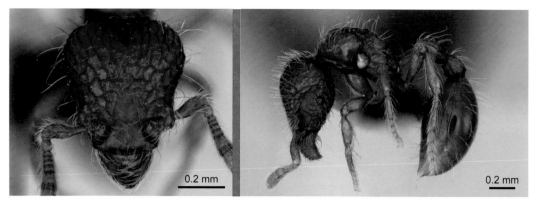

Pilotrochus

Pilotrochus besmerus CASENT0047617, Madagascar

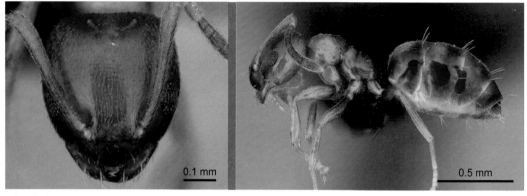

Plagiolepis

Plagiolepis MG01 CASENT0140938, Madagascar

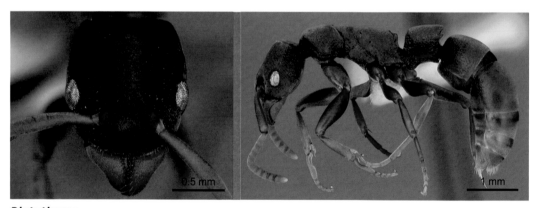

Platythyrea

Platythyrea mocquerysi CASENT0102996, Madagascar

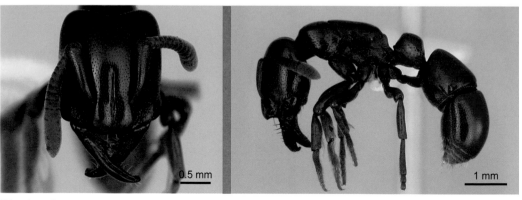

Plectroctena

Plectroctena ugandensis CASENT0003114, Democratic Republic of the Congo

PLATE 29

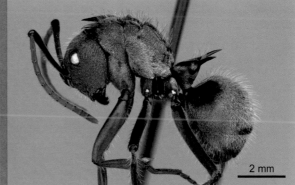

Polyrhachis

Polyrhachis UG01 CASENT0358740, Uganda

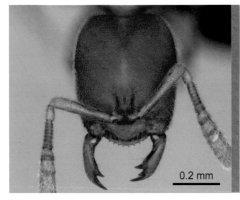

Ponera

Ponera petila CASENT0059796, Mauritius

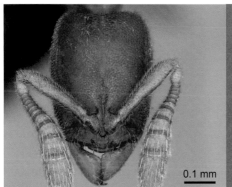

Prionopelta

Prionopelta subtilis CASENT0494610, Madagascar

Pristomyrmex

Pristomyrmex MU01 CASENT0060000, Mauritius

PLATE 30

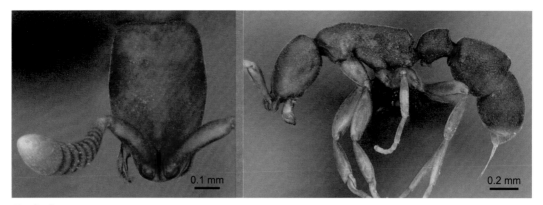

Probolomyrmex

Probolomyrmex tani CASENT0041507, Madagascar

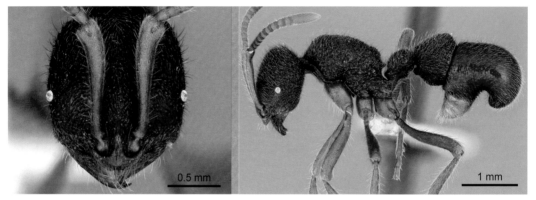

Proceratium

Proceratium diplopyx CASENT0100348, Madagascar

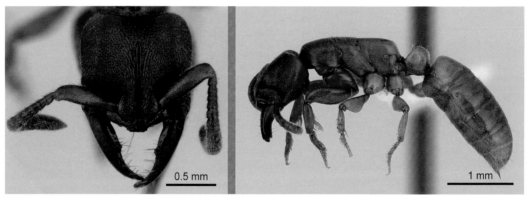

Promyopias

Promyopias silvestrii CASENT0178751, Cameroon

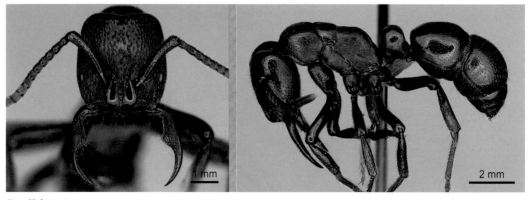

Psalidomyrmex

Psalidomyrmex UG02 CASENT0354816, Uganda

PLATE 31

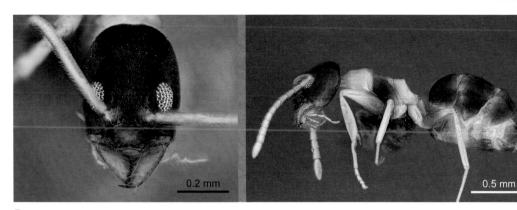

Ravavy

Ravavy MG01 CASENT0317901, Madagascar

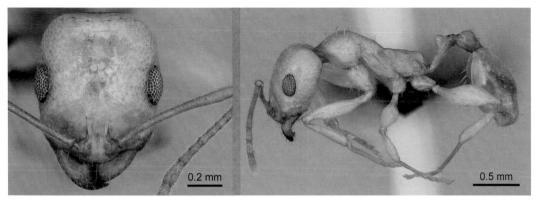

Royidris

Royidris notorthotenes CASENT0002219, Madagascar

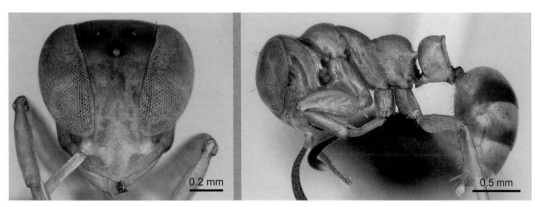

Santschiella

Santschiella kohli CASENT0004701, Gabon

Simopone

Simopone persculpta CASENT0173049, Mozambique

PLATE 32

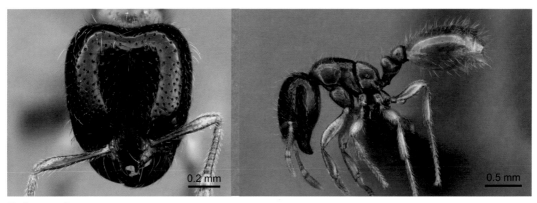

Solenopsis

Solenopsis punctaticeps CASENT0235999, South Africa

Stigmatomma

Stigmatomma MG06 CASENT0017556, Madagascar

Streblognathus

Streblognathus peetersi CASENT0173637, South Africa

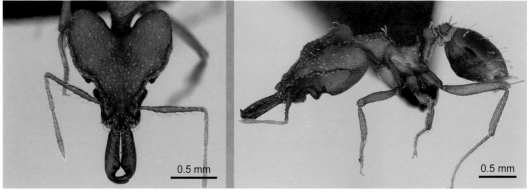

Strumigenys

Strumigenys dicomas CASENT0499800, Madagascar

PLATE 33

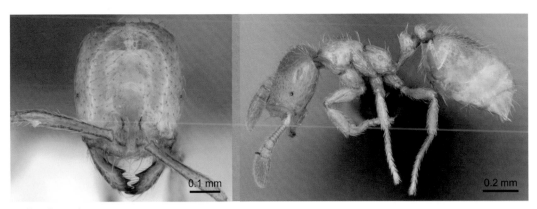

Syllophopsis

Syllophopsis cryptobia CASENT0906354, Saudi Arabia, head, CASENT0060392, Mauritius, profile

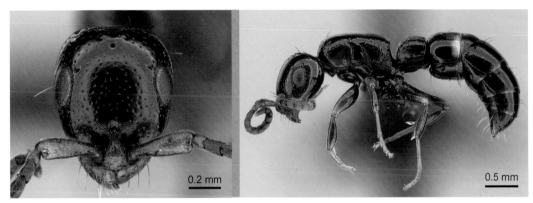

Tanipone

Tanipone varia CASENT0021184, Madagascar

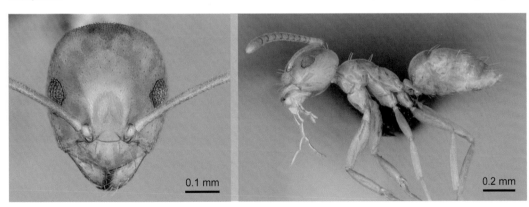

Tapinolepis

Tapinolepis MG02 CASENT0172773, Madagascar

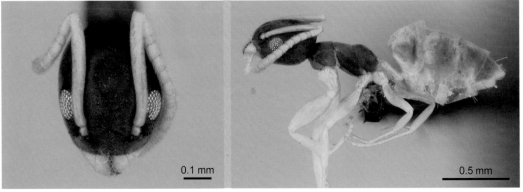

Tapinoma

Tapinoma melanocephalum CASENT0008659, Madagascar

PLATE 34

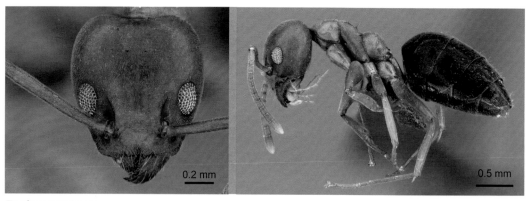

Technomyrmex

Technomyrmex hostilis CASENT0235734, Tanzania

Temnothorax

Temnothorax mpala CASENT0280870, Kenya

Terataner

Terataner FHG-nyx CASENT0104148, Madagascar

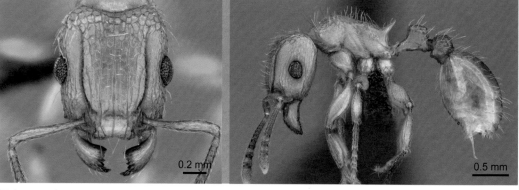

Tetramorium

Tetramorium notiale CASENT0249030, South Africa

PLATE 35

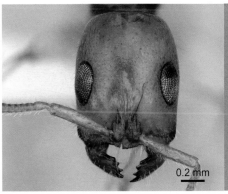

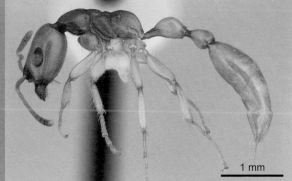

Tetraponera

Tetraponera fictrix CASENT0012854, Madagascar

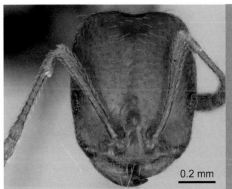

Trichomyrmex

Trichomyrmex destructor CASENT0481867, Madagascar

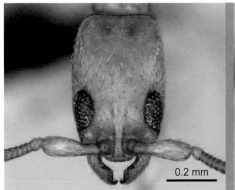

Vicinopone

Vicinopone conciliatrix CASENT0172777, Gabon

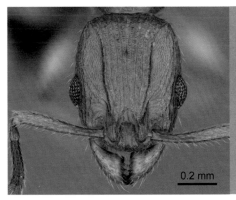

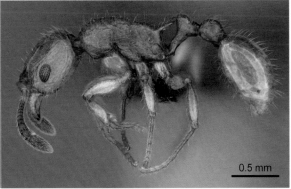

Vitsika

Vitsika suspicax CASENT0040798, Madagascar

PLATE 36

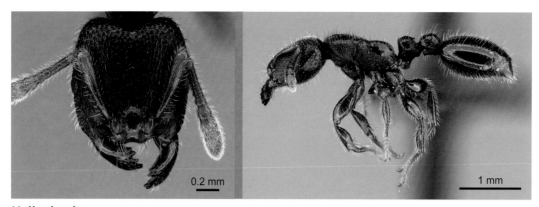

Vollenhovia

Vollenhovia oblonga CASENT0159709, Seychelles

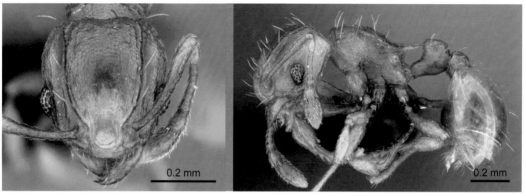

Wasmannia

Wasmannia auropunctata CASENT0097874, São Tomé and Principe

Xymmer

Xymmer MG02 CASENT0461545, Madagascar

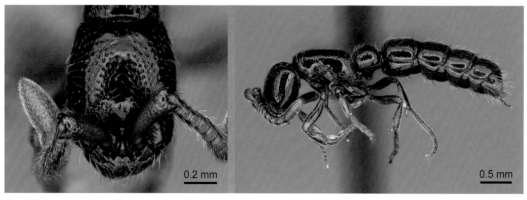

Zasphinctus

Zasphinctus UG01 CASENT0352813, Uganda

AFROTROPICAL AND MALAGASY GENERA

The identification of the 104 Afrotropical and 72 Malagasy genera analyzed here is provided in the form of two dichotomous keys to workers: the first to the Afrotropical fauna and the second to the Malagasy. The two keys will facilitate identification by students whose interest lies in one region, rather than having a single key that covers the genera of both regions. The characters chosen are, as far as possible, consistent in both keys for genera common to both regions, so that a species of *Tetramorium*, for example, will run through both keys, regardless of its region of origin.

Key to Afrotropical Genera (Workers)

1a	Body with a single reduced or isolated segment, the petiole (A2) between the mesosoma and the gaster **(A)** 2	
1b	Body with 2 reduced or isolated segments, the petiole and postpetiole (A2 and A3) between the mesosoma and the gaster **(AA)** 70	

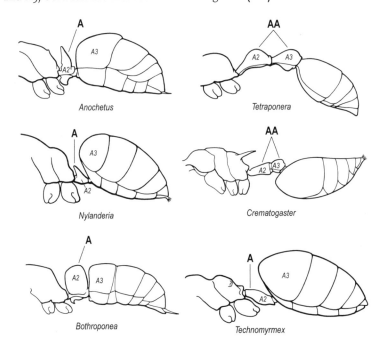

Anochetus

Tetraponera

Nylanderia

Crematogaster

Bothroponea

Technomyrmex

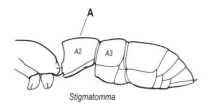

Stigmatomma

2a (1)   With head in ventral view and mouthparts closed, the prementum is not visible; it is entirely concealed by the labrum anteriorly and the maxillary stipites on each side, the stipites meet along the midline **(A)**. Lateral margins of the pygidium always with 1 or more short spines, teeth, or denticles on each side **(B)**. Lateral and posterior margins of the hypopygium always unarmed and unmodified 3

2b   With head in ventral view and mouthparts closed, the prementum is clearly visible; it is bounded by the labrum anteriorly and the maxillary stipites on each side of it **(AA)**. Lateral margins of the pygidium usually unarmed **(BB)**, only extremely rarely with short spines. Lateral and posterior margins of the hypopygium usually unarmed and unmodified but rarely a row of short teeth or spiniform setae present, which may project dorsally outside the pygidium; or the hypopygium terminates in an acidopore **(CC)**. 10

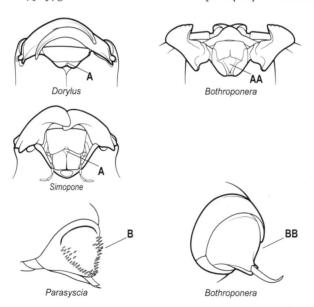

Dorylus   *Bothroponera*

Simopone

Parasyscia   *Bothroponera*

(Figure continues on next page)

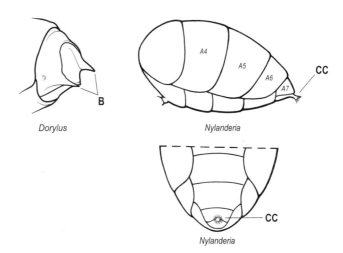

Dorylus

Nylanderia

Nylanderia

3a (2) With abdomen in profile or dorsal view, a deep girdling constriction is present between segments A3 and A4 (petiole = A2), between A4 and A5, and between A5 and A6 **(A)** **4**

3b With abdomen in profile or dorsal view, either girdling constrictions are entirely absent between the segments from A3 to the apex **(AA)**, or a constriction is present only between segments A3 and A4 **(AAA)** (petiole = A2) (when this constriction is very strongly developed, A3 is reduced and becomes isolated as the postpetiole) **(AAAA)**. Girdling constrictions are never present between segments A4 and A5, nor between A5 and A6 5

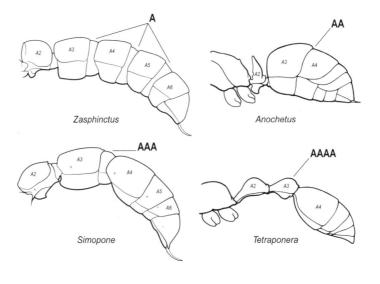

Zasphinctus

Anochetus

Simopone

Tetraponera

4a (3) Promesonotal suture absent from the dorsum of the mesosoma **(A)**. Propodeal spiracle low on the side and at or behind the midlength of the sclerite **(B)**. Apical gastral segment (A7) almost as large as preceding segment (A6) **(C)**. Propodeal lobes present **(D)** . . . . *Zasphinctus*

4b Promesonotal suture present across the dorsum of the mesosoma **(AA)**. Propodeal spiracle high on the side and in front of the midlength of the sclerite **(BB)**. Apical gastral segment (A7) very much smaller than preceding segment (A6) **(CC)**. Propodeal lobes absent **(DD)** . . . . *Aenictogiton*

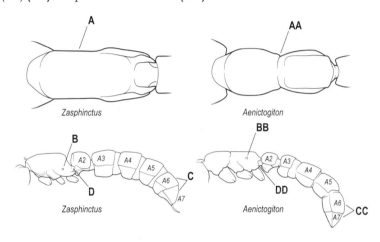

Zasphinctus | Aenictogiton | Zasphinctus | Aenictogiton

5a (3) Propodeal spiracle situated high up on the side and in front of the midlength of the propodeum **(A)**. Armament of pygidium consists of a single, posteriorly directed, short spine or tooth on each side **(B)**. Propodeal lobes absent **(C)**. Gastral segments 3–5 (A5–A7) with distinct presclerites. Eyes always absent. Anterior (free) margin of the labrum not cleft or indented medially **(D)**. Promesonotal suture strongly present but fused and immobile **(E)**. Sting very reduced **(F)**. . . . . *Dorylus*

5b Propodeal spiracle situated low down on the side and at or behind the midlength of the propodeum **(AA)**. Armament of pygidium a series of denticles or peg-like teeth, located laterally, posteriorly, or both; always several on each side **(BB)**. Propodeal lobes present **(CC)**. Gastral segments 3–5 (A5–A7) without differentiated presclerites. Eyes vary from large to absent. Anterior (free) margin of the labrum cleft or indented medially **(DD)**. Promesonotal suture usually absent, only rarely present or represented by a faint line **(EE)**. Sting strongly developed and usually clearly visible **(FF)** 6

(Continues on next page)

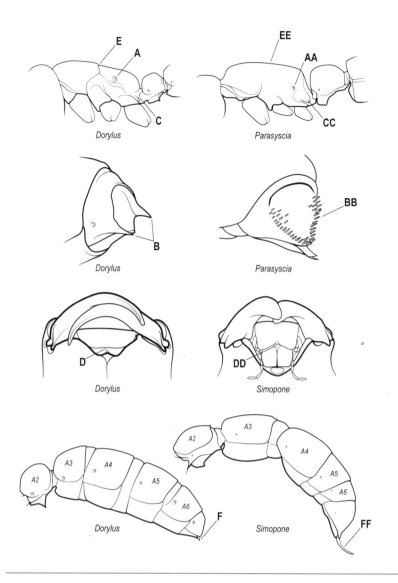

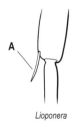

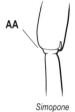

6a (5) Mesotibia always with at least 1 conspicuous spur **(A)**. Pretarsal claws of the metatarsus without a preapical tooth **(B)** 7

6b Mesotibia always lacks spurs **(AA)**. Pretarsal claws of the metatarsus with a preapical tooth **(BB)**. 9

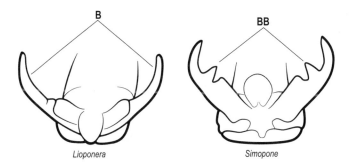

Lioponera   Simopone

7a (6)   Promesonotal suture present as an impressed groove across the dorsal mesosoma **(A)**. Eyes absent. Abdominal sternite 4, at the middle of its posterior
margin, with a conspicuous patch of pale cuticle **(B)** . . . . *Eburopone*

7b   Promesonotal suture absent, or at most represented by a vestigial line across
the dorsal mesosoma **(AA)**. Eyes present, may be very small. Abdominal sternite 4, at the middle of its posterior margin, unmodified, without a conspicu-
ous patch of pale cuticle **(BB)**. 8

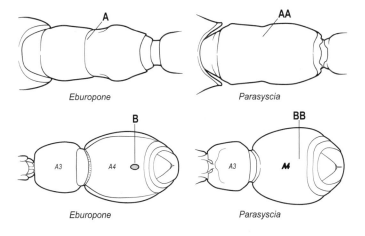

Eburopone   Parasyscia

Eburopone   Parasyscia

8a (7)   Petiole (A2) sharply marginate dorsolaterally **(A)**. Posterodorsal surface of
the metacoxa, just proximal of the trochanter, raised into a vertical cuticular flange or lamella **(B)**. Apical funicular segment large but not bulbous and
strongly swollen with respect to the preapical antennomere **(C)** . . . . *Lioponera*

8b   Petiole (A2) not marginate dorsolaterally, the sides rounding into the dorsum
(AA). Posterodorsal surface of the metacoxa, just proximal of the trochanter,
without a vertical cuticular flange or lamella **(BB)**. Apical funicular segment
very large, bulbous, and strongly swollen with respect to the preapical anten-
nomere **(CC)**. . . . . *Parasyscia* (part)

(Continues on next page)

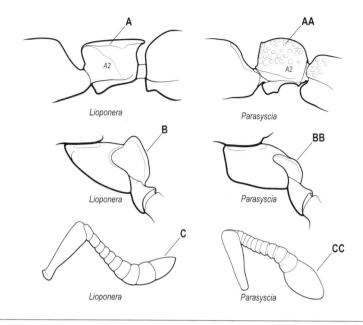

9a (6)   Antenna with 12 segments **(A)**. Ocelli absent **(B)**. Longitudinal glandular groove absent from the basal half of the ventral surface of the metabasitarsus. Tergite of the petiole (A2) not marginate laterally in dorsal view **(D)**. A vertical posterior surface present on the head above the occipital foramen **(E)**. With head in ventrolateral view the posterolateral marginal carina extends across the ventral surface to the midline **(F)**. Palp formula 3,2

. . . . *Vicinopone*

9b   Antenna with 11 segments **(AA)**. Ocelli present **(BB)**. Longitudinal glandular groove present on the basal half of the ventral surface of the metabasitarsus **(CC)**. Tergite of the petiole (A2) marginate laterally in dorsal view **(DD)**. No vertical posterior surface present on the head above the occipital foramen; instead, the vertex slopes evenly down to the foramen **(EE)**. With head in ventrolateral view, the posterolateral marginal carina does not extend across the ventral surface to the midline **(FF)**. Palp formula usually 6,4, rarely 5,3

. . . . ***Simopone***

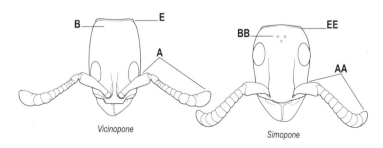

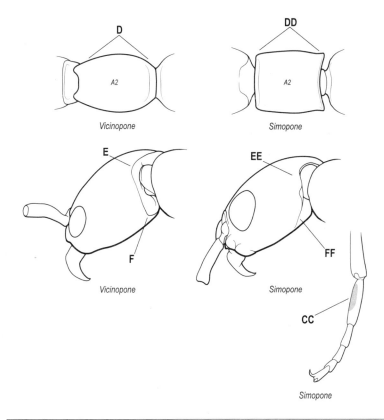

D

DD

A2   A2

Vicinopone   *Simopone*

E   EE

F   FF

Vicinopone   *Simopone*

CC

Simopone

10a (2) Apex of the gaster with a semicircular to circular acidopore formed from the apical portion of the hypopygium **(A)**. This structure often projects as a nozzle and is fringed with setae, but in some genera the acidopore has no nozzle or setae and is concealed by a projection of the pygidium (the acidopore becomes visible if the sclerites are separated); in these genera the antennal sockets are located well behind the posterior clypeal margin. **11**

10b Apex of the gaster with the hypopygium lacking an acidopore **(AA)**. Sting present or absent; when present, it is usually clearly visible, but when reduced or vestigial (or rarely when sting present but entirely retracted), the hypopygium has a smooth posterior margin, without trace of a circular or semicircular acidopore, and the antennal sockets are located close to or at the posterior clypeal margin **29**

(Continues on next page)

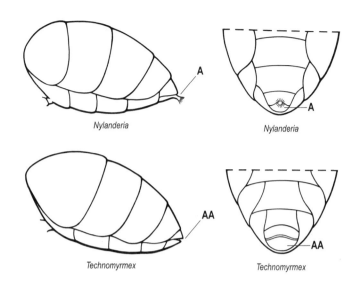

Nylanderia

Nylanderia

Technomyrmex

Technomyrmex

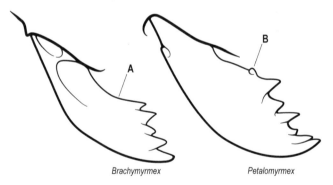

Brachymyrmex

Petalomyrmex

13a (12) Maxillary palp with 5 segments. Polymorphic species with roughly rectangular head capsule in full-face view **(A)**. In this view, the outer margin of the eye is situated a considerable distance in from the lateral margin of the head **(B)** .... *Aphomomyrmex*

13b Maxillary palp with 3 segments. Monomorphic species with heart-shaped head capsule in full-face view **(AA)**. In this view, the outer margin of the eye is situated very close to the lateral margin of the head **(BB)** .... *Petalomyrmex*

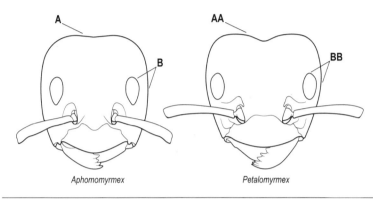

Aphomomyrmex   Petalomyrmex

14a (11) Antenna with 10 or 11 segments **15**
14b Antenna with 12 segments **20**

15a (14) Eyes minute to absent; if minutely present, the maximum diameter of the eye is very obviously less than the maximum width of the scape **(A)**. Maxillary palps short; with labiomaxillary complex fully retracted, the palps do not extend along the ventral surface of the head beyond the posterior margin of the buccal cavity .... *Acropyga*

15b Eyes present and conspicuous; maximum diameter of the eye is obviously greater than the maximum width of the scape **(AA)**. Maxillary palps long; with labiomaxillary complex fully retracted, the palps extend along the ventral surface of the head beyond the posterior margin of the buccal cavity **16**

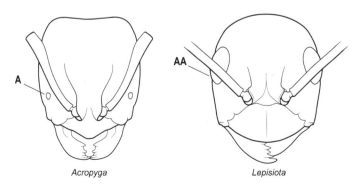

Acropyga   Lepisiota

16a (15) Antenna with 10 segments. . . . . *Agraulomyrmex*

16b Antenna with 11 segments 17

17a (16) Propodeum armed with a pair of spines, teeth, or tubercles **(A)**. Dorsal edge
of the petiole (A2) usually armed with a pair of teeth or spines **(A)** but some-
times only emarginate . . . . *Lepisiota*

17b Propodeum and petiole (A2) unarmed, without trace of spines, teeth, or
tubercles **(AA)** 18

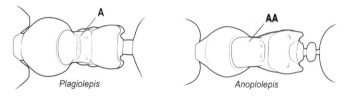

Lepisiota

Plagiolepis

18a (17) With the mesosoma in dorsal view, the mesonotum separated from the metanotum by a conspicuous transverse groove or impression, so that the metano-
tum forms a distinctly isolated sclerite on the dorsum **(A)** . . . . *Plagiolepis*

18b With the mesosoma in dorsal view, the mesonotum fused with the metanotum, the two not separated by a transverse groove or impression, the metanotum does not form an isolated sclerite on the dorsum **(AA)** 19

Plagiolepis

Anoplolepis

19a (18) Ventral apex of the metatibia with a divergent pair of coarse setae, between which is a large median spur **(A)**. Anterior arc of the torulus does not touch or indent the posterior clypeal margin **(B)**. Dorsum of the head behind the clypeus with numerous erect stout setae present **(C)**. Mandible with 6–9 teeth. Ocelli absent **(D)** except in largest workers of polymorphic species, where a single median ocellus may occur. . . . . *Anoplolepis*

19b Ventral apex of the metatibia with a divergent pair of coarse setae only, without a median spur between them **(AA)**. Anterior arc of the torulus touches and slightly indents the posterior clypeal margin **(BB)**. Dorsum of head behind the clypeus without erect setae **(CC)**. Mandible with 5 teeth. Ocelli present (all 3 distinct) **(DD)** . . . . *Tapinolepis*

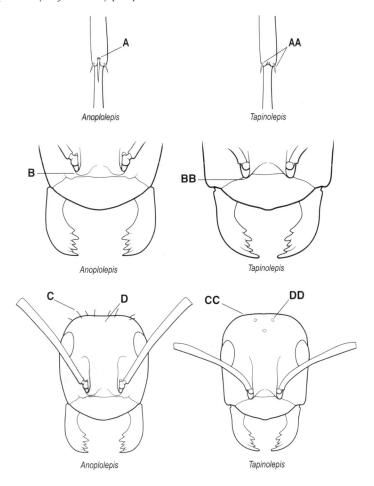

Anoplolepis

Tapinolepis

Anoplolepis

Tapinolepis

Anoplolepis

Tapinolepis

20a (14) Metapleuron with a distinct orifice for the metapleural gland, the orifice situated above the metacoxa and below the level of the propodeal spiracle **(A)**. Orifice of the metapleural gland protected and sometimes partially concealed by a line or tuft of guard-setae that are usually very conspicuous. Anteriormost point of the torulus indents, abuts **(B)**, or is immediately behind the clypeal suture **(C)**. If the last of these, the gap between clypeal suture and the anterior margin of the torulus is very narrow, distinctly less than the basal width of the scape shaft 21

20b Metapleural gland orifice absent, the surface of the metapleuron uninterrupted by a gland orifice above the metacoxa and below the level of the propodeal spiracle **(AA)**; guard-setae absent. Anteriormost point of the torulus far behind the clypeal suture, the gap between clypeal suture and the anterior margin of the torulus is at least equal to the basal width of the scape shaft and generally much more **(BB)** 26

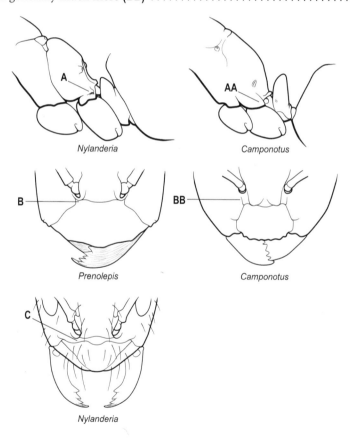

Nylanderia

Camponotus

Prenolepis

Camponotus

Nylanderia

21a (20) Antennal scape, when laid back in its natural resting position, passes below the eye. The eye enormous, in full-face view occupying almost all the side of the head **(A)**. In full-face view the antennal socket is on the line of the long axis of the eye **(B)** . . . . *Santschiella*

21b Antennal scape, when laid back in its natural resting position, passes above the eye. The eye smaller or (rarely) absent, when present in full-face view occupying less than one-half the side of the head **(AA)**. In full-face view the antennal socket is distinctly mesad of the line of the long axis of the eye **(BB)** **22**

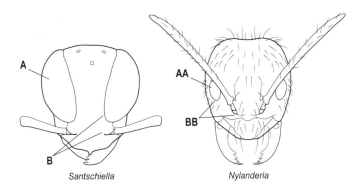

Santschiella   *Nylanderia*

22a (21) Propodeal spiracle an elongate vertical or near-vertical ellipse or slit **(A)**. First gastral sternite (sternite of A3) in ventral view with a transverse sulcus immediately behind the helcium **(B)**. With the mesosoma in ventral view the anterior margin of the foramen in which the petiole articulates does not extend forward to the level of a line that spans the anteriormost points of the metacoxal cavities **(C)**. Bases of metacoxae closely approximated when metacoxae are directed outward **(D)**. . . . . *Cataglyphis*

22b Propodeal spiracle circular to subcircular, usually small **(AA)**. First gastral sternite (sternite of A3) in ventral view without a transverse sulcus immediately behind the helcium **(BB)**. With the mesosoma in ventral view, the anterior margin of the foramen in which the petiole articulates extends forward at least to the level of a line that spans the anteriormost points of the metacoxal cavities and usually farther **(CC)**. Bases of metacoxae widely separated when metacoxae are directed outward **(DD)**. **23**

(Continues on next page)

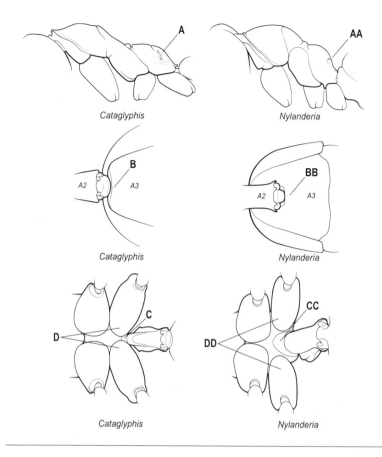

Cataglyphis · Nylanderia

Cataglyphis · Nylanderia

Cataglyphis · Nylanderia

23a (22) Maxillary palp with 3 segments **(A)**. Polymorphic species, eyes absent in minor workers, vestigial in major workers **(B)** . . . . *Paraparatrechina* (part)

23b Maxillary palp with 6 segments **(AA)**. Monomorphic species, always with large, conspicuous eyes **(BB)** **24**

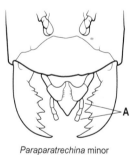

Paraparatrechina minor

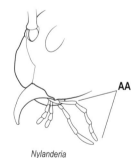

Nylanderia

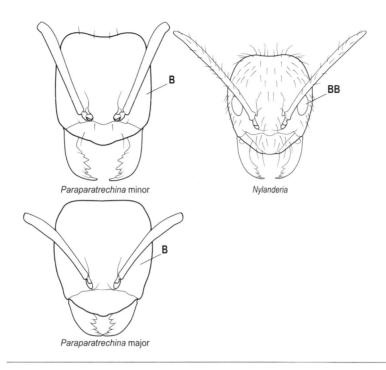

Paraparatrechina minor

Nylanderia

Paraparatrechina major

24a (23) Propodeal dorsum with erect setae present; usually a single pair but rarely with more **(A)**.................................... *Paraparatrechina* (part)

24b Propodeal dorsum without erect setae **(AA)**.............................25

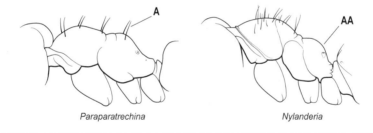

Paraparatrechina

Nylanderia

25a (24) Mandible unsculptured, with 6–7 teeth **(A)**; never with 5 or 8 teeth... *Nylanderia*

25b Mandible with light longitudinal striations and with 5 teeth **(AA)**, or distinctly striate and with 8 teeth **(AAA)**; never unsculptured and with 6–7 teeth .... *Paratrechina*

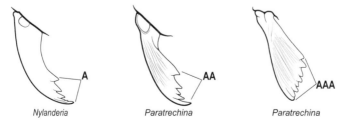

Nylanderia

Paratrechina

Paratrechina

26a (20) Maxillary palp with 5 segments. Mandible with 10 or more teeth or denticles altogether **(A)**. Apical tooth disproportionately large and the fourth tooth from the apex much larger than the third and the fifth tooth. Narrowly lamellate propodeal lobes present **(B)**. Petiole (A2) reduced to an elongate low node that allows the gaster to be reflexed over the mesosoma **(C)**. In profile the helcium projects from about the midheight of the first gastral segment (A3) **(D)** . . . .*Oecophylla*

26b Maxillary palp with 6 segments. Mandible usually with 7 teeth at most, sometimes fewer and only very rarely with more **(AA)**. Whatever the number of teeth, they decrease in size from the apex to base; the fourth tooth is not enlarged as above. Propodeal lobes absent **(BB)**. Petiole (A2) an erect node or scale, the gaster not capable of reflexion over the mesosoma **(CC)**. In profile the helcium projects from the base of the first gastral segment (A3) **(DD)** 27

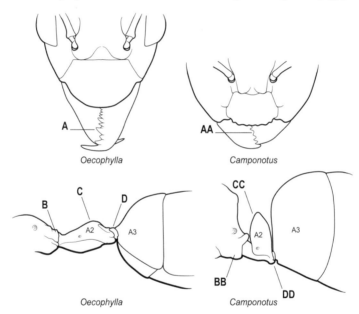

Oecophylla   *Camponotus*

Oecophylla   *Camponotus*

27a (26) Tergite of the first gastral segment (A3) large, accounting for at least half the length of the gaster **(A)**; in dorsal view the first gastral tergite distinctly much longer than the second. Pronotum usually, and petiole always, with spines or teeth **(B)** . . . . *Polyrhachis*

27b Tergite of the first gastral segment (A3) short, accounting for distinctly less than half the length of the gaster **(AA)**; in dorsal view the first gastral tergite at most only slightly longer than the second, often shorter. Both the pronotum and petiole usually without spines or teeth **(BB)**, rarely with such present in one or both places. 28

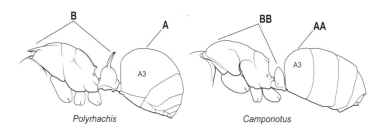

Polyrhachis   *Camponotus*

28a (27) Petiole (A2) a node or scale, never armed with teeth or spines **(A)**. Acidopore usually overlapped and concealed by the pygidium, usually not fringed with setae **(B)** . . . . *Camponotus*

28b Petiole (A2) armed with teeth, spines, or prominent angles **(AA)**. Acidopore exposed and usually fringed with setae; not overlapped and concealed by the pygidium **(BB)** . . . . *Phasmomyrmex*

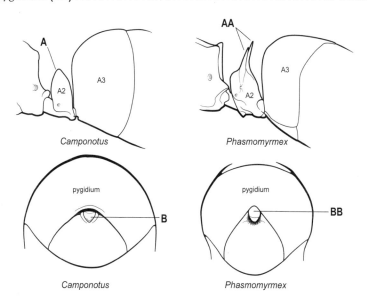

Camponotus   *Phasmomyrmex*

Camponotus   *Phasmomyrmex*

29a (10) Tergite of the helcium with an extensive U-shaped or V-shaped median notch or emargination dorsally in its anterior margin (easily visible if the gaster is slightly depressed with respect to the petiole) **(A)**. Sting vestigial or absent, never visibly projecting from apex of the gaster and its remnants not detectable without dissection 30

29b Tergite of the helcium entire, its anterior margin dorsally without a median U-shaped or V-shaped notch or emargination **(AA)**. Sting present and functional, usually projecting from apex of the gaster in dead specimens; in many species the sting shaft is visible through the cuticle of the ventral gastral apex even when sting fully retracted 34

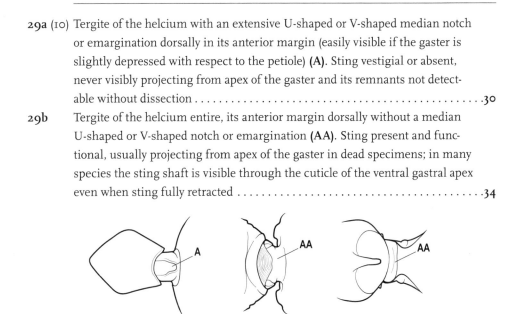

Technomyrmex   *Bothroponera*   *Odontomachus*

30a (29) Propodeum armed with a pair of teeth, tubercles or acute angles posterolaterally, these sometimes linked across by a narrow carina **(A)**. Between the tubercles, teeth or angles, or somewhat more dorsally, the propodeum at its midwidth with a prominent tubercle, projecting plate, longitudinal ridge, or tooth **(B)** . . . . *Axinidris*

30b Propodeum unarmed, without posterolateral teeth, tubercles or acute angles **(AA)**; usually without a median prominence of any description on the dorsum **(BB)**, but in a few species the posterior margin itself projects medially as a small angle or short tooth 31

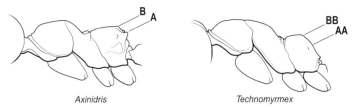

Axinidris   *Technomyrmex*

31a (30) Petiole (A2) a well-developed scale, distinct in profile **(A)**. Petiole scale somewhat inclined forward but not reduced and not overhung by an anterior
projection of the first gastral tergite (A3 tergite) above it. Anterior surface
of first gastral tergite without a longitudinal groove that accommodates the
petiole **(B)**. **32**

31b Petiole (A2) very reduced or vestigial, without a scale; in profile very small
and low or even more or less absent **(AA)**. Petiole overhung and concealed by
an anterior projection of the first gastral tergite (A3 tergite) above it. Anterior
surface of the first gastral tergite with a longitudinal groove that accommo-
dates the petiole **(BB)** **33**

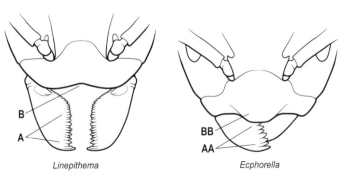

32a (31) Each mandible with 5–8 teeth plus 5–13 denticles **(A)**. Anterior clypeal margin concave in full-face view **(B)**. Fourth gastral (A6) sternite keel-shaped
posteriorly (introduced genus of Neotropical origin) . . . . *Linepithema*

32b Each mandible with 7 teeth and no denticles **(AA)**. Anterior clypeal margin
convex in full-face view **(BB)**. Fourth gastral (A6) sternite not keel-shaped
posteriorly . . . . *Ecphorella*

33a (31) In dorsal view only 4 gastral tergites (A3–A6) are visible **(A)**. The fifth gastral tergite (A7) is reflexed below the fourth, visible in ventral view, where it forms a transverse plate abutting the fifth sternite; the anal and associated orifices are thus situated ventrally. . . . . *Tapinoma*

33b In dorsal view 5 gastral tergites (A3–A7) are visible, the fifth gastral tergite (A7) is small but continues the line of the gaster and is not reflexed below the fourth **(AA)**; the anal and associated orifices are thus situated apically

. . . . *Technomyrmex*

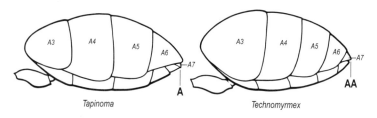

Tapinoma   Technomyrmex

34a (29) Promesonotal suture entirely absent from the dorsum of the mesosoma so that the pronotum and mesonotum are fused into a single unit **(A)** 35

34b Promesonotal suture conspicuously present across the dorsum of the mesosoma so that the pronotum and mesonotum are distinctly separated **(AA)** 37

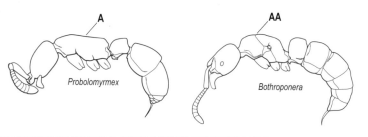

Probolomyrmex   Bothroponera

35a (34) Tergite of second gastral segment (A4) not arched and vaulted; the remaining segments directed posteriorly **(A)**. Sternite of second gastral segment large and longitudinal in profile, subrectangular to trapezoidal in shape. Eyes absent . . . . *Probolomyrmex*

35b Tergite of the second gastral segment (A4) strongly arched and vaulted so that the remaining segments point anteriorly **(AA)**. Sternite of the second gastral segment small to minute in profile, triangular in shape and with the apex of the triangle directed ventrally or anteriorly. Eyes usually present, often small 36

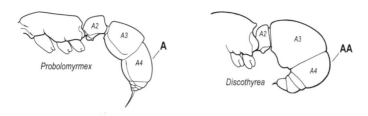

36a (35) Mandible edentate, overhung by a projecting frontoclypeal shelf that bears the antennal sockets **(A)**. Apical funicular segment strongly bulbous **(B)** .... *Discothyrea*

36b Mandible with 3 or more teeth, not overhung by a projecting frontoclypeal shelf **(AA)**. Apical funicular segment moderately enlarged but not strongly bulbous **(BB)** .... *Proceratium*

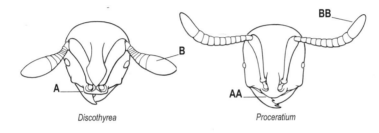

37a (34) Articulation of the petiole to the first gastral segment (A2 to A3) very broad in dorsal view **(A)**. In profile the petiole without a distinctly descending free posterior surface **(B)**. Helcium in profile with its upper margin projecting from very high on the anterior face of the first gastral segment **(C)**; above the helcium the first gastral segment has no free anterior surface, at most a short curve into the dorsum is present 38

37b Articulation of the petiole to the first gastral segment (A2 to A3) narrow in dorsal view **(AA)**. In profile the petiole has a distinctly descending posterior-free surface **(BB)**. Helcium in profile with its upper margin projecting from about the midheight or lower on the anterior face of the first gastral segment **(CC)**; above the helcium the first gastral segment has a free anterior surface below the curve into the dorsum 42

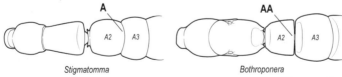

(Figure continues on next page)

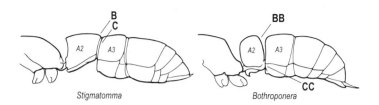

Stigmatomma   Bothroponera

38a (37) Mandible elongate and usually linear, multidentate, and not closing tightly against the clypeus, always with more than 3 teeth **(A)**..................... 39

38b Mandible short, triangular, or subtriangular, with relatively few teeth (1–3) and closing tightly against the clypeus **(AA)** 41

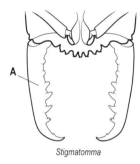

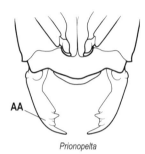

Stigmatomma   Prionopelta

39a (38) Mandible blunt at the apex and very long, longer than the head; tooth row on the inner margin of the mandible double **(A)**. Spatulate setae present on the head **(B)** ....*Mystrium*

39b Mandible pointed at the apex, not as long as the head; tooth row on the inner margin of the mandible single **(AA)**. Spatulate setae absent from the head **(BB)** 40

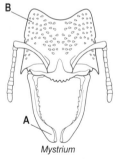

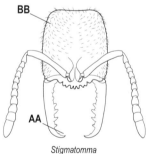

Mystrium   Stigmatomma

40a (39) Anterior margin of the median portion of the clypeus with a series of projecting, coarse, peg-like setae that appear as teeth under low magnification **(A)**.
Petiole sessile. Sternite of the petiole **(A2)** not reduced, usually extending the
length of the segment **(B)**. . . . . *Stigmatomma*

40b Anterior margin of the median portion of the clypeus smooth, without projecting peg-like setae **(AA)**. Petiole with a short anterior peduncle. Sternite of
the petiole **(A2)** reduced to a minute posteromedial sclerite **(BB)** . . . .*Xymmer*

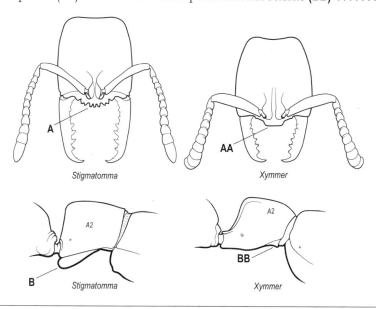

41a (38) Apical tooth of mandible followed by a short cleft, remainder of masticatory margin edentate **(A)**. Mesosoma in dorsal view marginate anteriorly and
strongly constricted from side to side in front of the metanotal groove **(B)**.
Antenna 9-segmented with a strongly defined 4-segmented club **(C)**. . . . *Concoctio*

41b Apical tooth of mandible followed by 2 more teeth, the second tooth small,
the third larger than the second **(AA)**. Mesosoma in dorsal view not marginate anteriorly nor constricted from side to side in front of the metanotal
groove **(BB)**. Antenna 8–12 segmented with a 3–4-segmented club **(CC)**
. . . . *Prionopelta*

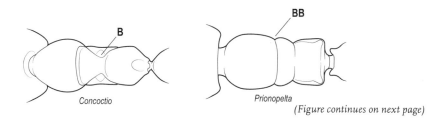

(Figure continues on next page)

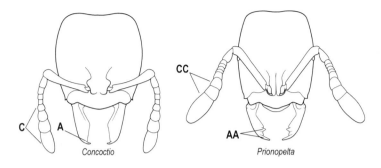

Concoctio

Prionopelta

42a (37) Frontal lobes absent; in full-face view antennal sockets entirely exposed and very close to the anterior margin of the head **(A)**. Sternite of the petiole (A2) reduced to a minute posteromedial sclerite **(B)**. Labrum with numerous peg-like setae present, arranged in transverse rows **(C)**. Sternite of the first gastral segment (A3), in ventral view, with a transverse sulcus posterior to the helcium **(D)**. . . . . *Apomyrma*

42b Frontal lobes present; in full-face view antennal sockets partially to entirely concealed and well behind the anterior margin of the head **(AA)**. Sternite of the petiole (A2) not reduced, usually extending the length of the segment **(BB)**. Labrum without transverse rows of peg-like setae **(CC)**. Sternite of the first gastral segment (A3), in ventral view, without a transverse sulcus posterior to the helcium **(DD)**. 43

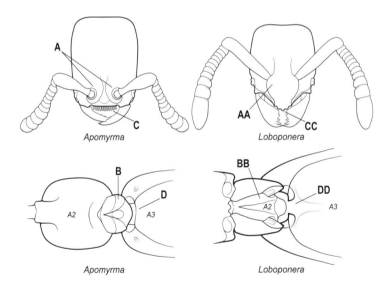

Apomyrma

Loboponera

Apomyrma

Loboponera

43a (42) Mandibles long and linear, in full-face view inserted in the middle of the anterior margin of the head, their bases closely approximated **(A)**. 44

43b Mandibles linear to triangular, in full-face view inserted at the anterolateral corners of the head, their bases conspicuously separated **(AA)**. 45

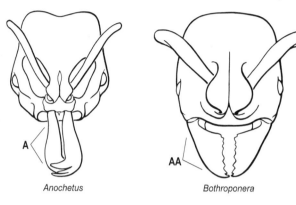

Anochetus   Bothroponera

44a (43) Nuchal carina (separates dorsal from posterior surfaces of the head) con-verges in a V-shape at the midline **(A)** and also receives a pair of prominent dark posterior apophyseal lines that converge to form the sharp median-dor-sal groove of the vertex **(B)**. Dorsalmost tooth of apical mandibular series truncated **(C)** . . . . *Odontomachus*

44b Nuchal carina forms a broad uninterrupted curve across the posterodorsal extremity of the head **(AA)**; posterior surface without paired dark apophyseal lines; on the vertex the median groove absent or ill-defined and shallow **(BB)**. Dorsalmost tooth of apical mandibular series acute **(CC)**. . . . . *Anochetus*

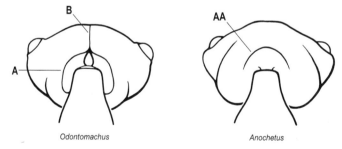

Odontomachus   Anochetus

(Figure continues on next page)

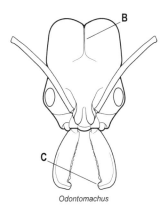

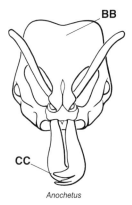

Odontomachus   Anochetus

45a (43) Dorsal (outer) surface of the metabasitarsus usually with simple setae and also equipped with strong, thickly spiniform or peg-like traction setae (**A**); similar traction setae also always present on the mesobasitarsus and mesotibia, together with simple setae. 46

45b Dorsal (outer) surface of the metabasitarsus with simple setae but without thickly spiniform or peg-like traction setae (**AA**); traction setae may very rarely occur on either the mesobasitarsus or mesotibia, together with simple setae, but if traction setae are present on either, they are absent from the metabasitarsus. 48

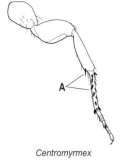

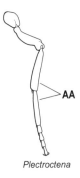

Centromyrmex   Plectroctena

46a (45) Mandible elongate and narrow, subfalcate (**A**). Apex of the mandible armed with a short vertical series of 3–4 small teeth. Apical half of the inner margin of the mandible concave; elongate basal margin shallowly convex (**B**). Prora a longitudinal, thick, bluntly convex crest that extends from just below the helcium almost to the apex of the first gastral sternite (A3) (**C**). Labial palp with 4 segments. . . . . *Promyopias*

46b Mandible triangular to elongate-triangular (**AA**); the apex of the mandible without a short vertical series of 3–4 small teeth. Masticatory and basal margins of the mandible both more or less straight (**BB**). Prora a transverse plate or a pair of weak longitudinal ridges on anterior face of the first gastral sternite (A3) **(CC)**. Labial palp with 3 segments **47**

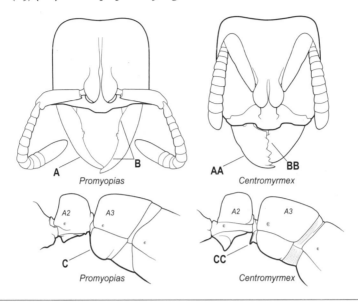

47a (46) Antennal funiculus gradually incrassate but without a distinct apical club **(A)**. Anterior clypeal margin without teeth that overhang the base of the mandible **(B)**. Maxillary palp with 4 segments. Orifice of the metapleural gland lateral **(C)**. Posterior surface of the metatibia without a depressed glandular area of pale cuticle above the spur. . . . .*Centromyrmex*

47b Antennal funiculus with a distinct 4-segmented apical club **(AA)**. Anterior clypeal margin with a strong triangular tooth on each side that projects over the base of the mandible **(BB)**. Maxillary palp with 2 segments. Orifice of the metapleural gland posteroventral **(CC)**. Posterior surface of the metatibia with a depressed glandular area of pale cuticle above the spur **(DD)**
. . . . *Feroponera*

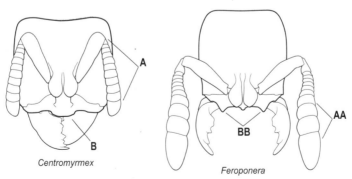

(Figure continues on next page)

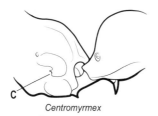

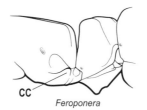

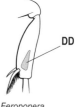

Centromyrmex   Feroponera   Feroponera

C   CC   DD

48a (45) Ventral apex of the metatibia, when viewed from in front with the metafemur at right-angles to the body, with a single large pectinate spur; without a second smaller spur in front of the pectinate main spur in the direction of observation **(A)**. 49

48b Ventral apex of the metatibia, when viewed from in front with the metafemur at right-angles to the body, with 2 spurs, consisting of a large pectinate spur and a second smaller spur that is in front of the pectinate main spur in the direction of observation **(AA)** 58

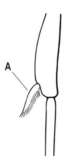
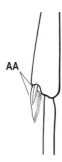

A   AA

49a (48) Mandible elongate, linear and weakly curved, the inner margin with 0–2 blunt teeth **(A)** 50

49b Mandible triangular to elongate-triangular, the masticatory margin sometimes edentate but usually with several to many teeth **(AA)** 51

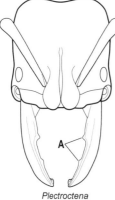
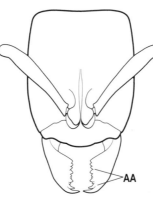
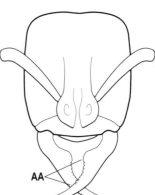

A   AA   AA

Plectroctena   Hypoponera   Psalidomyrmex

50a (49) Mandibular articulation associated with a marked semicircular excavation of the dorsal anterior margin of the head **(A)**. Mandible with a longitudinal groove on the inner half of the dorsal surface **(B)** ....*Plectroctena*

50b Mandibular articulation not associated with a semicircular excavation of the dorsal anterior margin of the head **(AA)**. Mandible without a longitudinal groove on the inner half of the dorsal surface **(BB)** .... ***Boloponera***

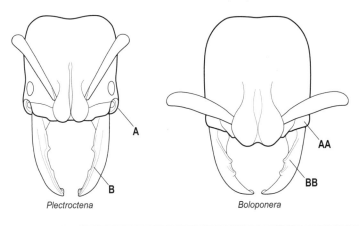

Plectroctena   Boloponera

51a (49) Basal portion of the mandible with a distinct circular or near-circular pit or fovea dorsolaterally **(A)** .... *Cryptopone*

51b Basal portion of the mandible without a dorsolateral pit or fovea **(AA)** 52

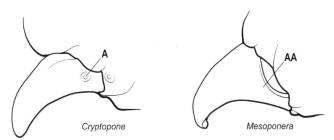

Cryptopone   Mesoponera

52a (51) Gaster (A3–A7) in profile and in dorsal view without an impression between the presclerites and the postsclerites of the second gastral segment (A4); the gaster without a girdling constriction **(A)** . . . . *Asphinctopone*

52b Gaster (A3–A7) in profile and in dorsal view with a distinct impression between the presclerites and the postsclerites of the second gastral segment (A4) that appears as a girdling constriction of the gaster **(AA)** 53

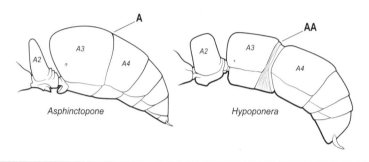

Asphinctopone   Hypoponera

53a (52) Mandible elongate-falcate, with an extremely long apical tooth so that the tips cross over at rest **(A)**. Masticatory margin edentate or crenulate. Labrum prominent, in full-face view it projects beyond the anterior clypeal margin as a striated lobe **(B)** . . . . *Psalidomyrmex*

53b Mandible short and triangular, lacking an extremely long apical tooth **(AA)**. Masticatory margin multidentate. The labrum does not project beyond the clypeus as a striated lobe in full-face view **(BB)** 54

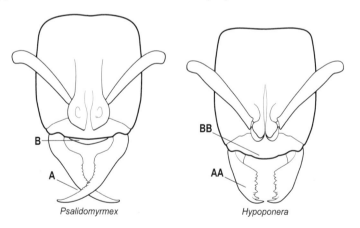

Psalidomyrmex   Hypoponera

54a (53) Metafemur mid-dorsally with a longitudinal groove on the basal half **(A)**. Tergite of the second gastral segment (A4) with dorsum vaulted, strongly arched and downcurved posteriorly **(B)**. Sternite of the second gastral segment much reduced and with a bluntly U-shaped outline in profile, very much smaller than the tergite . . . . *Loboponera*

54b Metafemur mid-dorsally without a longitudinal groove on the basal half. Tergite of the second gastral segment (A4) with the dorsum not vaulted, not arched nor strongly downcurved posteriorly **(BB)**. Sternite of the second gastral segment longitudinal, without a bluntly U-shaped outline in profile, only slightly smaller than the tergite 55

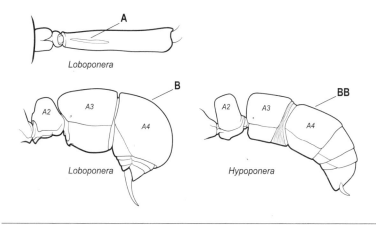

Loboponera

Loboponera

Hypoponera

55a (54) Petiole (A2) in profile subcylindrical, long and low, without an erect scale or node **(A)**. Prora absent from the first gastral sternite below the helcium **(B)**. Postsclerites of the second gastral segment (A4 posterior to gastral constriction) cylindrical, in profile very much longer than high and much longer than the first segment **(C)**. . . . . *Dolioponera*

55b Petiole (A2) in profile an erect scale or node **(AA)**. Prora present on the first gastral sternite below the helcium **(BB)**. Postsclerites of the second gastral segment (A4 posterior to gastral constriction) not cylindrical, in profile as high as long or nearly so and at most only slightly longer than the first segment **(CC)**. 56

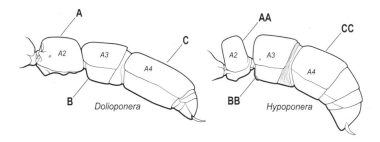

Dolioponera

Hypoponera

56a (55) With the mesosoma in dorsal view, the anterior half of the propodeum is so strongly constricted from side to side that its dorsum is reduced to a mere longitudinal crest, almost obliterated **(A)**. Mandible narrowly triangular and slender, with a total of 7 teeth and denticles, or fewer **(B)**. Exposed length of closed mandibles 0.45 × maximum length of head, or usually more

. . . . *Fisheropone*

56b With the mesosoma in dorsal view, the anterior half of the propodeum forms a distinctly defined transverse dorsal surface **(AA)**. Mandible broadly triangular and stout, usually with a total of more than 8 teeth and denticles **(BB)**. Exposed length of closed mandibles at most 0.35 × maximum length of head, usually less. 57

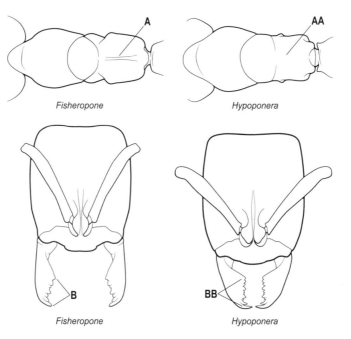

Fisheropone   Hypoponera

Fisheropone   Hypoponera

57a (56) Subpetiolar process in profile with a pair of teeth posteroventrally and with a fenestra or thin-spot anteriorly that is translucent **(A)**. Maxillary palp with 2 segments (a single known introduction in East Africa). . . . . *Ponera*

57b Subpetiolar process in profile rounded to acutely angulate posteroventrally but never with a pair of teeth; an anterior fenestra or thin-spot usually absent but present in some species **(AA)**. Maxillary palp with 0–1 segments

. . . . *Hypoponera*

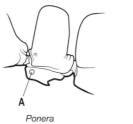
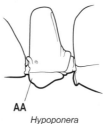

A
Ponera

AA
Hypoponera

58a (48) Helcium in profile located approximately at midheight on the front of the first gastral segment (A3), so that the first gastral segment does not have a long vertical anterior face in profile **(A)**. Mesotibia and metatibia each with 2 pectinate spurs, the anterior spur smaller than the posterior **(B)** . . . . . *Platythyrea*

58b Helcium in profile located very low on the front of the first gastral segment (A3) so that the first gastral segment has a long vertical anterior face **(AA)**. Mesotibia and metatibia each with 1 large pectinate spur and 1 small simple spur **(BB)** 59

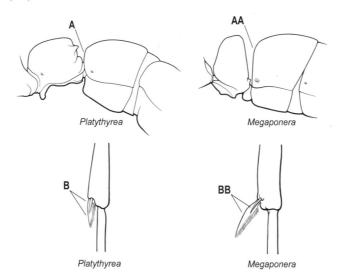

A

AA

Platythyrea

Megaponera

B

BB

Platythyrea

Megaponera

59a (58) Hypopygium with a series of dorsally directed teeth or spiniform setae on its
dorsal margin posteriorly (may be concealed among the regular setae of the
hypopygial apex) **(A)** 60

59b Hypopygium unarmed; usually with regular setae on its dorsal margin posteriorly, but without a series of dorsally directed teeth or spiniform setae **(AA)** . . . 61

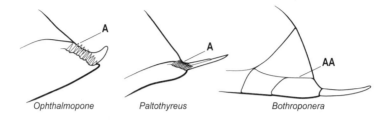

Ophthalmopone   Paltothyreus   Bothroponera

60a (59) Clypeus with a bluntly rectangular, conspicuously projecting median lobe,
the dorsal surface of which is transversely concave **(A)**. Metapleural gland orifice followed by a vertical cuticular flange. In dorsal view the base of the first
gastral tergite (A3) is weakly marginate and distinctly angulate on either side
(B). In full-face view the eyes are located in front of the midlength of the head
capsule **(C)**. Mandible with basal groove present **(D)**. . . . . *Paltothyreus*

60b Clypeus simple, its anterior margin convex, without a rectangular projecting
median lobe **(AA)**. Metapleural gland orifice not followed by a vertical cuticular flange. In dorsal view the base of the first gastral tergite (A3) is not marginate and is evenly rounded on either side **(BB)**. In full-face view the eyes are
large, located at or behind the midlength of the head capsule **(CC)**. Mandible
with basal groove absent **(DD)** . . . . *Ophthalmopone*

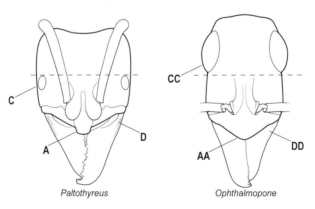

Paltothyreus   Ophthalmopone

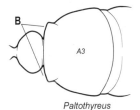

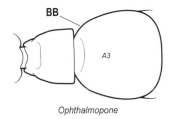

Paltothyreus   Ophthalmopone

61a (59) Pretarsal claws of metatarsus usually pectinate, extremely rarely with only
1–2 small teeth behind the apex **(A)**. Mandible with only 1–3 teeth (usually
2) **(B)**, and frontal lobes distinctly fail to cover the entire antennal sockets in
full-face view **(C)** . . . . *Leptogenys*

61b Pretarsal claws of the metatarsus never pectinate, the claws simple **(AA)** or
at most with teeth confined to the basal third or less **(AAA)**. If preapical teeth
present on claw, then mandible with 4 or more teeth **(BB)**, and the frontal
lobes fully conceal the antennal sockets in full-face view **(CC)**. 62

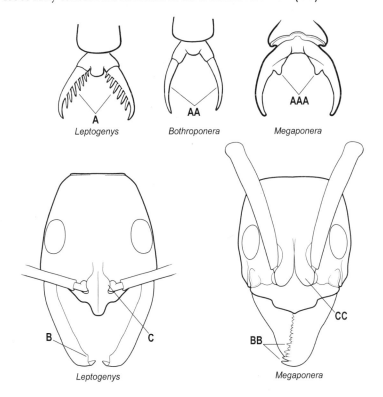

Leptogenys   Bothroponera   Megaponera

Leptogenys   Megaponera

62a (61) Head with a distinct carina on each side that extends from the anterior margin of the eye to the clypeal margin **(A)**. Polymorphic species, the antennal scapes conspicuously flattened **(B)**. Posterior portion of the clypeus broadly inserted between the frontal lobes **(C)** . . . .*Megaponera*

62b Head without a carina on each side that extends from the anterior margin of the eye to the clypeal margin **(AA)**. Monomorphic species, the antennal scapes not flattened **(BB)**. Posterior portion of the clypeus narrowly inserted between the frontal lobes **(CC)**. 63

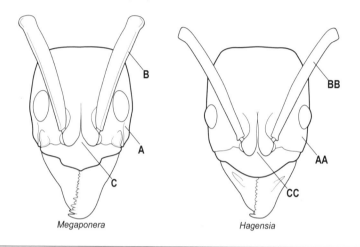

Megaponera   Hagensia

63a (62) Petiole (A2) dorsally with a comb of 5 long spines that curve backward over the base of the first gastral segment (A3) **(A)**. Propodeum with a pair of stout spines **(B)** . . . . *Phrynoponera*

63b Petiole (A2) dorsally without a comb of 5 spines **(AA)**. Propodeum unarmed or at most with a pair of small teeth **(BB)** 64

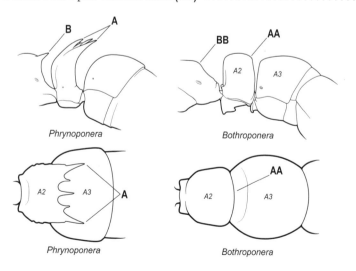

Phrynoponera   Bothroponera

Phrynoponera   Bothroponera

64a (63) Petiole (A2) with its sides convergent dorsally into a sharp longitudinal crest that runs the length of the segment **(A)**. Posterolateral margins of the petiole also sharply angulate in the dorsal half; these sharp angles meet the dorsal crest at its posterior end. Anterior clypeal margin broadly concave, the concavity terminates at each side in a prominent angle or tooth-like projection **(B)**. Propodeum with a pair of short teeth **(C)** . . . . *Streblognathus*

64b Petiole (A2) scale-like to nodiform but without a sharp longitudinal crest that runs the length of the dorsum **(AA)**. The clypeus usually prominent but if shallowly concave medially, then the concavity does not terminate in prominent angles or teeth **(BB)**. Propodeum unarmed **(CC)** 65

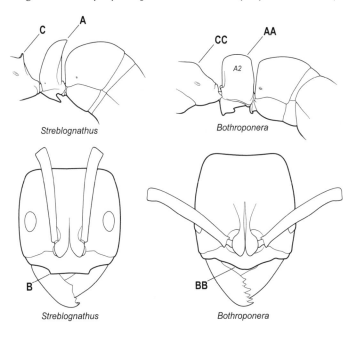

Streblognathus

Bothroponera

Streblognathus

Bothroponera

65a (64) Basal portion of the mandible either with an oblique dorsal groove **(A)** or with a small circular or near-circular pit or fovea dorsolaterally **(AA)** **66**

65b Basal portion of the mandible without an oblique dorsal groove, without a dorsolateral pit or fovea **(AAA)** **68**

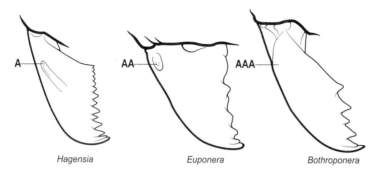

Hagensia   Euponera   Bothroponera

66a (65) Pretarsal claw of the metatarsus with a single, conspicuous preapical tooth on the inner margin **(A)**. Basal portion of the mandible with a distinct oblique dorsal groove **(B)**. Pronotum marginate laterally **(C)** . . . . *Hagensia*

66b Pretarsal claw of the metatarsus unarmed **(AA)**, or at most with an inconspicuous basal angle on the inner margin. Basal portion of the mandible with a dorsolateral pit or fovea **(BB)**. Pronotum not marginate laterally **(CC)** **67**

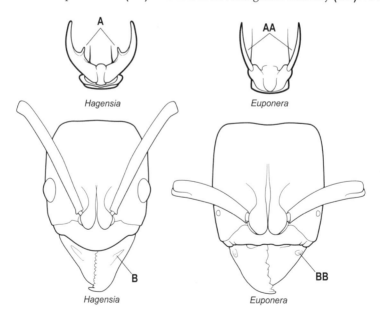

Hagensia   Euponera

Hagensia   Euponera

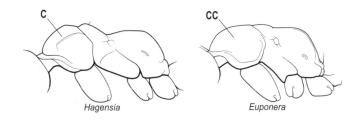

Hagensia   Euponera

67a (66) Prora absent from anteroventral angle of the first gastral sternite (A3) below the helcium **(A)**. Metanotal groove conspicuous across the dorsal mesosoma **(B)**; in profile the propodeal dorsum considerably depressed below the level of the mesonotal dorsum **(C)** . . . . *Brachyponera*

67b   Prora present at anteroventral angle of the first gastral sternite (A3) below the helcium **(AA)**. Metanotal groove vestigial to absent across the dorsal mesosoma **(BB)**; in profile the propodeal dorsum more or less continues at the same level as the mesonotal dorsum **(CC)** . . . . *Euponera*

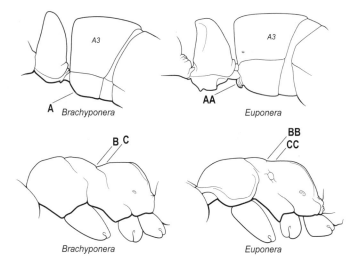

A *Brachyponera*   *Euponera*

Brachyponera   *Euponera*

68a (65) Petiole (A2) in profile surmounted by a thick node **(A)**. Metanotal groove usually entirely absent across the dorsum of the mesosoma **(B)**. . . . . *Bothroponera*

68b   Petiole (A2) in profile surmounted by a scale **(AA)**. Metanotal groove conspicuous across the dorsum of the mesosoma **(BB)** **69**

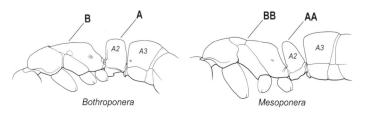

Bothroponera   *Mesoponera*

69a (68) Eyes usually absent **(A)**, very rarely a single ommatidium present. . . . *Parvaponera*

  69b Eyes always present **(AA)**, with more than 3 ommatidia . . . . *Mesoponera*

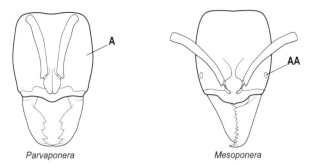

Parvaponera   Mesoponera

70a (1) Promesonotal suture fully and strongly developed across the dorsum of the mesosoma (posterior margin of the pronotum overlaps the anterior margin of the mesonotum) **(A)**; suture fully articulated and flexible, so that the pronotum and mesonotum are capable of movement relative to each other in fresh specimens. 71

  70b Promesonotal suture either entirely absent across the dorsum of the mesosoma, or at most vestigially present as a line or feeble impression (posterior margin of the pronotum does not overlap the anterior margin of the mesonotum) **(AA)**; the pronotum and mesonotum are fully fused, not capable of movement relative to each other in fresh specimens. 72

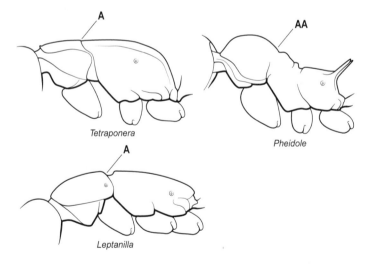

Tetraponera   Pheidole

Leptanilla

71a (70) Eyes present **(A)**. Antennal sockets inclined upward toward the midline of the head, at least their inner margins covered by frontal lobes **(B)**. Pretarsal claws each with a single preapical tooth **(C)**. Propodeal spiracle situated high on the side of the sclerite and far forward **(D)**. Propodeal lobes present **(E)**. Orifice of the metapleural gland simple, not overhung by a cuticular flap. Petiole and postpetiole (A2 and A3) without tergosternal fusion. Larger ants, total length of head and body greater than 2.00 mm. . . . . *Tetraponera*

71b Eyes absent **(AA)**. Antennal sockets horizontal and fully exposed, frontal lobes absent **(BB)**. Pretarsal claws without a preapical tooth **(CC)**. Propodeal spiracle situated at about the midheight of the side of the sclerite or lower, and behind the midlength **(DD)**. Propodeal lobes absent **(EE)**. Orifice of the metapleural gland overhung by a cuticular flap. Petiole and postpetiole (A2 and A3) with tergosternal fusion. Minute ants, total length of head and body less than 2.00 mm . . . . *Leptanilla*

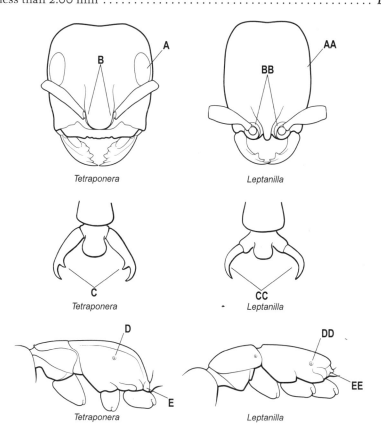

Tetraponera

Leptanilla

Tetraponera

Leptanilla

Tetraponera

Leptanilla

72a (70) Pygidium large and flattened dorsally, its lateral margins with a series of short spines, teeth, or denticles on each side **(A)**. With head in ventral view and mouthparts closed, the prementum is not visible; it is entirely concealed by the labrum anteriorly and the maxillary stipites on each side; the stipites meet along the midline **(B)**. . . . . *Parasyscia* (part)

72b Pygidium small and convex dorsally, its lateral margins unarmed **(AA)**. With head in ventral view and mouthparts closed, the prementum is visible; it is bounded by the labrum anteriorly and the maxillary stipites on each side of it **(BB)**. 73

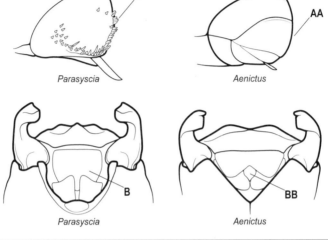

Parasyscia   *Aenictus*

Parasyscia   *Aenictus*

73a (72) Metatibial gland present: apical half of the ventral surface of the metatibia with an elongate impression or an area of extremely thin translucent cuticle **(A)**. Metapleural gland orifice overhung from above by a flap of cuticle. Eyes always absent. Antennal sockets horizontal, fully exposed; frontal lobes absent **(B)**. Antenna with 10 segments . . . . *Aenictus*

73b Metatibial gland absent: ventral surface of the metatibia without an elongate impression or an area of extremely thin translucent cuticle. Metapleural gland orifice a longitudinal slit or narrow crescent that opens dorsally to posterodorsally (may be difficult to see), or simple and opening posterolaterally. Eyes usually (but not always) present. Antennal sockets strongly incline upward toward the midline of the head; sockets usually (but not always) partially to entirely concealed by the frontal lobes **(BB)**. Antenna with 4–12 segments 74

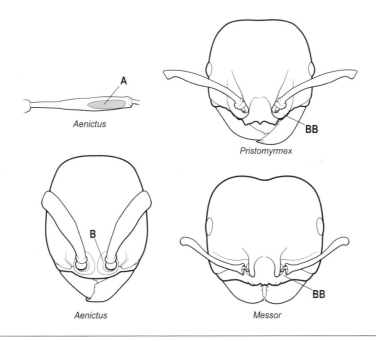

Aenictus

Pristomyrmex

Aenictus

Messor

74a (73)   Postpetiole (A3) articulated on the dorsal surface of the first gastral segment (A4) **(A)**. Helcium enlarged, in profile its height subequal to, or greater than, the height of the postpetiole **(B)**. Gaster in dorsal view roughly heart-shaped and capable of reflexion over the mesosoma. Petiole (A2) dorsoventrally flattened and without a node **(C)** . . . . *Crematogaster*

74b   Postpetiole (A3) articulated on the anterior face of the first gastral segment (A4) **(AA)**. Helcium small, in profile its height distinctly less than the height of the postpetiole **(BB)**. Gaster in dorsal view not roughly heart-shaped, not capable of reflexion over the mesosoma. Petiole (A2) not dorsoventrally flattened, with a node of some form **(CC)**. 75

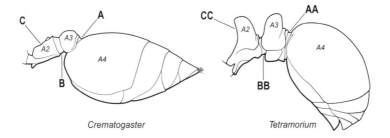

Crematogaster

Tetramorium

76a (75) Antenna with 7 segments. Gastral shoulder present: the tergosternal junction of the first gastral segment (A4) curves abruptly inward near the base, so that the sternite projects forward on each side of the postpetiole **(A)**. Sting spatulate. . . . . *Myrmicaria*

76b Antenna with 4–6 segments. Gastral shoulder absent: the tergosternal junction of the first gastral segment (A4) curves evenly to the base, the sternite does not project forward on each side of the postpetiole **(AA)**. Sting simple or extremely reduced, not spatulate 77

A / A2 / A3 / A4 — *Myrmicaria*

AA / A2 / A3 / A4 — *Microdaceton*

AA / A2 / A3 / A4 — *Strumigenys*

77a (76) Procoxa much smaller than the massively developed mesocoxa and metacoxa **(A)**. Lateral portion of the pronotum reduced to a narrow down-pointing triangle **(B)**. Frontal lobes confluent mid-dorsally; the clypeus does not project back between the frontal lobes **(C)**. Spongiform or lamellate appendages absent from waist segments **(D)**. Postpetiole (A3) very broadly articulated to the first gastral segment (A4) and the latter without presclerites **(E)**. Propodeal lobes absent. Mesobasitarsus and metabasitarsus equipped apically with a circlet of short traction spines. . . . . *Melissotarsus*

77b Procoxa as large as or larger than the mesocoxa and metacoxa **(AA)**. Lateral portion of the pronotum not reduced to a narrow down-pointing triangle **(BB)**. Frontal lobes widely separated; the clypeus projects back between the frontal lobes **(CC)**. Spongiform or lamellate appendages present on one or both waist segments **(DD)**. Postpetiole (A3) narrowly articulated to the first gastral segment (A4) and the latter with presclerites **(EE)**. Propodeal lobes present. Mesobasitarsus and metabasitarsus without an apical circlet of traction spines 78

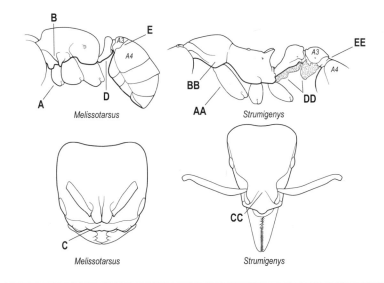

Melissotarsus · Strumigenys

Melissotarsus · Strumigenys

78a (77) Mesonotum and petiole node (A2) each with a pair of tubercles or teeth **(A)**. With head in profile the eye is not situated at the ventrolateral margin **(B)**. Antennal scrobes absent **(C)**. Base of the first gastral tergite (A4) without a transverse lip or flange of cuticle (limbus) behind the postpetiole **(D)**. Mandible linear, with an apical fork of 3 spiniform teeth arranged in a vertical series; without preapical dentition **(E)**. Spongiform tissue absent from the postpetiole **(F)**. Palp formula 3,2 . . . .*Microdaceton*

78b Mesonotum and petiole node (A2) without tubercles or teeth **(AA)**. With head in profile the eye is situated at the ventrolateral margin **(BB)**. Antennal scrobes usually present and conspicuous **(CC)**. Base of the first gastral tergite (A4) with a transverse lip or flange of cuticle (limbus) behind the postpetiole **(DD)**. Mandible triangular to linear; if linear then apical fork of 2 spiniform teeth at maximum and preapical dentition usually present **(EE)**. Spongiform tissue usually present on postpetiole **(FF)**. Palp formula 0,1 or 1,1 . . . . .***Strumigenys***

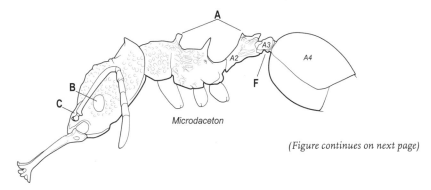

Microdaceton

(Figure continues on next page)

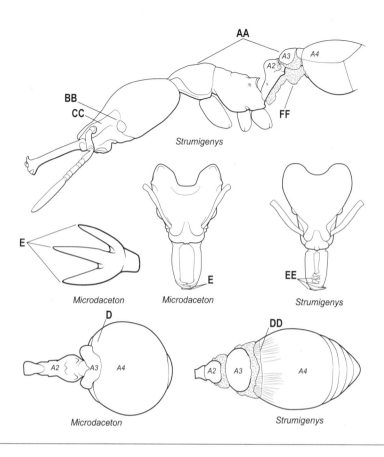

Strumigenys

Microdaceton   Microdaceton   Strumigenys

Microdaceton   Strumigenys

79a (75) Apical and preapical antennal segments are much larger than preceding funicular segments and form a conspicuous and usually very distinctive 2-segmented club **(A)** 80

79b Antenna never terminates in a conspicuous 2-segmented club. Either apical plus 2 preapical segments of antenna are enlarged and form a 3-segmented club **(AA)**, or less commonly the club has more than 3 segments; rarely the funiculus is filiform and lacks a developed apical club 84

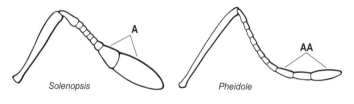

Solenopsis   Pheidole

80a (79) Antenna with 8–11 segments 81
80b Antenna with 12 segments 83

81a (80) Head with strongly developed sinuate frontal carinae that extend almost to the posterior margin **(A)**. Broad, deep antennal scrobes present that extend above the eyes. Anterior margin of the clypeus with a thin cuticular apron that fits tightly over the basal margins of the mandibles **(B)**. Lateral portions of the clypeus form a sharply raised transverse ridge in front of the antennal insertions **(C)** (introduced genus of Neotropical origin) . . . . *Wasmannia*

81b Head without frontal carinae and usually without antennal scrobes **(AA)**, the latter only extremely rarely present. Anterior margin of the clypeus without a cuticular apron that fits tightly over the basal margins of the mandibles. Lateral portions of the clypeus do not form a sharp transverse ridge in front of the antennal insertions **(CC)** 82

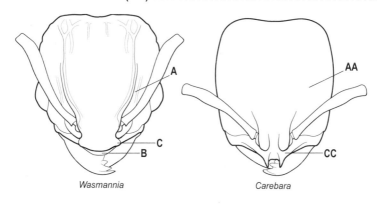

Wasmannia   *Carebara*

82a (81) Anterior clypeal margin with a single long, anteriorly projecting, unpaired median seta at the midpoint of the margin **(A)**. Propodeum always unarmed and rounded **(B)**. Antenna almost always with 10 segments, only extremely rarely with 11 . . . . *Solenopsis*

82b Anterior clypeal margin lacks a single unpaired median seta, instead usually with a pair of setae, one on each side of the midpoint of the margin **(AA)**. Propodeum sometimes unarmed and rounded but usually with spines or teeth, or sharply angulate **(BB)**. Antenna with 8–11 segments . . . . *Carebara*

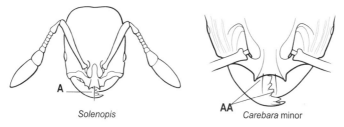

Solenopis   *Carebara* minor

(Figure continues on next page)

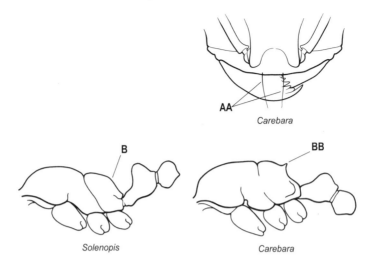

Carebara

Solenopis

Carebara

83a (80) Frontal lobes separated and median portion of the clypeus broadly inserted between them **(A)**. Lateral portions of the clypeus flattened and prominent, fused to the raised projecting median portion of the clypeus to form a shelf that projects forward over the mandibles **(B)**. Propodeal lobes low and rounded, not connected to propodeal spines (when present) by broad projecting lamellae **(C)**. Palp formula 5,3 . . . .***Cardiocondyla*** (part)

83b   Frontal lobes closely approximated and median portion of the clypeus reduced to an extremely narrow strip between them **(AA)**. Lateral portions of the clypeus not prominent, not fused to median portion, and not forming a shelf; instead, the median portion of the clypeus is sharply raised centrally and in the form of a narrow longitudinal ridge **(BB)**. Propodeal lobes large and prominent, connected to propodeal spines by broad conspicuous lamellae **(CC)**. Palp formula 2,2 . . . . ***Baracidris***

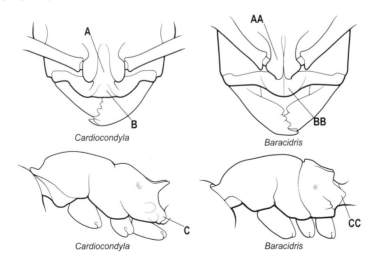

Cardiocondyla

Baracidris

Cardiocondyla

Baracidris

84a (79) Antenna with 9 segments . . . . *Meranoplus*

84b Antenna with 10–12 segments. 85

85a (84) Median portion of the clypeus vertical above the mandibles, with a conspicuous anteriorly projecting bilobed appendage at the top of the vertical face
(the clypeal fork), which projects out over the mandibles from about the same
level as the frontal lobes **(A)** 86

85b Median portion of the clypeus not vertical, without a bilobed appendage that
projects out over the mandibles from about the same level as the frontal lobes
(AA) 87

Calyptomyrmex   *Tetramorium*

86a (85) Antenna with 11 segments. Peduncle of the petiole (A2) short and very thick
in profile **(A)**, with a massive triangular anteroventral process that is inclined
anteriorly **(B)**. Propodeal lobe thin and lamellate in dorsal view; in profile
without a fenestra or thin-spot. Side of the petiolar peduncle, near base, without a carina that extends almost the height of the peduncle. All body setae
simple, without bizarre pilosity **(C)** . . . . *Dicroaspis*

86b Antenna with 12 segments. Peduncle of the petiole (A2) elongate and narrow
in profile **(AA)**, without an anteroventral process **(BB)**. Propodeal lobe thick
and coarse in dorsal view; in profile with a median fenestra or thin-spot. Side
of the petiolar peduncle, near base, with a sharp oblique carina that extends
almost the height of the peduncle (surface anterior to the carina highly polished, posterior to it sculptured). Body setae bizarre; either spatulate, squamate, clavate, star-shaped, or very short, thick, and stubbly with abruptly
tapered points **(CC)** . . . .*Calyptomyrmex*

(Continues on next page)

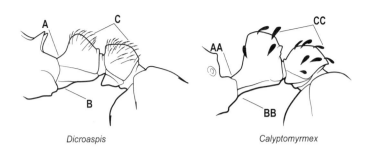

Dicroaspis   Calyptomyrmex

87a (85) Antenna with 10 segments. 88
87b   Antenna with 11 or 12 segments 90

88a (87) Eyes absent. Propodeum unarmed, without spines or teeth **(A)**. Posterior portion of the clypeus extremely narrowly inserted between the frontal carinae
(B). Postpetiole attached very high on the anterior face of the first gastral segment (A4) **(C)**. Palp formula 2,1. . . . .*Anillomyrma*
88b   Eyes present. Propodeum armed with a pair of spines or teeth **(AA)**. Posterior portion of the clypeus broadly inserted between the frontal carinae **(BB)**.
Postpetiole attached at midheight of the first gastral segment (A4) **(CC)**. Palp
formula 4,3 or 5,3. 89

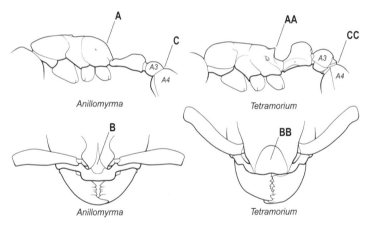

Anillomyrma   Tetramorium

Anillomyrma   Tetramorium

89a (88)   Sting shaft with an apicodorsal triangular to pennant-shaped lamelliform appendage **(A)**. Propodeal spiracle low on the side and behind the midlength, abutting the metapleural gland bulla, widely separated from the dorsal outline in profile **(B)**. Petiole (A2) with a long anterior peduncle **(C)**. Tergite of the first gastral segment (A4) does not broadly overlap the sternite on the ventral surface of the gaster **(D)**. Palp formula 4,3; a transverse crest present on stipes of maxilla. Monomorphic species . . . . *Tetramorium* (part)

89b   Sting shaft without an apicodorsal lamelliform appendage **(AA)**. Propodeal spiracle very high on the side and slightly in front of the midlength, widely separated from the metapleural gland bulla, very close to the dorsal outline in profile **(BB)**. Petiole (A2) subsessile, with an extremely short inconspicuous anterior peduncle **(CC)**. Tergite of the first gastral segment (A4) very broadly overlaps the sternite on the ventral surface of the gaster **(DD)**. Palp formula 5,3; without a transverse crest on the stipes of the maxilla. Polymorphic species . . . . *Atopomyrmex* (part)

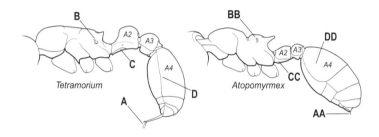

90a (87)   Antenna with 11 segments 91
90b   Antenna with 12 segments 99

91a (90)   Antennal scrobe present that extends below the eye **(A)**. Antennal insertions extremely widely separated so that in full-face view the outer margins of the frontal lobes form the visible lateral margin of the head in front of the eyes **(B)**. Dorsum of the gaster consists entirely of the expanded first tergite (A4), the remaining tergites visible in profile below the posterior margin of the first **(C)** . . . . *Cataulacus*

91b   Antennal scrobe either absent or present, but when present it extends above the eye **(AA)**. Antennal insertions more closely approximated so that in full-face view the outer margins of the frontal lobes are distinctly inside the visible lateral margin of the head in front of the eyes **(BB)**. Dorsum of the gaster does not consist entirely of the first tergite, the remaining tergites continue the line of the first and are visible in dorsal view **(CC)** 92

(Continues on next page)

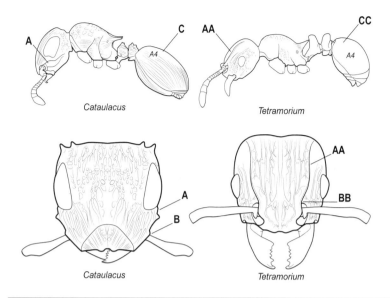

Cataulacus   Tetramorium

Cataulacus   Tetramorium

92a (91) Frontal lobes vestigial or absent so that in full-face view the antennal sockets are exposed, and the depressed area that contains the antennal sockets is clearly visible **(A)**. Anterior clypeal margin usually armed with a few blunt denticles or crenulae **(B)**. Dorsum of labrum, just behind its anterior margin, with a raised cuticular ridge, a transverse row of 2–3 denticles, or both of these **(C)** . . . . ***Pristomyrmex***

92b Frontal lobes present, in full-face view covering most or all of the antennal sockets; the antennal sockets are not fully visible in dorsal view **(AA)**. Anterior clypeal margin unarmed or with a pair of small teeth **(BB)**. Dorsum of labrum, just behind its anterior margin, without a raised cuticular ridge or a transverse row of 2–3 denticles **93**

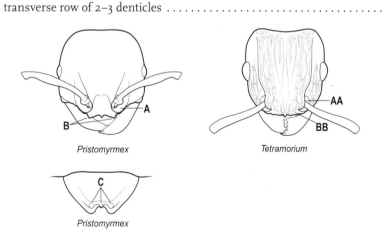

Pristomyrmex   Tetramorium

Pristomyrmex

93a (92) Median portion of the clypeus raised and produced forward as a large shield-like lobe that projects strongly over the mandibles **(A)**. Tibiae and basitarsi of middle and hind legs terminate in a number of peg-like stout spines **(B)**. Metatibia with a single, pectinate, apical spur **(C)**. Procoxa distinctly smaller than the mesocoxa and metacoxa **(D)** . . . .*Metapone*

93b Median portion of the clypeus not produced forward as a large shield-like lobe that projects strongly over the mandibles **(AA)**. Tibiae and basitarsi of middle and hind legs do not terminate in peg-like stout spines **(BB)**. Metatibia without a spur, or with a single, slender, simple spur **(CC)**. Procoxa larger than the mesocoxa and metacoxa **(DD)** 94

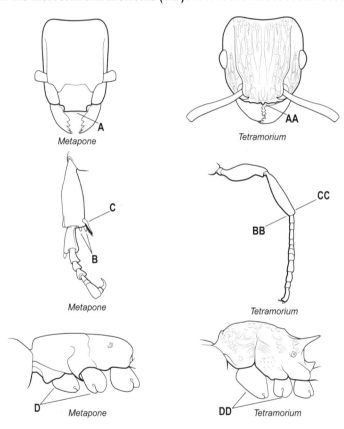

A
Metapone

AA
Tetramorium

C

B
Metapone

CC

BB
Tetramorium

D
Metapone

DD
Tetramorium

94a (93) Propodeum rounded to angulate, never armed with differentiated teeth or spines **(A)**. Anterior clypeal margin with a single unpaired seta at the midpoint **(B)**. Antennal scrobes always absent **(C)**. Mandible with only 4 teeth **(D)**. Maxillary palp with 1 or 2 segments 95

94b Propodeum bidentate or bispinose **(AA)**. Anterior clypeal margin without a single unpaired seta at the midpoint, usually with a pair of setae, one on each side of the midpoint **(BB)**. Antennal scrobes frequently, but not always, present **(CC)**. Mandible usually with 5 or more teeth **(DD)**. Maxillary palp with 3– 5 segments 97

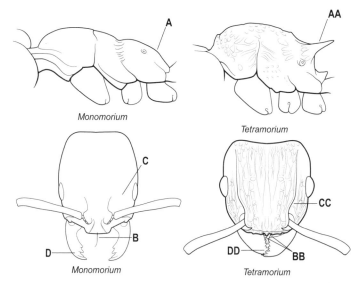

Monomorium

Tetramorium

Monomorium

Tetramorium

95a (94) Eyes absent. Propodeal spiracle a very large circular orifice **(A)**. Frontal lobes very closely approximated, and median portion of the clypeus narrow between the lobes **(B)**. Procoxa enormously enlarged when compared to the mesocoxa and metacoxa **(C)** .... *Bondroitia*

95b Eyes present. Propodeal spiracle small, usually pinhole-like **(AA)**. Frontal lobes widely separated, and median portion of the clypeus broad between the lobes **(BB)**. Procoxa larger than the mesocoxa and metacoxa but not enormously so **(CC)**. 96

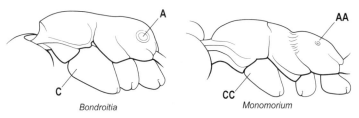

Bondroitia

Monomorium

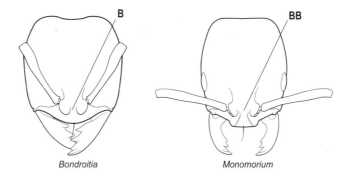

Bondroitia   Monomorium

96a (95)   Median portion of the clypeus distinctly raised, strongly to weakly longitudinally bicarinate **(A)**. Postpetiole node (A3) less voluminous than the petiole
node in profile and narrowly attached to the gaster **(B)** . . . . *Monomorium* (part)

96b   Median portion of the clypeus evenly transversely convex, not distinctly
raised nor longitudinally bicarinate **(AA)**. Postpetiole node (A3) much more
voluminous than the petiole node in profile and very broadly attached to the
gaster **(BB)** . . . . *Diplomorium*

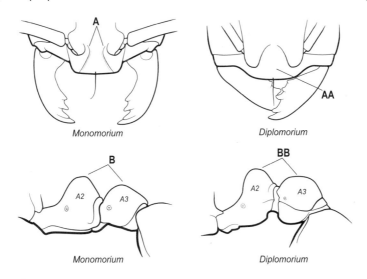

Monomorium   Diplomorium

Monomorium   Diplomorium

97a (94) Petiole and postpetiole each with a pair of spines laterally **(A)**. Mandible with 4–5 teeth. Gastral shoulder present: the tergosternal junction of the first gastral segment (A4) curves abruptly inward near the base, so that in dorsal view the sternite projects forward slightly beyond the tergite on each side of the postpetiole **(B)**. Sting simple, without a spatulate appendage preapically or apically. Maxillary palp with 5 segments . . . . *Nesomyrmex* (part)

97b Petiole and postpetiole unarmed, without spines **(AA)**. Mandible with more than 5 teeth. Gastral shoulder absent: the tergosternal junction of the first gastral segment (A4) curves evenly to the base, so that in dorsal view the sternite does not project beyond the tergite on each side of the postpetiole **(BB)**. Sting with a spatulate appendage preapically or apically. Maxillary palp with 3 or 4 segments 98

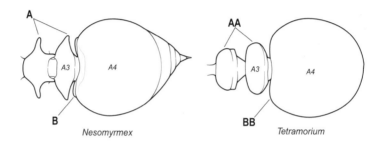

Nesomyrmex *Tetramorium*

98a (97) Lateral portions of the clypeus not raised into a narrow ridge or wall in front of the antennal insertions **(A)**. Median portion of the clypeus narrow and bicarinate, narrowly inserted between frontal lobes **(B)**. Mandible armed with 10–14 teeth that decrease in size from apex to base **(C)**. Promesonotum in profile with a swollen and dome-like outline with respect to the propodeum **(D)** . . . . *Cyphoidris*

98b Lateral portions of the clypeus raised into a narrow ridge or wall in front of the antennal insertions **(AA)**. Median portion of the clypeus broad, not bicarinate, broadly inserted between frontal lobes **(BB)**. Mandible armed with 2–3 enlarged teeth apically, followed by a row of at least 4 smaller denticles, sometimes more **(CC)**. Promesonotum in profile without a swollen and dome-like outline with respect to the propodeum **(DD)** . . . . *Tetramorium* (part)

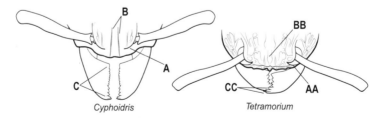

Cyphoidris *Tetramorium*

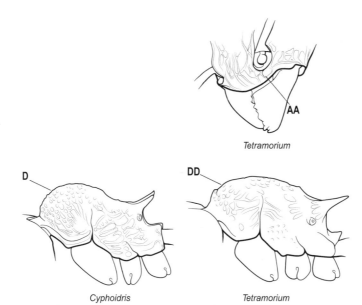

Tetramorium

Cyphoidris   *Tetramorium*

99a (90)   Dorsum of the petiole node (A2) armed with a pair of sharp spines **(A)** **100**
99b   Dorsum of the petiole node (A2) unarmed or indented medially, lacking
               sharp spines **(AA)** **102**

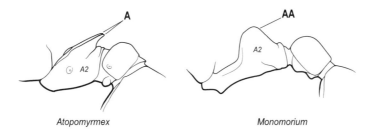

Atopomyrmex   *Monomorium*

100a (99) All of the visible portion of the gaster consists of the first tergite (A4), which
               is massively enlarged and ball-like, with an anteroventral orifice within which
               the remaining gastral segments are telescoped **(A)**. Eyes at extreme posterior
               corners of the head **(B)**. A broad lamellate crest is present across the posterior
               margin of the head **(C)**. The clypeus projects far forward and almost conceals
               the mandibles **(D)**. The metapleural gland orifice is a simple hole at the ven-
               trolateral corner of the metapleuron that opens posterolaterally ..... *Ankylomyrma*
100b   Gaster composed of 4 visible tergites and sternites that decrease in size pos-
               teriorly, the gaster with the first tergite (A4) not massively enlarged, not ball-
               like **(AA)**. Eyes not at extreme posterior corners of the head **(BB)**. Posterior
               margin of the head without a lamellate crest **(CC)**. The clypeus does not proj-

ect far forward over the mandibles **(DD)**. Metapleural gland orifice is a longitudinal slit or narrow crescent, well above the ventrolateral corner of the metapleuron, that opens dorsally to posterodorsally (may be very narrow and difficult to see) 101

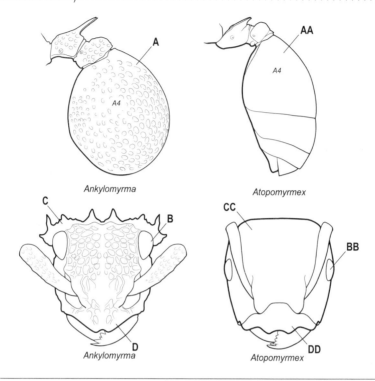

Ankylomyrma

Atopomyrmex

Ankylomyrma

Atopomyrmex

101a (100) Posterior corners of the head are evenly and broadly rounded in full-face view **(A)**. Stipes of the maxilla without a transverse crest. Ventral margin of the sides of the metapleuron eroded in front of the metapleural gland bulla. Ventral surface of the mesosoma with a very deep broad pit between the metacoxae. Polymorphic species, the propodeum armed with a pair of long spines . . . . ***Atopomyrmex*** (part)

101b Posterior corners of the head angulate to denticulate in full-face view **(AA)**. Stipes of the maxilla with a transverse crest **(BB)**. Ventral margin of sides of the metapleuron not eroded in front of the metapleural gland bulla but with a conspicuous broad groove running forward to the mesopleuron. Ventral surface of the mesosoma without a broad deep pit between the metacoxae. Monomorphic species, the propodeum short-bispinose at most . . . . ***Terataner*** (part)

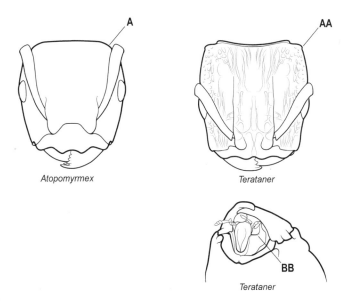

Atopomyrmex

Terataner

Terataner

102a (99) Mandible with the third tooth from the apex, and usually also the fourth tooth, double-ranked (i.e., with a tooth on the inner mandibular margin, internal to and concealed by the normally visible one when the mandibles are closed). Mesothoracic spiracles open on the dorsum of the mesosoma **(A)**. Propodeal spiracle extremely elongate, its orifice a long slit **(B)**. Eyes well behind the midlength of the head capsule in full-face view, and ventral surface of the head with a large psammophore **(C)** . . . . *Ocymyrmex*

102b Mandible without double-ranked teeth (i.e., the inner mandibular margin, internal to the visible teeth, is unarmed). Mesothoracic spiracles concealed by a pronotal flap on the sides of the mesosoma **(AA)**. Propodeal spiracle usually circular to subcircular **(BB)**, extremely rarely a short slit, in which case the eyes are in front of the midlength of the head capsule in full-face view, and the ventral surface of the head lacks a psammophore **103**

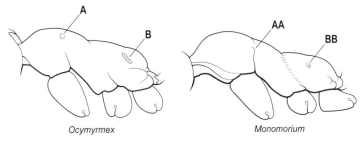

Ocymyrmex

Monomorium

(Figure continues on next page)

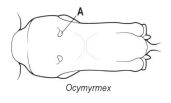

Ocymyrmex

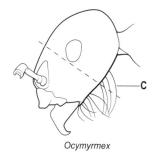

Ocymyrmex

103a (102) Lateral portions of the clypeus raised up into a sharp-edged ridge or shield-wall in front of the antennal insertions **(A)**. Sting terminates in an apical or apicodorsal lamellate, spatulate, or dentiform appendage **(B)**
. . . . ***Tetramorium*** (part)

103b Lateral portions of the clypeus not raised up into a sharp-edged ridge or shield-wall in front of the antennal insertions **(AA)**. Sting without an apical or apicodorsal lamellate, spatulate, or dentiform appendage **(BB)** **104**

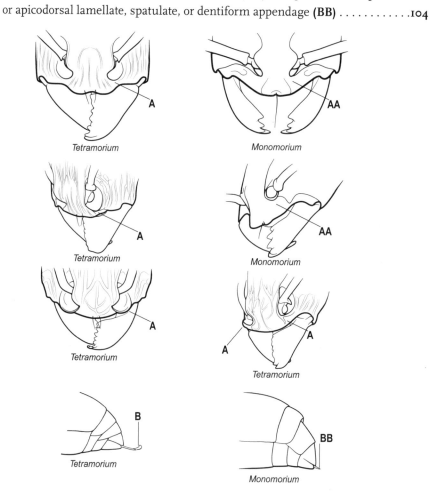

Tetramorium   *Monomorium*

Tetramorium   *Monomorium*

Tetramorium   *Tetramorium*

Tetramorium   *Monomorium*

104a (103) Posterior corners of the head distinctly tuberculate to sharply denticulate in full-face view **(A)**. Pronotum marginate laterally **(B)** . . . . *Terataner* (part)

104b Posterior corners of the head usually rounded, rarely angular, but never tuberculate or denticulate **(AA)**. Pronotum usually lacks lateral margination **(BB)**
. . . .**105**

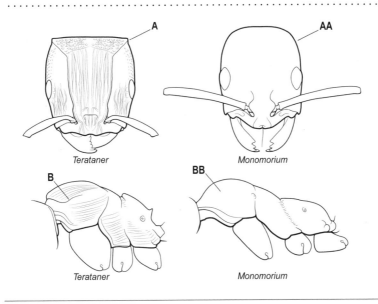

Terataner   Monomorium

Terataner   Monomorium

105a (104) Masticatory margin of the mandible with more than 5 teeth or denticles in total, usually with 7 or more altogether **(A)**; or very rarely the masticatory margin may be worn down and entirely edentate **(B)** **106**

105b Masticatory margin of the mandible with 3–5 teeth or denticles altogether, never with more; masticatory margin never entirely edentate **(AA)**. **107**

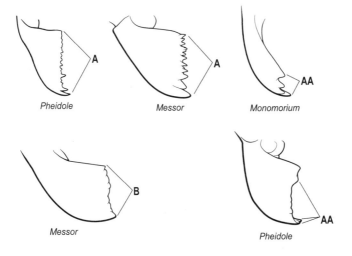

Pheidole   Messor   Monomorium

Messor   Pheidole

106a (105) Masticatory margin of mandible with the third tooth (counting from the apex) smaller than the fourth, or the reduced third tooth followed by a minute denticle before the larger fourth tooth; margin never edentate **(A)**. Antennal funiculi terminate in a strongly defined 3-segmented club **(B)**. Ventral surface of the head without a psammophore **(C)**. Palp formula 2,2 or more rarely 3,2. . . . .*Pheidole* (minor workers)

106b Masticatory margin of the mandible with the third tooth (counting from the apex) larger than the fourth **(AA)**, or margin edentate **(AAA)**. Antennal funiculi terminate in a weakly defined 4-segmented club or without a differentiated club, the segments gradually increasing in size toward the apex **(BB)**. Ventral surface of head usually (but not always) with a psammophore, composed of elongate J-shaped setae **(CC)**. Palp formula 4,3 or 5,3. Polymorphic species . . . . *Messor*

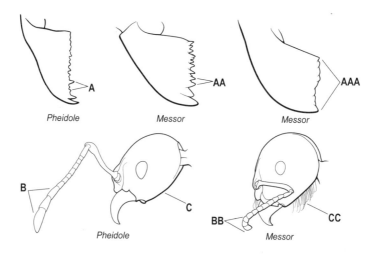

Pheidole   Messor   Messor

Pheidole   Messor

107a (105) Mandible powerfully constructed, armed with 2 large apical teeth followed by a long diastema and then 1 or 2 (rarely 3) basal teeth **(A)**. 2–4 hypostomal teeth usually present on the posterior margin of the buccal cavity **(B)**. Palp formula 2,2 or 3,2 and the clypeus lacks a long unpaired seta at the midpoint of the anterior margin. . . . . *Pheidole* (major workers)

107b Mandible delicately constructed, armed with 3–5 teeth, serially dentate and decreasing in size from apex to base, not arranged as above **(AA)**. Hypostomal teeth absent from the posterior margin of the buccal cavity **(BB)**. Palp formula variable, but if 2,2 then the clypeus with a long unpaired seta at the midpoint of the anterior margin 108

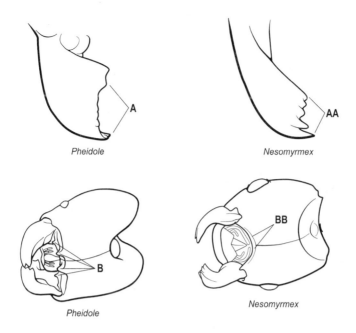

Pheidole

Nesomyrmex

Pheidole

Nesomyrmex

108a (107) Midpoint of the anterior clypeal margin with a single, unpaired, elongate seta that projects forward over the mandibles and is usually very conspicuous **(A)** **109**

108b Midpoint of the anterior clypeal margin without a single, unpaired elongate seta; instead, usually with a pair of short setae, one on each side of the midpoint **(AA)** **114**

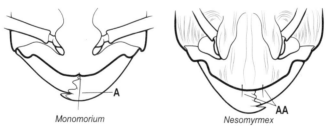

Monomorium

Nesomyrmex

109a (108) Maxillary palp with 5 segments. Median portion of the clypeus raised and projecting, not bicarinate, fused to the flattened prominent lateral portions of the clypeus to form a shelf that projects forward over the mandibles **(A)**. Propodeum usually bidentate or bispinose, only extremely rarely unarmed **(B)**. Mandible with 5 teeth . . . .*Cardiocondyla* (part)

109b Maxillary palp with 1–3 segments. Median portion of the clypeus concave to prominent anteriorly and weakly to acutely bicarinate **(AA)**; lateral portions of the clypeus not expanded forward nor fused with the median portion to form a broad projecting shelf. Propodeum usually unarmed and rounded **(BB)**, rarely angulate or bidenticulate. Mandible with 3–5 teeth (most commonly 3–4, only 1 species with 5). 110

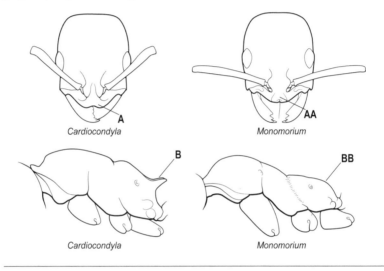

Cardiocondyla   Monomorium

Cardiocondyla   Monomorium

110a (109) With the head in profile the unpaired median clypeal seta arises from a vertical, or concave-vertical, surface well below the anteriormost point of the clypeal projection **(A)**. Eyes minute, of only 1–2 ommatidia. Propodeal dorsum angulate to denticulate between the dorsum and declivity **(B)**. . . . *Syllophopsis*

110b With the head in profile the unpaired median clypeal seta arises from the anteriormost point of the clypeal projection **(AA)**. Eyes larger, conspicuously of more than 2 ommatidia. Propodeal dorsum usually rounded between the dorsum and declivity, only rarely angulate or denticulate **(BB)** 111

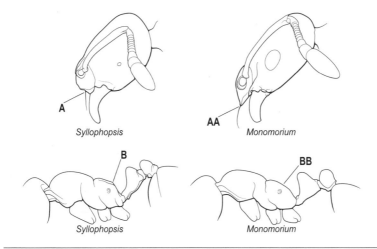

Syllophopsis   Monomorium

Syllophopsis   Monomorium

111a (110) Propodeal spiracle elongate-elliptical to short slit-shaped, oriented vertically **(A)**. Petiolar spiracle at, or close to, the midlength of the peduncle, not at the node **(B)**. Antenna without a strongly defined, 3-segmented apical club **(C)**. Anterior clypeal margin strongly bidentate, the apices of the 2 teeth widely separated (distance between the dental apices is much greater than the maximum width across the frontal lobes) **(D)**. Strongly polymorphic species . . . . ***Trichomyrmex*** (part)

111b Propodeal spiracle circular or very nearly circular **(AA)**. Petiolar spiracle at the node **(BB)**. Antenna with a strongly defined, 3-segmented apical club **(CC)**. Anterior clypeal margin rounded, biangulate, or bidentate medially; when biangulate or bidentate the apices of the 2 prominences close together (distance between their apices is less than the maximum width across the frontal lobes) **(DD)**. Monomorphic or polymorphic species 112

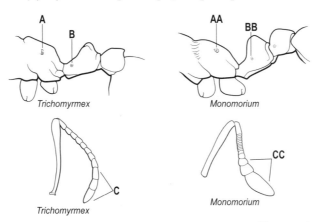

Trichomyrmex   Monomorium

Trichomyrmex   Monomorium

(Figure continues on next page)

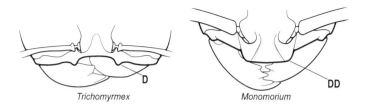

Trichomyrmex   Monomorium

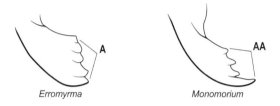

Erromyrma   Monomorium

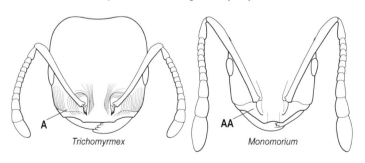

Trichomyrmex   Monomorium

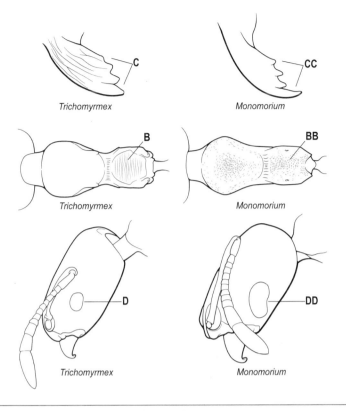

Trichomyrmex — C

Monomorium — CC

Trichomyrmex — B

Monomorium — BB

Trichomyrmex — D

Monomorium — DD

114a (108) Median portion of the clypeus forms an anteriorly projecting lobe or apron that fits tightly over the mandibular dorsum and distinctly overlaps the basal portion of the mandible **(A)**. In profile the anterior clypeal margin is closely adherent to the dorsal surface of the mandible . . . . ***Nesomyrmex*** (part)

114b Median portion of the clypeus does not form an anteriorly projecting lobe or apron that fits tightly over the mandibular dorsum and does not overlap the basal portion of the mandible **(AA)**. In profile the anterior clypeal margin is elevated slightly away from the dorsal surface of the mandible . . . . . . ***Temnothorax***

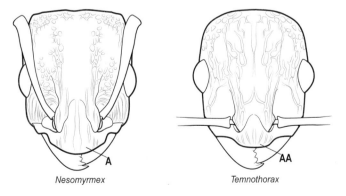

Nesomyrmex — A

Temnothorax — AA

Key to Malagasy Genera (Workers)

1a   Body with a single reduced or isolated segment, the petiole (A2) between the mesosoma and the gaster **(A)**. 2

1b   Body with 2 reduced or isolated segments, the petiole and postpetiole (A2 and A3) between the mesosoma and the gaster **(AA)**. **42**

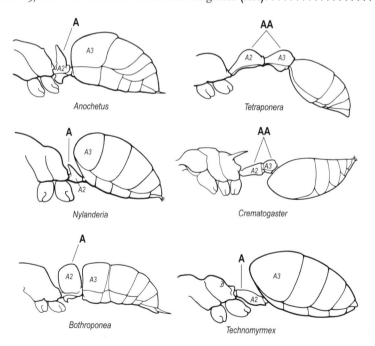

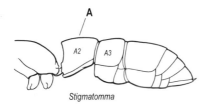

Stigmatomma

2a (1) With head in ventral view and mouthparts closed, the prementum is not visible; it is entirely concealed by the labrum anteriorly and the maxillary stipites on each side; the stipites meet along the midline **(A)**. Lateral margins of the pygidium always with a pair of distally converging rows of spines or peg-like teeth along the lateral and posterior margins **(B)**. 3

2b With head in ventral view and mouthparts closed, the prementum is clearly visible; it is bounded by the labrum anteriorly and the maxillary stipites on each side of it **(AA)**. Lateral and posterior margins of the pygidium unarmed, without spines, teeth, or denticles **(BB)**. 10

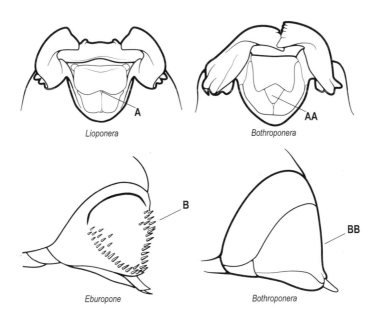

Lioponera   *Bothroponera*

Eburopone   *Bothroponera*

3a (2)   Sculpture predominantly strongly costate over head and body **(A)**. . . . . *Chrysapace*

3b   Sculpture predominantly punctuate on a smooth or finely sculptured background **(AA)**. 4

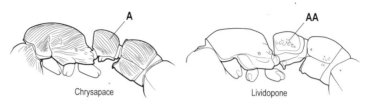

4a (3)   Mesotibia always with at least 1 conspicuous spur **(A)**. Ocelli absent **(B)**. 5

4b   Mesotibia always lacks spurs **(AA)**. Ocelli present **(BB)**. 9

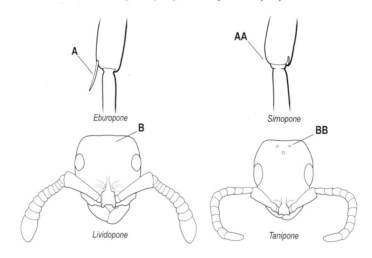

5a (4)   Helcium projects from very high on the anterior surface of A3 (the first gastral segment); in profile the dorsal surface of the helcium arises from immediately below the anterodorsal angle of A3 **(A)** . . . . *Lividopone*

5b   Helcium projects from the anterior surface of A3 (the first gastral segment) at about its midheight; in profile the dorsal surface of the helcium arises some distance below the anterodorsal angle of A3 **(AA)** 6

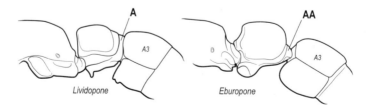

6a (5)   9 antennal segments **(A)**. . . . . *Ooceraea*

6b   12 antennal segments **(AA)** 7

7a (6)   Petiole node with sharp dorsolateral margins **(A)**. Posterodorsal surface of the metacoxa, just proximal of the trochanter, raised into a vertical cuticular flange or lamella **(B)** . . . .*Lioponera*

7b   Petiole node without sharp dorsolateral margins, the dorsal surface rounding evenly into the sides **(AA)**. Posterodorsal surface of metacoxa, just proximal of the trochanter, not raised into a vertical cuticular flange **(BB)**. 8

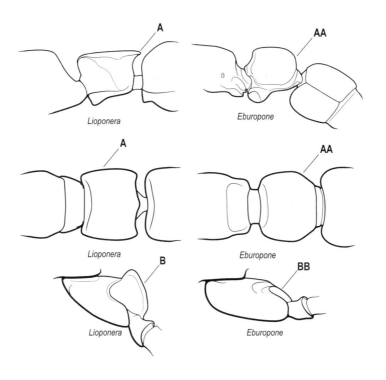

8a (7) Apical antennal segment bulbous, much larger in width than the preapical segment **(A)**. In profile and dorsal view, the promesonotal suture absent or indistinct **(B)**. Posterior margin of the abdominal sternite 4 without a glandular opening or pale glandular patch **(C)**. . . . . *Parasyscia* (part)

8b Apical antennal segment moderately enlarged but not strongly bulbous, more or less the same width as the preapical segment **(AA)**. In profile and dorsal view, the promesonotal suture present and impressed **(BB)**. Posterior margin of the abdominal sternite 4 with glandular opening **(CC)** or pale yellow patch **(CCC)** . . . . *Eburopone*

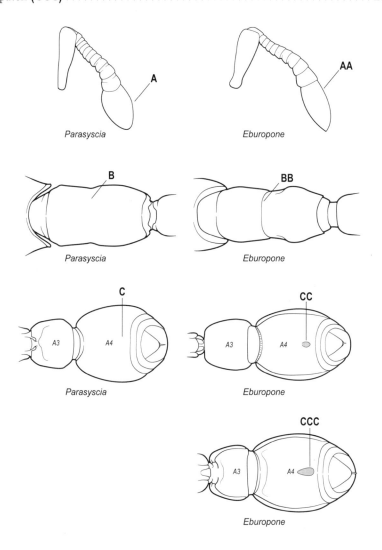

9a (4) Antenna with 11 segments **(A)**. Metabasitarsus ventrally with a longitudinal glandular groove that occupies at least the basal half of the tarsomere length **(B)**. Abdominal tergite 3 (the first gastral tergite) without subovate glandular patches on the posterior half; posterior margin of the A3 tergite without a transverse band of pale cuticle, without a pair of pale patches **(C)**. . . . . . . *Simopone*

9b Antenna with 12 segments **(AA)**. Metabasitarsus ventrally without a longitudinal glandular groove. Abdominal tergite 3 (the first gastral tergite) with a pair of subovate glandular patches on the posterior half (absent in 1 species); posterior margin of the A3 tergite with a transverse band of pale cuticle, or a pair of pale patches **(CC)**. . . . . *Tanipone*

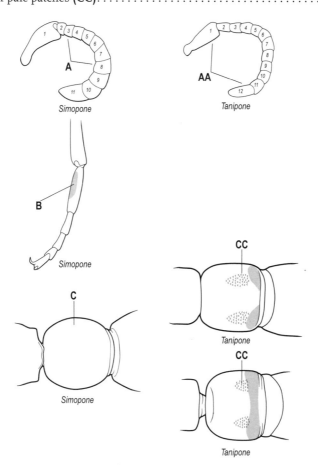

A

Simopone

AA

Tanipone

B

Simopone

C

Simopone

CC

Tanipone

CC

Tanipone

10a (2) Apex of the gaster with a semicircular to circular acidopore formed from the apical portion of the hypopygium **(A)**. This structure often projects as a nozzle and is fringed with setae, but in 1 genus the acidopore may have no nozzle or setae and may be concealed by a projection of the pygidium (the acidopore becomes visible if the sclerites are separated); in this genus the antennal sockets are located well behind the posterior clypeal margin. 11

10b Apex of the gaster with the hypopygium lacking an acidopore **(AA)**. Sting present or absent; when present it is usually clearly visible, but when reduced or vestigial (or rarely when sting present but entirely retracted), the hypopygium has a smooth posterior margin, without trace of a circular or semicircular acidopore, and the antennal sockets are located close to or at the posterior clypeal margin 19

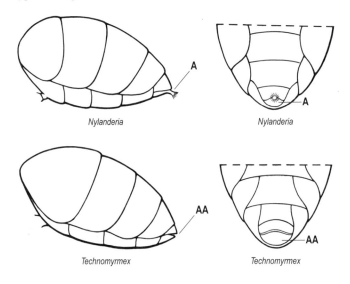

Nylanderia | Nylanderia

Technomyrmex | Technomyrmex

11a (10) Antenna with 9 segments **(A)** . . . . *Brachymyrmex*

11b Antenna with 11 **(AA)** or 12 segments **(AAA)** 12

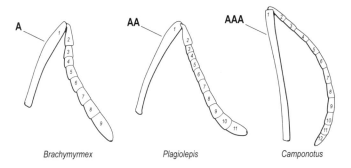

Brachymyrmex | Plagiolepis | Camponotus

| **12a** (11) | Antenna with 11 segments . . . .**13** |
| **12b** | Antenna with 12 segments. . . . .**16** |

13a (12) Propodeum armed with a pair of spines, teeth, or tubercles **(A)**. Dorsal edge
of the petiole **(A2)** usually armed with a pair of teeth or spines **(B)** but sometimes only emarginated . . . . *Lepisiota*

13b Propodeum and petiole **(A2)** unarmed, without trace of spines, teeth, or
tubercles **(AA, BB)** **14**

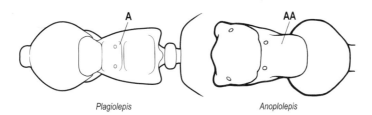

Lepisiota   Plagiolepis

14a (13) With the mesosoma in dorsal view, the mesonotum separated from the metanotum by a conspicuous transverse groove or impression, so that the metano-
tum forms a distinctly isolated sclerite on the dorsum **(A)**. . . . . *Plagiolepis*

14b With the mesosoma in dorsal view, the mesonotum fused with the metanotum, the two not separated by a transverse groove or impression, the metano-
tum does not form an isolated sclerite on the dorsum **(AA)**. **15**

Plagiolepis   Anoplolepis

15a (14) Ventral apex of the metatibia with a divergent pair of coarse setae, between which is a large median spur **(A)**. Scapes surpass the posterior margin of the head by two-thirds or more of their length **(B)**. Anterior arc of the torulus does not touch or indent the posterior clypeal margin **(C)**. Dorsum of the head behind the clypeus with numerous erect stout setae present. Mandible with 8 teeth . . . . *Anoplolepis*

15b Ventral apex of the metatibia with a divergent pair of coarse setae only, without a median spur between them **(AA)**. Scapes surpass the posterior margin of the head by less than one-quarter of their length **(BB)**. Anterior arc of the torulus indents the posterior clypeal margin **(CC)**. Dorsum of the head behind the clypeus with at most a single pair of erect setae. Mandible with 5–7 teeth. . . . . *Tapinolepis*

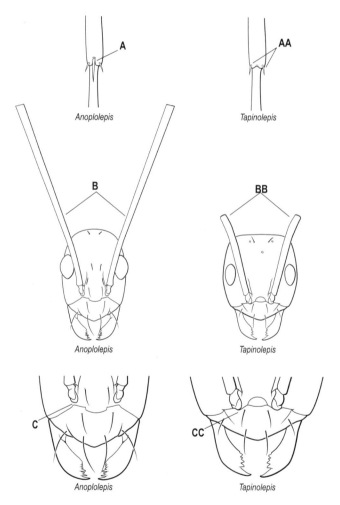

16a (12) Anteriormost point of the torulus far behind the clypeal suture, the gap between clypeal suture and the anterior margin of the torulus is at least equal to the basal width of the scape shaft and generally much more **(A)**. Metapleural gland orifice usually absent, the surface of the metapleuron uninterrupted by a gland orifice above the metacoxa and below the level of the propodeal spiracle; guard setae absent **(B)** .... *Camponotus*

16b Anteriormost point of the torulus indents, abuts, or is immediately behind the clypeal suture **(AA)**. If the last of these, then the gap between the clypeal suture and the anterior margin of the torulus is very narrow, distinctly less than the basal width of the scape shaft. Metapleuron with a distinct orifice for the metapleural gland, the orifice situated above the metacoxa and below the level of the propodeal spiracle **(BB)**. Orifice of metapleural gland protected and sometimes partially concealed by a line or tuft of guard setae that are usually very conspicuous **17**

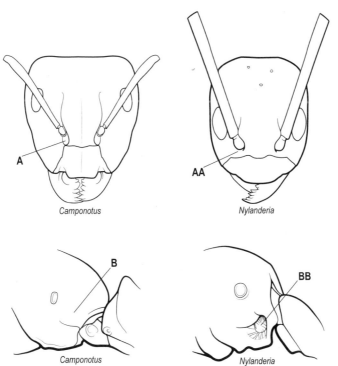

Camponotus

Nylanderia

Camponotus

Nylanderia

17a (16) Mandible with 6–7 teeth **(A)**. . . . . *Nylanderia*
17b Mandible with 5 teeth **(AA)**. 18

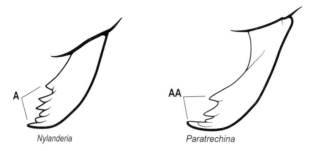

Nylanderia   *Paratrechina*

18a (17) Erect setae (1 pair) present on the propodeum **(A)**. Erect setae on the head form a pattern of 4 setae along the posterior margin and 6 or 7 rows from the posterior margin to the clypeal margin **(B)**. Femora and tibiae lacking large erect setae **(C)** . . . .*Paraparatrechina*
18b Erect setae absent from the propodeum **(AA)**. Erect setae on the front of head scattered across surface and not forming pairs **(BB)**. Legs with numerous erect setae **(CC)** . . . . *Paratrechina*

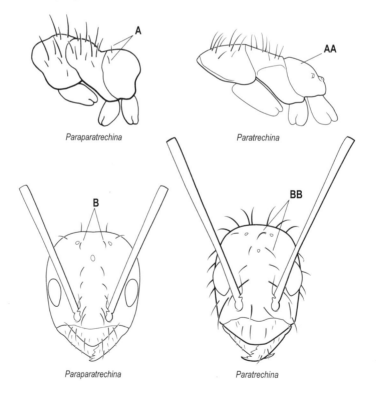

Paraparatrechina   *Paratrechina*

Paraparatrechina   *Paratrechina*

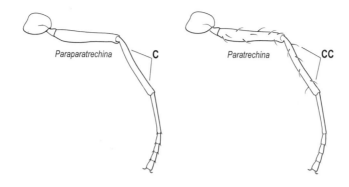

Paraparatrechina C   *Paratrechina* CC

19a (10) Tergite of the helcium with an extensive U-shaped or V-shaped median notch or emargination dorsally in its anterior margin (easily visible if the gaster is slightly depressed with respect to the petiole) **(A)**. Sting vestigial or absent, never visibly projecting from the apex of the gaster and its remnants not detectable without dissection **(B)**. Petiolar levator process with a lacuna **(C)** ... **20**

19b Tergite of the helcium entire, its anterior margin dorsally without a median U-shaped or V-shaped notch or emargination **(AA)**. Sting present and functional, usually projecting from the apex of the gaster in dead specimens; in many species the sting shaft is visible through the cuticle of the ventral gastral apex even when sting fully retracted **(BB)**. Petiolar levator process complete **24**

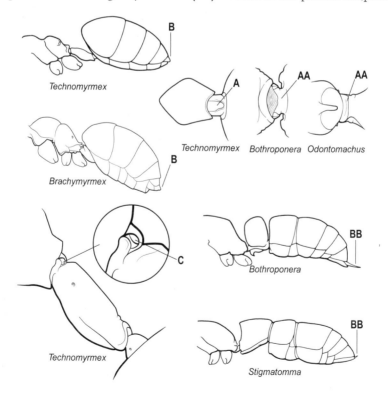

Technomyrmex B

Brachymyrmex B

A

Technomyrmex AA   *Bothroponera*   *Odontomachus* AA

Technomyrmex C

Bothroponera BB

Stigmatomma BB

20a (19) In profile the outline of the propodeal declivity distinctly concave **(A)**. Petiole (A2) in profile with a conspicuous, high-standing thin scale with distinct anterior and posterior faces **(B)** . . . . *Ochetellus*

20b In profile the outline of the propodeum more or less flat or convex **(AA)**. Petiole (A2) in profile greatly reduced, without a standing node or scale, the anterior face absent or at most indistinct **(BB)**. **21**

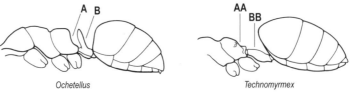

Ochetellus   Technomyrmex

21a (20) In dorsal view, 5 gastral tergites (A3–A7) are visible, the fifth gastral tergite (A7) is small but continues the line of the gaster and is not reflexed below the fourth; the anal and associated orifices are thus situated apically **(A)**. Pronotum commonly with 2–10 erect setae . . . . *Technomyrmex*

21b In dorsal view, only 4 gastral tergites (A3–A6) are visible. Fifth gastral tergite (A7) is reflexed below the fourth, visible in ventral view, where it forms a transverse plate abutting the fifth sternite; the anal and associated orifices are thus situated ventrally **(AA)**. Pronotum generally lacking erect setae **22**

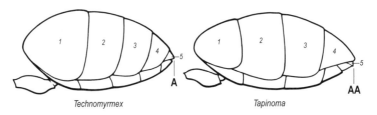

Technomyrmex   Tapinoma

22a (21) Dimorphic, with major and minor workers **(A)**. In profile the petiole (A2) in the form of node with a low, standing anterior face. . . . . *Aptinoma*

22b Monomorphic **(AA)**. In profile the petiole (A2) narrow and subcylindrical, without an anterior face **23**

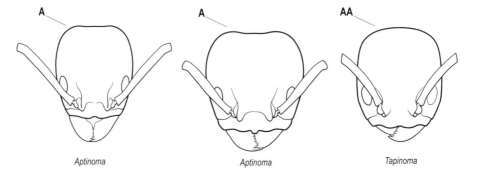

Aptinoma   Aptinoma   Tapinoma

23a (22) Propodeum in profile with a peak at the midposterodorsal point **(A)**. Mandible with a distinctly enlarged denticle at the junction of the basal and masticatory margins **(B)**. In full-face view, the eyes well in from the side of the head **(C)**. Scapes longer than the head **(D)** . . . . *Ravavy*

23b Propodeum in profile with the posterodorsal margin rounded, without a median peak **(AA)**. Mandible with denticles gradually and uniformly reduced in size from the apex to the base of the mandible **(BB)**. In full-face view, the eyes usually at margin of the head **(CC)**. Scapes either shorter **(DD)** or longer than the head **(DDD)**. . . . . **Tapinoma**

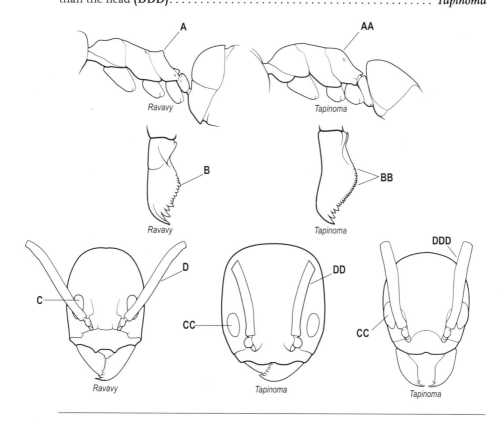

24a (19) Promesonotal suture entirely absent from the dorsum of the mesosoma so that the pronotum and mesonotum are fused into a single unit **(A)** 25

24b Promesonotal suture conspicuously present across the dorsum of the mesosoma so that the pronotum and mesonotum are distinctly separated **(AA)** 27

(Continues on next page)

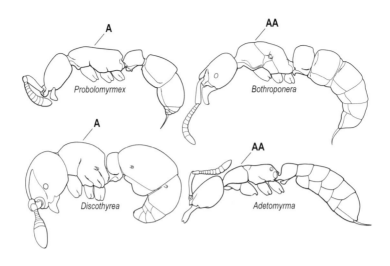

25a (24) Tergite of the second gastral segment (A4) not arched and vaulted, the remaining segments directed posteriorly **(A)**. Sternite of second gastral segment large and longitudinal in profile, subrectangular to trapezoidal in shape. Eyes absent..*Probolomyrmex*

25b Tergite of the second gastral segment (A4) strongly arched and vaulted so that the remaining segments point anteriorly **(AA)**. Sternite of the second gastral segment small to minute in profile, triangular in shape, and with the apex of the triangle directed ventrally or anteriorly. Eyes usually present, often small 26

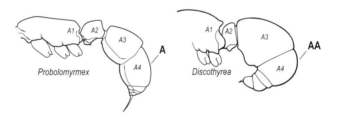

26a (25) Mandible edentate **(A)**, overhung by a projecting frontoclypeal shelf that bears the antennal sockets **(B)**. Apical funicular segment strongly bulbous **(C)** ....*Discothyrea*

26b Mandible with 3 or more teeth **(AA)**, not overhung by a projecting frontoclypeal shelf **(BB)**. Apical funicular segment moderately enlarged but not strongly bulbous **(CC)**..*Proceratium*

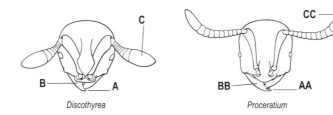

Discothyrea   Proceratium

27a (24) Articulation of the petiole to the first gastral segment (A2 to A3) very broad in dorsal view **(A)**. In profile the petiole without a distinctly descending free posterior surface **(B)**. Helcium in profile with its upper margin projecting from very high on anterior face of the first gastral segment **(C)**; above the helcium the first gastral segment has no free anterior surface—at most a short curve into the dorsum is present 28

27b Articulation of the petiole to first gastral segment (A2 to A3) narrow in dorsal view **(AA)**. In profile the petiole has a distinctly descending posterior free surface **(BB)**. Helcium in profile with its upper margin projecting from about the midheight or lower on anterior face of the first gastral segment **(CC)**; above the helcium the first gastral segment has a free anterior surface below the curve into the dorsum 32

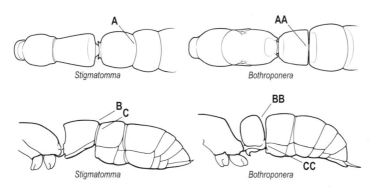

Stigmatomma   Bothroponera

Stigmatomma   Bothroponera

28a (27) Mandible short and slender, triangular or subtriangular, with 3 teeth (sometimes with a few very small additional denticles); the 3 teeth grouped together near the tip of the mandible, the middle tooth the smallest **(A)** . . . . *Prionopelta*

28b Mandibles short to long, slender, with 5 or more teeth that are distributed along the entire inner surface **(AA)**. 29

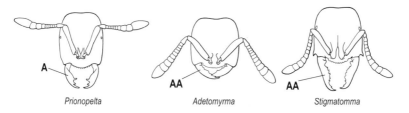

Prionopelta   Adetomyrma   Stigmatomma

29a (28) In profile or in dorsal view, the abdomen without an impression or girdling constriction between the petiole and the first gastral segment (A2 and A3) **(A)** and also without a constriction between the first and second gastral segments (A3 and A4) **(B)**. In profile the abdomen is greatly expanded posteriorly, the apical segment (A7) distinctly larger than A3, and the pygidium distinctly the longest tergite **(C)**. Sting enormous . . . . *Adetomyrma*

29b In profile or in dorsal view, the abdomen with an impression or girdling constriction between the petiole and the first gastral segment (A2 and A3) **(AA)**, and also with a constriction between the first and second gastral segments (A3 and A4) **(BB)**. In profile the abdomen reducing posteriorly, the apical segment (A7) distinctly smaller than A3, and the pygidium not the longest tergite **(CC)**. Sting normal 30

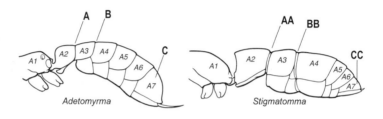

Adetomyrma   *Stigmatomma*

30a (29) Apex of the mandible rounded and with very short teeth, which may be blunt or sharp; mandible longer than the head, with 2 separate rows of teeth on the inner margin **(A)**. Setae on the head and body usually spatulate, rarely setae thin and tapering to points **(B)**. . . . .*Mystrium*

30b Apex of the mandible acute, with a single distinct, sharp tooth; mandible not as long as head, with a single row of teeth along the inner margin, teeth may be bicarinate but are never arranged in 2 separate rows **(AA)**. Setae on the head and body always long, thin, and tapering to points **(BB)** 31

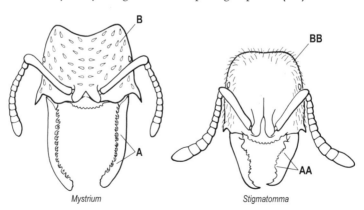

Mystrium   *Stigmatomma*

31a (30) Anterior clypeal margin with a row of specialized stout, dentiform setae **(A)**. Sternite of the petiole (A2) not reduced, usually extending the length of the segment **(B)** . . . . *Stigmatomma*

31b Anterior clypeal margin unarmed, without a row of specialized dentiform setae **(AA)**. Sternite of the petiole (A2) reduced to a minute posteromedian sclerite **(BB)** . . . .*Xymmer*

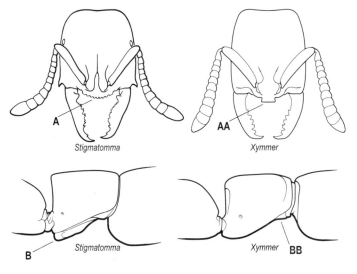

32a (27) Mandibles long and linear, in full-face view inserted close together in the middle of the anterior margin of the head, their bases closely approximated; 2 or 3 large teeth near the apex; inner margin often without teeth **(A)**. Petiole node (A2) often with at least 1 tooth or spine. Without a visible constriction between the first and second gastral segments (A3 and A4) **(B)** 33

32b Mandibles linear to triangular, short to long, in full-face view inserted at the anterolateral corners of the head, their bases conspicuously separated; teeth (when present) located along the inner margin **(AA)**. Petiole node (A2) without spines or teeth. With a visible constriction between the first and second gastral segments (A3 and A4) **(BB)** 34

(Continues on next page)

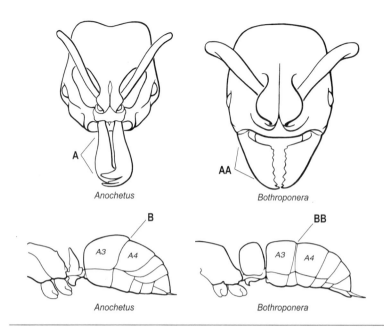

Anochetus

Bothroponera

Anochetus

Bothroponera

33a (32) Nuchal carina (separates dorsal from posterior surfaces of the head) converges in a V-shape at the midline **(A)** and also receives a pair of prominent dark posterior apophyseal lines that converge to form the sharp median-dorsal groove of the vertex **(B)** .... *Odontomachus*

33b Nuchal carina forms a broad uninterrupted curve across the posterodorsal extremity of the head **(AA)**; posterior surface without paired dark apophyseal lines; on vertex the median groove absent or ill-defined and shallow **(BB)**
.... *Anochetus*

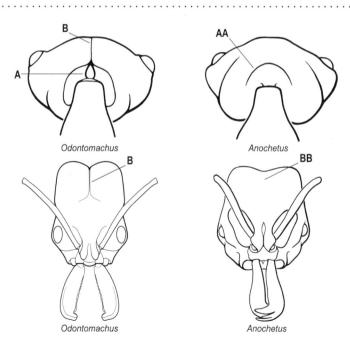

Odontomachus

Anochetus

Odontomachus

Anochetus

34a (32) Ventral apex of the metatibia, when viewed from in front with the metafemur at right-angles to the body, with a single large pectinate spur **(A)**; without a second smaller spur in front of the pectinate main spur in the direction of observation. 35

34b Ventral apex of the metatibia, when viewed from in front with the metafemur at right-angles to the body, with 2 spurs, consisting of a large pectinate spur and a second smaller spur that is in front of the pectinate main spur in the direction of observation **(AA)**. 36

35a (34) Subpetiolar process in profile with a pair of teeth posteroventrally **(A)** and with a fenestra or thin-spot anteriorly that is translucent **(B)**. Maxillary palp with 2 segments . . . . *Ponera*

35b Subpetiolar process in profile rounded to acutely angulate posteroventrally **(AA)** but never with a pair of teeth; an anterior fenestra or thin-spot usually absent but present in some species. **(BB)** Maxillary palp with 0–1 segments . . . . *Hypoponera*

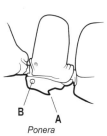

Ponera

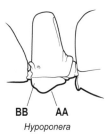

Hypoponera

36a (34) Helcium in profile located approximately at midheight on the anterior surface of the first gastral segment (A3), so that in profile the first gastral segment does not have a long vertical anterior face above the helcium **(A)**. Mesotibia and metatibia each with 2 pectinate spurs, the anterior spur smaller than the posterior **(B)**. Posterodorsal margin of the petiole node often with 2 or 3 blunt or triangular teeth that are directed posteriorly **(C)**. Anterior section of the frontal lobes and antennal sockets widely separated broadly by a rounded or triangular section of the clypeus, which extends between them **(D)** ... *Platythyrea*

36b Helcium in profile located very low on the anterior surface of the first gastral segment (A3), so that in profile the first gastral segment has a long vertical anterior face above the helcium **(AA)**. Mesotibia and metatibia each with 1 large pectinate spur and 1 small simple spur **(BB)**. Posterodorsal margin of the petiole node always without teeth **(CC)**. Anterior sections of the frontal lobes and antennal sockets very close together and at most separated by a very narrow triangle of the clypeus, which extends posteriorly between them **(DD)** 37

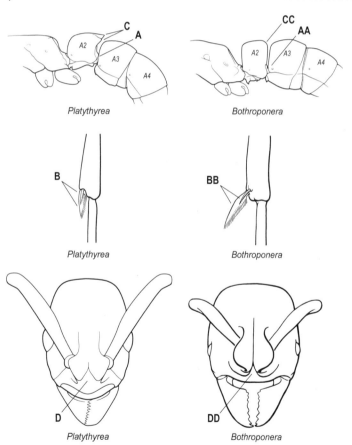

Platythyrea

Bothroponera

Platythyrea

Bothroponera

Platythyrea

Bothroponera

37a (36) Mandibles subtriangular **(A)** or slender and elongate **(B)**, often without teeth on the inner margins, at most armed with only 1–3 teeth. With head in full-face view, the frontal lobes cover only the inner margins of the antennal sockets, so that most of the sockets are exposed **(C)**. Pretarsal claws of the metatarsus usually pectinate, rarely with only 1–2 small teeth behind the apical point **(D)**. . . . . *Leptogenys*

37b Mandibles always subtriangular and armed with 5 or more teeth **(AA)**. With head in full-face view, the frontal lobes cover the entire antennal sockets **(CC)**. Pretarsal claws of the metatarsus never pectinate, never toothed behind the apical point **(DD)**. 38

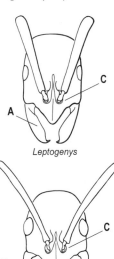

Leptogenys

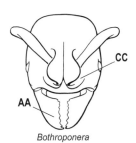

Bothroponera

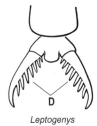

Leptogenys

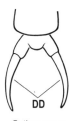

Leptogenys   *Bothroponera*

38a (37) Prora absent from anteroventral angle of the first gastral sternite (A3) below the helcium **(A)** (Mauritius) . . . . *Brachyponera*

38b Prora present at anteroventral angle of the first gastral sternite (A3) below the helcium **(AA)** 39

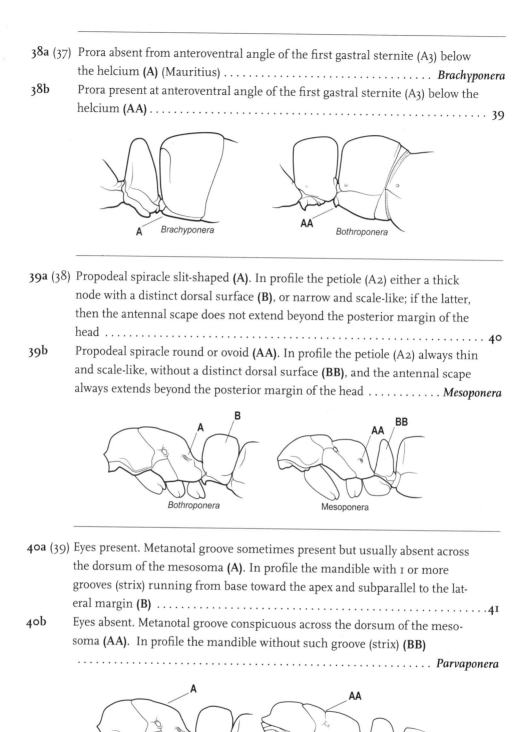

A   *Brachyponera*   AA   *Bothroponera*

39a (38) Propodeal spiracle slit-shaped **(A)**. In profile the petiole (A2) either a thick node with a distinct dorsal surface **(B)**, or narrow and scale-like; if the latter, then the antennal scape does not extend beyond the posterior margin of the head 40

39b Propodeal spiracle round or ovoid **(AA)**. In profile the petiole (A2) always thin and scale-like, without a distinct dorsal surface **(BB)**, and the antennal scape always extends beyond the posterior margin of the head . . . . *Mesoponera*

Bothroponera   *Mesoponera*

40a (39) Eyes present. Metanotal groove sometimes present but usually absent across the dorsum of the mesosoma **(A)**. In profile the mandible with 1 or more grooves (strix) running from base toward the apex and subparallel to the lateral margin **(B)** 41

40b Eyes absent. Metanotal groove conspicuous across the dorsum of the mesosoma **(AA)**. In profile the mandible without such groove (strix) **(BB)**
. . . . *Parvaponera*

Bothroponera   *Parvaponera*

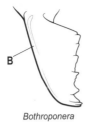

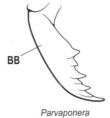

Bothroponera   *Parvaponera*

41a (40) Base of the mandible without a dorsolateral pit or fovea **(A)**; laterobasal margin of the mandible, at junction with the head capsule, without conspicuous concave, smooth impression **(B)**. Metanotal groove always absent
. . . . *Bothroponera*

41b Base of the mandible with a dorsolateral pit or fovea **(AA)**; laterobasal margin of the mandible, at junction with the head capsule, with a concave, glossy smooth impression **(BB)**. Metanotal groove sometimes present . . . . *Euponera*

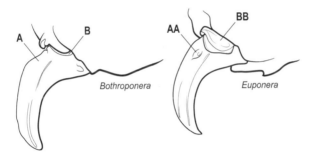

Bothroponera   *Euponera*

42a (1) With head in ventral view and mouthparts closed, the prementum is not visible **(A)**; it is entirely concealed by the labrum anteriorly and the maxillary stipites on each side; the stipites meet along the midline. Lateral margins of the pygidium always with a pair of distally converging rows of spines or peglike teeth along the lateral and posterior margins **(B)** 43

42b With head in ventral view and mouthparts closed, the prementum is clearly visible **(AA)**; it is bounded by the labrum anteriorly and the maxillary stipites on each side of it. Lateral and posterior margins of the pygidium unarmed, without spines, teeth, or denticles **(BB)**. 45

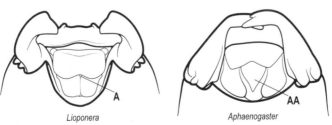

Lioponera   *Aphaenogaster*

(Figure continues on next page)

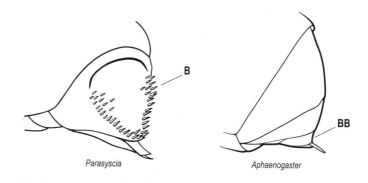

Parasyscia

Aphaenogaster

43a (42) 9 antennal segments **(A)**. . . . . *Ooceraea*

43b 12 antennal segments **(AA)** **44**

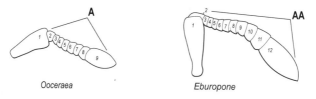

Ooceraea

Eburopone

44a (43) Apical antennal segment bulbous, much larger in width than the preapical segment **(A)**. In profile and dorsal view, the promesonotal suture absent or indistinct **(B)**. Posterior margin of abdominal sternite 4 without a glandular opening or pale glandular patch **(C)** . . . . *Parasyscia* (part)

44b Apical antennal segment moderately enlarged but not strongly bulbous, more or less the same width as the preapical segment **(AA)**. In profile and dorsal view, the promesonotal suture present and impressed **(BB)**. Posterior margin of abdominal sternite 4 with glandular opening **(CC)** or pale yellow patch **(CCC)** . . . . *Eburopone*

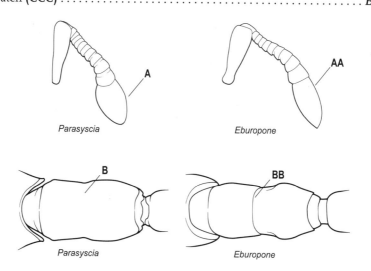

Parasyscia

Eburopone

Parasyscia

Eburopone

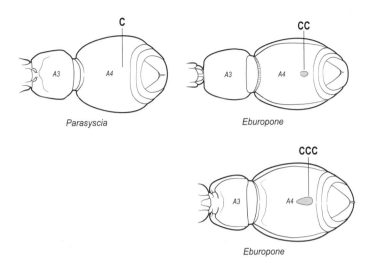

Parasyscia

Eburopone

Eburopone

45a (42) Promesonotal suture fully and strongly developed across the dorsum of the mesosoma (the posterior margin of the pronotum overlaps the anterior margin of the mesonotum); suture fully articulated and flexible, so that the pronotum and mesonotum are capable of movement relative to each other in fresh specimens **(A)**. Pretarsal claws each with a single preapical tooth **(B)**. Metatibia always with 2 spurs, a large pectinate spur **(C)** and a second spur that may be simple or pectinate **(D)**. Eyes very large and elongate **(E)**. Ocelli sometimes present . . . . ***Tetraponera***

45b Promesonotal suture either entirely absent across the dorsum of the mesosoma, or at most vestigially present as a line or feeble impression (the posterior margin of the pronotum does not overlap the anterior margin of the mesonotum); the pronotum and the mesonotum are fully fused, not capable of movement relative to each other in fresh specimens **(AA)**. Pretarsal claws simple, without a preapical tooth **(BB)**. Metatibia without spurs or with a single simple spur **(CC)**, only extremely rarely with 2 spurs present. Eyes sometimes absent, generally small and more or less round **(EE)**. Ocelli always absent **46**

Tetraponera

Tetramorium

(Figure continues on next page)

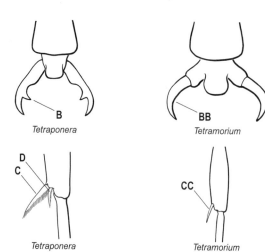

Tetraponera   Tetramorium

Tetraponera   Tetramorium

46a (45)   Postpetiole (A3) articulated on dorsal surface of the first gastral segment (A4) **(A)**. Helcium enlarged, in profile its height subequal to, or greater than, the height of the postpetiole **(B)**. Gaster in dorsal view roughly heart-shaped and capable of reflexion over the mesosoma. Petiole (A2) dorsoventrally flattened and without a node **(C)** . . . . *Crematogaster*

46b   Postpetiole (A3) articulated on anterior face of the first gastral segment (A4) **(AA)**. Helcium small, in profile its height distinctly less than the height of the postpetiole **(BB)**. Gaster in dorsal view not roughly heart-shaped, not capable of reflexion over the mesosoma. Petiole (A2) not dorsoventrally flattened, with a node of some form **(CC)**. 47

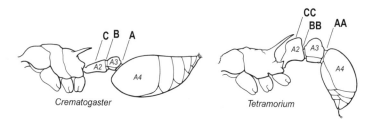

Crematogaster   Tetramorium

47a (46) Antennal scrobes present that run below the eye **(A)** **48**

47b Antennal scrobes absent **(AA)** or present but running above the eye. In some genera, both eye and scrobe absent **49**

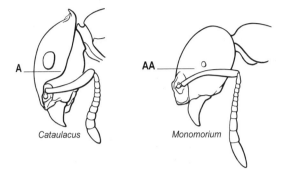

Cataulacus Monomorium

48a (47) Antenna with 11 segments, with an apical club of 3 segments. Petiole (A2) sessile, without an anterior peduncle **(A)**. First gastral tergite (A4) greatly expanded, obscuring the remaining tergites in dorsal view; tergites A5–A7 visible in profile below rear margin of the first **(B)**. Head posteroventrally with a triangular tooth or ridge on each side that fits into a corresponding groove on the anterolateral pronotum when the head is flexed downward **(C)**. Palp formula 5,3. .... *Cataulacus*

48b Antenna with 7 segments, with an apical club of 2 segments. Petiole (A2) with an anterior peduncle **(AA)**. First gastral tergite (A4) not obscuring remaining tergites in dorsal view **(BB)**. Head posteroventrally without a triangular tooth or ridge on each side, without a corresponding groove to receive it on the anterolateral pronotum **(CC)**. Palp formula 1,2. .... *Eurhopalothrix*

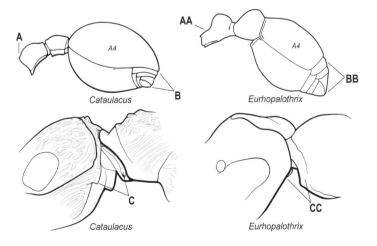

Cataulacus Eurhopalothrix

Cataulacus Eurhopalothrix

49a (47) Apical and preapical antennal segments are much larger than preceding funicular segments and form a conspicuous and usually very distinctive 2-segmented club **(A)** 50

49b Antenna never terminates in a conspicuous 2-segmented club. Either apical plus 2 preapical segments of antenna are enlarged and form a 3-segmented club, or less commonly the club has more than 3 segments; rarely the funiculus is filiform and lacks a developed apical club **(AA)** 56

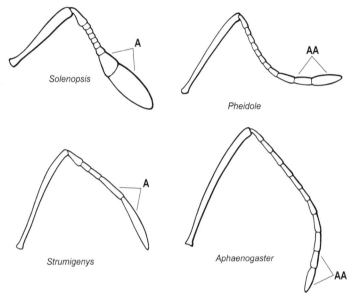

50a (49) Antenna 4–6 segments 51
50b Antenna 8–12 segments 52

51a (50) Procoxa much smaller than massively developed mesocoxa and metacoxa **(A)**. Lateral portion of the pronotum narrowed to a down-pointing triangle **(B)**. Frontal lobes confluent mid-dorsally **(C)**; the clypeus does not project back between the frontal lobes. Spongiform tissue absent from waist segments **(D)**. Postpetiole (A3) very broadly articulated to the first gastral segment (A4) **(E)**. Propodeal lobes absent. Mesobasitarsus and metabasitarsus equipped apically with a circlet of short traction spines **(F)**. . . . . *Melissotarsus*
51b Procoxa as large as or larger than the mesocoxa and metacoxa **(AA)**. Lateral portion of the pronotum not so reduced **(BB)**. Frontal lobes widely separated; the clypeus projects back between the frontal lobes **(CC)**. Spongiform tissue present on one or both waist segments **(DD)**. Postpetiole (A3) narrowly articu-

lated to the first gastral segment (A4) **(EE)**. Propodeal lobes present. Mesobasitarsus and metabasitarsus without an apical circlet of traction spines **(FF)**
....*Strumigenys*

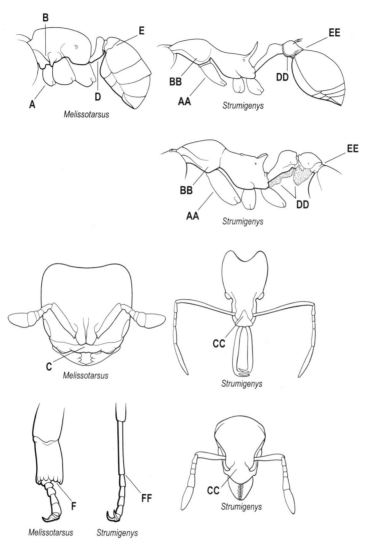

52a (50) Mandible with 7 large teeth that increase in size from apex to base; between each tooth is a minute denticle **(A)**. Mesopleuron with a depressed circular pit filled with fine radially arranged setae **(B)**. Antennal scrobes present **(C)**; antenna with 8 segments . . . .*Pilotrochus*

52b Mandible with 4–6 teeth that decrease in size from apex to base **(AA)**, or with only 2 teeth apically that are followed by an edentate oblique margin; without denticles between teeth. Mesopleuron lacking a seta-filled circular pit **(BB)**. Antennal scrobes usually absent **(CC)**; antenna generally with 9–12 segments, only extremely rarely with 8 53

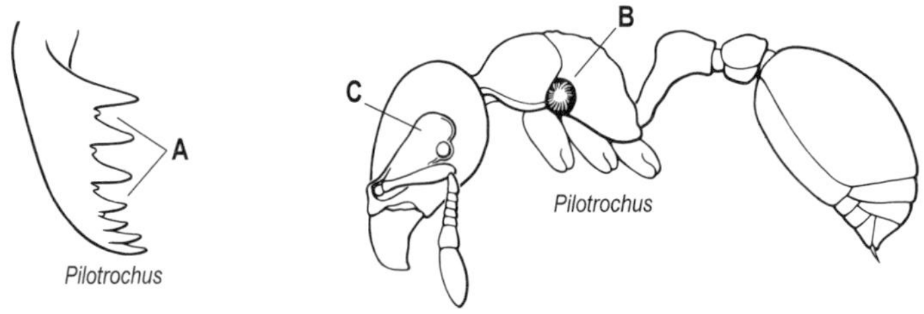

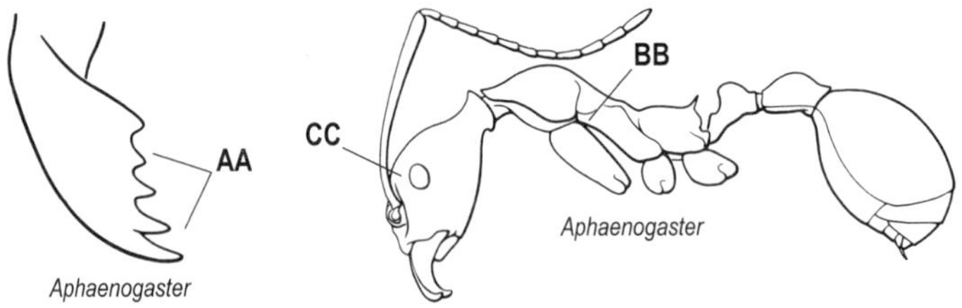

Pilotrochus

Pilotrochus

Aphaenogaster

Aphaenogaster

53a (52) Antenna with 12 segments. 54
53b Antenna with 8–11 segments. 55

54a (53) Anterior margin of the clypeus usually with a pair of teeth **(A)** and basal margin of the mandible often with a tooth close to its midlength **(B)**; if teeth absent on the clypeus, then tooth clearly visible on basal margin of the mandible. Postpetiole (A3) not dorsoventrally flattened in profile **(C)**, not broad in dorsal view **(D)**. Palp formula 1,1 **(E)** . . . . *Adelomyrmex*

54b Anterior margin of the clypeus without a pair of teeth **(AA)**. Basal margin of the mandible unarmed **(BB)**. Postpetiole (A3) dorsoventrally flattened in profile **(CC)**, in dorsal view **(DD)** very broad. Palp formula 5,3 **(EE)** . . . . . ***Cardiocondyla***

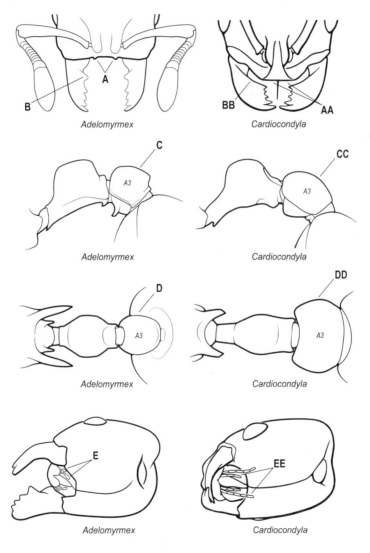

Adelomyrmex   Cardiocondyla

Adelomyrmex   Cardiocondyla

Adelomyrmex   Cardiocondyla

Adelomyrmex   Cardiocondyla

55a (53) Anterior margin of the clypeus with a single, unpaired elongate seta at its midpoint **(A)**. Propodeum rounded, never with teeth, spines, or thin flanges **(B)**. Worker caste polymorphic with intermediates. Antenna always with 10 segments .... *Solenopsis*

55b Anterior margin of the clypeus with a pair of elongate setae that straddle midpoint, without an isolated, unpaired, median seta **(AA)**. Propodeum armed with a pair of teeth or spines **(BB)**, which vary from short and angular to long and thin, or occasionally propodeum with a pair of thin, elongate, vertical flanges. Dimorphic, with minors **(CC)** and majors **(DD)** but without intermediates. Antenna with 8–11 segments .... *Carebara*

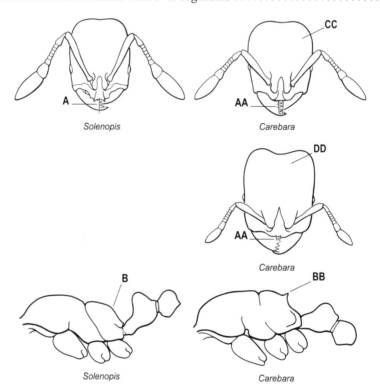

Solenopis

Carebara

Carebara

Solenopis

Carebara

56a (49) Median portion of the clypeus vertical above the mandibles, with a conspicuous anteriorly projecting bilobed appendage at the top of the vertical face
(the clypeal fork), which projects out over the mandibles from about the same
level as the frontal lobes **(A)** 57

56b Median portion of the clypeus not vertical, without a bilobed appendage
which projects out over the mandibles from about the same level as the fron-
tal lobes **(AA)** 58

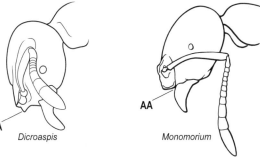

A
Dicroaspis

AA
Monomorium

57a (56) Antenna with 11 segments. Peduncle of the petiole (A2) short and very thick
in profile, with a massive triangular anteroventral process that is inclined
anteriorly **(A)**. All body setae simple, without spatulate or short and stubbly
pilosity **(B)** . . . . *Dicroaspis*

57b Antenna with 12 segments. Peduncle of the petiole (A2) elongate and narrow in profile **(AA)**. Body setae spatulate or short and stubbly with abruptly
tapered points **(BB)** . . . . *Calyptomyrmex*

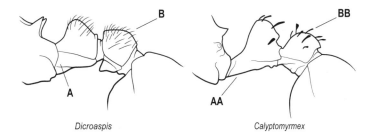

B

BB

A
Dicroaspis

AA
Calyptomyrmex

58a (56) Antenna with 9 segments. Dorsum of the mesosoma forming a broad flange with thin lateral edges that project outward over the sides of the mesosoma **(A)**. Petiole (A2) sessile, without an anterior peduncle **(B)** . . . . *Meranoplus*

58b Antenna with 11 or 12 segments. Dorsum of the pronotum and mesonotum usually rounding gradually into sides of the mesosoma, or at most separated by a distinct angle or ridge, but never a thin projecting flange **(AA)**. Petiole usually with an anterior peduncle **(BB)**. 59

Meranoplus   **B**   *Tetramorium*

59a (58) Antenna with 11 segments 60

59b Antenna with 12 segments. 64

60a (59) Frontal lobes vestigial or absent so that in full-face view the antennal sockets are exposed and the depressed area that contains the antennal sockets is clearly visible **(A)**. Anterior clypeal margin armed with denticles **(B)**. Dorsum of labrum, just behind its anterior margin, with a raised cuticular ridge, a transverse row of 2–3 denticles, or both of these **(C)** (Mauritius) . . . . . *Pristomyrmex*

60b Frontal lobes present, in full-face view covering most or all of the antennal sockets, which are not fully visible in dorsal view **(AA)**. Anterior clypeal margin unarmed or with a pair of small teeth **(BB)**. Dorsum of labrum, just behind its anterior margin, without a raised cuticular ridge or a transverse row of 2–3 denticles. 61

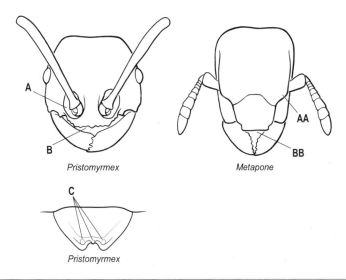

Pristomyrmex   Metapone

Pristomyrmex

61a (60) Frontal lobes massively expanded laterally, in full-face view covering most
of the sides of the head anterior of the eye **(A)**. With the head in profile, a
sharply differentiated diagonal carina is present that extends up from the
mandible insertion to above the eye **(B)**. Mesonotum with cuticular tubercles
present (Réunion) . . . . *Cyphomyrmex*

61b Frontal lobes not expanded laterally to cover the sides of the head anterior of
the eye **(AA)**. With the head in profile, lacking a diagonal carina that extends
from the mandible insertion to above the eye **(BB)**. Mesonotum without
cuticular tubercles. 62

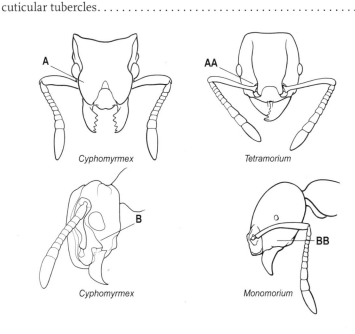

Cyphomyrmex   Tetramorium

Cyphomyrmex   Monomorium

62a (61) Eyes located behind the midlength of the sides of the head **(A)**. Central portion of the clypeus raised and produced forward as a large shield-like lobe, which projects strongly over the mandibles **(B)**. Tibiae and basitarsi of middle and hind legs terminate in a number of peg-like stout spines **(C)**. Procoxa distinctly smaller than the mesocoxa and metacoxa **(D)**. Metatibia with a single, pectinate, apical spur . . . .*Metapone*

62b Eyes located at or in front of the midlength of sides of the head, or eyes absent **(AA)**. Median portion of the clypeus not produced forward as a large shield-like lobe **(BB)**. Tibiae and basitarsi of middle and hind legs do not terminate in peg-like stout spines **(CC)**. Procoxa larger than the mesocoxa and metacoxa **(DD)**. Metatibia without a spur, or with a single, slender, simple spur. 63

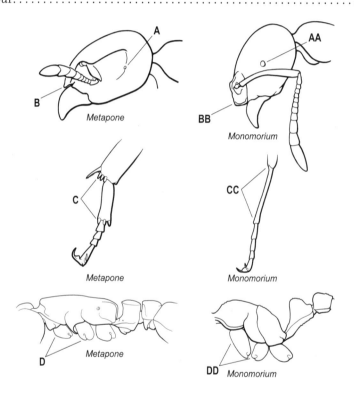

63a (62) Propodeum smoothly rounded, never armed with differentiated teeth or spines **(A)**. Area of the clypeus immediately in front of antennal sockets more or less flat **(B)**. Antennal scrobes always absent. Mandible usually with 4 teeth, rarely with 5. Apex of sting thin and pointed, lacking a triangular lamellate extension **(C)**. Anterior margin of the clypeus with an unpaired single, median elongate seta that is often flanked by paired setae on each side **(D)**. Maxillary palp with 1 or 2 segments . . . . *Monomorium* (part)

63b Propodeum bidentate or bispinose **(AA)**. Area of the clypeus immediately in front of antennal sockets raised up into a narrow ridge or shield wall so that the sockets appear to be placed within deep pits **(BB)**. Antennal scrobes usually but not always present. Mandible armed with 6 or more teeth: 2–3 enlarged teeth apically, followed by a row of at least 4 smaller denticles, sometimes more. Apex of sting with a triangular to pennant-shaped lamellate extension projecting upward from shaft (visible only when sting is extended) **(CC)**. Anterior margin of the clypeus without an unpaired single, median elongate seta; usually with pairs of elongate setae that straddle the midpoint **(DD)**. Maxillary palp with 3 or 4 segments . . . . *Tetramorium* (part)

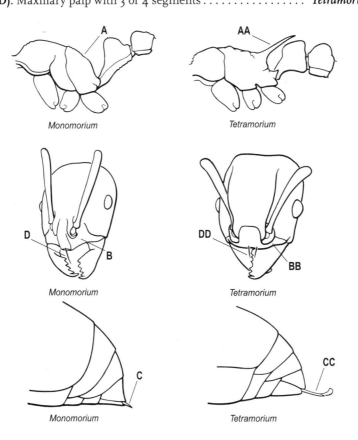

Monomorium

Tetramorium

Monomorium

Tetramorium

Monomorium

Tetramorium

64a (59) Petiole (A2) sessile, lacking an anterior peduncle; the petiole ventrally with a large projecting plate-like keel **(A)**. Central portion of the clypeus longitudinally bicarinate **(B)** (Seychelles) . . . . *Vollenhovia*

64b Petiole (A2) pedunculate to subsessile, ventrally without a large plate-like keel **(AA)** (often a small anteroventral process present on the peduncle). If the petiole subsessile, then the median portion of the clypeus is not bicarinate **(BB)**
. 65

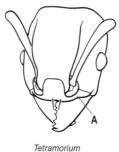

Vollenhovia **A**

Tetramorium **AA**

Vollenhovia **B**

Tetramorium **BB**

65a (64) Lateral portions of the clypeus, immediately in front of the antennal sockets, raised up into a narrow ridge or shield wall on each side, so that the sockets appear to be placed within deep pits **(A)** 66

65b Lateral portions of the clypeus, immediately in front of antennal sockets, usually flat but occasionally in the form of a shallow ridge **(AA)** 67

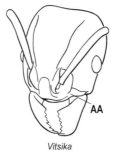

Tetramorium **A**

Vitsika **AA**

66a (65) Apex of sting thin and pointed, lacking a triangular lamellate extension **(A)**. Anterior margin of the clypeus with a small triangular point centrally **(B)**. Mandible armed with 6–8 teeth that decrease regularly in size from apical to basal . . . . *Eutetramorium*

66b Apex of sting with a triangular to pennant shaped lamellate extension projecting upward from the shaft (visible only when sting is extended) **(AA)**. Anterior margin of the clypeus without a triangular point centrally **(BB)**. Mandible armed with 6–11 teeth (only very rarely 6, usually 7 or more), arranged with 2–3 larger teeth apically, followed by at least 4 smaller teeth or denticles. . . . . *Tetramorium* (part)

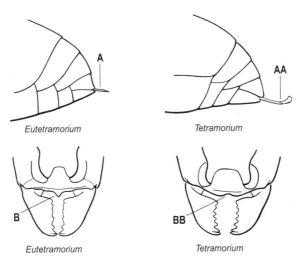

Eutetramorium   Tetramorium

Eutetramorium   Tetramorium

67a (65) In full-face view, posterior corners of the head often distinctly tuberculate to sharply denticulate **(A)**. Pronotum often marginate laterally and also anteriorly. In dorsal view pronotal shoulder angulate, denticulate, or tuberculate **(B)**. Petiole (A2) node dorsally with a transverse crest, or indented medially, or with a pair of long, sharp spines or teeth **(C)**. Propodeum short-bispinose at most, never with long sharp spines **(D)** . . . . *Terataner*

67b In full-face view, posterior corners of the head evenly broadly rounded, rarely angular, but never tuberculate or denticulate **(AA)**. Pronotum usually lacking lateral margination but occasionally denticulate **(BB)**. Petiole (A2) node dorsally often unarmed and without medial indention **(CC)**; if the petiole node is armed with long spines, then long spines are also present on the propodeum **(DD)**. 68

(Continues on next page)

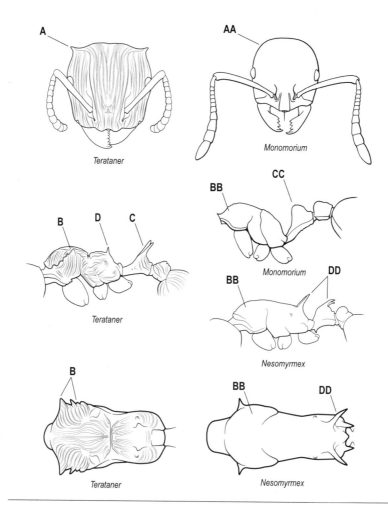

Terataner

Monomorium

Terataner

Monomorium

Nesomyrmex

Terataner

Nesomyrmex

68a (67)   Mandible with 6 or more teeth or denticles in total **(A)**. 69

68b   Mandible with 3–5 teeth or denticles in total, never with more **(AA)**. 73

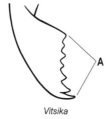

Vitsika

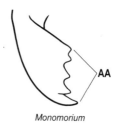

Monomorium

69a (68) Sting strongly developed and usually clearly visible **(A)**. Anterior clypeal margin with a single long, anteriorly projecting, unpaired median seta **(B)**. Very rarely, a second seta present **70**

69b Sting very reduced, never visible **(AA)**. Anterior clypeal margin without a single unpaired median seta; instead, usually with a pair of setae, one on each side of the midpoint of the margin **(BB)** **72**

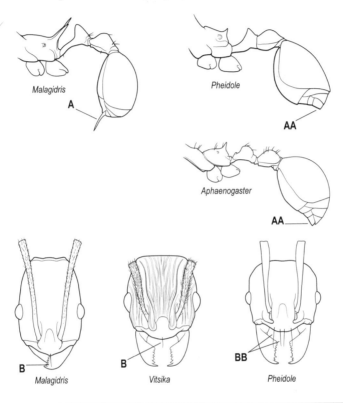

70a (69) Frontal carinae extend to close to the posterior margin of the head and form the upper margins of the antennal scrobes (**A**). With the mesosoma in profile the dorsal outline simple, more or less evenly shallowly convex from front to back **(B)** . . . . ***Vitsika*** (part)

70b Frontal carinae restricted to frontal lobes; antennal scrobes absent **(AA)**. With the mesosoma in profile, the dorsal outline complex; the pronotum or pronotum plus anterior mesonotum forms a dome-like or markedly convex arc; the propodeum is on a much lower level and forms a separate shallow convexity or flat plateau **(BB)**. **71**

(Continues on next page)

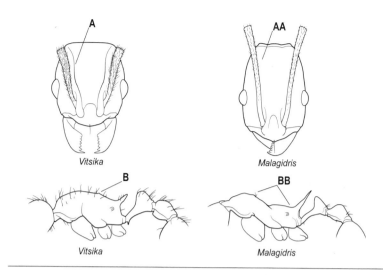

71a (70) Anterior clypeal margin without a median notch; with a single, unpaired, stout seta at the midpoint of the margin **(A)**. In full-face view, eyes obviously located in front of the midlength of the head capsule **(B)**. Antennal scape with appressed setae only **(C)**; promesonotum with sparse, short, stiff setae; propodeal dorsum with 0–2 pairs of setae; dorsal (outer) surface of the metatibia without standing setae **(D)**. Antennal club of 3 segments. With the head in full-face view, occipital corners usually narrow and elongated **(E)**. . . . . . .*Malagidris*

71b Anterior clypeal margin with a small median notch; usually with a pair of short, slender setae, 1 on each side of the midpoint within the notch **(AA)**. In full-face view, eyes located at about the midlength of the side of the head capsule **(BB)**. Antennal scape with numerous long, fine, erect, and semierect setae **(CC)**; promesonotum and propodeal dorsum with numerous long, fine, flexuous erect, and semierect setae; dorsal (outer) surface of the metatibia with standing setae present **(DD)**. Antennal club of 3–5 segments. With the head in full-face view, occipital corners broadly rounded **(EE)** . . . .*Vitsika* (part)

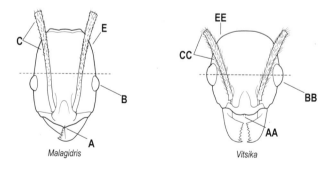

Malagidris

Vitsika

72a (69) Antenna terminating in a strongly defined 3-segmented club **(A)**. The posterior portion of the head usually rounded but the neck and collar may be present **(B)**. The workers' size variable. Mandible with third tooth (counting from the apex) smaller than fourth **(C)**, or reduced third tooth followed by a minute denticle before larger fourth tooth; margin never edentate. If third tooth not smaller than fourth (*P. lucida* complex), then the head will be without a neck or collar. Palp formula 2,2 or more rarely 3,2 **(D)**. . . . .*Pheidole* (minor workers)

72b Antenna terminating in a weakly defined 4-segmented club or with apical 4 segments gradually increasing in size toward the apex **(AA)**. The posterior portion of the head is always drawn out and tapers to a strongly constricted neck, beyond which the head flares out into a pronounced collar **(BB)**. The workers are large, elongate, with long spindly legs. Mandible with third tooth (counting from the apex) larger than fourth, or margin edentate **(CC)**. Palp formula 4,3 or 5,3 **(DD)** . . . .*Aphaenogaster*

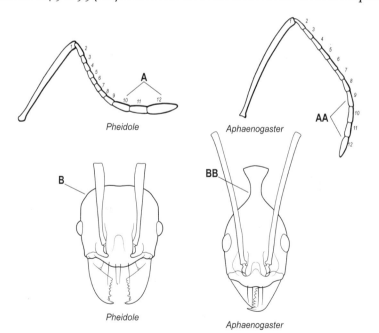

Pheidole

Aphaenogaster

Pheidole

Aphaenogaster

(Figure continues on next page)

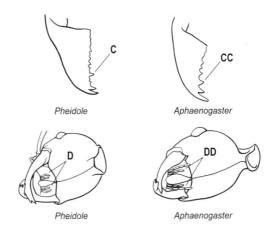

Pheidole   Aphaenogaster

Pheidole   Aphaenogaster

73a (68) Midpoint of the anterior clypeal margin without a single, unpaired elongate seta; instead, usually with a pair of short setae, 1 on each side of the midpoint **(A)** . . . . ·74

73b Midpoint of the anterior clypeal margin with a single, unpaired, elongate seta that projects forward over the mandibles and is usually very conspicuous **(AA)** . . . . ·75

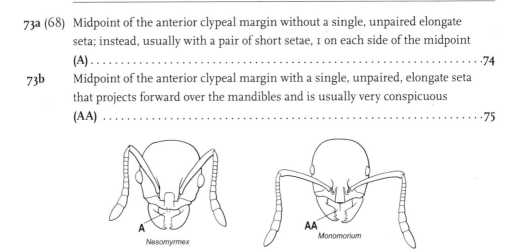

A   AA
Nesomyrmex   Monomorium

74a (73) Mandible powerfully constructed, armed with 2 large apical teeth followed by a long gap without teeth (diastema) and then 1 or 2 (rarely 3) basal teeth **(A)**. 2–5 teeth usually present on the posterior margin of the buccal cavity **(B)** . . . . *Pheidole* (major workers)

74b Mandible delicately constructed, armed with an uninterrupted series of 3–5 teeth (serially dentate) and decreasing in size from the apex to base, not arranged as above **(AA)**. Teeth absent from the posterior margin of the buccal cavity **(BB)** . . . . *Nesomyrmex*

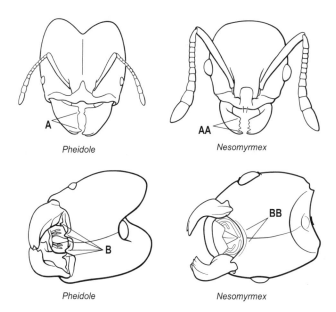

A — Pheidole   AA — Nesomyrmex

B — Pheidole   BB — Nesomyrmex

75a (73)   Palp formula 5,3 **(A)**. Mandible armed with 5 distinct teeth **(B)**. Occipital carina conspicuous along posterior margin of the head between vertex and occiput **(C)**. First gastral tergite (tergite of A4) does not overlap the sternite on the ventral surface of the gaster **(D)**. . . . . *Royidris*

75b   Palp formula 3,3 or less **(AA)**. Mandible usually armed with 3–4 teeth, only extremely rarely with 5 **(BB)**. Occipital carina vestigial or short, not conspicuous along posterior margin of the head between vertex and occiput **(CC)**. First gastral tergite (tergite of A4) strongly overlaps the sternite on the ventral surface of the gaster **(DD)**. 76

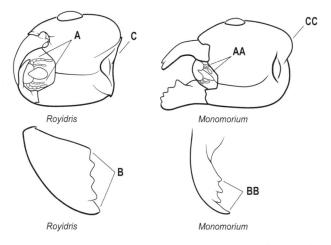

A, C — Royidris   AA, CC — Monomorium

B — Royidris   BB — Monomorium

(Figure continues on next page)

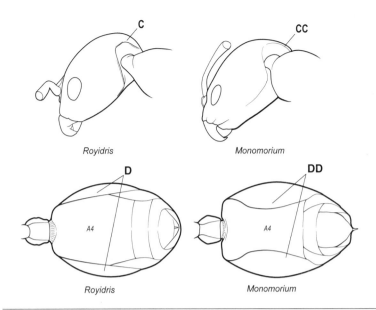

Royidris

Monomorium

Royidris

Monomorium

76a (75) Circular striolae around antennal insertions (A). Mandible distinctly longitudinally striate (B); maximum number of teeth 4 (C). . . . . *Trichomyrmex*

76b   Circular striolae absent around antennal insertions (AA). Mandible usually
smooth and shining (BB), except for seta-filled pits; with 3 to 5 teeth (CC). 77

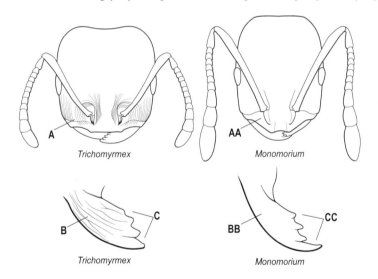

Trichomyrmex

Monomorium

Trichomyrmex

Monomorium

77a (76) Mandible armed with 5 teeth (A). Propodeum elongate, its dorsum and sides
finely and densely transversely striolate (B). Palp formula 3,3 . . . . *Erromyrma*

77b   Mandible armed with 3 or 4 teeth (AA) (5 teeth very rarely present in *M.
chnodes*). Propodeum not as above, usually smooth on dorsum and sides (BB).
Palp formula 3,2, 2,2 or 1,2. 78

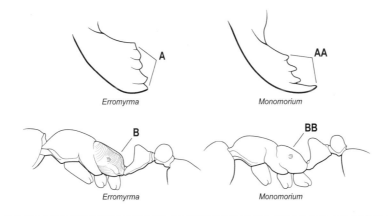

Erromyrma | Monomorium

Erromyrma | Monomorium

78a (77) With the head in profile, the unpaired, median clypeal seta arises from the anteriormost point of the clypeal projection **(A)**. Propodeal dorsum rounding into declivitous face without distinct angle **(B)**. Eyes large and with more than 2 ommatidia. Underside of the petiolar peduncle and node without transverse ridges **(C)** . . . . *Monomorium* (part)

78b In profile the unpaired, median clypeal seta arises from a vertical, or concave-vertical, surface well below the anteriormost point of the clypeal projection **(AA)**. Propodeal dorsum angulate to denticulate **(BB)**. Eyes usually small and with 1–2 ommatidia. If eyes large and with more than 2 ommatidia, then several to many fine, transverse ridges present on underside of the petiolar peduncle and node (effaced in very small workers) **(CC)** . . . . . *Syllophopsis*

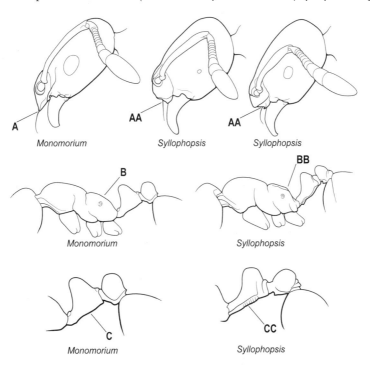

Monomorium | Syllophopsis | Syllophopsis

Monomorium | Syllophopsis

Monomorium | Syllophopsis

Genus Accounts

The genera are discussed in turn, in alphabetical order. The subfamily to which the genus belongs is noted to the right of its name, and the number of species and subspecies in each region is given immediately below its name. The entry for each genus gives notes that cover diagnostic remarks, distribution and ecology, and references to papers for the identification of its species, if such exist. In some instances there is an additional section, taxonomic comment, where points of particular taxonomic interest are noted. These sections are followed by a formal morphological definition of the worker of the genus as well as a list of all its available nominal taxa that occur in the regions under consideration. Each genus is illustrated by both photographs and line drawings, and distribution maps are provided. For genera with numerous widespread collection records, the distribution is indicated in gray with diagonal pattern. For genera with fewer known collection records, specimen localities are mapped (black circles) and the predicted range is indicated in solid gray. Specialized anatomical terms are defined in the glossary of morphology at the end of the book.

ACROPYGA Roger, 1862 — Formicinae

3 species, Afrotropical — PLATE 6

DIAGNOSTIC REMARKS: Among the formicines, *Acropyga* is immediately characterized by its combination of 10- or 11-segmented antennae and minute eyes; the maximum diameter of the eye is always considerably less than the maximum width of the scape.

DISTRIBUTION AND ECOLOGY: This genus is distributed throughout the world's tropics, and has some subtropical species, but it is absent from Madagascar. Although only 3 species occur in the Afrotropical region, they are very widely distributed and are hypogaeic, living in leaf litter and the soil. Nests are constructed in and under rotten wood, under stones, and 1 was found in the wall of an abandoned termitary. The species appear to be highly dependent upon mealybug secretions for food; mealybugs are tended by the workers on the roots

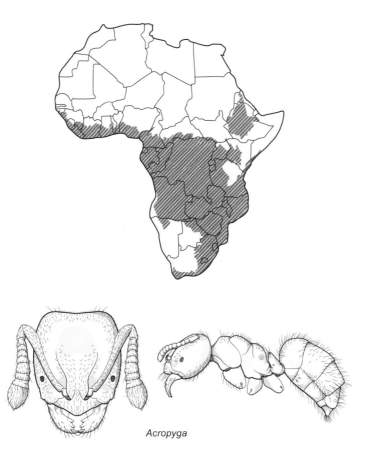

Acropyga

of plants. This dependency is so extensive that the queens of many species (including 1 Afrotropical species, *A. arnoldi*) have been observed carrying a mealybug in their mandibles during their nuptial flight.

IDENTIFICATION OF SPECIES: LaPolla, 2004a; LaPolla and Fisher, 2005.

DEFINITION OF THE GENUS (WORKER): Characters of subfamily FORMICINAE, plus:

Mandible with 5–9 teeth; tooth 3 (counting from the apical) is smaller than tooth 4.
    Mandibles at full closure not mostly concealed by the clypeus.
Palp formula 5,4, or 4,3, or 3,3.
Frontal carinae terminate posteriorly just behind the apices of the toruli.
Antenna with 10–11 segments.
Antennal sockets close to the posterior clypeal margin; anterior arc of torulus touches or
    indents the clypeal suture.
Eyes present, small to minute, located in front of the midlength of the sides of the head;
    ocelli absent or the median ocellus present (1 species).
Metapleural gland present.
Propodeal spiracle circular.

Propodeal foramen long: in ventral view, the foramen extends anterior of a line that spans the anteriormost points of the metacoxal cavities.

Metacoxae widely separated: in ventral view, with the mesocoxae and metacoxae directed at right-angles to the long axis of the mesosoma; the distance between the bases of the mesocoxae is markedly less than the distance between the bases of the metacoxae.

Tibial spurs: mesotibia o; metatibia o.

Abdominal segment 2 (petiole) a scale, erect or slightly inclined forward, without spines or teeth.

Petiole (A2) with its ventral margin U-shaped in section.

Abdominal segment 3 (first gastral segment) without tergosternal fusion on each side of the helcium.

Abdominal sternite 3 without a transverse sulcus across the sclerite immediately behind the helcium sternite (i.e., presternite and poststernite of A3 not separated by a sulcus).

Abdominal segment 3 in ventral view with the tergosternal suture curving forward (and usually upward) on each side of the helcium; the suture then narrowly arches around and runs posteriorly.

Acidopore with a seta-fringed nozzle.

AFROTROPICAL *ACROPYGA*

arnoldi Santschi, 1926
  = *rhodesiana* Santschi, 1928
bakwele LaPolla and Fisher, 2005
silvestrii Emery, 1915

ADELOMYRMEX Emery, 1897 Myrmicinae

2 species, Malagasy (Seychelles Islands) PLATE 6

DIAGNOSTIC REMARKS: Among the Malagasy myrmicines, *Adelomyrmex* is characterized by 12-segmented antennae that terminate in a distinct 2-segmented club and the presence of a tooth on the basal margin of the mandible. The clypeus posteriorly is very narrowly inserted between the frontal lobes, behind which there are no frontal carinae. The anterior clypeal margin has a triangular tooth on each side of its raised central portion.

DISTRIBUTION AND ECOLOGY: *Adelomyrmex* is predominantly a Neotropical genus, but a few species are found in the Pacific Islands and New Guinea; it does not occur in the Afrotropical region. Present in the Seychelles are 2 unidentified leaf-litter species; it is uncertain if they are endemics or introductions.

TAXONOMIC COMMENT: The Afrotropical *Baracidris* also has 12-segmented antennae with a 2-segmented club, but *Baracidris* lacks teeth on the basal mandibular margin and on the anterior clypeus. Compare the formal definitions for other differences.

IDENTIFICATION OF SPECIES: Fernández, 2003; Longino, 2012.

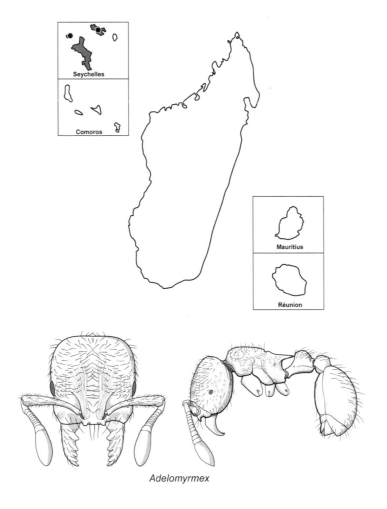

Adelomyrmex

DEFINITION OF THE GENUS (WORKER): Characters of subfamily MYRMICINAE, plus:

Mandible triangular, narrow; masticatory margin with 5–7 teeth and the basal margin with a tooth followed by a notch. Inner surface of mandible, just behind masticatory margin, with a row of translucent, lamelliform setae (not visible with mandibles closed).

Palp formula 2,2, or 1,2, or 1,1 (palp formula 1,1 in Seychelles species).

Stipes of the maxilla without a transverse crest.

Hypostoma with a median small tooth.

Clypeus with median portion narrow from side to side and sharply raised centrally. Clypeus posteriorly very narrowly inserted between the frontal lobes. Anterior clypeal margin with a triangular tooth on each side of the raised central portion.

Clypeus with an unpaired seta at the midpoint of the anterior margin.

Frontal carinae represented only by the frontal lobes.

Antennal scrobes absent.

Antenna with 12 segments, with an apical club of 2 segments.

Torulus partially or entirely concealed by frontal lobe in full-face view.

Eyes present, located slightly in front of the midlength of the head capsule.

Promesonotal suture absent from the dorsum of the mesosoma; promesonotum forms a single convexity in profile.

Propodeum short, with a pair of triangular teeth or spines.

Propodeal spiracle well in front of the margin of the declivity.

Metasternal process absent.

Tibial spurs: mesotibia 0; metatibia 0.

Abdominal segment 2 (petiole) with an anterior peduncle.

Abdominal tergite 4 (first gastral) does not broadly overlap the sternite on the ventral gaster; gastral shoulders absent.

Sting present, simple, strongly developed.

Main pilosity of dorsal head and body: simple.

MALAGASY *ADELOMYRMEX*

2 unidentified species, Seychelles Islands

ADETOMYRMA Ward, 1994 — Amblyoponinae

9 species, Malagasy

PLATE 6

DIAGNOSTIC REMARKS: Among the amblyoponines, *Adetomyrma* is characterized by short, blade-like mandibles that close tightly against the clypeus, the absence of a constriction separating the petiole (A2) from the first gastral segment (A3) in dorsal view, the lack of a girdling constriction on the second gastral segment (A4) that separates presclerites from postsclerites, and abdominal segments that increase in size posteriorly, so that A7 (the apical segment) is conspicuously larger than A3. Bizarre setae are absent from the head and body.

DISTRIBUTION AND ECOLOGY: This genus is restricted to Madagascar, where it nests in and under rotten logs in various forest zones, from montane rainforest to spiny forest. Tenebrionid beetle larvae have been recorded as prey, and both alate and ergatoid queens are known.

IDENTIFICATION OF SPECIES: Yoshimura and Fisher, 2012.

DEFINITION OF THE GENUS (WORKER): Characters of subfamily AMBLYOPONINAE, plus:

Mandible short, narrowly blade-like, and gently curved, closing against the clypeal margin; with 5–6 teeth.

Palp formula 3,3.

Clypeus narrow from front to back, so that antennal sockets are close to the anterior margin of the head.

Frontal carinae expanded into narrow frontal lobes that only partially cover the antennal sockets.

Eyes absent.

Antenna with 12 segments, funiculus gradually incrassate apically.

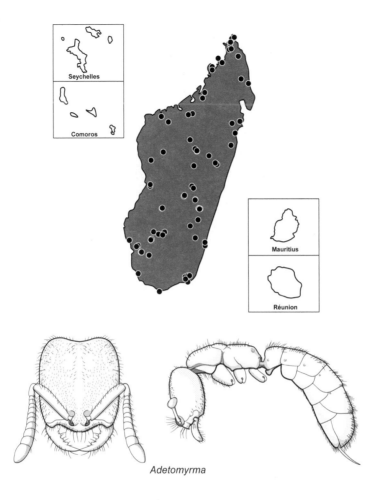

Adetomyrma

Metanotal groove vestigial.

Tibial spurs: mesotibia 1; metatibia 2.

Abdominal segment 2 (petiole) in dorsal view not separated from A3 (first gastral segment) by a constriction; segments A2–A6 form an uninterrupted series that gradually increases in width.

Petiole (A2) in profile with a steep anterior face, a weakly convex dorsum, and no trace of a posterior face; petiole very broadly attached to abdominal segment 3 (first gastral segment).

Helcium tergite not set off from posttergite 3 by a constriction; A3 without differentiated presclerites.

Helcium attached high on the anterior face of the first gastral segment (A3), so that in profile A3 above the helcium has no free anterior face.

Abdominal segment 3 without tergosternal fusion.

Prora absent.

Abdominal segment 4 without presclerites, without tergosternal fusion.

Pygidium hypertrophied; sting very large.

aureocuprea Yoshimura and Fisher, 2012

bressleri Yoshimura and Fisher, 2012

caputleae Yoshimura and Fisher, 2012

cassis Yoshimura and Fisher, 2012

caudapinniger Yoshimura and Fisher, 2012

cilium Yoshimura and Fisher, 2012

clarivida Yoshimura and Fisher, 2012

goblin Yoshimura and Fisher, 2012

venatrix Ward, 1994

AENICTOGITON Emery, 1901 — Dorylinae

7 species, Afrotropical   PLATE 6

DIAGNOSTIC REMARKS: Among the dorylines, *Aenictogiton* has the promesonotal suture present and conspicuous. The waist is of a single segment, the petiole (A2). Deep girdling constrictions are present between the second and third gastral segments (A4 and A5), and between the third and fourth gastral segments (A5 and A6), as well as between the first and second (A3 and A4), giving the gaster a distinctly irregular appearance. Only 1 other doryline genus shares this gastral morphology, *Zasphinctus*, but in that genus the promesonotal suture is absent.

DISTRIBUTION AND ECOLOGY: Restricted to the central Afrotropical region, the worker caste of this genus is known only from leaf-litter samples taken in Ugandan forest. All earlier records refer to males, taken at light or by sweeping.

TAXONOMIC COMMENT: The 7 described species listed below are all based on males. The worker caste is now known but has not yet been formally described.

IDENTIFICATION OF SPECIES: Santschi, 1924 (males); no recent review exists.

DEFINITION OF THE GENUS (WORKER): Characters of subfamily DORYLINAE, plus:

Prementum not visible when mouthparts closed.

Labrum with distal (free) margin convex, not indented or cleft medially.

Palp formula apparently 1,1 (awaits confirmation).

Clypeus with an unpaired seta at the midpoint of its anterior margin.

Eyes absent; ocelli absent.

Frontal carinae present as a pair of extremely closely approximated low ridges between the antennal sockets.

Antennal sockets fully exposed in full-face view.

Parafrontal ridges present but very reduced.

Antenna with 12 segments, the 3 apical segments forming a weak club.

Antennal scrobes absent.

Promesonotal suture strongly present on the dorsum of the mesosoma but fused and inflexible.

Pronotal-mesopleural suture present on side of mesosoma.

Metanotal groove absent.

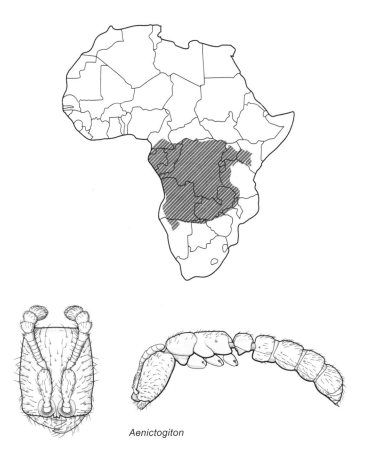

Aenictogiton

Propodeal spiracle situated high on the side and at about the midlength of the sclerite, not subtended by a longitudinal impression.

Propodeal lobes absent.

Tibial spurs: mesotibia 1, pectinate; metatibia 1, pectinate.

Metatibial gland present.

Pretarsal claws simple.

Waist of 1 segment (A2 = petiole); not marginate laterally.

Petiole (A2) sessile.

Helcium projects from anterior surface of A3 (first gastral segment) just above its midheight.

Prora a simple U-shaped margin on anterior surface of the sternite of A3.

Abdominal segments 4, 5, and 6 all with strongly developed presclerites; junction of presclerite and postsclerite on each segment a deep girdling constriction.

Pygidium (tergite of A7) of moderate size (but distinctly smaller than tergite of A6), slightly flattened dorsally, unarmed except for lateral marginal tubercles from which setae arise.

Sting present.

attenuatus Santschi, 1919	*fossiceps* Emery, 1901
bequaerti Forel, 1913	*schoutedeni* Santschi, 1924
elongatus Santschi, 1919	*sulcatus* Santschi, 1919
emeryi Forel, 1913	

AENICTUS Shuckard, 1840

Dorylinae

34 species (+ 17 subspecies), Afrotropical

PLATE 7

DIAGNOSTIC REMARKS: Among the dorylines, *Aenictus* species are monomorphic ants in which the promesonotal suture is absent. The waist is of 2 conspicuously reduced segments (A2 and A3, unique in Afrotropical dorylines). The antenna has 10 segments in all known Afrotropical species (some extralimital species have fewer). Eyes are always absent, and the spiracle of the postpetiole (A3) is situated at or usually behind the midlength of the tergite. In addition, this is the only doryline genus to exhibit a very reduced, unarmed pygidium. In all other genera the pygidium is large, conspicuous, and armed with teeth or stout spiniform setae.

DISTRIBUTION AND ECOLOGY: This genus occurs throughout the Old World tropics, except for Madagascar and its associated islands. All species are terrestrial to subterranean and frequently encountered in the topsoil, among the roots of trees and shrubs, in leaf litter, and sometimes in and under rotten wood. They are nomadic mass predators and do not make permanent nests; their narrow marching columns are sometimes encountered on the ground or in the leaf litter, and their prey includes other ants. The large, dichthadiigyne queens of *Aenictus* are only very rarely discovered, but the conspicuous males are commonly attracted to lights.

TAXONOMIC COMMENT: The species-rank taxonomy of Afrotropical *Aenictus* is confused, mostly by the early development of a dual system, with 1 taxonomy based on the worker caste and the other on the very conspicuous male sex. The 2 are strikingly different and, unless captured together, could not be associated before the advent of DNA sequencing. Thus, of the 51 nominal taxa in the species-group, the majority are described only from males. It is apparent that many taxa currently bear 2 names, 1 based on workers and the other on males, of what are actually the same species.

IDENTIFICATION OF SPECIES: No review exists.

DEFINITION OF THE GENUS (WORKER): Characters of subfamily DORYLINAE, plus:

Prementum partially to mostly exposed when mouthparts closed.

Palp formula 2,2.

Eyes absent; ocelli absent.

Frontal carinae present as a pair of closely approximated raised ridges (sometimes almost fused into a single median crest).

Antennal sockets fully exposed in full-face view.

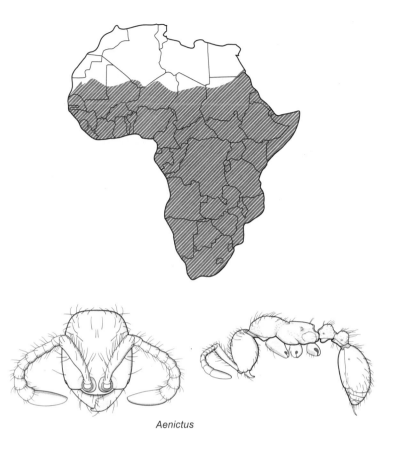

Aenictus

Parafrontal ridges present.

Antenna with 10 segments; funiculus gradually incrassate apically or with a weak club of 2–3 segments.

Antennal scrobes absent.

Promesonotal suture absent on the dorsum of the mesosoma.

Pronotal-mesopleural suture partially to entirely fused on side of mesosoma.

Metanotal groove often weakly present, sometimes vestigial or absent.

Propodeal spiracle situated high on side and in front of the midlength of the sclerite, not subtended by a longitudinal impression.

Propodeal lobes present.

Tibial spurs: mesotibia 0–2 (usually 1); metatibia 0–2 (usually 1).

Metatibial gland present, usually conspicuous, located distally on ventral surface.

Pretarsal claws simple.

Waist of 2 segments (A2 and A3 = petiole and postpetiole).

Petiole (A2) sessile to subsessile.

Abdominal segment 3 (postpetiole) with the spiracle situated at or usually behind the midlength of the tergite.

Prora present as an anterior subpostpetiolar process.

Abdominal segment 4 (first gastral segment) with strongly developed presclerites; postsclerites of A4 constricted to a narrow neck immediately behind the presclerites.
Abdominal segments 5–7 without presclerites.
Pygidium (tergite of A7) very small, reduced to a narrow U-shaped sclerite.
Sting large and functional.

AFROTROPICAL AENICTUS

alluaudi Santschi, 1910
    alluaudi falcifer Santschi, 1924
anceps Forel, 1910
asantei Campione, Novak, and
        Gotwald, 1983
asperivalvus Santschi, 1919
bayoni Menozzi, 1933
bottegoi Emery, 1899
    bottegoi noctivagus Santschi, 1913
brazzai Santschi, 1910
buttgenbachi Forel, 1913
congolensis Santschi, 1911
crucifer Santschi, 1914
    crucifer tuberculatus Arnold, 1915
decolor (Mayr, 1879)
    = *batesi* Forel, 1911
    = *bidentatus* Donisthorpe, 1942
eugenii Emery, 1895
    = *eugeniae kenyensis* Santschi, 1933
    eugenii caroli Forel, 1910
    eugenii henrii Santschi, 1924
foreli Santschi, 1919
furculatus Santschi, 1919
    furculatus andrieui Santschi, 1930
furibundus Arnold, 1959
fuscovarius Gerstäcker, 1859
    fuscovarius magrettii Emery, 1892
    fuscovarius laetior Forel, 1910

hamifer Emery, 1896
humeralis Santschi, 1910
    humeralis chevalieri Santschi, 1910
    humeralis viridans Santschi, 1915
inconspicuus Westwood, 1845
luteus Emery, 1892
    luteus moestus Santschi, 1930
mariae Emery, 1895
    mariae natalensis Forel, in Emery,
        1901
mentu Weber, 1942
moebii Emery, 1895
    moebii sankisianus Forel, 1913
mutatus Santschi, 1913
    mutatus pudicus Santschi, 1919
pharao Santschi, 1924
raptor Forel, 1913
rixator Forel, in Emery, 1901
rotundatus Mayr, 1901
    rotundatus guineensis Santschi, 1924
    rotundatus merwei Santschi, 1932
soudanicus Santschi, 1910
    soudanicus brunneus Forel, 1913
steindachneri Mayr, 1901
togoensis Santschi, 1915
vagans Santschi, 1924
villiersi Bernard, 1953
weissi Santschi, 1910

AGRAULOMYRMEX Prins, 1983

Formicinae

2 species, Afrotropical

PLATE 7

DIAGNOSTIC REMARKS: Among the formicines, *Agraulomyrmex* is the only genus that combines 10-segmented antennae with eyes that are relatively large; their maximum diameter is conspicuously greater than the maximum width of the scape.

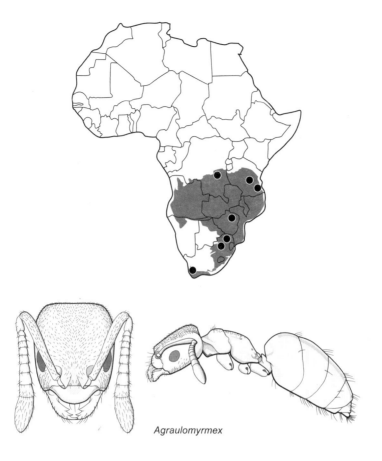

Agraulomyrmex

DISTRIBUTION AND ECOLOGY: Restricted to the southern half of the Afrotropical region, where undescribed species are also known. The nest sites of *Agraulomyrmex* species are in the ground, but workers forage on weeds and low vegetation, where they tend aphids. Replete workers are present in nests, suggesting a dependence on homopterous sugar secretions during adverse conditions.

TAXONOMIC COMMENT: The taxonomy of this genus and its relationship with species in supposedly related genera, such as those Afrotropical species currently referred to as *Brachymyrmex*, is not properly understood. In addition, there are a few undescribed formicine species that are currently unplaced at the genus-level but are apparently closely related to *Agraulomyrmex*. These species will reach couplet 19 in the African key but fail to key either to *Anoplolepis* or *Tapinolepis*, sharing only some characters with each of these genera.

IDENTIFICATION OF SPECIES: Prins, 1983.

DEFINITION OF THE GENUS (WORKER): Characters of subfamily FORMICINAE, plus:

Mandible with 4–6 (usually 5) teeth; tooth 3 (counting from the apical) is smaller than tooth 4.

Palp formula 6,4, or 5,3.

Frontal carinae terminate posteriorly just behind the apices of the toruli.

Antenna with 10 segments.

Antennal sockets close to the posterior clypeal margin; anterior arc of the torulus indents the clypeal suture.

Eyes present, large, located in front of the midlength of the sides of the head; ocelli absent.

Metapleural gland present.

Metanotum not present as a discrete sclerite on the dorsum of the mesosoma.

Propodeal spiracle circular.

Propodeal foramen long: in ventral view, the foramen extends anterior of a line that spans the anteriormost points of the metacoxal cavities (not confirmed but present in related genera).

Metacoxae widely separated: in ventral view, with the mesocoxae and metacoxae directed at right-angles to the long axis of the mesosoma, the distance between the bases of the mesocoxae is markedly less than the distance between the bases of the metacoxae (not confirmed but present in related genera).

Tibial spurs: mesotibia 0; metatibia 0.

Abdominal segment 2 (petiole) a low scale that is slightly inclined forward, without spines or teeth; petiole partially to entirely overhung by gaster when the mesosoma and gaster are aligned.

Petiole (A2) with its ventral margin U-shaped in section.

Abdominal segment 3 (first gastral) basally with complete tergosternal fusion for some distance on each side of the helcium; the line of fusion follows the edges of the impression in the anterior face of A3 and the unfused portions of the tergite and sternite commence some distance above the helcium, at the dorsolateral apices of the impression.

Abdominal sternite 3 without a transverse sulcus across the sclerite immediately behind the helcium sternite (i.e., presternite and poststernite of A3 not separated by a sulcus).

Acidopore with a seta-fringed nozzle.

AFROTROPICAL AGRAULOMYRMEX

meridionalis Prins, 1983
wilsoni Prins, 1983

ANILLOMYRMA Emery, 1913 — Myrmicinae

2 species, Afrotropical — PLATE 7

DIAGNOSTIC REMARKS: Among the myrmicines, *Anillomyrma* species are eyeless and have 10-segmented antennae that terminate in an apical club of 3 segments. The procoxae are greatly enlarged, conspicuously much larger than the mesocoxae and metacoxae; the propodeum is unarmed; and in profile the postpetiole (A3) is seen to be attached very high on the anterior surface of the first gastral segment (A4).

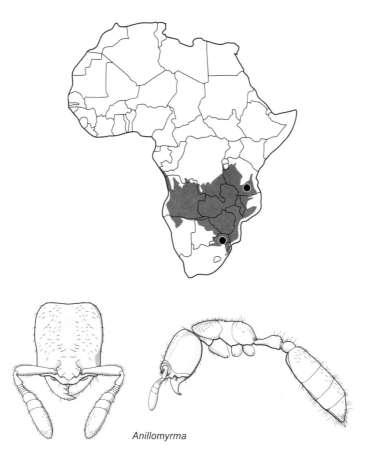

Anillomyrma

DISTRIBUTION AND ECOLOGY: Samples of this genus, from South Africa and Tanzania, appear to represent 2 undescribed species. Previously described forms inhabit the Oriental and Malesian regions, but it remains unknown in the Malagasy region.

TAXONOMIC COMMENT: What is known of the genus is summarized in Bolton, 1987, and Eguchi et al., 2009.

IDENTIFICATION OF SPECIES: The Afrotropical species are as yet undescribed.

DEFINITION OF THE GENUS (WORKER): Characters of subfamily MYRMICINAE, plus:

Mandible narrowly triangular and with masticatory margin very oblique, so that the mandibles cross over at full closure; with 3–4 teeth.

Palp formula 2,2.

Stipes of the maxilla without a transverse crest.

Clypeus with median portion narrowed from side to side, posteriorly very narrowly inserted between the small frontal lobes. Median portion of the clypeus not longitudinally bicarinate.

Clypeus with an unpaired seta at the midpoint of the anterior margin.

Frontal carinae restricted to closely approximated, small frontal lobes.

Antennal scrobes absent.

Antenna with 10 segments, with an apical club of 3 segments; funiculus segments 2–6 reduced to narrow annuli.

Torulus with upper lobe visible in full-face view; maximum width of torulus lobe is posterior to the point of maximum width of the frontal lobes.

Eyes absent.

Promesonotal suture absent from the dorsum of the mesosoma; promesonotum flat. Metanotal groove present but not impressed.

Propodeum unarmed; propodeal lobe vestigial to absent.

Propodeal spiracle small, approximately at the midlength of the sclerite.

Metasternal process absent.

Procoxa enlarged, very much larger than mesocoxa and metacoxa.

Tibial spurs: mesotibia 0; metatibia 0.

Abdominal segment 2 (petiole) without an anteroventral process on the peduncle.

Abdominal segment 3 (postpetiole) in profile attached very high on the anterior surface of A4.

Abdominal tergite 4 (first gastral) overlaps the sternite on the ventral gaster; gastral shoulders absent.

Sting strongly developed.

Main pilosity of dorsal head and body: simple.

AFROTROPICAL ANILLOMYRMA

2 undescribed species, in Tanzania and South Africa

ANKYLOMYRMA Bolton, 1973 Agroecomyrmecinae

1 species, Afrotropical PLATE 7

DIAGNOSTIC REMARKS: *Ankylomyrma* is unique in its possession of a massive, ball-shaped gaster that consists almost entirely of the hypertrophied first gastral (A4) tergite. In addition, antennal scrobes are present; there is a transverse cuticular lamella posteriorly on the head; the eyes are at the extreme posterolateral corners of the head (full-face view); the pronotum is swollen and has 4 pairs of tubercles; the metapleural gland orifice is a simple round to elliptical hole; and the waist is 2-segmented, with a pair of sharp spines on the petiole (A2).

DISTRIBUTION AND ECOLOGY: The single bizarre species is known only from a few collections made in the rainforest zone of West and Central Africa (Cameroon, Central African Republic, Gabon, and Ghana). It is entirely arboreal, but beyond that nothing is known of its biology.

IDENTIFICATION OF SPECIES: Bolton, 1973b; Bolton, 1981b.

DEFINITION OF THE GENUS (WORKER): Characters of subfamily AGROECOMYRMECINAE, plus:

Mandible triangular, with 5 teeth that occupy the entire margin.

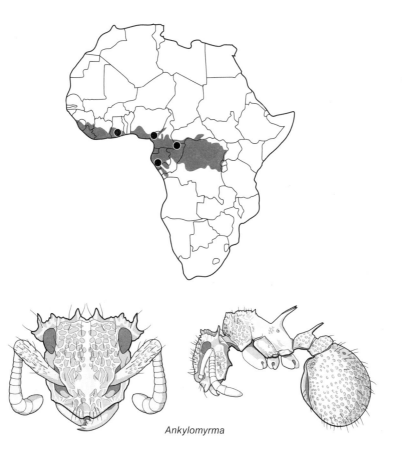

Ankylomyrma

Palp formula 6,4, the maxillary palp extremely long; when laid back, it extends posteriorly beyond the level of the occipital foramen.

Frontal carinae well in from the lateral margins of the head in full-face view.

Clypeus with median portion overhanging the mandibles.

Antennal sockets and frontal lobes in full-face view not strongly migrated laterally.

Eyes at extreme posterolateral corners of the head.

Posterior margin of the head with a broad transverse lamella that projects into a series of dentiform processes.

Mesosoma elongate; promesonotum domed.

Pronotum with 4 pairs of tubercles or teeth.

Propodeum and petiole (A2) each bispinose.

Abdominal segment 2 (petiole) short-pedunculate, without a ventral process, spiracle opens laterally.

Petiole (A2) with fused tergite and sternite in posterior view equally convex, their inner margins forming a circle; the tergite and sternite meet edge to edge and do not overlap.

Pygidium extremely reduced, simple.

AFROTROPICAL ANKYLOMYRMA

coronacantha Bolton, 1973

ANOCHETUS Mayr, 1861   Ponerinae

19 species, Afrotropical; 5 species, Malagasy   PLATE 8

DIAGNOSTIC REMARKS: Among the ponerines are only 2 genera that have linear mandibles that are articulated very close together near the midpoint of the anterior margin of the head: *Anochetus* and *Odontomachus*. The 2 are easily separated because in *Anochetus* the petiole (A2) never terminates in a single dorsal spine. In addition, the carina that separates the dorsal from the posterior surface of the head (the nuchal carina) forms a broad uninterrupted curve across the posterodorsal extremity of the head. The posterior surface of the head lacks paired dark apophyseal lines, and on the vertex there is either no median groove, or the groove is shallow and poorly defined.

DISTRIBUTION AND ECOLOGY: *Anochetus* is a pantropical genus, with a couple of species present in the southern Palaearctic. Numerous species are present in the regions under consideration here, found in all vegetation zones, and some may be locally quite common. Nests are usually constructed in the ground, directly or under stones, or in and under rotten wood, and sometimes in the walls of termitaries. A couple of species are arboreal, nesting in rot cavities or rotten branches in trees, and foraging entirely arboreally. All species are carnivorous, taking a wide range of insect prey, and several species are known to be nocturnal.

IDENTIFICATION OF SPECIES: Brown, 1978 (Afrotropical + Malagasy); Fisher and Smith, 2008 (Malagasy).

DEFINITION OF THE GENUS (WORKER): Characters of subfamily PONERINAE, plus:

Mandibles linear and nearly parallel when closed, articulated close to the midline of the anterior margin of the head. Mandible equipped apically with a vertical series of 3 teeth, the uppermost (apicodorsal) tooth acute. A pair of trigger setae present below the mandibles.

Palp formula 4,4, or 4,3.

Clypeus with anterior margin projecting on each side of the mandibles.

Frontal lobes small; in full-face view, their anterior margins do not overhang the anterior clypeal margin.

Eyes present, minute to large, in front of the midlength of the head capsule; each eye located on a broad, rounded cuticular prominence.

Head capsule with eye separated from frons by a broad impression in the dorsum.

Vertex without a median furrow, with a nuchal carina posteriorly but without a V-shaped pair of dark apophyseal lines that extend down to the occipital foramen.

Metanotum usually absent from dorsal mesosoma, rarely visible as a narrow, poorly delimited, transverse strip.

Epimeral sclerite present.

Metapleural gland orifice opens posteriorly.

Propodeal spiracle with orifice circular to short-elliptical.

Tibial spurs: mesotibia 0–1; metatibia 1, pectinate. Mesotibial spur, when present, is very small and simple.

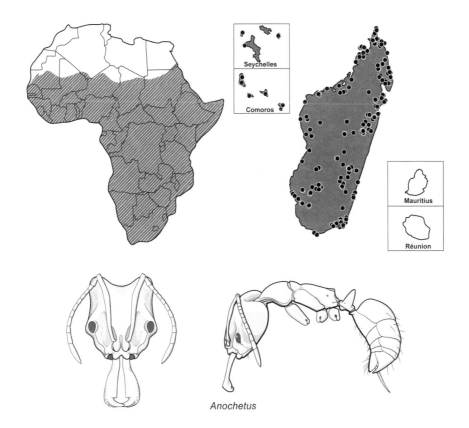

Anochetus

Pretarsal claws simple.

Abdominal segment 2 (petiole) dorsally variously shaped but never surmounted by a single spine.

Helcium located at the base of the anterior face of A3 (first gastral segment).

Prora present, small.

Abdominal segment 4 (second gastral segment) with presclerites present but sometimes only very weakly differentiated and without a girdling constriction.

Stridulitrum present or absent on the pretergite of A4.

AFROTROPICAL ANOCHETUS

africanus (Mayr, 1865)
  = *africanus camerunensis* Mayr, 1896
  = *pasteuri* Santschi, 1923
angolensis Brown, 1978
bequaerti Forel, 1913
  = *bequaerti abstracta* Santschi, 1914
  = *estus* Wheeler, W.M., 1922
  = *opaciventris* Wheeler, W.M., 1922
faurei Arnold, 1948
fuliginosus Arnold, 1948

jonesi Arnold, 1926
katonae Forel, 1907
  = *parvus* Santschi, 1914
  = *parvus longiceps* Santschi, 1914
  = *punctatus* Santschi, 1914
  = *punctatus occidentalis* Santschi, 1914
  = *concinnus* Santschi, 1920
  = *punctatus durbanensis* Arnold, 1926
  = *gnomulus* Bernard, 1953
  = *lamottei* Bernard, 1953

levaillanti Emery, 1895

maynei Forel, 1913

natalensis Arnold, 1926

obscuratus Santschi, 1911

    = *schoutedeni* Santschi, 1923

    = *schoutedeni ustus* Santschi, 1923

pellucidus Emery, 1902

    = *pellucidus aurifrons* Santschi, 1910

pubescens Brown, 1978

punctaticeps Mayr, 1901

rothschildi Forel, 1907

sedilloti Emery, 1884

siphneus Brown, 1978

talpa Forel, 1901

traegaordhi Mayr, 1904

    = *gracilicornis* Viehmeyer, 1923

    = *sudanicus* Weber, 1942

    = *angusticornis* Arnold, 1946

    = *silvaticus* Bernard, 1953

MALAGASY *ANOCHETUS*

boltoni Fisher, 2008

goodmani Fisher, 2008

grandidieri Forel, 1891

    = *madecassus* Santschi, 1928

madagascarensis Forel, 1887

    = *africanus friederichsi* Forel, 1918

pattersoni Fisher, 2008

ANOPLOLEPIS Santschi, 1914   Formicinae

9 species (+ 4 subspecies), Afrotropical; 1 species (+ 1 subspecies), Malagasy, introduced   PLATE 8

DIAGNOSTIC REMARKS: Among the formicines, *Anoplolepis* can be identified by its combination of 11-segmented antennae, absence of ocelli, lack of a differentiated metanotum on the mesosoma, lack of spines or teeth on the propodeum and petiole, and the presence of a stout spur on the metatibia. In addition, the eyes are located well behind the midlength of the head, and the dorsum of the head, behind the clypeus, has numerous stout, standing setae present.

DISTRIBUTION AND ECOLOGY: Almost entirely Afrotropical but with 1 species (*A. gracilipes*) that is very widespread in the Old World tropics. Malagasy records appear to represent introductions. *A. steingroeveri gertrudae* appears to be nothing more than a casual introduction on Réunion Island and is probably a synonym of the South African *A. steingroeveri*. *A. gracilipes* has been recorded from Réunion Island, Mauritius, and the Seychelles but not from Madagascar itself. Species of this genus are ground nesters, either directly or under surface objects; rainforest species of the *A. gracilipes* group have also been found nesting in the compacted walls of termite colonies. The larger, polymorphic species of the *A. custodiens* group forage freely on the surface of the ground. They are aggressive, fast-moving ants that are predaceous on a wide range of arthropods, but they also tend homopterous insects, even ascending trees for honeydew. In arid areas, excess sugars are stored by repletes within the nest.

IDENTIFICATION OF SPECIES: Prins, 1982 (partial review of South Africa species); no complete review exists.

DEFINITION OF THE GENUS (WORKER): Characters of subfamily FORMICINAE, plus:

Worker caste polymorphic in *A. custodiens* group, monomorphic in *A. gracilipes* group.

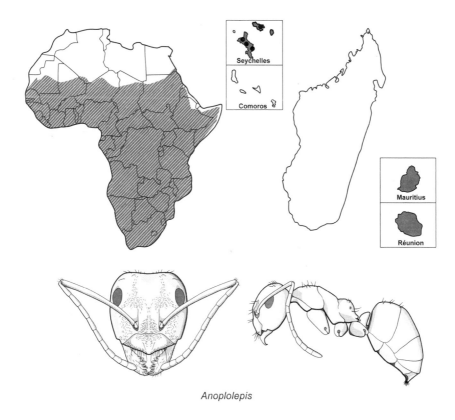

Anoplolepis

Mandible with 6–9 teeth; tooth 3 (counting from the apical) is smaller than tooth 4; basal margin of mandible may have 1–2 small teeth close to the basal angle.

Palp formula 6,4.

Frontal carinae extend posteriorly far beyond the apices of the toruli.

Antenna with 11 segments.

Antennal sockets close to the posterior clypeal margin but anterior arc of the torulus does not indent the clypeal suture.

Eyes present, located behind the midlength of the sides of the head; ocelli usually absent (a median ocellus may be present in largest workers of some polymorphic species).

Cephalic dorsum behind the clypeus with erect stout setae.

Metanotum does not form a distinct isolated sclerite on the dorsum of the mesosoma.

Metapleural gland present.

Propodeal spiracle circular to subcircular.

Propodeal foramen long: in ventral view, the foramen extends anterior of a line that spans the anteriormost points of the metacoxal cavities.

Metacoxae widely separated: in ventral view, with the mesocoxae and metacoxae directed at right-angles to the long axis of the mesosoma, the distance between the bases of the mesocoxae is markedly less than the distance between the bases of the metacoxae.

Tibial spurs: mesotibia 1; metatibia 1, metatibial spur flanked by a pair of stout setae.

Abdominal segment 2 (petiole) a scale, without spines or teeth.

Petiole (A2) with ventral margin U-shaped in section.

Abdominal segment 3 (first gastral) without a sharply margined longitudinal concavity into which the petiole fits.

Abdominal segment 3 without tergosternal fusion on each side of the helcium.

Abdominal sternite 3 without a transverse sulcus across the sclerite immediately behind the helcium sternite (i.e., presternite and poststernite of A3 not separated by a sulcus).

With abdominal segment 3 in ventral view, the tergosternal suture extends laterally or curves forward on each side of the helcium; the suture then arches around and runs posteriorly.

Acidopore with a seta-fringed nozzle.

AFROTROPICAL *ANOPLOLEPIS*

carinata (Emery, 1899)	*gracilipes* (Smith, F., 1857)
custodiens (Smith, F., 1858)	*nuptialis* (Santschi, 1917)
= *hendecarthrus* (Roger, 1863)	*opaciventris* (Emery, 1899)
= *berthoudi* (Forel, 1876)	*rufescens* (Santschi, 1917)
custodiens detrita (Emery, 1892)	*steingroeveri* (Forel, 1894)
custodiens hirsuta (Emery, 1892)	= *braunsi* (Forel, 1913)
custodiens pilipes (Emery, 1892)	*steingroeveri parsonsi* Santschi, 1937
fallax (Mayr, 1865)	*tenella* (Santschi, 1911)

MALAGASY *ANOPLOLEPIS*

gracilipes (Smith, F., 1857)

steingroeveri (Forel, 1894)

  steingroeveri gertrudae (Forel, 1900)

APHAENOGASTER Mayr, 1853   Myrmicinae

3 species (+ 3 subspecies), Malagasy   PLATE 8

DIAGNOSTIC REMARKS: Among the Malagasy myrmicines, *Aphaenogaster* has 12-segmented antennae that either terminate in a 4-segmented club or the apical 4 segments gradually increase in size apically. The posterior portion of the head capsule is drawn out into a strongly constricted neck, behind which the head capsule flares out again and forms a pronounced collar. The first gastral tergite (the tergite of A4) broadly overlaps the sternite on the side of the gaster, and the sting is very reduced, never visible. The workers are large, elongate, with long, spindly legs.

DISTRIBUTION AND ECOLOGY: *Aphaenogaster* has an almost worldwide distribution, with a large number of species in the Holarctic. A few species occur in Madagascar, but the genus is absent from the Afrotropical region and absent from the Indian Ocean islands.

TAXONOMIC COMMENT: The Malagasy species all appear to belong to a group that is typically Oriental and Malesian, sometimes called the subgenus *Deromyrma* Forel, whose main distinguishing feature is the cephalic constriction noted above.

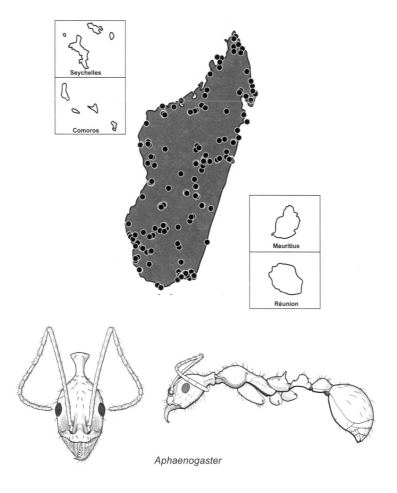

Aphaenogaster

IDENTIFICATION OF SPECIES: No review exists.

DEFINITION OF THE GENUS (WORKER): Characters of subfamily MYRMICINAE, plus:

Mandible triangular, relatively stout. Masticatory margin of mandible with 6–16 teeth, the third tooth (counting from the apex) larger than the fourth; sometimes with 1 or more denticles on the basal margin.

Palp formula 5,3, or 4,3.

Stipes of the maxilla usually without a transverse crest, uncommonly with a crest feebly present.

Clypeus large, posteriorly broadly inserted between the frontal lobes; median portion of the clypeus not bicarinate.

Clypeus without an unpaired seta at the midpoint of the anterior margin.

Frontal carinae short, restricted to well defined but narrow frontal lobes.

Antennal scrobes absent.

Antenna with 12 segments, without an apical club or with a club of 4 segments.

Torulus with upper lobe visible in full-face view; maximum width of torulus lobe is posterior to the point of maximum width of the frontal lobes.

Eyes present, located at, or slightly in front of, the midlength of the head capsule.

Promesonotal suture absent or represented across the dorsum by a narrow line.

Pronotum, or pronotum plus anterior mesonotum, swollen and convex in profile, usually dome-like; posteriorly the mesonotum slopes steeply to the metanotal groove.

Propodeum bidentate to bispinose; in profile the propodeal dorsum usually on a much lower level than the apex of the convex pronotum.

Propodeal spiracle high on the side of the sclerite and close to its midlength.

Metasternal process at most minute, usually absent.

Tibial spurs: mesotibia 0 or 1; metatibia 0 or 1.

Abdominal segment 2 (petiole) with a distinct anterior peduncle, without a subpetiolar process.

Stridulitrum present on the pretergite of A4.

Abdominal tergite 4 (first gastral) broadly overlaps the sternite on the side of the gaster; gastral shoulders usually absent.

Sting only weakly developed.

Main pilosity of dorsal head and body simple.

MALAGASY APHAENOGASTER

friederichsi Forel, 1918

gonacantha (Emery, 1899)

swammerdami Forel, 1886

   swammerdami curta Forel, 1891

   swammerdami spinipes Santschi, 1911

   swammerdami clara (Santschi, 1915)

APHOMOMYRMEX Emery, 1899   Formicinae

1 species, Afrotropical   PLATE 8

DIAGNOSTIC REMARKS: Among the formicines, *Aphomomyrmex* has only 9 antennal segments, has a tooth on the basal margin of the mandible, and has a tall, narrow petiolar scale. Only 1 other genus shares this suite of characters, *Petalomyrmex*. But *Aphomomyrmex* is polymorphic and has eyes that are located high on the head capsule, so that in full-face view the eyes appear quite widely separated from the lateral margins of the head. In addition, the single species of *Aphomomyrmex* is black and has 5 maxillary palp segments, while that of the monomorphic *Petalomyrmex* is brownish yellow and has 3 maxillary palp segments.

DISTRIBUTION AND ECOLOGY: This monotypic Afrotropical genus is known from several collections in the Central African rainforest and from a dubious record from South Africa, which may be mislabeled. It is an uncommon arboreal ant and may be associated with myrmecophytes.

IDENTIFICATION OF SPECIES: Snelling, 1979.

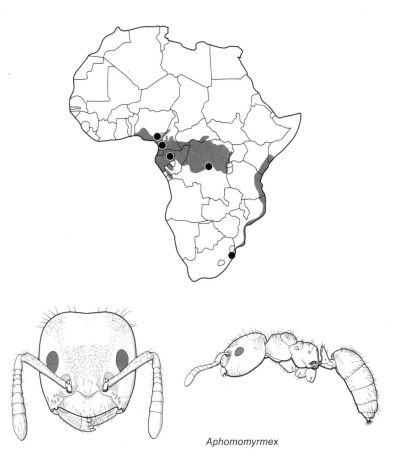

Aphomomyrmex

DEFINITION OF THE GENUS (WORKER): Characters of subfamily FORMICINAE, plus:

Worker caste polymorphic.

Mandible with 5–6 teeth; tooth 3 (counting from the apical) is smaller than tooth 4; basal tooth small and on the basal margin.

Palp formula 5,3.

Frontal triangle broad and shallow, not depressed and only weakly differentiated from the clypeus and inner margins of the frontal carinae, the latter very weakly represented and obtuse.

Antenna with 9 segments.

Antennal sockets close to the posterior clypeal margin; anterior arc of the torulus touches or slightly indents the clypeal suture.

Eyes present, slightly in front of the midlength of the sides of the head; ocelli present, minute in smallest workers.

Mesothorax not constricted in its anterior half.

Metapleural gland present.

Propodeal spiracle circular, large.

Propodeal foramen long: in ventral view the foramen extends anterior of a line that spans the anteriormost points of the metacoxal cavities.

Metacoxae widely separated: in ventral view, with the mesocoxae and metacoxae directed at right-angles to the long axis of the mesosoma, the distance between the bases of the mesocoxae is markedly less than the distance between the bases of the metacoxae.

Tibial spurs: mesotibia 0; metatibia 0.

Abdominal segment 2 (petiole) a tall, erect slender scale, its dorsal surface without teeth or spines.

Petiole (A2) with a short posterior peduncle that is attached low down on A3 (the first gastral segment).

Petiole (A2) with its ventral margin U-shaped in section.

Abdominal segment 3 (the first gastral) with a longitudinal concavity, into which the posterior peduncle of the petiole fits when the mesosoma and gaster are aligned.

Abdominal segment 3 basally with complete tergosternal fusion for some distance on each side of the helcium; the line of fusion follows the edges of the impression in the anterior face of A3, and the unfused portions of the tergite and sternite commence some distance above the helcium, at the dorsolateral apices of the impression.

Abdominal sternite 3 without a transverse sulcus across the sclerite immediately behind the helcium sternite (i.e., presternite and poststernite of A3 not separated by a sulcus).

Acidopore with a seta-fringed nozzle.

AFROTROPICAL APHOMOMYRMEX

afer Emery, 1899

= *muralti* Forel, 1910

APOMYRMA Brown, Gotwald, and Lévieux, 1971 — Apomyrminae

1 species, Afrotropical — PLATE 9

DIAGNOSTIC REMARKS: In *Apomyrma* a double row of anteriorly directed, peg-like setae are present on the labrum (but such setae are absent from the clypeus); eyes and frontal carinae are absent; the antennal sockets are entirely exposed and very close to the anterior margin of the head; the promesonotal suture is very deeply impressed and articulated; the waist is 1-segmented and nodiform, with the petiole (A2) short-pedunculate anteriorly and having a well-defined posterior surface, and the helcium is attached low down on the anterior face of the first gastral segment (A3).

DISTRIBUTION AND ECOLOGY: Known from only a few collections in West and Central African rainforests and gallery forests (Benin, Cameroon, Central African Republic, Ghana, Ivory Coast, and Nigeria), the single species may occur at great depth in the soil, where it apparently both nests and forages. Geophilomorph centipedes form at least part of the diet, as dismembered individuals have been found in nests. The genus is absent from Madagascar.

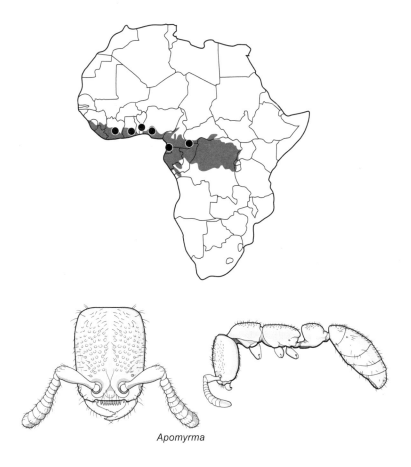

Apomyrma

IDENTIFICATION OF SPECIES: Brown, Gotwald, and Lévieux, 1971; Bolton, 1990a; Bolton, 2003.

DEFINITION OF THE GENUS (WORKER): Characters of subfamily APOMYRMINAE, plus:

Mandible short and curved, narrowly blade-like, denticulate in distal half, terminating in an apical tooth.

Palp formula 2,2.

Antenna terminates in a weakly 4-segmented club.

Epimeral sclerite present, small.

Abdominal segment 2 (petiole) nodiform, with a short anterior peduncle; in profile petiole with a steep anterior face, a weakly convex dorsum, and a conspicuous posterior face; petiole narrowly attached to abdominal segment 3 (first gastral segment).

Prora absent.

Abdominal sternite 3 (first gastral sternite) with a transverse sulcus across the width of the sclerite close behind the helcium.

AFROTROPICAL APOMYRMA

stygia Brown, Gotwald, and Lévieux, 1971

DIAGNOSTIC REMARKS: Among the dolichoderines, *Aptinoma* is very similar to *Tapinoma*, but the worker caste of *Aptinoma* is dimorphic and the petiole in profile takes the form of a low node that has a short, near-vertical anterior face and a flat dorsum. This contrasts with the extremely reduced low segment, lacking a dorsal node, that is typical of *Tapinoma*.

DISTRIBUTION AND ECOLOGY: Known only from Madagascar, where its species nest and forage arboreally. Nest sites may be in dead vines or twigs that are still attached to the tree (*A. mangabe*) or under canopy litter and moss on branches (*A. antongil*).

IDENTIFICATION OF SPECIES: Fisher, 2009.

DEFINITION OF THE GENUS (WORKER): Characters of subfamily DOLICHODERINAE, plus:

Worker caste dimorphic, with distinct major and minor workers.

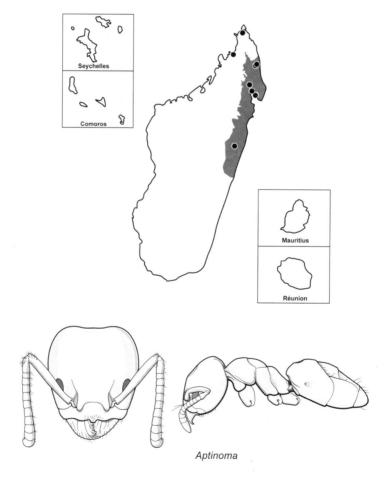

Aptinoma

Masticatory margin of mandible with a combination of teeth and denticles (about 8–14 in total).

Palp formula 6,3 or 6,4.

Anterior clypeal margin transverse.

Antenna with 12 segments.

Metathoracic spiracles dorsolateral, visible in dorsal view.

Metanotal groove present.

Propodeum unarmed.

Tibial spurs: mesotibia 1; metatibia 1, pectinate.

Abdominal segment 2 (petiole) a reduced, low node, with a short, near-vertical anterior face and a flat dorsum.

Abdominal segment 3 (first gastral segment) projects forward over the reduced petiole and entirely conceals the petiole in dorsal view when the gaster and mesosoma are aligned.

Abdominal segment 3, on the ventral surface of projecting portion, with a longitudinal impression or groove that accommodates the entire petiole when the mesosoma and gaster are aligned.

Abdominal tergites 3–6 visible in dorsal view (i.e., 4 gastral tergites visible in dorsal view); tergite of A7 (pygidium) is reflexed onto the ventral surface of the gaster, where it abuts abdominal sternite 7 (hypopygium).

Abdominal sternite 6 (fourth gastral) not keel-shaped posteriorly.

MALAGASY APTINOMA

antongil Fisher, 2009
manganbe Fisher, 2009

ASPHINCTOPONE Santschi, 1914   Ponerinae

3 species, Afrotropical   PLATE 9

DIAGNOSTIC REMARKS: Among the ponerines, *Asphinctopone* combines the characters of presence of a single spur on the metatibia and absence of traction setae on the metabasitarsus, with the following: The clypeus is specialized, as described below. The promesonotal suture and the metanotal groove are both conspicuously impressed. The petiole (A2) is a high, unarmed scale, with a stout posterior peduncle that has 3–4 strong transverse carinae dorsally. The second gastral segment (A4) does not have differentiated presclerites (i.e., a gastral girdling constriction is absent).

DISTRIBUTION AND ECOLOGY: Only 3 species of this uniquely Afrotropical genus are known; all are uncommon and only rarely discovered. The most common species (*A. silvestrii*) is widespread throughout the rainforest zone of West and Central Africa, where it has been found in leaf litter, topsoil, pieces of rotten wood, and rotting vegetation on the forest floor and once in an abandoned termitary. The other 2 species are known only from their type-collections, in rainforest litter in the Central African Republic (*A. differens*) and by hand collection in primary forest in Tanzania (*A. pilosa*).

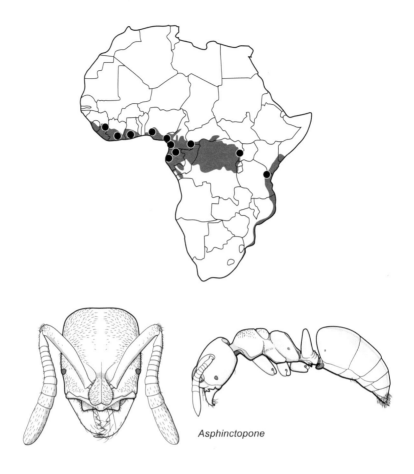

Asphinctopone

IDENTIFICATION OF SPECIES: Bolton and Fisher, 2008a; Hawkes, 2010.

DEFINITION OF THE GENUS (WORKER): Characters of subfamily PONERINAE, plus:

Mandible oblique, not triangular, with a gap between the mandibles basally when they are closed; with a weakly developed basal groove; masticatory margin with 5–6 teeth. Basalmost tooth is at the rounded basal angle. Basal margin shallowly convex, near the articulation of the inner margin usually with a small tooth-like process (overlapped by anterolateral angles of the clypeal lobe when mandibles closed).

Palp formula 3,3; second maxillary palpomere elongate and slender.

Clypeus complex: in full-face view the median portion projects anteriorly as a broad lobe that terminates in a distinct angle on each side; these angles overlap the basal margins of the mandibles; the anterior clypeal margin has a small median rounded projection, on each side of which the anterior margin is shallowly concave; above the median projection the central portion of the clypeus forms a narrow ridge to the frontal lobes.

Frontal lobes small; in full-face view their anterior margins do not overhang the anterior clypeal margin.

Eyes present, small, in front of midlength of the head capsule.

Antenna with the terminal 3 or 4 segments forming a weak club; apical antennomere hypertrophied, longer than the 5 preceding segments together.

Promesonotal suture moderately to deeply impressed, cross-ribbed on extreme anterior mesonotum.

Metanotal groove present, moderately to deeply impressed and cross-ribbed; mesonotum conspicuously isolated and convex between promesonotal and metanotal impressions.

Epimeral sclerite present.

Mesopleuron with a transverse sulcus that divides it into anepisternum and katepisternum.

Metapleural gland orifice opens posterolaterally.

Propodeal spiracle very small, its sclerite almost circular but the orifice itself a small ellipse.

Tibial spurs: mesotibia 1; metatibia 1; each pectinate.

Pretarsal claws simple.

Abdominal segment 2 (petiole) a high, unarmed scale that is narrow in profile and broad in dorsal view.

Petiole (A2) with a short, stout posterior peduncle that is traversed dorsally by 3–4 strong transverse carinae.

Petiole sternite, when detached and in posterior view, complex: in the posterior third of its length the sternite bifurcates into an externally visible ventral plate and a slightly shorter, internally projecting sclerite that is completely concealed by the external plate in normal view.

Helcium located at the base of the anterior face of A3 (first gastral segment); with a narrow transverse crest of cuticle between the ventral apices of the tergal arch.

Prora present, long in profile, broad in anterior view with a thick outer annulus and a deep central concavity; prora in profile projects anteriorly to about the midlength of the helcium.

Abdominal segment 4 (second gastral segment) without differentiated presclerites.

Stridulitrum absent from the pretergite of A4.

AFROTROPICAL ASPHINCTOPONE

differens Bolton and Fisher, 2008

pilosa Hawkes, 2010

silvestrii Santschi, 1914

  = *lucidus* Weber, 1949

  = *lamottei* (Bernard, 1953)

ATOPOMYRMEX André, 1889   Myrmicinae

3 species, Afrotropical   PLATE 9

DIAGNOSTIC REMARKS: Among the Afrotropical myrmicines, *Atopomyrmex* species have a polymorphic worker caste in which the frontal carinae and antennal scrobes are variably developed among the different sizes of worker. The antennae are of 10 or 12 segments, terminating in a club of 3 segments, and the posterior corners of the head capsule

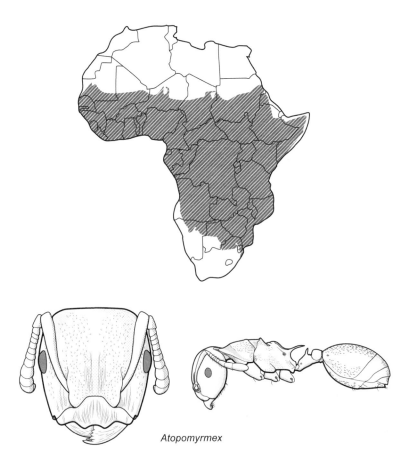

Atopomyrmex

are broadly, evenly rounded in full-face view. The metapleuron seems strange in profile, because its lower margin appears to have been eroded away, so that the metacoxa seems to rest upon the bulla of the metapleural gland. The petiole (A2) is subsessile and usually has a pair of dorsal spines.

DISTRIBUTION AND ECOLOGY: A small genus, restricted to the Afrotropical region. The 2 commonest species, *A. cryptoceroides* and *A. mocquerysi*, are widespread in forested areas. The former is more confined to rainforest, but the latter can apparently live anywhere trees are present. They form extensive nests within the standing trunks of dead trees, or in dead parts of standing tree trunks, and forage arboreally. The third species, *A. calpocalycola*, has been found only once, in the internodes of a myrmecophyte in the Cameroon rainforest.

TAXONOMIC COMMENT: At different places in the original description of *A. calpocalycola* Snelling (1992) states that the species has 9- or 8-segmented antennae. All workers in the type-series actually have 10-segmented antennae.

IDENTIFICATION OF SPECIES: Bolton, 1981b; Snelling, 1992.

DEFINITION OF THE GENUS (WORKER): Characters of subfamily MYRMICINAE, plus:

Worker caste polymorphic.

Mandible short, triangular, and stout; masticatory margin with 4–7 teeth. Number of teeth tends to increase with body size, but in largest workers all teeth may be worn down and rounded.

Palp formula 5,3, or 4,3.

Stipes of the maxilla without a transverse crest.

Clypeus large, median portion shield-like, posteriorly broadly inserted between the frontal lobes. Anterior clypeal margin extends forward as a lobe that fits tightly over the closed mandibles and overlaps their bases.

Clypeus without an unpaired seta at the midpoint of the anterior margin.

Frontal carinae development varies with size: in smallest workers short, terminating a short distance behind the frontal lobes; with increasing size, the carinae become longer and more strongly defined; in largest workers, the frontal carinae extend well beyond the level of the posterior margins of the eyes.

Antennal scrobes absent in smallest workers; developing, extending, and becoming deeper with increased size; conspicuous and capable of accommodating the scape in largest workers.

Antenna with 10 or 12 segments, with an apical club of 3 segments.

Torulus concealed by frontal lobe in full-face view.

Eyes well developed, located behind the midlength of the head capsule.

Head capsule with posterior corners broadly, evenly rounded in full-face view.

Pronotum more or less flat to shallowly transversely concave, usually bluntly marginate laterally.

Promesonotal suture usually vestigial to absent on dorsum (visible in minors of 1 species); metanotal groove present.

Mesonotum in profile usually broadly and bluntly bituberculate.

Metapleuron in profile with its lower margin appearing to have been eroded away, so that the metacoxa appears to rest upon the bulla of the metapleural gland.

Metathorax ventrally with a deep pit between the apices of the metacoxae, behind the metasternal pit; metasternal process absent.

Propodeum bispinose.

Propodeal spiracle large, well in front of the margin of the declivity.

Tibial spurs: mesotibia 0; metatibia 0.

Pretarsal claws large.

Abdominal segment 2 (petiole) dorsally usually with a pair of short, stout spines (absent on 1 species).

Stridulitrum present on the pretergite of A4.

Abdominal tergite 4 (first gastral) broadly overlaps the sternite on the ventral gaster; gastral shoulders present.

Sting reduced.

Main pilosity of dorsal head and body: usually absent, rarely present and simple.

calpocalycola Snelling, 1992

cryptoceroides Emery, 1892

  = *deplanatus* Mayr, 1895

  = *mocquerysi curvispina* Forel, 1911

  = *cryptoceroides melanoticus* Santschi,

    1925

mocquerysi André, 1889

  = *mocquerysi australis* Santschi, 1914

  = *mocquerysi arnoldi* Santschi, 1923

  = *mocquerysi obscura* Santschi, 1923

  = *mocquerysi opaca* Santschi, 1923

  = *mocquerysi erigens* Santschi, 1924

AXINIDRIS Weber, 1941

21 species, Afrotropical

Dolichoderinae

PLATE 10

DIAGNOSTIC REMARKS: Among the dolichoderines, *Axinidris* is the only Afrotropical genus that has the propodeum armed. In all others the propodeum is evenly rounded and unadorned, or at most has a low cuticular transverse crest between the propodeal dorsum and the declivity. The propodeal armament varies in different *Axinidris* species; spines, teeth, tubercles, or angles may occur, and usually a longitudinal median carina is also present, or uncommonly a median longitudinal flange.

DISTRIBUTION AND ECOLOGY: Restricted to forested areas of the Afrotropical region and most common in rainforest zones. All known species are arboreal, nesting within hollow plant stems, both living and dead, and also in rot holes and rotten branches on standing trees. A couple of species have been recovered from myrmecophytes. Foraging is mostly arboreal, but occasionally workers descend to the litter layer. One West African species, *A. palligastrion*, is found in association with *Crematogaster clariventris* Mayr, sharing its forage trails on tree trunks and running in the same manner, with the pale gaster elevated.

IDENTIFICATION OF SPECIES: Shattuck, 1991; Snelling, 2007.

DEFINITION OF THE GENUS (WORKER): Characters of subfamily DOLICHODERINAE, plus:

Masticatory margin of mandible with a combination of teeth and denticles (about 7–12
    in total).

Palp formula 6,4.

Anterior clypeal margin with a median notch.

Antenna with 12 segments.

Metathoracic spiracles dorsal.

Metanotal groove present.

Propodeum armed with a pair of spines, teeth, tubercles, or angles; usually also with a
    longitudinal median carina or uncommonly a median flange.

Propodeal spiracle dorsolateral and at about the midlength of the sclerite.

Tibial spurs: mesotibia 1; metatibia 1, pectinate.

Abdominal segment 2 (petiole) with a low scale that is narrowly rounded to subacute
    dorsally and inclined anteriorly; the anterior face of scale much shorter than the
    posterior face.

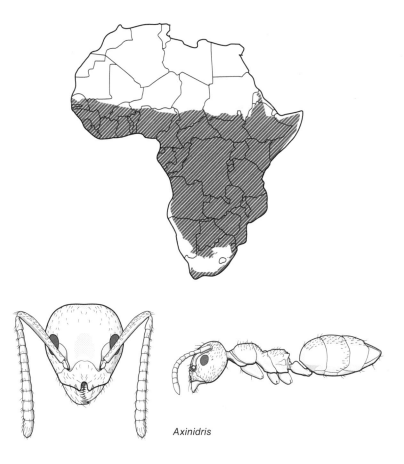

Axinidris

Abdominal segment 3 (first gastral) projects forward but does not conceal the entire petiole in dorsal view.

Abdominal segment 3, on ventral surface of projecting portion, with a longitudinal impression or groove that accommodates the basal portion, or at most the posterior face, of the petiole when the mesosoma and gaster are aligned.

Abdominal tergites 3–6 visible in dorsal view (i.e., 4 gastral tergites visible in dorsal view); tergite of A7 (pygidium) is vertical or reflexed onto the ventral surface of the gaster, not visible in dorsal view.

Abdominal sternite 6 (fourth gastral sternite) not keel-shaped posteriorly.

AFROTROPICAL AXINIDRIS

acholli Weber, 1941	= *parvus* Shattuck, 1991
bidens Shattuck, 1991	*icipe* Snelling, 2007
denticulata (Wheeler, W.M., 1922)	*kakamegensis* Shattuck, 1991
gabonica Snelling, 2007	*kinoin* Shattuck, 1991
ghanensis Shattuck, 1991	*lignicola* Snelling, 2007
hylekoites Shattuck, 1991	*luhya* Snelling, 2007
hypoclinoides (Santschi, 1919)	*mlalu* Snelling, 2007

murielae Shattuck, 1991	*okekai* Snelling, 2007
namib Snelling, 2007	*palligastrion* Shattuck, 1991
nigripes Shattuck, 1991	*stageri* Snelling, 2007
occidentalis Shattuck, 1991	*tridens* (Arnold, 1946)

BARACIDRIS Bolton, 1981   Myrmicinae

3 species, Afrotropical   PLATE 10

DIAGNOSTIC REMARKS: Among the Afrotropical myrmicines, *Baracidris* is the only genus to combine 12-segmented antennae that terminate in a distinct 2-segmented club, with very closely approximated frontal lobes that almost touch anteromedially. The clypeus between the frontal lobes is reduced to an extremely narrow strip. The median portion of the clypeus is sharply raised centrally and takes the form of a narrow longitudinal ridge, and the large propodeal lobes are connected to the propodeal spines by conspicuous lamellae that ascend the declivity.

DISTRIBUTION AND ECOLOGY: *Baracidris* is known only from the Afrotropical region, where its 3 species have been discovered fewer than a dozen times, in samples of leaf litter from the rainforest zone.

TAXONOMIC COMMENT: The Malagasy (Seychelles Islands) *Adelomyrmex* species also have 12-segmented antennae with a 2-segmented club, but *Adelomyrmex* has teeth on the basal mandibular margin and on the anterior clypeus. Compare the formal definitions for other differences.

IDENTIFICATION OF SPECIES: Bolton, 1981b; Fernández, 2003.

DEFINITION OF THE GENUS (WORKER): Characters of subfamily MYRMICINAE, plus:

Mandible triangular, narrow; masticatory margin with 5 teeth and without a tooth on the basal margin. Mandibles, when closed, enclose a space between their basal margins and the anterior clypeal margin. Inner surface of mandible, just behind masticatory margin, with a row of translucent, lamelliform setae (not visible with mandibles closed).

Palp formula 2,2.

Stipes of the maxilla without a transverse crest.

Hypostoma without a median small tooth.

Clypeus with median portion sharply raised centrally and in the form of a narrow, longitudinal crest from the anterior margin to the frontal lobes. Clypeus posteriorly very narrowly inserted between the frontal lobes.

Clypeus without an unpaired seta at the midpoint of the anterior margin.

Frontal carinae represented only by small, very closely approximated, frontal lobes that almost touch anteriorly.

Antennal scrobes absent.

Antenna with12 segments, with a strongly developed apical club of 2 segments.

Torulus with upper lobe not concealed by frontal lobe in full-face view.

Eyes minute and inconspicuous, located approximately at the midlength of the head capsule.

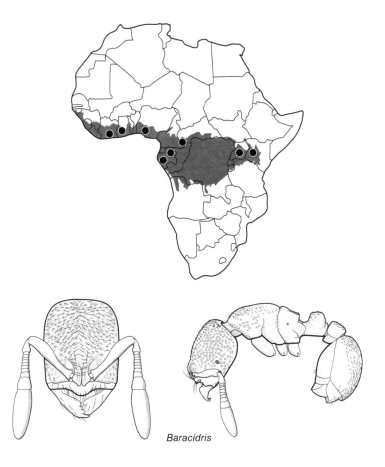

Baracidris

Promesonotal suture absent from the dorsum of the mesosoma; promesonotum forms a single long shallow convexity in profile.

Metanotal groove impressed.

Propodeum short, with a pair of triangular teeth. Propodeal lobes broad and rounded, linked to the propodeal teeth by a lamina.

Propodeal spiracle well in front of the margin of the declivity.

Metasternal process absent.

Tibial spurs: mesotibia 0; metatibia 0.

Abdominal segment 2 (petiole) with a stout anterior peduncle.

Abdominal tergite 4 (first gastral) does not broadly overlap the sternite on the ventral gaster; gastral shoulders absent.

Sting present, simple.

Main pilosity of dorsal head and body: usually absent, sometimes simple.

AFROTROPICAL BARACIDRIS

meketra Bolton, 1981
pilosa Fernández, 2003
sitra Bolton, 1981

DIAGNOSTIC REMARKS: Among the ponerines, *Boloponera* combines the characters of presence of a single spur on the metatibia and absence of traction setae on the metabasitarsus, with the following: The mandibles are linear and somewhat curved, and the mandibular inner margin has 2 small, blunt teeth. The anterior margin of the head lacks a semicircular excavation at the outer base of the mandible. The frontal lobes are large, in full-face view they distinctly overhang the anterior clypeal margin. The mesofemur does not have a mid-dorsal longitudinal groove such as is present on the metafemur.

DISTRIBUTION AND ECOLOGY: This genus remains known only from the holotype worker, which was retrieved from a litter sample taken in the rainforest of Cameroon.

IDENTIFICATION OF SPECIES: Fisher, 2006; Schmidt and Shattuck, 2014.

DEFINITION OF THE GENUS (WORKER): Characters of subfamily PONERINAE, plus:

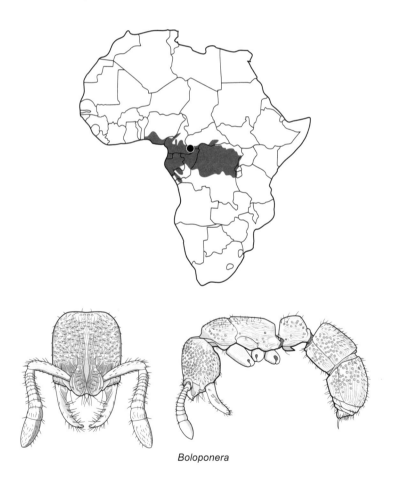

Boloponera

Mandible linear and shallowly curved, acute apically and the inner margin with 2 small, blunt teeth; without a longitudinal groove or trench on the dorsal surface.

Palp formula unknown.

Clypeus simple; median portion of the anterior margin without a lobe.

Frontal lobes very large; in full-face view their anterior margins distinctly overhang the anterior clypeal margin.

Eyes absent.

Antenna terminates in a 2-segmented club.

Mesopleural transverse sulcus present; anepisternum fused to the mesonotum.

Metanotal groove absent.

Epimeral sclerite absent.

Metapleural gland orifice opens laterally.

Propodeal spiracle circular or very nearly so.

Propodeal dorsum without a median longitudinal groove or impression.

Metafemur with a longitudinal groove mid-dorsally along its entire length; mesofemur without such a groove.

Tibial spurs: mesotibia 1; metatibia 1; each pectinate.

Pretarsal claws simple.

Abdominal segment 2 (petiole) a node.

Helcium arises near midheight on the anterior face of A3 (first gastral segment).

Prora present, small.

Abdominal segment 4 (second gastral segment) with very strongly differentiated presclerites.

Stridulitrum: presence/absence unknown.

AFROTROPICAL BOLOPONERA

vicans Fisher, 2006

BONDROITIA Forel, 1911   Myrmicinae

2 species, Afrotropical   PLATE 10

DIAGNOSTIC REMARKS: Among the Afrotropical myrmicines, *Bondroitia* is eyeless and has 11-segmented antennae that terminate in a club of 3 segments. The procoxae are enormously enlarged compared to the mesocoxae and metacoxae. The propodeal spiracle has a very large circular orifice, and the propodeum is unarmed. In addition, the clypeus is only narrowly inserted between the very closely approximated small frontal lobes, and the mandible has only 4 teeth.

DISTRIBUTION AND ECOLOGY: Only 2 species of this exclusively Afrotropical genus are known. *B. saharensis* is known only from the queen and male, from a single collection made in Niger. The other, *B. lujae*, has been recorded only 3 times, once each from Angola, the Democratic Republic of Congo, and Zambia.

IDENTIFICATION OF SPECIES: Bolton, 1987.

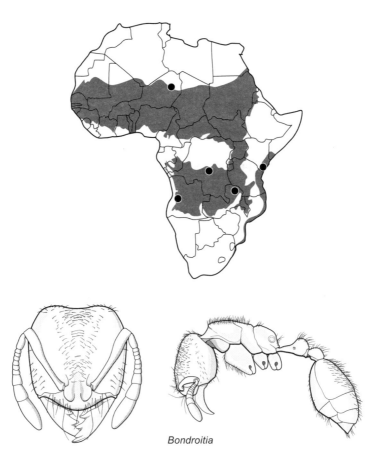

Bondroitia

DEFINITION OF THE GENUS (WORKER): Characters of subfamily MYRMICINAE, plus:

Worker caste size-variable but not polymorphic.

Mandible narrowly triangular and with masticatory margin very oblique, so that the mandibles cross over at full closure; with 4 teeth.

Palp formula 2,2.

Stipes of the maxilla without a transverse crest.

Clypeus posteriorly narrowly inserted between the small frontal lobes. Median portion of the clypeus not longitudinally bicarinate. Extreme lateral portion of the clypeus projects as a low, broadly triangular tooth over the outer basal border of the mandible.

Clypeus with an unpaired seta at the midpoint of the anterior margin.

Frontal carinae restricted to very closely approximated small frontal lobes.

Antennal scrobes absent.

Antenna with 11 segments, with an apical club of 3 segments.

Torulus with upper lobe visible in full-face view; maximum width of the torulus lobe is posterior to the point of maximum width of the frontal lobes.

Eyes absent.

Promesonotal suture absent from the dorsum of the mesosoma; promesonotum flat.

Metanotal groove impressed.

Propodeum unarmed; propodeal lobe small but distinct.

Propodeal spiracle enormous, its posterior margin close to the margin of the declivity and low on the side.

Metasternal process absent.

Procoxa enormously enlarged, very much larger than mesocoxa and metacoxa.

Tibial spurs: mesotibia 0; metatibia 0.

Abdominal segment 2 (petiole) without an anteroventral process on the peduncle.

Abdominal segment 3 (postpetiole) small, in profile distinctly less voluminous than petiole (A2), not attached high on the anterior surface of A4.

Abdominal tergite 4 (first gastral) overlaps the sternite on the ventral gaster; gastral shoulders present.

Sting small and inconspicuous.

Main pilosity of dorsal head and body: simple.

AFROTROPICAL BONDROITIA

lujae (Forel, 1909)

  = *coecum* (Forel, 1911)

saharensis (Santschi, 1923)

BOTHROPONERA Mayr, 1862 — Ponerinae

22 species (+ 14 subspecies), Afrotropical; 8 species, Malagasy — PLATE 11

DIAGNOSTIC REMARKS: Among the ponerines, *Bothroponera* combines the characters of 2 spurs on the metatibia, absence of traction setae on the metabasitarsus, and location of the helcium at the base of the anterior face of the first gastral segment (A3), with the following: The mandible does not have a basal pit. The frontal lobes are large. The metanotal groove is usually absent, and the propodeal spiracle is slit-shaped. The petiole (A2) is always a node, never a scale.

DISTRIBUTION AND ECOLOGY: *Bothroponera* is predominantly a genus of the Afrotropical and Malagasy regions, but it also has some species in the Oriental region, and 1 in the Malesian. Its species nest in the ground, either directly in compacted soil, including the walls of termite nests, or under surface objects, or in rotten wood. Foraging takes place in the leaf litter and topsoil, and some species are nocturnal. Most species have alate queens, but at least 1 has lost the caste and reproduces by gamergates.

TAXONOMIC COMMENT: The name of a large and recently described large South African species, *B. umgodikulula* Joma and Mackay, 2013, was listed in Schmidt and Shattuck (2014) but does not yet appear to have been published in an available manner.

IDENTIFICATION OF SPECIES: Wheeler, W.M., 1922 (Afrotropical); Arnold, 1952 (South Africa); Rakotonirina and Fisher, 2013a (Malagasy).

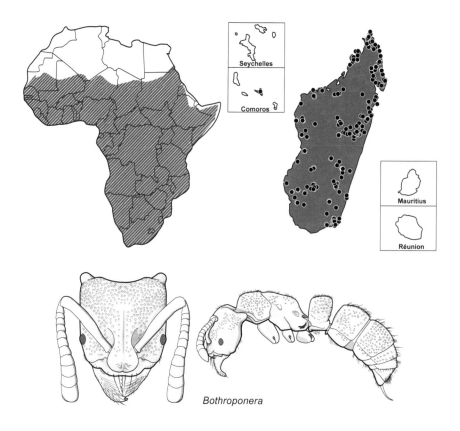

Bothroponera

DEFINITION OF THE GENUS (WORKER): Characters of subfamily PONERINAE, plus:

Mandible stoutly triangular; with or without a basal groove; without a basal pit; masticatory margin with 6–10 teeth. Basal angle angulate.

Palp formula 4,4, or 4,3.

Clypeus simple, without projecting lobes or teeth.

Frontal lobes large; in full-face view, their anterior margins well behind the anterior clypeal margin.

Eyes present, small to moderate in size, well in front of the midlength of the head capsule.

Metanotal groove usually absent, sometimes a faint and feeble vestige discernible.

Epimeral sclerite present.

Metapleural gland orifice opens laterally or posteriorly.

Propodeal spiracle with orifice slit-shaped.

Tibial spurs: mesotibia 2; metatibia 2; on each tibia the anterior spur small and simple, the posterior spur larger and pectinate.

Metatibial posterior surface in species of the *B. crassa/soror* group with a slightly depressed oval area of pale, very finely granular cuticle above the base of the spur, that appears to be glandular; feature absent outside this group.

Pretarsal claws simple or inner margin with an enhanced basal angle or a small basal tooth.

Abdominal segment 2 (petiole) a node.

Helcium located at the base of the anterior face of A3 (first gastral segment).

Prora present, varying from very small to large and conspicuous.

Abdominal segment 4 (second gastral segment) with presclerites strongly differentiated.

Stridulitrum absent or present on the pretergite of A4.

AFROTROPICAL BOTHROPONERA

cariosa Emery, 1895

cavernosa (Roger, 1860)

  cavernosa montivaga Arnold, 1947

crassa (Emery, 1877)

  crassa crassior Santschi, 1930

  crassa ilgii (Forel, 1910)

  = *crassa gamzea* (Özdikmen, 2010)

    (redundant replacement name)

cribrata (Santschi, 1910)

fugax (Forel, 1907)

granosa (Roger, 1860)

kenyensis Santschi, 1937

kruegeri (Forel, 1910)

  kruegeri asina (Santschi, 1912)

  kruegeri rhodesiana (Forel, 1913)

laevissima (Arnold, 1915)

laevissima aspera Arnold, 1962

lamottei Bernard, 1953

mlanjiensis Arnold, 1946

pachyderma (Emery, 1901)

  pachyderma attenata (Santschi, 1920)

pachyderma postsquamosa (Santschi, 1920)

pachyderma funerea Wheeler, W.M., 1922

picardi (Forel, 1901)

pumicosa (Roger, 1860)

  pumicosa sculpturata (Santschi, 1912)

rubescens Santschi, 1937

sanguinea (Santschi, 1920)

silvestrii (Santschi, 1914)

  silvestrii nimba Bernard, 1953

soror (Emery, 1899)

  soror ancilla (Emery, 1899)

  soror suturalis (Forel, 1907)

strigulosa Emery, 1895

  = *berthoudi* (Forel, 1901)

talpa André, 1890

  = *clavicornis* (Bernard, 1953)

  talpa variolata (Santschi, 1912)

variolosa Arnold, 1947

zumpti Santschi, 1937

MALAGASY BOTHROPONERA

cambouei Forel, 1891

  = *kipyatkovi* (Dubovikoff, 2013) syn. n.

comorensis (André, 1887)

masoala (Rakotonirina and Fisher, 2013)

perroti Forel, 1891

  = *perroti admista* Forel, 1892

planicornis (Rakotonirina and Fisher, 2013)

tavaratra (Rakotonirina and Fisher, 2013)

vazimba (Rakotonirina and Fisher, 2013)

wasmannii Forel, 1887

incertae sedis: jonesii (Forel, 1891)

BRACHYMYRMEX Mayr, 1868

Formicinae

1 species, Afrotropical; 1 species, Malagasy, introduced

PLATE 11

DIAGNOSTIC REMARKS: Among the formicines, *Brachymyrmex* has only 9 antennal segments, is monomorphic, does not have a tooth on the basal margin of the mandible, and has a petiole (A2) that is a small, low scale, overhung from behind by the first gastral segment (A3).

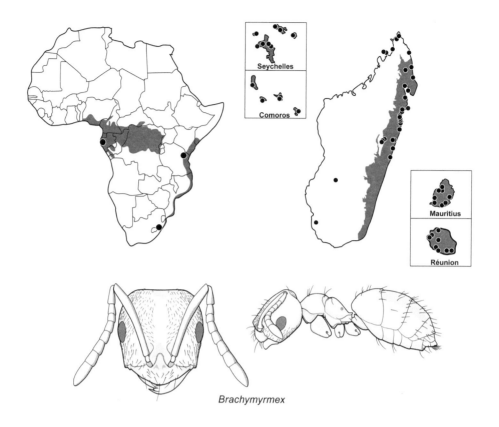

Brachymyrmex

DISTRIBUTION AND ECOLOGY: *Brachymyrmex* is a poorly understood Neotropical genus, currently with 41 extant species, and a large number of unresolved subspecies. There is at least 1 undescribed Afrotropical representative that is tentatively included in this genus. The identity of the Malagasy species is possibly *B. cordemoyi*, but this awaits confirmation.

TAXONOMIC COMMENT: The undescribed Afrotropical species may be incorrectly included in the genus. The relationships of Afrotropical species currently placed in *Agraulomyrmex*, and a number of undescribed taxa, await genus-rank resolution with respect to *Brachymyrmex*.

IDENTIFICATION OF SPECIES: No review exists.

DEFINITION OF THE GENUS (WORKER): Characters of subfamily FORMICINAE, plus:

Mandible with 5 teeth; tooth 3 (counting from the apical) is smaller than tooth 4; without a tooth on the basal margin.

Palp formula 6,4.

Frontal triangle small, depressed, its extent sharply differentiated by the clypeus anteriorly and the inner margins of the frontal carinae laterally.

Antenna with 9 segments.

Antennal sockets close to the posterior clypeal margin; anterior arc of the torulus indents the clypeal suture.

Eyes present, at or in front of the midlength of the sides of the head; ocelli present.

Mesothorax not constricted in its anterior half.

Metanotum does not form a distinct sclerite on the dorsum of the mesosoma.

Metapleural gland present.

Propodeum unarmed, with a very short dorsal surface and a long, shallowly sloped declivity; propodeal spiracle circular and near the midlength of the declivity.

Propodeal foramen long: in ventral view the foramen extends anterior of a line that spans the anteriormost points of the metacoxal cavities.

Metacoxae widely separated: in ventral view, with the mesocoxae and metacoxae directed at right-angles to the long axis of the mesosoma, the distance between the bases of the mesocoxae is markedly less than the distance between the bases of the metacoxae.

Tibial spurs: mesotibia 0; metatibia 0.

Abdominal segment 2 (petiole) a small, low scale, its dorsal surface without teeth or spines.

Petiole (A2) with a short posterior peduncle that is attached low down on A3 (first gastral segment).

Petiole (A2) with its ventral margin U-shaped in section.

Abdominal segment 3 (first gastral) with a longitudinal concavity, into which the posterior peduncle of the petiole fits when the mesosoma and gaster are aligned.

Abdominal segment 3 (first gastral) basally with complete tergosternal fusion for some distance on each side of the helcium; the line of fusion follows the edges of the impression in the anterior face of A3 and the unfused portions of the tergite and sternite commence some distance above the helcium, at the dorsolateral apices of the impression.

Abdominal sternite 3 without a transverse sulcus across the sclerite immediately behind the helcium sternite (i.e., presternite and poststernite of A3 not separated by a sulcus).

Acidopore with a seta-fringed nozzle.

AFROTROPICAL BRACHYMYRMEX

1 species, possibly incorrectly placed here

MALAGASY BRACHYMYRMEX

cordemoyi Forel, 1895

BRACHYPONERA Emery, 1900 Ponerinae

1 species (+ 2 subspecies), Afrotropical; 1 species, Malagasy (Mauritius) PLATE 11

DIAGNOSTIC REMARKS: Among the ponerines, *Brachyponera* combines the characters of 2 spurs on the metatibia, absence of traction setae on the metabasitarsus, and location of the helcium at the base of the anterior face of the first gastral segment (A3), with the following: The mandible has a basal pit present but lacks a basal groove. The metanotal groove is present and impressed, and the propodeal spiracle is not slit-shaped. The petiole (A2) is a high, unarmed scale. There is no prora at the base of the first gastral segment (A3).

DISTRIBUTION AND ECOLOGY: *Brachyponera* species occur throughout the Old World tropics. *B. sennaarensis* is extremely widespread in Africa outside the rainforest zones and

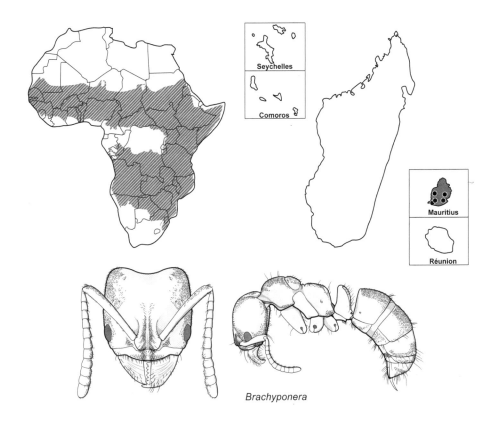

Brachyponera

can also be found along roads and in clearings within forests. It nests directly in the ground, usually in sunlit spots, and forages on the surface. It is primarily an opportunist omnivore but, remarkably for a ponerine, it will also collect and eat grass seeds. The workers of *B. sennaarensis* are size polymorphic, and the queen is much larger than her workers. The single Malagasy species, *B. obscurans*, is of Sri Lankan origin and has been found only on the Indian Ocean island of Mauritius, where it probably represents an introduction.

IDENTIFICATION OF SPECIES: No recent review exists.

DEFINITION OF THE GENUS (WORKER): Characters of subfamily PONERINAE, plus:

Mandible stoutly triangular; without a basal groove but with a basal pit close to the lateral margin; masticatory margin with 6–8 teeth. Basal angle angulate.

Palp formula 3,3.

Clypeus simple, without projecting lobes or teeth.

Frontal lobes small; in full-face view, their anterior margins well behind the anterior clypeal margin.

Eyes present, of moderate size, well in front of the midlength of the head capsule.

Metanotal groove present, distinctly impressed.

Epimeral sclerite present.

Metapleural gland orifice opens posteriorly.

Propodeal spiracle with orifice round to short-elliptical.

Tibial spurs: mesotibia 2; metatibia 2; on each tibia the anterior spur small and simple, the posterior spur larger and pectinate.

Metatibial posterior surface with a slightly depressed oval area of pale, very finely granular, cuticle above the base of the spur that appears to be glandular.

Pretarsal claws simple.

Abdominal segment 2 (petiole) a high, unarmed scale.

Petiole (A2) sternite, when detached and in posterior view, complex: in the posterior quarter of its length, the sternite bifurcates to produce a broad ventral flange that considerably overlaps, and partially conceals, the articulatory apex of the sclerite.

Helcium located at the base of the anterior face of A3 (first gastral segment).

Prora absent; in dissected specimens, the remains of the prora can be seen as a small transverse bar of cuticle that extends across between the posteroventral apices of the helcium tergite, below its true sternite.

Abdominal segment 4 (second gastral segment) with presclerites weakly differentiated.

Stridulitrum absent from the pretergite of A4.

AFROTROPICAL BRACHYPONERA

sennaarensis (Mayr, 1862)
  = *sorghi* (Roger, 1863)
  sennaarensis ruginota (Stitz, 1916)
  sennaarensis decolor (Santschi, 1921)

MALAGASY BRACHYPONERA

obscurans (Walker, 1859)

CALYPTOMYRMEX Emery, 1887 — Myrmicinae

16 species, Afrotropical; 1 species, Malagasy (Comoro Islands)   PLATE 11

DIAGNOSTIC REMARKS: Among the myrmicines, *Calyptomyrmex* shares a specialized clypeal morphology only with *Dicroaspis*. In both, the anterior face of the clypeus rises vertically from its anterior margin. At the top of the vertical surface, about at the level of the frontal lobes, the clypeus projects anteriorly as a bilobed appendage, the clypeal fork, which projects above the mandibles. Of the 2 genera with this structure, *Calyptomyrmex* has 12-segmented antennae, has the petiolar peduncle elongate and narrow in profile, and has bizarre setae on the head and body that may be spatulate, squamate, clavate, star-shaped, or very short, thick, and stubbly with abruptly tapered points; setae are never simple, fine and soft.

DISTRIBUTION AND ECOLOGY: *Calyptomyrmex* species are distributed throughout the Old World tropics and are characteristically species of the soil, humus, and leaf litter. Nests are usually constructed in the soil under logs or stones but have also been found in rotten wood and in the walls of termitaries; sometimes workers forage on bare soil on the forest floor. Workers frequently appear dirty, as they have specialized setae that hold fine soil particles close to the body as a form of camouflage.

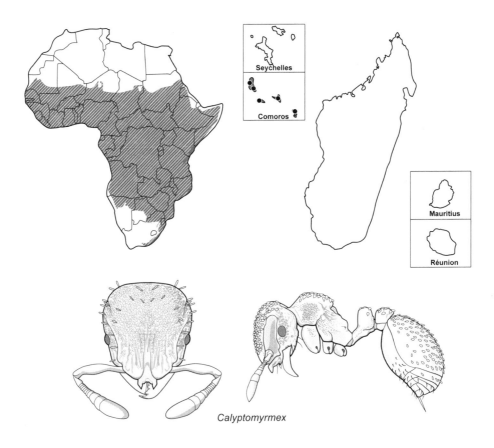

Calyptomyrmex

IDENTIFICATION OF SPECIES: Bolton, 1981a (Afrotropical).

DEFINITION OF THE GENUS (WORKER): Characters of subfamily MYRMICINAE, plus:

Mandible triangular and stoutly constructed; masticatory margin with 6–9 teeth.

Palp formula 2,2.

Stipes of the maxilla with a transverse crest.

Clypeus constricted posteriorly, narrowly and deeply inserted between the broad frontal lobes. Median portion of the clypeus with a narrow anterior apron, then near-vertical, terminating dorsally in a prominent bidentate cuticular fork that projects anteriorly from just below the level of the frontal lobes.

Clypeus without an unpaired seta at the midpoint of the anterior margin.

Frontal carinae long and form the dorsal margins of the deep scrobes. In full-face view the anterior margins of the frontal lobes (and the clypeal fork between them) overhang and conceal the anterior clypeal margin.

Antennal scrobes strongly developed, deep, extending above the eyes; scrobe capable of containing the entire antenna.

Antenna with 12 segments, with a stout apical club of 3 segments; funiculus segments 2–8 very narrow annuli, often difficult to count. Antennal scape slender basally, strongly broadened and incrassate in the apical half.

Torulus concealed by the frontal lobe in full-face view.

Eyes present, usually small, below the lower margin of the scrobe.

Promesonotal suture absent from the dorsum of the mesosoma. Promesonotum forms a single convexity in profile that is elevated above the level of the sloping propodeal dorsum.

Propodeum unarmed, or with a pair of denticles or small teeth, never with elongate spines.

Propodeal lobe thick and coarse in dorsal view, in profile with a median fenestra or thin spot.

Propodeal spiracle close to the margin of the declivity.

Metasternal process absent.

Tibial spurs: mesotibia 0; metatibia 0.

Abdominal segment 2 (petiole) with peduncle elongate and narrow in profile, without an anteroventral process; side of peduncle, near its base, with a sharp oblique carina that extends almost the height of the peduncle (surface anterior to the carina highly polished, posterior to it sculptured).

Abdominal segment 3 (postpetiole) with an anteroventral process that consists of a tooth on each side.

Abdominal tergite 4 (first gastral) does not broadly overlap the sternite on the ventral gaster; gastral shoulders absent; tergite not vaulted apically.

Sting present, simple.

Main pilosity of dorsal head and body: usually bizarre (clavate, spatulate, squamate, stellate), uncommonly all setae short and stout, simple and stubbly. Similar specialized setae usually also present on legs.

AFROTROPICAL CALYPTOMYRMEX

arnoldi (Forel, 1913)	*nedjem* Bolton, 1981
barak Bolton, 1981	*nummuliticus* Santschi, 1914
brevis Weber, 1943	= *reticulatus* Weber, 1952
brunneus Arnold, 1948	*piripilis* Santschi, 1923
clavatus Weber, 1952	= *cataractae* Arnold, 1926
claviseta (Santschi, 1914)	= *cataractae litoralis* Arnold, 1948
duhun Bolton, 1981	= *punctatus* Weber, 1952
foreli Emery, 1915	*rennefer* Bolton, 1981
= *emeryi* (Forel, 1910)	*shasu* Bolton, 1981
= *pusillus* Santschi, 1915	*stellatus* Santschi, 1915
= *arnoldi hartwigi* Arnold, 1948	*tensus* Bolton, 1981
kaurus Bolton, 1981	

MALAGASY CALYPTOMYRMEX

1 undescribed species, Comoro Islands

169 species (+ 116 subspecies), Afrotropical; 50 species (+ 33 subspecies), Malagasy   PLATE 12

DIAGNOSTIC REMARKS: Among the formicines, *Camponotus* species are dimorphic, with differentiated major and minor worker castes. Antennae with 12 segments, the sockets located well behind the posterior clypeal margin (a distance greater than the basal width of the scape). Eyes are present, located behind the midlength of the head. The metapleural gland orifice is usually absent (present in 1 small Madagascan species-group). Mandibular teeth decrease in size from apical to basal; tooth 3 is not strikingly reduced compared to tooth 4. The mesosoma and petiole are always unarmed and always lack spines or teeth, though the propodeum may be marginate. Abdominal tergite 3 (first gastral) is short, at most only slightly longer than the tergite of A4, often shorter; A3 tergite, in dorsal view or in profile, accounts for distinctly less than half the length of the gaster.

DISTRIBUTION AND ECOLOGY: *Camponotus* is one of the largest ant genera and has a worldwide distribution. Its hundreds of species occupy almost all terrestrial habitats, from harsh desert conditions to the tips of rainforest trees. Nests may be constructed under stones or directly into the earth, in and under rotten wood, in tree stumps, in hollow stems and branches, directly into standing trees, both living and dead, in rot holes in standing timber, or in termitaries, both active and abandoned. A number of species actively tunnel in sound wood, while others construct fibrous or silken nests on the undersides of leaves. Some species are entirely arboreal while others are purely nocturnal, but the majority are diurnal and terrestrial. Most species will avidly tend homopterous insects, but many are also opportunist predators and scavengers.

TAXONOMIC COMMENT: At present, *Camponotus* contains about 46 subgenera worldwide. Some of these probably represent monophyletic groups, but the majority appear artificial. A detailed taxonomic investigation of the status and validity of the subgenera is long overdue. For a synopsis of the subgenera as they currently stand, see Bolton, 2003. One species, the very widely distributed and variable *C. maculatus*, is a contender for the most oversplit species in myrmecology, as demonstrated by the long list of its regional infraspecific junior synonyms presented below.

IDENTIFICATION OF SPECIES: Robertson and Zachariades, 1997 (*C. fulvopilosus* group, Afrotropical); otherwise no review exists and the species-rank taxonomy is confused.

DEFINITION OF THE GENUS (WORKER): Characters of subfamily FORMICINAE, plus:

Worker caste dimorphic.

Mandible with 5–8 teeth that decrease in size from apical to basal; tooth 3 (counting
    from the apical) is larger than tooth 4.

Palp formula 6,4.

Frontal carinae strongly developed.

Antenna with 12 segments.

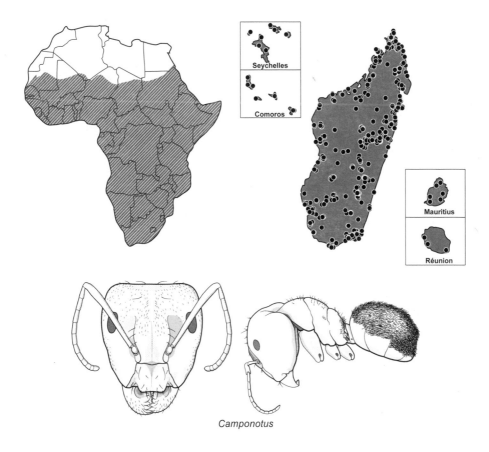

Camponotus

Antennal sockets located a considerable distance behind the posterior clypeal margin (a distance greater than the basal width of the scape).

Eyes present, located behind the midlength of the sides of the head; ocelli usually absent, but a single median ocellus may rarely occur in the largest workers of some species.

Mesosoma without spines or teeth on pronotum or propodeum; propodeum may be marginate laterally or posteriorly and sometimes overhangs the declivity posteriorly.

Metapleural gland usually absent, present in some Madagascan species.

Propodeal spiracle elliptical to slit-shaped, located posteriorly on the side and low down.

Propodeal foramen short: in ventral view the foramen does not extend anterior of a line that spans the anteriormost points of the metacoxal cavities.

Metacoxae closely approximated: in ventral view, with the mesocoxae and metacoxae directed at right-angles to the long axis of the mesosoma, the distance between the bases of the mesocoxae is about equal to the distance between the bases of the metacoxae.

Tibial spurs: mesotibia 0 or 1; metatibia 0 or 1.

Abdominal segment 2 (petiole) a node or a scale, without spines or teeth.

Petiole (A2) with its ventral margin V-shaped in section.

Abdominal segment 3 (first gastral) without tergosternal fusion on each side of the helcium.

Abdominal tergite 3 short, at most only slightly longer than the tergite of A4, often shorter; A3 tergite, in dorsal view or in profile, accounts for distinctly less than half the length of the gaster.

Abdominal sternite 3 with a transverse sulcus or mobile suture across the sclerite, immediately behind the helcium sternite (i.e., presternite and poststernite of A3 separated by a sulcus).

Acidopore normally concealed by pygidium when not in use, usually without a seta-fringed nozzle.

AFROTROPICAL CAMPONOTUS

abjectus Santschi, 1937

acvapimensis Mayr, 1862
  = *mombassae* Forel, 1886
  = *acwapimensis poultoni* Forel, 1913
  = *flavosetosus* Donisthorpe, 1945

aegyptiacus Emery, 1915
  = *maculatus infrasquameus* Santschi, 1926

aequatorialis Roger, 1863
  aequatorialis kohli Forel, 1915

aequitas Santschi, 1920

agonius Santschi, 1915

amphidus Santschi, 1926

angusticeps Emery, 1886

argus Santschi, 1935

arminius Forel, 1910
  arminius bicontractus Wheeler, W.M., 1922

arnoldinus Forel, 1914

atriscapus Santschi, 1926

aurofasciatus Santschi, 1915

auropubens Forel, 1894
  auropubens absalon Santschi, 1915
  auropubens argentopubens Santschi, 1915
  auropubens jacob Santschi, 1915

avius Santschi, 1926
  avius hertigi Santschi, 1937

barbarossa Emery, 1920
  barbarossa micipsa Wheeler, W.M., 1922
  barbarossa sulcatinasis Santschi, 1926

basuto Arnold, 1958

bayeri Forel, 1913

baynei Arnold, 1922

belligerus Santschi, 1920

benguelensis Santschi, 1911

berthoudi Forel, 1879

bertolonii Emery, 1895

bianconii Emery, 1895

bifossus Santschi, 1917

bituberculatus André, 1889

bottegoi Emery, 1895

braunsi Mayr, 1895
  braunsi erythromelus Emery, 1896
  braunsi epinotalis Santschi, 1916
  braunsi candidus Santschi, 1926

brevicollis Stitz, 1916

brevisetosus Forel, 1910

brookei Forel, 1914

brutus Forel, 1886

buchholzi Mayr, 1902

burgeoni Santschi, 1926

buttikeri Arnold, 1958

caesar Forel, 1886
  caesar imperator Emery, 1899

caffer Emery, 1895

callmorphus Stitz, 1923

carbo Emery, 1877
  carbo occidentalis Mayr, 1902

chapini Wheeler, W.M., 1922
  chapini ganzii Weber, 1943

chrysurus Gerstäcker, 1871
  chrysurus acutisquamis Mayr, 1902
  chrysurus securifer Emery, 1920
  chrysurus yvonnae Forel, 1920

chrysurus apellis Wheeler, W.M., 1922
cinctellus (Gerstäcker, 1859)
  = *venustus* Mayr, 1867
  cinctellus belliceps Santschi, 1939
cleobulus Santschi, 1919
cognatocompressus Forel, 1904
compressiscapus André, 1889
compressus (Fabricius, 1787)
  compressus probativus Santschi, 1921
confluens Forel, 1913
  confluens bequaerti Forel, 1913
  confluens trematogaster Santschi, 1915
congolensis Emery, 1899
  congolensis weissi Santschi, 1911
coniceps Santschi, 1926
conradti Forel, 1914
  conradti fimbriatipes Santschi, 1920
cosmicus (Smith, F., 1858)
crawleyi Emery, 1920
crepusculi Arnold, 1922
criniticeps Menozzi, 1939
crucheti Santschi, 1911
cubangensis Forel, 1901
  cubangensis dofleini Forel, 1911
  = *mayri ledieui* Forel, 1916
cuneiscapus Forel, 1910
debellator Santschi, 1926
desantii Santschi, 1915
detritus Emery, 1886
dewitzii Forel, 1886
dicksoni Arnold, 1948
diplopunctatus Emery, 1915
  diplopunctatus subconvexus
    Viehmeyer, 1923
donisthorpei Emery, 1920
druryi Forel, 1886
emarginatus Emery, 1886
empedocles Emery, 1920
erinaceus Gerstäcker, 1871
errabundus Arnold, 1949
etiolipes Bolton, 1995
  = *longipes* (Gerstäcker, 1859)
    (homonym)

eugeniae Forel, 1879
  eugeniae amplior Forel, 1913
favorabilis Santschi, 1919
ferreri Forel, 1913
  ferreri akka Forel, 1916
  ferreri cavisquamis Santschi, 1926
flavomarginatus Mayr, 1862
  = *micans albisectus* Emery, 1892
florius Santschi, 1926
foraminosus Forel, 1879
  foraminosus chrysogaster Emery, 1895
  foraminosus cuitensis Forel, 1901
  foraminosus honorus Forel, 1910
  foraminosus deductus Santschi, 1915
  foraminosus flavus Stitz, 1916
  foraminosus dorsalis Santschi, 1926
fornasinii Emery, 1895
fulvopilosus (De Geer, 1778)
  = *pilosa* (Olivier, 1792)
  = *rufiventris* (Fabricius, 1804)
  = *fulvopilosus flavopilosus* Emery, 1895
  = *fulvopilosus detritoides* Forel, 1910
furvus Santschi, 1911
gabonensis Santschi, 1926
galla Forel, 1894
  = *foraminosus latinotus* Forel, 1907
grandidieri Forel, 1886
  grandidieri ruspolii Forel, 1892
  grandidieri comorensis Santschi, 1915
  grandidieri eumendax Özdikmen, 2010
  = *foraminosus mendax* Emery, 1895
    (homonym)
guttatus Emery, 1899
  guttatus minusculus Viehmeyer, 1914
haereticus Santschi, 1914
hapi Weber, 1943
havilandi Arnold, 1922
heros Santschi, 1926
hova Forel, 1891
  hova pictiventris Mayr, 1901
ilgii Forel, 1894
immigrans Santschi, 1914
importunus Forel, 1911

jeanneli Santschi, 1914

kersteni Gerstäcker, 1871

klugii Emery, 1895

knysnae Arnold, 1922

kollbrunneri Forel, 1910

lamborni Donisthorpe, 1933

langi Wheeler, W.M., 1922

  langi jejunus Santschi, 1926

ligeus Donisthorpe, 1931

lilianae Forel, 1913

  lilianae cornutus Forel, 1913

limbiventris Santschi, 1911

liogaster Santschi, 1932

longipalpis Santschi, 1926

maculatus (Fabricius, 1782)

  = *cognata* (Smith, F., 1858)

  = *lacteipennis* (Smith, F., 1858)

  = *sexpunctatus liengmei* Forel, 1894

  = *maculatus atramentarius* Forel, 1904

  = *maculatus ballioni* Forel, 1904

  = *maculatus intonsus* Emery, 1905

  = *maculatus liocnemis* Emery, 1905

  = *maculatus manzer* Forel, 1910

  = *maculatus mathildae* Forel, 1910

  = *maculatus schereri* Forel, 1911

  = *maculatus cavallus* Santschi, 1911

  = *maculatus melanocnemis* Santschi, 1911

  = *maculatus schultzei* Forel, 1912

  = *maculatus cluisoides* Forel, 1913

  = *maculatus miserabilis* Santschi, 1914

  = *maculatus lohieri* Emery, 1915

  = *maculatus hieroglyphicus* Santschi, 1917

  = *maculatus hannae* Santschi, 1919

  = *maculatus conakryensis* Emery, 1920

  = *maculatus erythraea* Emery, 1920

  = *maculatus sarmentus* Emery, 1920

  = *maculatus semispicatus* Emery, 1920

  = *maculatus thomensis* Santschi, 1920

  = *maculatus flavominor* Emery, 1925

  = *ballioni boera* Santschi, 1925

  = *maculatus tuckeri* Santschi, 1932

  = *maculatus flavifemur* Santschi, 1937

  = *maculatus zumpti* Santschi, 1937

  = *maculatus nubis* Weber, 1943

  = *maculatus sudanicus* Weber, 1943

  = *maculatus proletaria* Baroni Urbani, 1971

  maculatus ugandensis Santschi, 1923

maguassa Wheeler, W.M., 1922

massinissa Wheeler, W.M., 1922

maynei Forel, 1916

mayri Forel, 1879

  mayri chimporensis Santschi, 1930

moderatus Santschi, 1930

monardi Santschi, 1930

mystaceus Emery, 1886

  mystaceus exsanguis Forel, 1910

  mystaceus kamae Forel, 1910

namacola Prins, 1973

nasutus Emery, 1895

  nasutus quinquedentatus Forel, 1910

  nasutus pretiosus Arnold, 1922

  nasutus subnasutus Arnold, 1922

  nasutus fenestralis Santschi, 1937

  nasutus quadridentatus Santschi, 1937

natalensis (Smith, F., 1858)

  natalensis corvus Forel, 1879

  natalensis diabolus Forel, 1879

  natalensis fulvipes Forel, 1914

  natalensis politiceps Santschi, 1914

niveosetosus Mayr, 1862

  niveosetosus irredux Forel, 1910

oculatior Santschi, 1935

olivieri Forel, 1886

  olivieri lemma Forel, 1886

  olivieri delagoensis Forel, 1894

  olivieri osiris Forel, 1911

  olivieri concordius Santschi, 1915

  olivieri infelix Santschi, 1915

  olivieri moshianus Santschi, 1915

  olivieri sorptus Santschi, 1915

  olivieri tenuipilis Santschi, 1915

  olivieri pax Menozzi, 1942

  olivieri patersoni Arnold, 1959

  olivieri zaireicus Özdikmen, 2010

= *olivieri nitidior* Santschi, 1926
(homonym)
orinobates Santschi, 1919
orites Santschi, 1919
= *orinodromus* Santschi, 1919
orthodoxus Santschi, 1914
ostiarius Forel, 1914
perrisii Forel, 1886
perrisii jucundus Santschi, 1911
perrisii nigeriensis Santschi, 1914
perrisii densipunctatus Stitz, 1916
perrisii insularis Stitz, 1916
= *foraminosus annobonensis* Santschi,
1920
petersii Emery, 1895
petersii janus Forel, 1911
pompeius Forel, 1886
pompeius marius Emery, 1899
pompeius cassius Wheeler, W.M., 1922
pompeius iota Santschi, 1926
posticus Santschi, 1926
postoculatus Forel, 1914
prosulcatus Santschi, 1935
puberulus Emery, 1897
pulvinatus Mayr, 1907
reevei Arnold, 1922
rhamses Santschi, 1915
rhamses completus Santschi, 1930
robecchii Emery, 1892
robecchii troglodytes Forel, 1894
robecchii rhodesianus Forel, 1913
robecchii var. abyssinicus Santschi, 1913
robecchii dispar Stitz, 1923
robertae Santschi, 1926
rotundinodis Santschi, 1935
roubaudi Santschi, 1911
rubripes (Latreille, 1802)
rufoglaucus (Jerdon, 1851)
= *redtenbacheri* Mayr, 1862
rufoglaucus zulu Emery, 1895
rufoglaucus zanzibaricus Forel, 1911
rufoglaucus controversus Santschi, 1916
= *rufoglaucus flavopilosus* Viehmeyer,
1913 (homonym)

rufoglaucus syphax Wheeler, W.M., 1922
rufoglaucus latericius Stitz, 1923
sacchii Emery, 1899
sankisianus Forel, 1913
scabrinodis Arnold, 1924
scalaris Forel, 1901
schoutedeni Forel, 1911
sellidorsatus Prins, 1973
sericeus (Fabricius, 1798)
= *aurulenta* (Latreille, 1802)
sericeus sulgeri Santschi, 1913
sericeus euchrous Santschi, 1926
sexpunctatus Forel, 1894
simulans Forel, 1910
simus Emery, 1908
simus manidis Forel, 1909
solon Forel, 1886
solon chiton Emery, 1925
solon jugurtha Emery, 1925
somalinus André, 1887
somalinus curtior Forel, 1894
somalinus pattensis Forel, 1907
storeatus Forel, 1910
tameri Bolton, 1995
= *tricolor* Weber, 1943 (homonym)
tauricollis Forel, 1894
thales Forel, 1910
thraso Forel, 1893
thraso negus Forel, 1907
thraso agricola Forel, 1910
thraso assabensis Emery, 1925
thraso montinanus Santschi, 1926
thraso nefasitensis Menozzi, 1931
tilhoi Santschi, 1926
traegaordhi Santschi, 1914
traegaordhi fumeus Santschi, 1926
transvaalensis Arnold, 1948
transvaalensis griqua Arnold, 1952
transvaalensis prinsi Özdikmen, 2010
= *transvaalensis arnoldi* Prins, 1965
(homonym)
trifasciatus Santschi, 1926
valdeziae Forel, 1879
varus Forel, 1910

vespertinus Arnold, 1960
vestitus (Smith, F., 1858)
  vestitus comptus Santschi, 1926
  vestitus intuens Santschi, 1926
  vestitus perpectitus Santschi, 1926
  vestitus strophiatus Santschi, 1926
  vestitus anthracinus Santschi, 1930
  vestitus bombycinus Santschi, 1930
  vestitus lujai Santschi, 1930
  vestitus pectitus Santschi, 1930
victoriae Arnold, 1959
viri Santschi, 1915
vividus (Smith, F., 1858)
  = *laboriosa* (Smith, F., 1858)

vividus meinerti Forel, 1886
vividus reginae Forel, 1901
vividus cato Forel, 1913
  = *vividus laevithorax* Menozzi, 1924
vividus semidepilis Wheeler, W.M.,
  1922
vulpus Santschi, 1926
wellmani Forel, 1909
  wellmani rufipartis Forel, 1916
  wellmani gamma Santschi, 1926
werthi Forel, 1908
  werthi skaifei Arnold, 1959
zimmermanni Forel, 1894
  zimmermanni pansus Santschi, 1926

MALAGASY CAMPONOTUS

aurosus Roger, 1863
batesii Forel, 1895
butteli Forel, 1905
cambouei Forel, 1891
cervicalis Roger, 1863
  cervicalis gaullei Santschi, 1911
christi Forel, 1886
  christi foersteri Forel, 1886
  christi ambustus Forel, 1892
  christi maculiventris Emery, 1895
  christi ferruginea Emery, 1899
concolor Forel, 1891
  = *gallienii* Forel, 1916
darwinii Forel, 1886
  darwinii rubropilosus Forel, 1891
  darwinii themistocles Forel, 1910
  descarpentriesi Santschi, 1926
dromedarius Forel, 1891
  dromedarius pulcher Forel, 1892
dufouri Forel, 1891
  dufouri imerinensis Forel, 1891
echinoploides Forel, 1891
edmondi André, 1887
  edmondi ernesti Forel, 1891
ellioti Forel, 1891
  ellioti relucens Santschi, 1911
ethicus Forel, 1897
gerberti Donisthorpe, 1949

gibber Forel, 1891
gouldi Forel, 1886
grandidieri Forel, 1886
  grandidieri atrabilis Santschi, 1915
hagensii Forel, 1886
heteroclitus Forel, 1895
hildebrandti Forel, 1886
  hildebrandti dichromothrix Emery, 1920
hova Forel, 1891
  hova boivini Forel, 1891
  hova radamae Forel, 1891
  hova hovahovoides Forel, 1892
  hova hovoides Dalla Torre, 1893
  hova mixtellus Dalla Torre, 1893
  hova fulvus Emery, 1894
  hova fairmairei Santschi, 1911
  hova becki Santschi, 1923
  hova obscuratus Emery, 1925
  hova cemeryi Özdikmen, 2010
  = *hova luteolus* Emery, 1925 (homonym)
imitator Forel, 1891
  imitator resinicola Santschi, 1911
immaculatus Forel, 1892
kelleri Forel, 1886
  kelleri invalidus Forel, 1897
legionarium Santschi, 1911
leveillei Emery, 1895
lubbocki Forel, 1886

lubbocki christoides Forel, 1891
lubbocki rectus Forel, 1891
maculatus (Fabricius, 1782)
  = maculatus radamoides Forel, 1891
  = maculatus lividior Santschi, 1911
mocquerysi Emery, 1899
nasicus Forel, 1891
niveosetosus Mayr, 1862
  niveosetosus madagascarensis Forel,
    1886
nosibeensis André, 1887
olivieri Forel, 1886
  olivieri freyeri Santschi, 1915
perroti Forel, 1897
  perroti aeschylus Forel, 1913
pictipes Forel, 1891
putatus Forel, 1892
  quadrimaculatus Forel, 1886
  quadrimaculatus sellaris Emery, 1895

quadrimaculatus opacatus Emery,
  1925
= quadrimaculatus opaca Emery,
  1899 (homonym)
radovae Forel, 1886
  radovae radovaedarwinii Forel, 1891
reaumuri Forel, 1892
repens Forel, 1897
  = madagascarensis (Forel, 1886)
    (homonym)
robustus Roger, 1863
roeseli Forel, 1910
sibreei Forel, 1891
sikorae Emery, 1920
strangulatus Santschi, 1911
  = hova maculatoides Emery, 1920
thomasseti Forel, 1912
ursus Forel, 1886
voeltzkowii Forel, 1894

CARDIOCONDYLA Emery, 1869

Myrmicinae

14 species, Afrotropical; 5 species (+ 1 subspecies), Malagasy

PLATE 12

DIAGNOSTIC REMARKS: Among the myrmicines, *Cardiocondyla* has a unique clypeal form. The lateral portions of the clypeus are dorsoventrally flattened and project forward. They are fused to the raised, prominent median portion of the clypeus and together form a shelf that projects over the mandibles (in some, the lateral portions project further forward than the median portion). Seen in profile, this shelf is elevated far above the dorsal surface of the mandible. In addition, the clypeus posteriorly is broadly inserted between small, narrow frontal lobes, the antennae have 12 segments, and the clypeus has an unpaired seta at the midpoint of the anterior margin. Eyes are present, located in front of the midlength of the head capsule, and the postpetiole (A3) is broad in dorsal view and dorsoventrally flattened.

DISTRIBUTION AND ECOLOGY: *Cardiocondyla* species are numerous and widely distributed in the Old World, but New World representation is sparse and consists almost entirely of introduced tramp species, of which the genus has several. Two of the more accomplished tramp species, *C. emeryi* and *C. wroughtonii*, occur fairly commonly in the regions under consideration here, together with a number of endemic species. Most species are small and inconspicuous, found nesting in the ground, often at the bases of trees, in compressed leaf litter, or in fallen twigs. Foraging is usually terrestrial, but workers will also ascend trees to tend homopterous insects. The males of *Cardiocondyla* species may be ergatoid or dimorphic, with a mixture of alate and ergatoid forms.

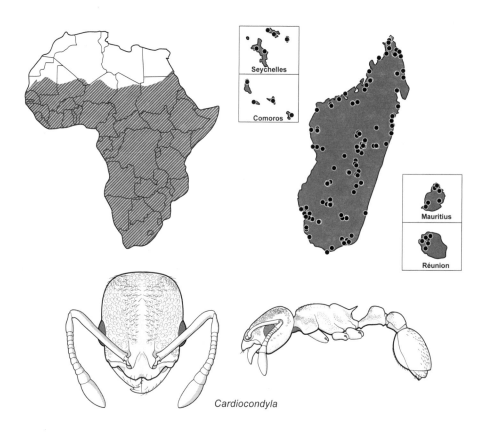

Cardiocondyla

IDENTIFICATION OF SPECIES: Bolton, 1982 (Afrotropical); Rigato, 2002 (Afrotropical); Seifert, 2003 (Afrotropical).

DEFINITION OF THE GENUS (WORKER): Characters of subfamily MYRMICINAE, plus:

Mandible short triangular; masticatory margin with 5 teeth.

Palp formula 5,3.

Stipes of the maxilla with a transverse crest.

Clypeus posteriorly broadly inserted between the small, narrow frontal lobes. Clypeus with lateral portions flattened and projecting, fused to the raised, prominent median portion to form a shelf that projects out over the mandibles (lateral portions may project further forward than median portion). In profile, this clypeal shelf elevated considerably away from the dorsal surface of the mandible.

Clypeus with an unpaired seta at the midpoint of the anterior margin.

Frontal carinae restricted to small frontal lobes.

Antennal scrobes absent.

Antenna with 12 segments, with an apical club of 2 or 3 segments.

Torulus with upper lobe concealed by the frontal lobe in full-face view.

Eyes present, usually large and conspicuous, located in front of the midlength of the head capsule in full-face view.

Promesonotal suture absent from the dorsum of the mesosoma.

Propodeum unarmed to bispinose.

Propodeal spiracle small, approximately at the midlength of the sclerite and often low down on the side.

Metasternal process absent.

Tibial spurs: mesotibia 0; metatibia 0.

Abdominal segment 3 (postpetiole) dorsoventrally flattened in profile; in dorsal view very broad.

Abdominal tergite 4 (first gastral) overlaps the sternite on the ventral gaster; gastral shoulders present.

Sting strongly developed.

Main pilosity of dorsal head and body: usually absent.

AFROTROPICAL CARDIOCONDYLA

emeryi Forel, 1881	= *fusca* Weber, 1952
= *emeryi mahdii* Karavaiev, 1911	= *wassmanni* Santschi, 1926
fajumensis Forel, 1913	= *wasmanni sculptior* Santschi, 1926
= *nilotica* Weber, 1952	*venustula* Wheeler, W.M., 1908
longinoda Rigato, 2002	= *badonei* Arnold, 1926
luciae Rigato, 2002	= *globinodis* Stitz, 1923
monardi Santschi, 1930	*weserka* Bolton, 1982
neferka Bolton, 1982	*wroughtonii* (Forel, 1890)
obscurior Wheeler, W.M., 1929	= *emeryi chlorotica* Menozzi, 1930
sekhemka Bolton, 1982	*yoruba* Rigato, 2002
shuckardi Forel, 1891	*zoserka* Bolton, 1982
= *brevispinosa* Weber, 1952	

MALAGASY CARDIOCONDYLA

cristata (Santschi, 1912)	*nuda sculptinodis* Santschi, 1913
emeryi (Forel, 1881)	*shuckardi* Forel, 1891
= *emeryi rasalamae* Forel, 1891	= *nuda shuckardoides* Forel, 1895
= *mauritia* Donisthorpe, 1946	*wroughtonii* (Forel, 1890)
nuda (Mayr, 1866)	

CAREBARA Westwood, 1840 — Myrmicinae

62 species (+ 11 subspecies), Afrotropical; 3 species, Malagasy — PLATE 12

DIAGNOSTIC REMARKS: Among the myrmicines, *Carebara* species have 8–11 segmented antennae that always terminate in a strong club of 2 segments (in some dimorphic species the major workers have 1 more antennal segment than the minors). The majority of species have dimorphic workers, but some are monomorphic and some polymorphic. The anterior clypeal margin has a pair of median setae, 1 on each side of the midpoint but none at the midpoint itself. The propodeum is usually angulate or armed with teeth or spines; less commonly it may be unarmed and rounded. The head usually lacks frontal carinae behind the frontal lobes and lacks antennal scrobes, but short scrobes are present in the major workers of 2 species.

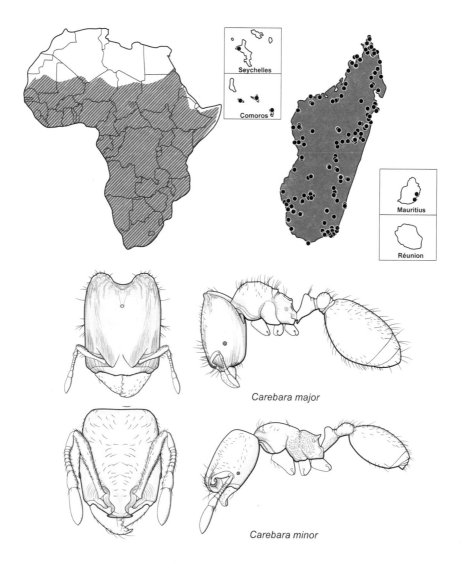

Carebara major

Carebara minor

DISTRIBUTION AND ECOLOGY: This large genus occurs throughout the tropics and sub-tropics of the world. The vast majority of species are minute and cryptic, nesting and foraging in the soil, in leaf litter, or in rotten wood, but some have been found in trees, either on the branches or in epiphytic mosses. In contrast, some of the larger, polymorphic species forage in swarms on the surface of the ground. A number of species can often be found using the compacted soil of termitary walls as a nest site. One such group of minute, monomorphic, entirely hypogaeic species appear always to nest in the walls of subterranean termitaries; this group may prey entirely on the termites themselves. Queens of *Carebara* species tend to be much larger than their workers, sometimes enormously so.

TAXONOMIC COMMENT: *Carebara* has a large number of genus-group names as synonyms. Most of the references that follow refer to various groups of *Carebara* as they were understood at the time of publication. A full list of current synonyms is given in the introduction.

IDENTIFICATION OF SPECIES: Santschi, 1913b (Afrotropical *Oligomyrmex*); Santschi, 1914 (Afrotropical *Carebara*); Wheeler, W.M., 1922 (Afrotropical *Carebara*); Weber, 1950 (Afrotropical *Oligomyrmex*); Weber, 1952 (Afrotropical *Aeromyrma*); Bolton and Belshaw, 1993 (*Paedalgus*); Fischer, Azorsa, and Fisher, 2014 (Afrotropical *C. polita* group); no full review exists, and the species-rank taxonomy is otherwise confused.

DEFINITION OF THE GENUS (WORKER): Characters of subfamily MYRMICINAE, plus:

Worker caste often strikingly dimorphic, with major and minor workers, the major workers always with disproportionately enlarged heads; a number of species monomorphic (apparently by loss of the major caste), a few species polymorphic, with all intermediate sizes between the smallest (minima) and largest (maxima) workers.

Mandible triangular; masticatory margin with 3–7 teeth (margin rarely edentate in some maxima workers of polymorphic species).

Palp formula 2,2, or 1,2.

Stipes of the maxilla without a transverse crest, except in some maxima workers of polymorphic species, where an oblique partial crest may be present.

Clypeus posteriorly narrowly to moderately broadly inserted between the frontal lobes. Median portion of the clypeus usually longitudinally bicarinate dorsally (carinae reduced, or even lost, in a very few species); each carina usually terminates in an angle or denticle at the anterior margin.

Clypeus without an unpaired seta at the midpoint of the anterior margin; usually a pair of median setae present, 1 on each side of the midpoint.

Frontal carinae almost always restricted to the small frontal lobes, extended posteriorly only in a few species.

Antennal scrobes almost universally absent; present only in the major workers of 2 species, which have extended frontal carinae and phragmotic heads.

Antenna with 8–11 segments, with a conspicuous apical club of 2 segments; in some dimorphic species the majors with 1 antennal segment more than the minors.

Torulus with upper lobe visible in full-face view except in some majors; maximum width of torulus lobe is posterior to the point of maximum width of the frontal lobes.

Eyes absent or present; when present, frequently small or very small and usually located in front of the midlength of the head capsule, more rarely at about the midlength; in some polymorphic species the eyes are progressively farther behind the midlength as worker size diminishes. One or more ocelli may be developed in major workers.

Promesonotal suture absent from the dorsum of the mesosoma in minors, frequently discernible in majors; in polymorphic species the definition of the suture usually increases with increased size. In dimorphic and polymorphic species the sclerites of the mesosoma also increase, so that in the largest workers the mesosoma may be gyne-like. Metanotal groove usually present and impressed but frequently reduced and not impressed.

Propodeum unarmed to bispinose.

Propodeal spiracle at, or behind, the midlength of the sclerite; sometimes close to the margin of the declivity.

Metasternal process absent.

Tibial spurs: mesotibia 0 or 1; metatibia 0 or 1.

Abdominal segment 4 (first gastral) tergite overlaps the sternite on the ventral gaster; gastral shoulders present.

Sting strongly developed to weak.

Main pilosity of dorsal head and body: simple; sometimes sparse, may be absent in places.

AFROTROPICAL CAREBARA

aberrans (Santschi, 1937)

acuta (Weber, 1952)

africana (Forel, 1910)

alluaudi (Santschi, 1913)
  = kenyensis (Ettershank, 1966)
  alluaudi cataractae (Santschi, 1919)

ampla Santschi, 1912
  ampla cincta Santschi, 1926
  ampla obscurithorax Santschi, 1926
  ampla rugosa Santschi, 1928

angolensis (Santschi, 1914)
  angolensis congolensis (Forel, 1916)

arnoldi (Forel, 1913)

arnoldiella (Santschi, 1919)

bartrumi Weber, 1943

convexa (Weber, 1950)

crigensis (Belshaw and Bolton, 1994)

debilis (Santschi, 1913)

diabolica (Baroni Urbani, 1969)

diabola (Santschi, 1913)

distincta (Bolton and Belshaw, 1993)

diversa (Jerdon, 1851)
  diversa standfussi (Forel, 1911)

donisthorpei (Weber, 1950)

elmenteitae (Patrizi, 1948)

erythraea (Emery, 1915)

frontalis (Weber, 1950)

guineana Fernández, 2006
  = silvestrii Santschi, 1914 (homonym)

hammoniae (Stitz, 1923)

hostilis (Smith, F., 1858)

incerta (Santschi, 1919)

infima (Santschi, 1913)

jeanneli (Santschi, 1913)

junodi Forel, 1904

khamiensis (Arnold, 1952)

kunensis (Ettershank, 1966)

= arnoldi (Santschi, 1928) (homonym)

langi Wheeler, W.M., 1922

latro (Santschi, 1937)

lucida (Santschi, 1917)

mayri (Santschi, 1928)

menozzii (Ettershank, 1966)
  = eidmanni (Menozzi, 1942) (homonym)

nana (Santschi, 1919)

octata (Bolton and Belshaw, 1993)

osborni Wheeler, W.M., 1922

paeta (Santschi, 1937)

patrizii Menozzi, 1927

perpusilla (Emery, 1895)
  perpusilla spinosa (Forel, 1907)
  perpusilla concedens (Santschi, 1914)
  perpusilla arnoldiana (Ettershank, 1966)
    = perpusillum arnoldi (Forel, 1914) (homonym)

petulca (Wheeler, W.M., 1922)

pisinna (Bolton and Belshaw, 1993)

polita (Santschi, 1914)
  polita nicotianae (Arnold, 1948)

punctata (Karavaiev, 1931)

rara (Bolton and Belshaw, 1993)

robertsoni (Bolton and Belshaw, 1993)

santschii (Weber, 1943)

sarita (Bolton and Belshaw, 1993)

semilaevis (Mayr, 1901)
  = hewitti (Santschi, 1919)

sicheli Mayr, 1862

silvestrii (Santschi, 1914)

solitaria (Stitz, 1910)

sudanensis (Weber, 1943)

sudanica Santschi, 1933

termitolestes (Wheeler, W.M., 1918)

thoracica (Weber, 1950)
traegaordhi (Santschi, 1914)
ugandana (Santschi, 1923)
vidua Smith, F., 1858
  = dux Smith, F., 1858
  = colossus Gerstäcker, 1859

= vidua abdominalis Santschi, 1912
  vidua fur Santschi, 1928
villiersi (Bernard, 1953)
volsellata (Santschi, 1937)
vorax (Santschi, 1914)

MALAGASY CAREBARA

grandidieri (Forel, 1891)
nosindambo (Forel, 1891)
voeltzkowi (Forel, 1907)

CATAGLYPHIS Foerster, 1850 — Formicinae

4 species (+ 4 subspecies), Afrotropical — PLATE 13

DIAGNOSTIC REMARKS: Among the Afrotropical formicines, *Cataglyphis* is easily isolated by its character combination of 12-segmented antennae, eyes that are well behind the midlength of the head, presence of ocelli, presence of a psammophore on the ventral head, and the possession of a large, elongate propodeal spiracle, the orifice of which is a near-vertical ellipse or slit.

DISTRIBUTION AND ECOLOGY: Primarily a genus of the Palaearctic region, where many species are present, the few Afrotropical forms represent a limited radiation from north of the Sahara. They are mostly restricted to the Sahelian zone, across the width of the continent just south of the great desert. At least 1 species occurs coastally from Ghana to western Nigeria, having colonized from the north via the forest gap in Benin. There are no records from the Malagasy region. All species are ground nesters and foragers, capable of considerable speed. They usually forage on the surface of the ground, seeking dead or heat-stressed arthropods, but honeydew may also form part of their diet.

IDENTIFICATION OF SPECIES: Santschi, 1929 (includes several Afrotropical forms, but some of the names are treated as infrasubspecific and therefore unavailable); otherwise no review exists.

DEFINITION OF THE GENUS (WORKER): Characters of subfamily FORMICINAE, plus:

Mandible with 5–7 teeth; tooth 3 (counting from the apical) is larger than tooth 4.
Palp formula 6,4; basal segment of the maxillary palp flat.
Frontal carinae strongly developed.
Antenna with 12 segments.
Antennal sockets close to or abutting the posterior clypeal margin.
Eyes present, located behind the midlength of the sides of the head; ocelli present.
Head anteroventrally usually with a distinct psammophore present.
Metapleural gland present.
Propodeal spiracle elliptical to slit-shaped, high on the side.

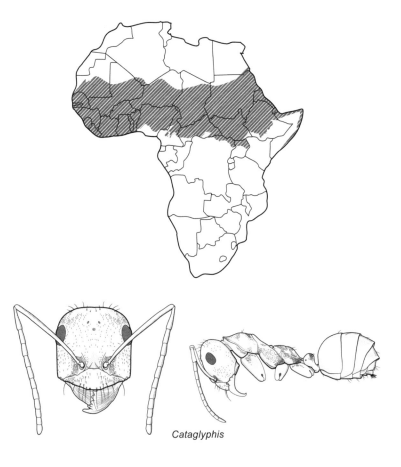

Cataglyphis

Propodeal foramen short: in ventral view the foramen does not extend anterior of a line
that spans the anteriormost points of the metacoxal cavities.

Metacoxae closely approximated: in ventral view, with the mesocoxae and metacoxae
directed at right-angles to the long axis of the mesosoma, the distance between
the bases of the mesocoxae is about equal to the distance between the bases of the
metacoxae.

Tibial spurs: mesotibia 1; metatibia 1.

Metatibia ventrally with a double longitudinal row of short, stout setae.

Abdominal segment 2 (petiole) a node or scale, without spines or teeth.

Petiole (A2) with its ventral margin V-shaped in section.

Abdominal segment 3 (first gastral) without tergosternal fusion on each side of the
helcium.

Abdominal sternite 3 with a transverse sulcus across the sclerite immediately behind
the helcium sternite (i.e., the presternite and poststernite of A3 separated by a
sulcus).

Acidopore with a seta-fringed nozzle.

abyssinica (Forel, 1904)

albicans (Roger, 1859)

   albicans franchettii Menozzi, 1931

aurata Menozzi, 1932

bicolor (Fabricius, 1793)

   bicolor seticornis (Emery, 1906)

   bicolor sudanica (Karavaiev, 1912)

   bicolor congolensis (Stitz, 1916)

CATAULACUS Smith, F., 1853   Myrmicinae

39 species, Afrotropical; 8 species, Malagasy   PLATE 13

DIAGNOSTIC REMARKS: Among the myrmicines, *Cataulacus* is very conspicuous. It has antennae with 11 segments that terminate in an apical club of 3 segments. Antennal scrobes are present that extend below the large eyes. The antennal insertions are extremely widely separated so that in full-face view the outer margins of the frontal lobes form the visible lateral margins of the head in front of the eyes. The dorsum of the gaster consists entirely of the expanded first tergite (the tergite of A4); the remaining tergites are visible in profile, below the posterior margin of the first. Bizarre pilosity is frequently present and all known species are mostly or entirely black.

DISTRIBUTION AND ECOLOGY: Present throughout the Old World tropics except for New Guinea and Australia, all species of *Cataulacus* are exclusively arboreal. Nests are usually made in rotten or hollow twigs or stems but are sometimes constructed in large rotten branches, or in rotten cavities in trunks. Some of the smaller species nest in shallow rot-holes that may be concealed by superficial debris or moss, or under patches of rotten bark. Several species nest in galls and acacia thorns; a number are associated with myrmecophytes; and occasionally some species utilize the walls of termitaries. Foraging usually takes place upon trees or shrubs where the nests are situated, but some species will cross the ground and ascend adjacent plants. Homopterous insects are tended for sugars, but some are also predaceous, with 1 species recorded as breaking into termite tunnels to attack the colony. The males of *Cataulacus* species are alate, but their appearance is the most worker-like of any ant genus.

IDENTIFICATION OF SPECIES: Bolton, 1974a (Afrotropical + Malagasy); Bolton, 1982 (Afrotropical).

DEFINITION OF THE GENUS (WORKER): Characters of subfamily MYRMICINAE, plus:

Worker caste monomorphic in most species, some with very size-variable workers.

Mandible short and thick, short-triangular, or with subparallel basal and outer margins. Masticatory margin with 1–3 teeth apically, followed by a denticulate or edentate margin.

Palp formula 5,3.

Stipes of the maxilla with a transverse crest close to the apex.

Clypeus large, posteriorly broadly inserted between the frontal lobes.

Clypeus without an unpaired seta at the midpoint of the anterior margin.

Frontal carinae (and antennal sockets) very widely separated; frontal carinae overhang the sides of the head in front of the eyes and terminate at, or just in front of, the

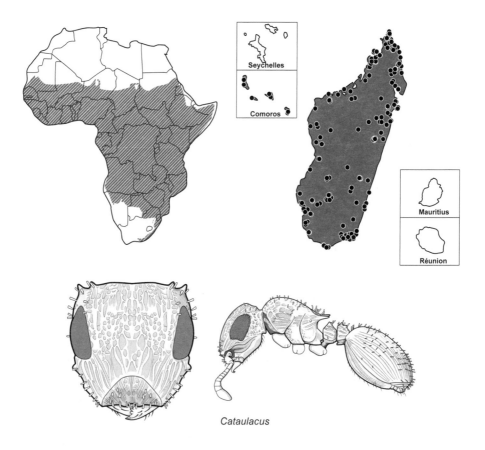

Cataulacus

anterior margin of the eye; in full-face view usually with a small preocular tooth or tubercle at the apex of each frontal carina.

Antennal scrobe present that extends below the eye.

Antenna with 11 segments, with an apical club of 3 segments.

Torulus concealed by frontal lobe in full-face view.

Eyes dorsolateral, large.

Head posteroventrally with a triangular tooth on each side that locks into a corresponding groove on the anterolateral pronotum when the head is flexed downward.

Promesonotal suture absent from dorsum, or a remnant present as a weak, shallow impression. Pronotum usually marginate laterally.

Propodeum usually bidentate to bispinose, only rarely unarmed.

Propodeal spiracle high on the side, at or slightly behind the midlength of the sclerite.

Metasternal process very small to absent.

Procoxa no larger than metacoxa, often slightly smaller; profemur can be tightly applied to concave side of pronotum.

Tibial spurs: mesotibia 0; metatibia 0.

Abdominal segment 2 (petiole) sessile.

Abdomnal segment 3 (postpetiole) with the tergite and sternite fused.

Abdominal tergite 4 (first gastral) hypertrophied and in dorsal view comprising the entire gastral dorsum. Abdominal segments 5–7 very reduced but in profile visible apically, below the posterior rim of A4 tergite.

Abdominal tergite 4 (first gastral) does not broadly overlap the sternite on the ventral gaster; gastral shoulders absent.

Sting reduced to vestigial, apparently nonfunctional.

Main pilosity of dorsal head and body: absent in some species; when present, setae usually short, blunt and stubbly, or bizarre (spatulate, squamate, stud-like, clavate, stalked-suborbicular); setae only uncommonly simple and long.

AFROTROPICAL CATAULACUS

adpressus Bolton, 1974

bequaerti Forel, 1913

boltoni Snelling, 1979

brevisetosus Forel, 1901
    = *meduseus* Santschi, 1939

centrurus Bolton, 1982

cestus Bolton, 1982

difficilis Santschi, 1916

egenus Santschi, 1911
    = *egenus simplex* Santschi, 1914

elongatus Santschi, 1924

erinaceus Stitz, 1910
    = *erinaceus crassispina* Santschi, 1917

fricatidorsus Santschi, 1914

greggi Bolton, 1974

guineensis Smith, F., 1853
    = *parallelus* Smith, F., 1853
    = *guineensis sulcinodis* Emery, 1892
    = *sulcatus* Stitz, 1910
    = *sulcatus alenensis* Stitz, 1910
    = *sulcatus fernandensis* Stitz, 1910

huberi André, 1890
    = *huberi longispinus* Stitz, 1910
    = *huberi herteri* Forel, 1913
    = *huberi guilelmi* Wheeler, W.M., 1925

impressus Bolton, 1974

inermis Santschi, 1924

intrudens (Smith, F., 1876)
    = *hararicus* Forel, 1894
    = *intrudens rugosus* Forel, 1894
    = *baumi* Forel, 1901
    = *baumi batonga* Forel, 1913
    = *rugosus subrugosus* Santschi, 1914
    = *intrudens intermedius* Santschi, 1917

    = *baumi bulawayensis* Arnold, 1917
    = *johannae densipunctatus* Stitz, 1923
    = *baumi pseudotrema* Santschi, 1926
    = *baumi gazanus* Santschi, 1928
    = *foveosquamosus* Santschi, 1937
    = *umbilicatus* Santschi, 1937

jacksoni Bolton, 1982

jeanneli Santschi, 1914
    = *pygmaeus degener* Santschi, 1916
    = *jeanneli loveridgei* Santschi, 1926

kenyensis Santschi, 1935

kohli Mayr, 1895
    = *kohli brazzavillensis* Santschi, 1910
    = *foveolatus* Stitz, 1910
    = *latipes* Menozzi, 1933

lobatus Mayr, 1895

lujae Forel, 1911
    = *lujae gilviventris* Forel, 1913

mckeyi Snelling, 1979

micans Mayr, 1901
    = *intrudens tristiculus* Santschi, 1919

mocquerysi André, 1889
    = *mocquerysi nainei* Forel, 1918

moloch Bolton, 1982

pilosus Santschi, 1920

pullus Santschi, 1910
    = *coriaceus* Stitz, 1910
    = *pullus orientalis* Santschi, 1914

pygmaeus André, 1890
    = *pygmaeus chariensis* Santschi, 1911
    = *pygmaeus bakusuensis* Forel, 1913

satrap Bolton, 1982

striativentris Santschi, 1924
    = *donisthorpei* Santschi, 1937

tardus Santschi, 1914

    = *schoutedeni* Santschi, 1919

taylori Bolton, 1982

theobromicola Santschi, 1939

traegaordhi Santschi, 1914

    = *traegaordhi ugandensis* Santschi,

        1914

    = *marleyi* Forel, 1914

    = *pygmaeus suddensis* Weber, 1943

vorticus Bolton, 1974

weissi Santschi, 1913

    = *traegaordhi plectroniae* Wheeler,

        W.M., 1922

    = *jeanneli aethiops* Santschi, 1924

wissmannii Forel, 1894

    = *wissmannii otii* Forel, 1901

    = *micans durbanensis* Forel, 1914

    = *wissmannii linearis* Santschi, 1914

MALAGASY CATAULACUS

ebrardi Forel, 1886

intrudens (Smith, F., 1876)

    = *johannae* Forel, 1895

oberthueri Emery, in Forel, 1891

porcatus Emery, 1899

regularis Forel, 1892

tenuis Emery, 1899

voeltzkowi Forel, 1907

wasmanni Forel, 1897

CENTROMYRMEX Mayr, 1866   Ponerinae

10 species, Afrotropical   PLATE 13

DIAGNOSTIC REMARKS: Among the ponerines, *Centromyrmex* combines the following characters. Stout, thickly spiniform, or peg-like traction setae are present on the mesotibia, mesobasitarsus, and metabasitarsus (in 1 species also on the metatibia). Eyes are absent. The helcium arises close to the midheight of the anterior face of A3 (the first gastral segment), not at its base. The metapleural gland orifice is located well above the ventral margin of the metapleuron and far anterior of the posteroventral angle of the metapleuron. The mesotibia and metatibia may each have 2 spurs, or the mesotibia lacks spurs and the metatibia has only 1.

DISTRIBUTION AND ECOLOGY: *Centromyrmex* species are widespread in the Afrotropics and also occur in the Oriental, Malesian, and Neotropical regions but are absent from the Malagasy region and from Australia. Species are subterranean and can be found nesting in or near active subterranean termitaries or in the walls of mound-building termites. Nests also occur in and under termite-infested rotten logs that are embedded in the ground. Outside of termitaries, foragers may also be found under rotten wood or in the root mat below the leaf-litter layer; specimens are known from leaf-litter samples but are uncommon. Termites appear to be the sole prey of all species.

IDENTIFICATION OF SPECIES: Bolton and Fisher, 2008c.

DEFINITION OF THE GENUS (WORKER): Characters of subfamily PONERINAE, plus:

Worker caste of *C. bequaerti* and *C. secutor* are polymorphic; all others are monomorphic.

Mandible triangular to elongate-triangular, with a distinct basal groove.

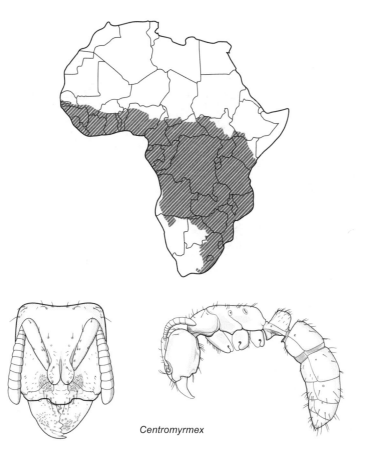

Centromyrmex

Palp formula 4,3.

Clypeus with the anterior margin convex, without a projecting tooth on each side.

Frontal lobes in full-face view with their anterior margins not overhanging the anterior clypeal margin.

Eyes absent.

Antenna with scape strongly dorsoventrally flattened in its basal half, the leading edge extremely thin.

Metanotal groove absent (though the dorsum of the mesosoma may be strongly longitudinally concave).

Epimeral sclerite present.

Metapleural gland orifice a small pore or short slit that opens laterally, located well above the ventral margin of the metapleuron and far anterior of the posteroventral angle of the metapleuron.

Propodeal spiracle with orifice almost circular to distinctly slit-shaped.

Mesotibia, mesobasitarsus, and metabasitarsus with strongly sclerotized spiniform or peg-like traction setae (also on metatibia in 1 species).

Tibial spurs: either mesotibia 2; metatibia 2, the posterior spur on each pectinate (*bequaerti* group); or mesotibia 0; metatibia 1, pectinate (*feae* group).

Pretarsal claws simple.

Abdominal segment 2 (petiole) a node.

Helcium located close to midheight on the anterior face of A3 (first gastral segment).

Prora present but sometimes very weak.

Abdominal segment 4 (second gastral segment) with differentiated presclerites.

Stridulitrum absent from the pretergite of A4.

AFROTROPICAL CENTROMYRMEX

angolensis Santschi, 1937	*raptor* Bolton and Fisher, 2008
bequaerti (Forel, 1913)	*secutor* Bolton and Fisher, 2008
= *rufigaster* (Arnold, 1916)	*sellaris* Mayr, 1896
decessor Bolton and Fisher, 2008	= *constanciae* Arnold, 1915
ereptor Bolton and Fisher, 2008	= *arnoldi* Santschi, 1919
fugator Bolton and Fisher, 2008	= *congolensis* Weber, 1949
longiventris Santschi, 1919	= *arnoldi guineensis* Bernard, 1953
praedator Bolton and Fisher, 2008	

CHRYSAPACE Crawley, 1924 — Dorylinae

1 species Malagasy, undescribed   PLATE 13

DIAGNOSTIC REMARKS: Among the dorylines, *Chrysapace* is immediately characterized by its unique sculpture. The dorsum and sides of the head, mesosoma, petiole (A2), and first gastral tergite (tergite of A3) are all strikingly strongly costate to sulcate. In addition, both the mesotibiae and metatibiae each have 2 spurs present, and the pretarsal claws each have a single preapical tooth.

DISTRIBUTION AND ECOLOGY: A single undescribed species of *Chrysapace* is known from the forests of Madagascar; the genus is otherwise represented by a few species in the Oriental and Malesian regions.

TAXONOMIC COMMENT: *Chrysapace* is currently recorded as a junior synonym of *Cerapachys*. It will be formally reinstated by Marek Borowiec, in a genus-rank revision of Dorylinae that is in preparation.

IDENTIFICATION OF SPECIES: The Malagasy species is undescribed.

DEFINITION OF THE GENUS (WORKER): Characters of subfamily DORYLINAE, plus:

Prementum not visible when mouthparts closed.

Palp formula 5,3.

Eyes present, distinctly behind midlength of the head capsule in full-face view; ocelli present.

Frontal carinae present as a pair of closely approximated raised ridges, without horizontal frontal lobes.

Antennal sockets fully exposed in full-face view.

Parafrontal ridges present.

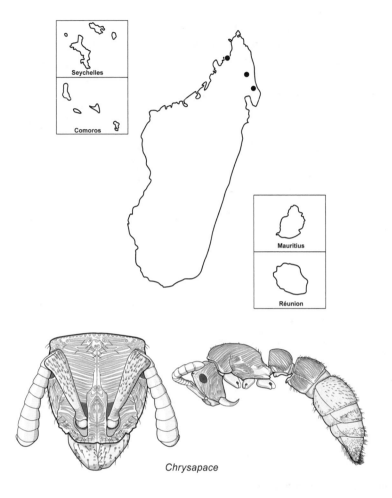

Chrysapace

Antenna with 12 segments; funiculus gradually incrassate, apical segment not swollen and bulbous.

Antennal scrobes absent.

Promesonotal suture absent on the dorsum of the mesosoma.

Pronotal-mesopleural suture present on side of mesosoma.

Metanotal groove absent.

Propodeal spiracle low on side and situated at or behind midlength of the sclerite, not subtended by a longitudinal impression.

Propodeal lobes present.

Tibial spurs: mesotibia 2, pectinate; metatibia 2, pectinate.

Metacoxa without a posterodorsal vertical cuticular flange or lamella.

Pretarsal claws each with a single preapical tooth.

Waist of 1 segment (petiole = A2).

Petiole (A2) sessile; not marginate dorsolaterally, the dorsum rounds into the sides.

Helcium projects from anterior surface of A3 (first gastral segment) at about its midheight.

Prora present as a strongly projecting cuticular flange.

Abdominal segment 4 with strongly developed presclerites.

Abdominal segments 5 and 6 without presclerites.

Pygidium (tergite of A7) armed dorsolaterally with a series of denticles or peg-like spiniform setae and terminating apically in a short cuticular fork.

Sting present, large and fully functional.

Sculpture characteristic: dorsum and sides of head, mesosoma, petiole (A2), and abdominal tergite 3 strikingly strongly costate to sulcate.

MALAGASY CHRYSAPACE

1 species, undescribed

CONCOCTIO Brown, 1974 Amblyoponinae

1 species, Afrotropical PLATE 14

DIAGNOSTIC REMARKS: Among the amblyoponines, *Concoctio* is characterized by the possession of triangular mandibles that are edentate except for an apical tooth and close tightly against the clypeus; the antennae are 9-segmented, with the 4 apical antennomeres forming a club; the pronotum is marginate anteriorly.

DISTRIBUTION AND ECOLOGY: The single described species of this genus has been collected only a couple of times, in rainforest areas of West Africa (Cameroon and Gabon), in soil and leaf-litter samples. One or more undescribed species are known from Tanzania and South Africa.

IDENTIFICATION OF SPECIES: Brown, 1974a.

DEFINITION OF THE GENUS (WORKER): Characters of subfamily AMBLYOPONINAE, plus:

Mandible triangular, with a distinct apical tooth that is followed by a small cleft; remainder of masticatory margin edentate. When mandibles are fully closed, there is no space between the masticatory margins nor between basal margins of the mandibles and the clypeus.

Palp formula unknown.

Clypeus with a small, rounded median lobe.

Frontal lobes small and closely approximated, their anterior margins well behind the anterior clypeal margin.

Eyes minutely present, located slightly behind the midlength of the head capsule.

Antenna with 9 segments, with the apical 4 antennomeres forming a club.

Pronotum marginate anteriorly.

Mesonotal dorsum strongly narrowed posteriorly; in dorsal view the lateral mesonotal margins converge strongly from promesonotal suture to metanotal groove.

Metanotal groove present as a short, transverse line.

Tibial spurs: mesotibia 0; metatibia 0.

Abdominal segment 2 (petiole) with a short anterior peduncle; in profile petiole with a

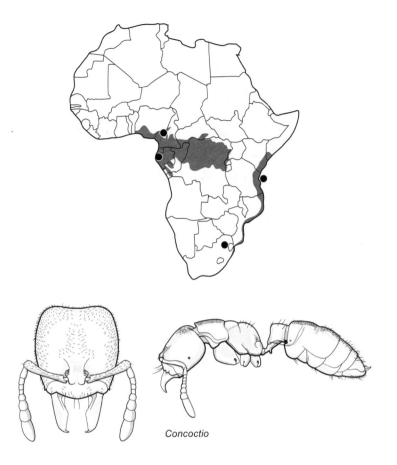

Concoctio

steep anterior face and weakly convex dorsum but without a differentiated posterior face; petiole broadly attached to abdominal segment 3 (first gastral segment).

Helcium with its dorsal surface arising very high on the anterior face of the first gastral segment (A3), so that in profile A3 above the helcium has no free anterior face (at most a short, rounded angle present between anterior and dorsal surfaces).

Prora present but small.

Abdominal segment 4 with distinctly differentiated presclerites.

AFROTROPICAL CONCOCTIO

concenta Brown, 1974

CREMATOGASTER Lund, 1831   Myrmicinae

135 species (+ 173 subspecies), Afrotropical; 37 species, Malagasy   PLATE 14

DIAGNOSTIC REMARKS: Among the myrmicines, *Crematogaster* may be recognized instantly by the unique articulation of the postpetiole (A3), which posteriorly is attached on the dorsal surface of the first gastral segment (A4). In addition, the head is usually broader than long, the antennae usually have 11 segments, more rarely 10, and the clypeus is broadly

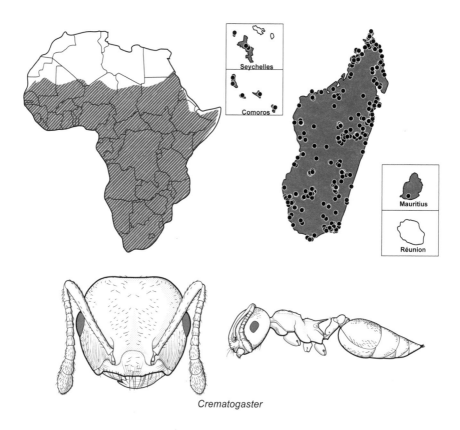

Crematogaster

inserted between short frontal lobes. The petiole (A2) is flattened dorsally and lacks any sort of a node (in life the petiole is capable of extreme elevation so that its dorsal surface lies against the propodeal declivity). The helcium is enlarged; in profile its height is sub-equal to, or greater than, the height of the postpetiole. The gaster in dorsal view is roughly heart-shaped.

DISTRIBUTION AND ECOLOGY: With a worldwide distribution, *Crematogaster* is one of the most common and species-rich genera in the regions under consideration here. Its species-rank taxonomy in the Afrotropical region is in poor condition. A number of *Crematogaster* species nest in the ground, either directly or under a surface object; others nest in rotten wood on the ground, and some nest in fallen twigs in the leaf litter. Of these ground-nesting species, some forage entirely on the surface or in the leaf litter, but others ascend plants, including large trees, in search of food. A considerable number of species are entirely arboreal, some being specialist inhabitants of galls and thorns, and species of the *C. depressa* group construct large, conspicuous nests of carton high on the trunks of large trees. Some species will nest in termitaries, both active and abandoned, either on the ground or in trees.

TAXONOMIC COMMENT: Throughout its taxonomic history, *Crematogaster* accumulated about 15 subgeneric names, which have now been reduced to 2 by Blaimer (2012c),

C. (Crematogaster) and *C. (Orthocrema)*. Many of the earlier subgenera did not occur in the Afrotropical and Malagasy regions but the subgenera that were present, whose names are often encountered in the earlier literature, are given in the synonymic list of family-group and species-group names in the introduction.

IDENTIFICATION OF SPECIES: Blaimer, 2010, 2012a, 2012b (Malagasy groups); Blaimer and Fisher, 2013a, 2013b (Malagasy groups); no review exists for Afrotropical taxa, and the species-rank taxonomy is very confused.

DEFINITION OF THE GENUS (WORKER): Characters of subfamily MYRMICINAE, plus:

Worker caste polymorphic in some species, many with considerable worker size variation; otherwise monomorphic.

Mandible short-triangular or with basal and outer margins subparallel (almost falcate in major workers of at least 1 polymorphic species). Masticatory margin with 4–5 teeth.

Palp formula 5,3, or 4,3, or very rarely 3,2.

Stipes of the maxilla without a transverse crest.

Clypeus posteriorly broadly inserted between the frontal lobes.

Frontal carinae short, restricted to well-defined but narrow frontal lobes that are relatively widely separated.

Antennal scrobes absent.

Antenna with 10–11 segments, with an apical club of 2–3 segments.

Torulus with upper lobe visible in full-face view.

Eyes small to well developed, usually located at, or slightly behind, the midlength of the head capsule.

Head capsule usually broader than long in full-face view.

Promesonotal suture often represented by an impression or weak line across the dorsum; promesonotum usually convex; metanotal groove usually conspicuous.

Propodeal spiracle at margin of the declivity.

Propodeal lobes absent; propodeum unarmed to strongly bidentate.

Metasternal process absent.

Tibial spurs: mesotibia 0; metatibia 0.

Abdominal segment 2 (petiole) subsessile, flattened dorsally and without a node; petiole capable of extreme elevation so that its dorsal surface lies against the propodeal declivity.

Helcium much enlarged, its depth in profile at least subequal to the depth of the postpetiole.

Abdominal segment 3 (postpetiole) attached to the dorsal surface of A4 (first gastral segment).

Abdominal tergite 4 strongly overlaps the sternite on the ventral surface of the gaster; gastral shoulders absent.

Stridulitrum present on the pretergite of A4.

Gaster (A4–A7) in profile flat dorsally and convex ventrally.

Sting spatulate.

Main pilosity of dorsal head and body: usually simple, may be sparse.

acaciae Forel, 1892
  acaciae gloriosa Santschi, 1914
  acaciae victoriosa Santschi, 1916
  acaciae generosa Santschi, 1919
aegyptiaca Mayr, 1862
  aegyptiaca robusta Emery, 1877
  aegyptiaca turkanensis Santschi, 1935
affabilis Forel, 1907
africana Mayr, 1895
  africana schumanni Mayr, 1895
  africana variegata Mayr, 1902
  africana biemarginata Forel, 1910
  africana alligatrix Forel, 1911
  africana thoracica Santschi, 1921
  africana tibialis Santschi, in Wheeler,
    W.M., 1922
  africana camena Wheeler, W.M., 1922
  africana stanleyi Wheeler, W.M., 1922
  africana stolonis Santschi, 1937
  africana polymorphica Weber, 1943
aloysiisabaudiae Menozzi, 1930
alulai Emery, 1901
  alulai scrutans Forel, 1910
amabilis Santschi, 1911
  amabilis retiaria Santschi, 1933
ambigua Santschi, 1926
amita Forel, 1913
  amita matabele Arnold, 1920
  amita bushimana Santschi, 1926
  amita caffra Santschi, 1926
  amita makololo Santschi, 1926
ancipitula Forel, 1917
angusticeps Santschi, 1911
arnoldi Forel, 1914
  arnoldi loveridgei Santschi, 1926
arthurimuelleri Forel, 1894
batesi Forel, 1911
bequaerti Forel, 1913
  bequaerti ludia Forel, 1913
  bequaerti mutabilis Santschi, 1914
  bequaerti gerardi Santschi, 1915
  bequaerti atraplex Santschi, in
    Wheeler, W.M., 1922

  bequaerti modica Santschi, 1926
  = bequaerti saga Santschi, 1937
  bequaerti semiclara Santschi, 1928
boera Santschi, 1926
breviventris Santschi, 1920
brunneipennis André, 1890
  brunneipennis omniparens Forel, 1914
  brunneipennis yorubosa Santschi, 1933
buchneri Forel, 1894
  = halli Donisthorpe, 1945
  buchneri graeteri Forel, 1916
  buchneri composita Santschi, 1933
  buchneri uasina Santschi, 1935
capensis Mayr, 1862
  capensis calens Forel, 1910
  capensis tropicorum Forel, 1910
castanea Smith, F., 1858
  = tricolor decolor Forel, 1891
  castanea arborea Smith, F., 1858
  = castanea decolorata Santschi, 1926
  castanea ferruginea Forel, 1892
  castanea hararica Forel, 1894
  castanea rufonigra Emery, 1895
  castanea busschodtsi Emery, 1899
  castanea aquila Forel, 1907
  castanea inversa Forel, 1907
  castanea mediorufa Forel, 1907
  castanea analis Santschi, 1910
  castanea simia Forel, 1910
  castanea museisapientiae Forel, 1911
  castanea ulugurensis Forel, 1911
  castanea yambatensis Forel, 1913
  castanea rufimembrum Santschi, 1914
  castanea durbanensis Forel, 1914
  castanea insidiosa Santschi, 1920
  castanea bruta Santschi, 1926
  castanea adusta Santschi, 1935
censor Forel, 1910
  censor junodi Forel, 1916
chiarinii Emery, 1881
  chiarinii taediosa Forel, 1894
  chiarinii cincta Emery, 1896
  chiarinii aethiops Forel, 1907

chiarinii nigra Forel, 1910

chiarinii v-nigra Forel, 1910

chiarinii sellula Santschi, 1914

chiarinii subsulcata Santschi, 1914

chiarinii bayeri Santschi, 1926

chlorotica Emery, 1899

chopardi Bernard, 1950

cicatriculosa Roger, 1863

clariventris Mayr, 1895

   clariventris biimpressa Mayr, 1895

coelestis Santschi, 1911

   coelestis mercatori Santschi, 1939

concava Emery, 1899

constructor Emery, 1895

   constructor kirbyella Emery, 1922

   = *constructor kirbyi* Mayr, 1895

    (homonym)

cuvierae Donisthorpe, 1945

delagoensis Forel, 1894

   delagoensis rhodesiana Arnold, 1920

   delagoensis merwei Santschi, 1932

   delagoense acutidens Arnold, 1944

depressa (Latreille, 1802)

   = *platygnatha* (Roger, 1863)

   = *mandibularis* André, 1889

   = *buchneri foreli* Mayr, 1895

   depressa fuscipennis Emery, 1899

   depressa adultera Santschi, 1915

desperans Forel, 1914

dolens Forel, 1910

donisthorpei Santschi, 1934

edentula Santschi, 1914

excisa Mayr, 1895

   excisa lacustris Santschi, 1914

   excisa cavinota Stitz, 1916

   excisa bomaella Santschi, 1935

flaviventris Santschi, 1910

foraminiceps Santschi, 1913

   foraminiceps staitchi Forel, 1915

   foraminiceps mirmillo Santschi, 1935

gabonensis Emery, 1899

   gabonensis fuscitatis Forel, 1913

gallicola Forel, 1894

   gallicola rauana Forel, 1907

   gallicola latro Forel, 1910

gambiensis André, 1889

   = *gambiensis longiruga* Forel, 1907

   = *fulva* Donisthorpe, 1945

   gambiensis krantziana Forel, 1914

   gambiensis transversiruga Santschi, 1914

   gambiensis sejuncta Stitz, 1916

gerstaeckeri Dalla Torre, 1892

   = *cephalotes* Gerstäcker, 1871

    (homonym)

   gerstaeckeri sjostedti Mayr, 1907

   gerstaeckeri oraclum Forel, 1913

   = *bulawayensis* Santschi, 1916

   = *godfreyi foraminicipoides* Arnold, 1920

   gerstaeckeri godefreyi Forel, 1914

   gerstaeckeri maledicta Forel, 1914

   = *neuvillei carininotum* Santschi, 1917

   gerstaeckeri infaceta Santschi, 1916

   gerstaeckeri kohliella Santschi, 1918

   gerstaeckeri pulla Wheeler, W.M., 1922

   gerstaeckeri rufescens Wheeler, W.M.,

    1922

   gerstaeckeri zulu Wheeler, W.M., 1922

   gerstaeckeri inquieta Santschi, 1928

   gerstaeckeri pudica Santschi, 1930

gratiosa Santschi, 1926

gutenbergi Santschi, 1914

hemiceros Santschi, 1926

homeri Forel, 1913

hottentota Emery, 1899

   hottentota bassuto Santschi, 1925

ilgii Forel, 1910

impressa Emery, 1899

   = *excisa andrei* Forel, 1911

   impressa brazzai Santschi, 1910

   impressa maynei Forel, 1913

   impressa sapora Forel, 1916

impressiceps Mayr, 1902

   impressiceps lujana Forel, 1915

   impressiceps longiscapa Stitz, 1916

   impressiceps frontalis Santschi, in

    Wheeler, W.M., 1922

inconspicua Mayr, 1896

incorrecta Santschi, 1917

jeanneli Santschi, 1914
jullieni Santschi, 1910
juventa Santschi, 1926
kachelibae Arnold, 1954
kasaiensis Forel, 1913
kneri Mayr, 1862
  kneri dakarensis Santschi, 1914
  kneri pronotalis Santschi, 1914
  kneri behanzini Santschi, 1915
  kneri funerea Santschi, 1915
kohli Forel, 1909
  kohli winkleri Forel, 1909
laestrygon Emery, 1869
  laestrygon airensis Santschi, 1937
lamottei Bernard, 1953
lango Weber, 1943
latuka Weber, 1943
laurenti Forel, 1909
  laurenti zeta Emery, 1922
libengensis Stitz, 1916
  libengensis rufula Santschi, 1926
liengmei Forel, 1894
  liengmei weitzeckeri Emery, 1895
  = *coelestis kloofensis* Forel, 1914
  liengmei caculata Forel, 1915
litoralis Arnold, 1955
longiceps Santschi, 1910
lotti Weber, 1943
luctans Forel, 1907
  luctans liebknechti Forel, 1915
  luctans sordidoides Santschi, 1930
magitae Forel, 1910
margaritae Emery, 1895
  margaritae lujae Forel, 1913
  margaritae brevarmata Forel, 1915
  margaritae cupida Santschi, 1935
melanogaster Emery, 1895
  melanogaster homonyma Emery, 1922
menilekii Forel, 1894
  menilekii occidentalis Mayr, 1902
  menilekii spuria Forel, 1913
  menilekii satan Forel, 1916
  menelikii proserpina Santschi, 1919
  menilekii viehmeyeri Santschi, 1921

menilekii completa Santschi, 1928
  menilekii suddensis Weber, 1943
microspina Menozzi, 1942
mimosae Santschi, 1914
  mimosae tenuipilis Santschi, 1937
misella Arnold, 1920
monticola Arnold, 1920
mottazi Santschi, 1928
muralti Forel, 1910
  muralti ugandensis Santschi, 1914
  muralti livingstonei Santschi, 1919
natalensis Forel, 1910
  natalensis braunsi Emery, 1922
  natalensis dulcis Santschi, 1926
neuvillei Forel, 1907
  neuvillei cooperi Forel, 1914
nigeriensis Santschi, 1914
  nigeriensis wilnigra Forel, 1916
nigrans Forel, 1915
nigriceps Emery, 1897
  nigriceps prelli Forel, 1911
  nigriceps saganensis Consani, in
    Menozzi and Consani, 1952
nigronitens Santschi, 1917
ochraceiventris Stitz, 1916
opaciceps Mayr, 1901
  opaciceps defleta Forel, 1910
  opaciceps clepens Forel, 1913
  opaciceps cacodaemon Forel, 1914
orobia Santschi, 1919
oscaris Forel, 1910
painei Donisthorpe, 1945
paolii Menozzi, 1930
pauciseta Emery, 1899
  pauciseta grossulior Forel, 1916
peringueyi Emery, 1895
  peringueyi cacochyma Forel, 1914
  peringueyi angustior Arnold, 1920
  peringueyi gedeon Arnold, 1920
  peringueyi dentulata Stitz, 1923
petiolidens Forel, 1916
phoenix Santschi, 1921
pseudinermis Viehmeyer, 1923
  pseudinermis muellerianus Finzi, 1939

pulchella Bernard, 1953

rectinota Forel, 1913

resulcata Bolton, 1995

  = *subsulcata* Santschi, 1932 (homonym)

retifera Santschi, 1926

rivai Emery, 1897

  rivai luctuosa Menozzi, 1930

rufigena Arnold, 1958

rugosa André, 1895

  rugosa nigriventris Santschi, 1919

  rugosa rugaticeps Santschi, 1919

rugosior Santschi, 1910

ruspolii Forel, 1892

  ruspolii atriscapis Forel, 1915

rustica Santschi, 1935

santschii Forel, 1913

  santschii clymene Forel, 1915

schultzei Forel, 1910

senegalensis Roger, 1863

  senegalensis goliathula Forel, 1922

sewellii Forel, 1891

  sewellei marnoi Mayr, 1895

  sewellii acis Forel, 1913

similis Stitz, 1911

solenopsides Emery, 1899

  solenopsides flavida Mayr, 1907

  solenopsides costeboriensis Santschi, 1919

  solenopsides mandonbii Santschi, 1926

solers Forel, 1910

sordidula (Nylander, 1849)

  sordidula molognori Weber, 1943

stadelmanni Mayr, 1895

  stadelmanni angustata Mayr, 1895

  stadelmanni intermedia Mayr, 1896

  stadelmanni dolichocephala Santschi, 1911

  stadelmanni scnereri Forel, 1911

stadelmanni anguliceps Stitz, 1916

stadelmanni ovinodis Stitz, 1916

stadelmanni gracilenta Viehmeyer, 1922

stadelmanni spissata Santschi, 1937

stenocephala Emery, 1922

stigmata Santschi, 1914

striatula Emery, 1892

  striatula benitensis Santschi, 1910

  striatula obstinata Santschi, 1911

  striatula langi Santschi, 1926

  striatula omega Santschi, 1935

  striatula horatii Santschi, 1937

theta Forel, 1911

togoensis Donisthorpe, 1945

transiens Forel, 1913

transvaalensis Forel, 1894

  transvaalensis hammi Arnold, 1920

trautweini Viehmeyer, 1914

tricolor Gerstäcker, 1859: 263

vidua Santschi, 1928

vulcania Santschi, 1913

wasmanni Santschi, 1910

wellmani Forel, 1909

  = *wellmani retusa* Santschi, 1916

  = *boxi* Donisthorpe, 1945

  = *dewasi* (Bruneau de Miré, 1966)

  wellmani weissi Santschi, 1910

  wellmani luciae Forel, 1913

werneri Mayr, 1907

  werneri cacozela Santschi, 1914

  werneri pasithea Santschi, 1915

wilwerthi Santschi, 1910

  wilwerthi fauconneti Forel, 1910

  wilwerthi confusa Santschi, 1911

zavattarii Menozzi, 1926

  zavattarii edmeeae Santschi, 1937

zonacaciae Weber, 1943

MALAGASY CREMATOGASTER

agnetis Forel, 1892

alafara Blaimer, in Blaimer and Fisher, 2013

bara Blaimer, in Blaimer and Fisher, 2013

castanea Smith, F., 1858

  = *tricolor decolor* Forel, 1891

degeeri Forel, 1886

  = *degeeri lunaris* Santschi, 1928

dentata Dalla Torre, 1893

  = *sewelli improba* Forel, 1907

= *sewellii mauritiana* Forel, 1907
ensifera Forel, 1910
grevei Forel, 1891
hafahafa Blaimer, in Blaimer and
    Fisher, 2013
hazolava Blaimer, in Blaimer and
    Fisher, 2013
hova Forel, 1887
    = *hova latinoda* Forel, 1892
kelleri Forel, 1891
    = *gibba* Emery, 1894
    = *adrepens* Forel, 1897
lobata Emery, 1895
    = *lobata pacifica* Santschi, 1919
madagascariensis André, 1887
madecassa Emery, 1895
mafybe Blaimer, in Blaimer and Fisher,
    2013
mahery Blaimer, 2010
maina Blaimer, in Blaimer and Fisher,
    2013
malahelo Blaimer, in Blaimer and
    Fisher, 2013
malala Blaimer, 2010
marthae Forel, 1892
masokely Blaimer, in Blaimer and
    Fisher, 2013
mpanjono Blaimer, 2012

nosibeensis Forel, 1891
ramamy Blaimer, in Blaimer and
    Fisher, 2013
ranavalonae Forel, 1887
    = *emmae* Forel, 1891
    = *emmae laticeps* Forel, 1892
    = *inops* Forel, 1892
    = *ranovalonae paulinae* Forel, 1892
    = *ranavalonae pepo* Forel, 1922
    = *descarpentriesi* Santschi, 1928
rasoherinae Forel, 1891
    = *rasoherinae brunneola* Emery, 1922
    = *rasoherinae brunnea* Forel, 1907
    (homonym)
    = *voeltzkowi* Forel, 1907
razana Blaimer, 2012
sabatra Blaimer, 2010
schenki Forel, 1891
sewellii Forel, 1891
sisa Blaimer, 2010
tavaratra Blaimer, in Blaimer and
    Fisher, 2013
telolafy Blaimer, 2012
tricolor Gerstäcker, 1859: 263
tsisitsilo Blaimer, in Blaimer and
    Fisher, 2013
volamena Blaimer, 2012

CRYPTOPONE Emery, 1893   Ponerinae

1 species, Afrotropical   PLATE 14

DIAGNOSTIC REMARKS: Among the ponerines, *Cryptopone* combines the characters of presence of a single spur on the metatibia and the absense of traction setae on the metabasitarsus, with the following: The eyes are absent. The base of the mandible has a laterally placed pit. The sides of the propodeum strongly converge dorsally, so that the dorsum is very narrow, especially just behind the much broader mesonotum. The mesotibia has sparse, short, stout traction setae scattered among the normal setae, and the helcium is not at the base of the anterior face of the first gastral segment (A3).

DISTRIBUTION AND ECOLOGY: *Cryptopone* currently contains 22 species, with a mainly pantropical distribution, but is also represented in the southern portions of the holarctic. It is absent from the Malagasy region. The sole Afrotropical species, *C. hartwigi*, is known from southern Angola and South Africa, and appears to be rare.

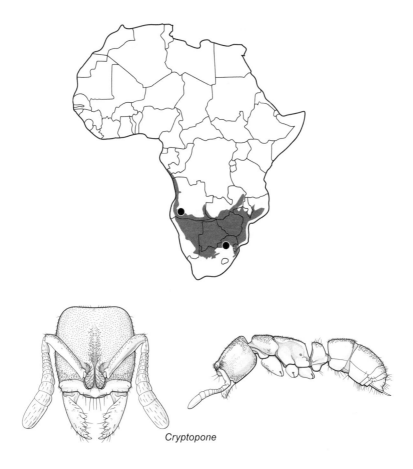

Cryptopone

TAXONOMIC COMMENT: The inclusion of *C. hartwigi* in *Cryptopone* may be incorrect as analysis of its DNA has indicated affinity with *Fisheropone*, which implies that its *Cryptopone*-like morphological characters may have been acquired by convergence. See the discussion in Schmidt and Shattuck, 2014. Resolution of its taxonomy will depend on a full investigation of the whole genus.

IDENTIFICATION OF SPECIES: No review exists.

DEFINITION OF THE GENUS (WORKER): Characters of subfamily PONERINAE, plus:

Mandible triangular, with a basal pit laterally, without a basal groove; masticatory margin with 5–7 teeth.
Palp formula unknown.
Clypeus simple, without lobes or lateral teeth.
Frontal lobes small, their anterior margins behind the anterior clypeal margin.
Eyes absent.
Antenna with the 4 apical segments forming a club.
Metanotal groove absent.

Propodeum with sides strongly convergent dorsally, so that the dorsum is very narrow, especially just behind the much broader mesonotum.

Epimeral sclerite absent.

Metapleural gland orifice opens posteriorly.

Propodeal spiracle small, circular.

Tibial spurs: mesotibia 1; metatibia 1; each spur pectinate.

Mesotibia with sparse, short, stout traction setae among the normal setae; traction setae absent from mesobasitarsus and metabasitarsus.

Pretarsal claws small and simple.

Abdominal segment 2 (petiole) a low, thick scale.

Helcium not at base of the anterior face of A3 (first gastral segment).

Prora rudimentary, at most a narrow transverse rim.

Abdominal segment 4 (second gastral segment) with strongly differentiated presclerites.

Stridulitrum presence/absence unknown.

AFROTROPICAL CRYPTOPONE

hartwigi Arnold, 1948

CYPHOIDRIS Weber, 1952 Myrmicinae

4 species, Afrotropical PLATE 14

DIAGNOSTIC REMARKS: Among the Afrotropical myrmicines, *Cyphoidris* has 11-segmented antennae that terminate in a 3-segmented club. The mandibles are stoutly triangular and always have more than 9 teeth. The median portion of the clypeus is narrow and bicarinate, posteriorly narrowly inserted between the frontal lobes; the lateral portions of the clypeus are not raised into a ridge in front of the antennal sockets. Long frontal carinae and strong antennal scrobes are present that extend above the eyes. The promesonotum in profile appears swollen and dome-like with respect to the propodeum, and the propodeal spiracle is at the margin of the declivity.

DISTRIBUTION AND ECOLOGY: The 4 known species of this purely Afrotropical genus are found in leaf litter, particularly in the rainforest zone. Collections are uncommon, and nest sites remain unknown but are most probably in rotten wood on the forest floor.

IDENTIFICATION OF SPECIES: Bolton, 1981b.

DEFINITION OF THE GENUS (WORKER): Characters of subfamily MYRMICINAE, plus:

Mandible triangular, stoutly constructed; masticatory margin elongate, with 10–14 teeth.

Palp formula 4,3, or 3,2.

Stipes of the maxilla with a transverse crest.

Clypeus with median portion narrowed from side to side and elevated, longitudinally bicarinate on the dorsal surface of elevated section, the carinae commence between the frontal lobes; the clypeus posteriorly narrowly inserted between the frontal lobes.

Clypeus with lateral portions not forming a raised ridge in front of the antennal sockets.

Clypeus without an unpaired seta at the midpoint of the anterior margin.

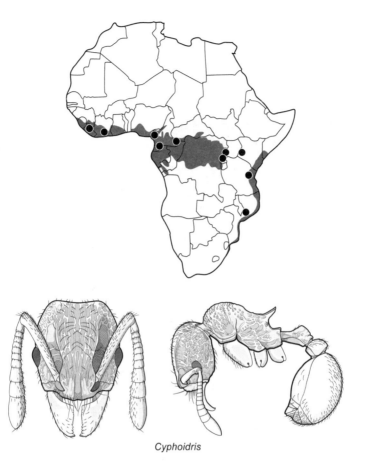

Cyphoidris

Frontal carinae long, anteriorly expanded into frontal lobes that cover the antennal sockets, posteriorly forming the dorsal margins of the scrobes.

Antennal scrobes present, broad, extending back above the eye almost to the posterior margin.

Torulus concealed by the frontal lobe in full-face view.

Antenna with 11 segments, with an apical club of 3 segments.

Eyes well developed, located in front of the midlength of the head capsule.

Promesonotal suture absent; promesonotum swollen, in profile its dorsum dome-like and strongly convex.

Propodeum bispinose; in profile the propodeal dorsum on a much lower level than the apex of the convex promesonotum.

Propodeal spiracle located at the margin of the declivity.

Metasternal process absent.

Tibial spurs: mesotibia 0; metatibia 0.

Abdominal segment 4 (first gastral) tergite does not broadly overlap the sternite on the ventral gaster; gastral shoulders absent.

Sting terminates in a linear, narrowly spatulate appendage.

Main pilosity of dorsal head and body: simple, may be sparse.

exalta Bolton, 1981

parissa Bolton, 1981

spinosa Weber, 1952

werneri Bolton, 1981

CYPHOMYRMEX Mayr, 1862   Myrmicinae

1 species, Malagasy (Réunion Island), introduced   PLATE 15

DIAGNOSTIC REMARKS: Among the Malagasy myrmicines, *Cyphomyrmex* is instantly recognizable by its massively laterally expanded frontal lobes, which in full-face view overlap and conceal the entire side of the head in front of the eyes. In addition, the frontal carinae, behind the lobes, extend to the posterior margin of the head and are subtended by scrobes that extend above the eyes; the antennae have 11 segments and terminate in a weak 2-segmented club.

DISTRIBUTION AND ECOLOGY: A single species of this New World genus, which currently contains 39 extant species, has been introduced on Réunion Island in the Indian Ocean. It is found under stones, often in cultivated land.

IDENTIFICATION OF SPECIES: Snelling and Longino, 1992.

DEFINITION OF THE GENUS (WORKER): Characters of subfamily MYRMICINAE, plus:

Mandible triangular; masticatory margin with 5 teeth.

Palp formula 4,2.

Stipes of the maxilla without a transverse crest.

Clypeus broadly inserted between the massive frontal lobes. Median portion of the
    clypeus with an anterior apron that fits tightly over the basal mandibular margins.

Clypeus with an unpaired seta at the midpoint of the anterior margin.

Frontal lobes enormously expanded laterally; in full-face view the frontal lobe overlap-
    ping and concealing the entire sides of the head in front of the eye.

Frontal carinae extend to the posterior margin of the head.

Antennal scrobe present above the eye; scrobe shallow and much shorter than the
    length of the scape.

Antenna with 11 segments, with a weak apical club of 2 segments.

Torulus concealed by frontal lobe in full-face view.

Eyes present; side of the head immediately in front of the eye with an oblique carina;
    side of the head separated from ventral surface by a blunt angle.

Promesonotal suture absent or at most a vestigial impression; metanotal groove deep.

Mesonotum with cuticular tubercles.

Propodeum without spines or sharp teeth.

Propodeal spiracle close to the margin of the declivity.

Metasternal process absent.

Tibial spurs: mesotibia 0; metatibia 0.

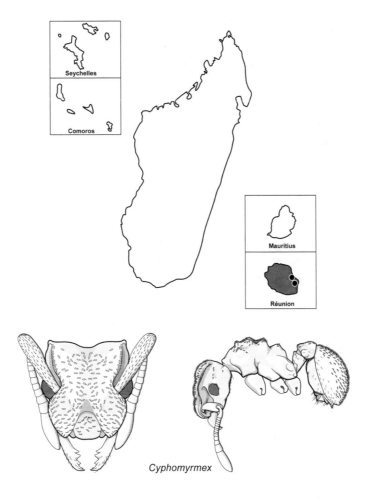

Cyphomyrmex

Abdominal segment 2 (petiole) subsessile, with a short, thick anterior peduncle and lacking a differentiated node.

Abdominal tergite 4 (first gastral tergite) does not broadly overlap the sternite on the ventral gaster; gastral shoulders absent.

Abdominal sternite 4 (first gastral sternite) truncated basally.

Sting reduced.

Main pilosity of dorsal head and body: standing setae absent; minute appressed setae or pubescence present.

MALAGASY CYPHOMYRMEX

rimosus (Spinola, 1851)

DICROASPIS Emery, 1908

Myrmicinae

2 species, Afrotropical; 1 species, Malagasy (Comoro Islands)

PLATE 15

DIAGNOSTIC REMARKS: Among the myrmicines, *Dicroaspis* shares a specialized clypeal morphology only with *Calyptomyrmex*. In both, the anterior face of the clypeus rises vertically from its anterior margin. At the top of the vertical surface, about at the level of the frontal lobes, the clypeus projects anteriorly as a bilobed appendage, the clypeal fork, which projects out above the mandibles. Of the 2 genera with this structure, *Dicroaspis* has 11-segmented antennae, the petiolar peduncle is short and very thick in profile, and all setae on the head and body are simple, fine, and soft.

DISTRIBUTION AND ECOLOGY: This genus is restricted to the Afrotropical and Malagasy regions but has not been found on the island of Madagascar itself. Nests are made in rotten logs, and workers have also been retrieved from samples of leaf litter in wet forest areas.

IDENTIFICATION OF SPECIES: Bolton, 1981a (Afrotropical).

DEFINITION OF THE GENUS (WORKER): Characters of subfamily MYRMICINAE, plus:

Mandible triangular and stoutly constructed; masticatory margin with 7–9 teeth.
Palp formula 2,2.
Stipes of the maxilla with a transverse crest.

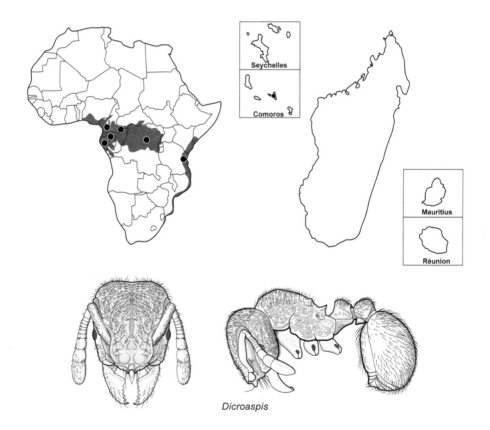

Dicroaspis

Clypeus constricted posteriorly, narrowly inserted between the broad frontal lobes. Median portion of the clypeus with a narrow anterior apron, then near-vertical, terminating dorsally in a prominent bidentate cuticular fork that projects anteriorly from just below the level of the frontal lobes.

Clypeus without an unpaired seta at the midpoint of the anterior margin.

Frontal carinae long and form the dorsal margins of the deep scrobes. In full-face view the anterior margins of the frontal lobes (and the clypeal fork between them) overhang and conceal the anterior clypeal margin.

Antennal scrobes strongly developed, deep, extending above the eyes; the scrobe capable of containing the entire antenna.

Torulus entirely concealed by the large frontal lobe in full-face view.

Antenna with 11 segments, with an apical club of 3 segments. Antennal scape not incrassate in the apical half.

Eyes present, small, below the lower margin of the scrobe.

Promesonotal suture absent from the dorsum of the mesosoma.

Propodeum with dorsum sloping to a pair of short, stout spines or teeth.

Propodeal lobe thin in dorsal view, in profile without a median fenestra or thin spot.

Propodeal spiracle well in front of the margin of the declivity.

Metasternal process absent.

Tibial spurs: mesotibia 0 or 1; metatibia 0 or 1.

Abdominal segment 2 (petiole) with peduncle short and very thick in profile, with a massive triangular anteroventral process that is inclined anteriorly; side of peduncle, near base, without a carina that extends almost to the height of the peduncle.

Abdominal segment 3 (postpetiole) with an anteroventral process that consists of a single strong transverse ridge that appears as a tooth in profile.

Abdominal tergite 4 (first gastral) does not broadly overlap the sternite on the ventral gaster; gastral shoulders absent; tergite weakly vaulted apically.

Sting present, simple.

Main pilosity of dorsal head and body: simple, fine and soft.

AFROTROPICAL DICROASPIS

cryptocera Emery, 1908
  = *wheeleri* (Menozzi, 1924)
laevidens (Santschi, 1919)

MALAGASY DICROASPIS

1 undescribed species, Comoro Islands

DIPLOMORIUM Mayr, 1901   Myrmicinae

1 species, Afrotropical   PLATE 15

DIAGNOSTIC REMARKS: Among the Afrotropical myrmicines, *Diplomorium* has 11-segmented antennae that terminate in a club of 3 segments. Superficially it resembles the few species of *Monomorium* in which the antennae have 11 segments, but *Diplomorium*

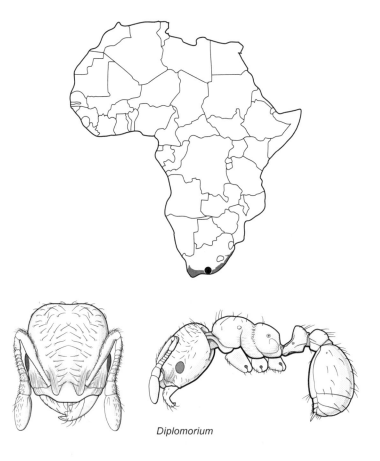

Diplomorium

is distinguished by its clypeal structure, as the median portion of the clypeus is swollen and transversely convex; is not longitudinally bicarinate; and posteriorly is broadly inserted between the small but widely separated frontal lobes. In addition, the postpetiole node (A3) is more voluminous than the petiole node in profile and is broadly attached to the gaster.

DISTRIBUTION AND ECOLOGY: This strange little monotypic genus is known from only a very few collections made in eastern Cape Province, South Africa; it is not known to occur anywhere else.

IDENTIFICATION OF SPECIES: Bolton, 1987.

DEFINITION OF THE GENUS (WORKER): Characters of subfamily MYRMICINAE, plus:

Worker caste with some size variation but not polymorphic.

Mandible short triangular; masticatory margin with 4–5 teeth.

Palp formula 2,2.

Stipes of the maxilla without a transverse crest.

Clypeus posteriorly broadly inserted between the small but widely separated frontal lobes. Median portion of the clypeus swollen and evenly convex, not longitudinally bicarinate.

Clypeus with an unpaired seta at the midpoint of the anterior margin.

Frontal carinae restricted to the small frontal lobes.

Antennal scrobes absent.

Antenna with 11 segments, with a weak apical club of 3 segments.

Torulus with upper lobe visible in full-face view; maximum width of torulus lobe is posterior to the point of maximum width of the frontal lobes.

Eyes present and conspicuous, in front of the midlength of the side of the head capsule.

Promesonotal suture absent from the dorsum of the mesosoma. Metanotal groove impressed.

Propodeum unarmed.

Propodeal spiracle small, low on the side, and at, or close to, the midlength of the sclerite.

Metasternal process absent.

Tibial spurs: mesotibia 0; metatibia 0.

Abdominal segment 3 (postpetiole) enlarged, in profile more voluminous than petiole (A2); very broadly attached to A4.

Abdominal tergite 4 (first gastral) overlaps the sternite on the ventral gaster; gastral shoulders present.

Sting strongly developed.

Main pilosity of dorsal head and body: simple.

AFROTROPICAL DIPLOMORIUM

longipenne Mayr, 1901

DISCOTHYREA Roger, 1863   Proceratiinae

7 species, Afrotropical; 1 species, Malagasy   PLATE 15

DIAGNOSTIC REMARKS: Among the proceratiines, *Discothyrea* species are compact ants in which the tergite of A4 (the second gastral tergite) is enlarged, arched, and strongly vaulted so that the remaining segments point forward. The clypeus and anterior portion of the frons form a shelf that projects out over the mandibles, and the latter are edentate behind the apical tooth. The antennal sockets are located anteriorly on the frontoclypeal shelf. Eyes are present but may be minute. The antennae are 6–12 segmented, and the apical segment is always greatly enlarged and bulbous.

DISTRIBUTION AND ECOLOGY: Species of this genus are distributed throughout the world's tropics, and a number of subtropical forms are also known. In the regions under consideration here, most species are found in the leaf litter and soil, including the compacted soil walls of termite nests, but some inhabit rotten wood on the ground, particularly in forested areas, and nests may also be found in small pieces of rotten wood below the surface. The prey is probably arthropod eggs. There are many undescribed Afrotropical species of this genus, some of them minute.

IDENTIFICATION OF SPECIES: No review exists.

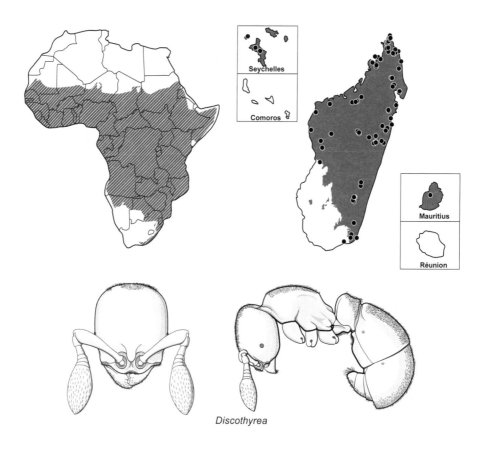

Discothyrea

DEFINITION OF THE GENUS (WORKER): Characters of subfamily PROCERATIINAE, plus:

Mandible small, masticatory margin edentate except for the apical tooth.

Palp formula 5,4, or 4,4, or 4,3, or 3,4, or 1,3.

Clypeus and the anterior portion of the frons forming a strongly projecting fron-toclypeal shelf that overhangs the mandibles.

Frontal carinae either reduced and fused to form a single median lamina, or elevated as a flat-topped platform behind the level of the antennal sockets.

Antennal sockets located anteriorly on the frontoclypeal shelf.

Antennal scrobes present in *D. oculata* species-group, where they are broad and shallow; absent in other species-groups.

Eyes minute to moderate, located at or just in front of the midlength of the head capsule.

Antenna with 6–12 segments, the apical antennomere greatly enlarged and bulbous.

Metanotal groove absent.

Tibial spurs: mesotibia 0 or 1; metatibia 1, pectinate.

Abdominal segment 2 (petiole) with the tergite and sternite fused.

Helcium very short, its dorsum located high on the anterior face of A3 (first gastral segment).

Helcium sternite visible in profile but inconspicuous in this view. With helcium detached and in anterior view, the tergite overlaps the sternite on each side; the sternite is convex and bulges ventrally below the level of the tergal apices.

Prora present, usually merely a transverse ridge.

Abdominal segment 4 (second gastral segment) with strongly developed presclerites, but the presclerites and the girdling constriction are usually completely concealed below the posterior margin of A3.

Abdominal segment 4 with the tergite enlarged and strongly vaulted, the tergite hypertrophied with respect to the sternite, which is reduced.

AFROTROPICAL DISCOTHYREA

hewitti Arnold, 1916	*poweri* (Arnold, 1916)
mixta Brown, 1958	*sculptior* Santschi, 1913
oculata Emery, 1901	*traegaordhi* Santschi, 1914
patrizii Weber, 1949	

MALAGASY DISCOTHYREA

berlita Fisher, 2005

DOLIOPONERA Brown, 1974 — Ponerinae

1 species, Afrotropical — PLATE 16

DIAGNOSTIC REMARKS: Among the ponerines, *Dolioponera* combines the characters of presence of a single spur on the metatibia, and absence of traction setae on the metabasitarsus, with the following: The clypeus has a projecting median lobe that is truncated apically. The petiole (A2) is large, long and barrel-shaped, and only fractionally shorter than the first gastral segment (A3). The helcium arises just below the midheight of the anterior face of A3 (not near its base). The second gastral segment (A4) is very long and cylindrical, more than 1.5 times the length of A3.

DISTRIBUTION AND ECOLOGY: The single recorded species of *Dolioponera* is known only from a very few collections, 1 each in the rainforest zones of Cameroon, Central African Republic, Gabon, Ghana, and Ivory Coast. It is probably more widespread but uncommon. The few specimens in collections were all retrieved from soil or leaf-litter samples.

IDENTIFICATION OF SPECIES: Brown, 1974b; Fisher, 2006; Schmidt and Shattuck, 2014.

DEFINITION OF THE GENUS (WORKER): Characters of subfamily PONERINAE, plus:

Mandible short-triangular, with the basal angle rounded. Masticatory margin with an apical and a preapical tooth, behind which the margin is edentate but has a series of short, stout, spiniform setae. Basal groove absent.

Palp formula 2,2.

Clypeus with a projecting median lobe that is truncated apically.

Frontal lobes broad; not overhanging the anterior clypeal margin.

Eyes absent or very small; when present, located low on the side and far forward.

Antenna with apical segment much enlarged.

Mesopleuron entirely fused to mesonotum, without trace of a suture between the 2 sclerites.

Metanotal groove absent.

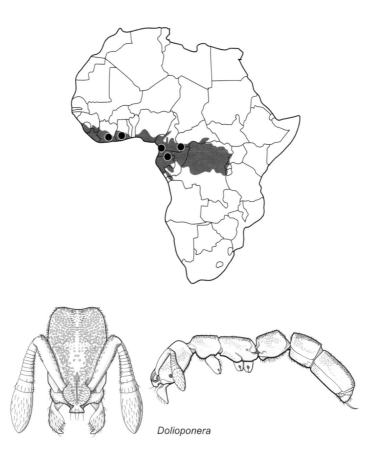

Dolioponera

Epimeral sclerite absent.

Metapleural gland orifice opens laterally.

Propodeal spiracle circular.

Tibial spurs: mesotibia 1; metatibia 1; each spur pectinate.

Pretarsal claws simple.

Abdominal segment 2 (petiole) large, elongated and barrel-shaped, only fractionally shorter than A3 (first gastral segment).

Helcium arises just below midheight on the anterior face of A3 (first gastral segment).

Prora absent.

Abdominal segment 4 (second gastral segment) with strongly differentiated presclerites, the segment cylindrical and very long, more than 1.5 times the length of A3.

Stridulitrum absent from the pretergite of A4.

AFROTROPICAL DOLIOPONERA

fustigera Brown, 1974

DORYLUS Fabricius, 1793

55 species (+ 63 subspecies), Afrotropical

Dorylinae

PLATE 16

DIAGNOSTIC REMARKS: Among the dorylines, *Dorylus* species are strikingly polymorphic ants in which the promesonotal suture is strongly present but entirely fused and immobile. Eyes are always absent, and the clypeus has an elongate, unpaired seta at the midpoint of its anterior margin. The propodeal spiracle is situated high up on the side and in front of the midlength of the propodeum, and propodeal lobes are absent. The waist is of a single segment, the petiole (A2). Gastral segments 3–5 (A5–A7) all have differentiated presclerites. The pygidium is armed with a single, posteriorly directed, short spine or tooth on each side. The sting is very reduced and not functional as a weapon.

DISTRIBUTION AND ECOLOGY: Primarily Afrotropical but also with a few species in the extreme southern Palaearctic as well as in the Oriental and Malesian regions, *Dorylus* is absent from the Malagasy region. In the Afrotropics the genus is best known from its group of highly conspicuous surface raiders, the driver ants (*D. nigricans* group), whose trails are often seen crossing paths by means of sunken, partially covered runways that shield them from the sun. The majority of species are smaller and usually never seen on the surface of the ground during the day. These species are mostly to entirely subterranean in their foraging behavior, being found in the earth, beneath stones and logs, in rotten logs, tree stumps, and termitaries. All species of *Dorylus* are nomadic mass predators that do not construct permanent nests. The driver ant group (subgenus *D. [Anomma]*) takes any animal it can catch as prey. Arthropods are the most common victims, but small vertebrates are also taken. The other groups, mainly subterranean, are often found raiding termitaries. Queens of *Dorylus* are enormous dichthadiigynes and are only rarely discovered, but their males, the large "sausage flies," are commonly attracted to lights.

TAXONOMIC COMMENT: The species-rank taxonomy of Afrotropical *Dorylus* is confused, not only by the extreme number of very poorly defined subspecies but also by the early development of a dual system, with one taxonomy based on the worker caste and the other on the very conspicuous male sex. The 2 are strikingly different and, unless captured together, could not be associated before the advent of DNA sequencing. Thus, of the 118 nominal taxa in the species-group, some taxa are described only from workers, others only from males. It is apparent that many taxa currently bear at least 2 names: one based on workers, the other on males, of what are actually the same species. Currently, there are 6 recognized subgenera of *Dorylus*, 5 of which occur in the Afrotropical region: *D. (Alaopone)* Emery, 1881; *D. (Anomma)* Shuckard, 1840; *D. (Dorylus)* Fabricius, 1793; *D. (Rhogmus)* Shuckard, 1840; *D. (Typhlopone)* Westwood, 1839.

IDENTIFICATION OF SPECIES: Emery, 1895 (*Dorylus*); Santschi, 1939 (*D. [Alaopone]* males); Raignier and van Boven, 1955 (*D. [Anomma]*); Gotwald, 1982 (key to subgenera of *Dorylus*). No recent review of Afrotropical species exists.

DEFINITION OF THE GENUS (WORKER): Characters of subfamily DORYLINAE, plus:

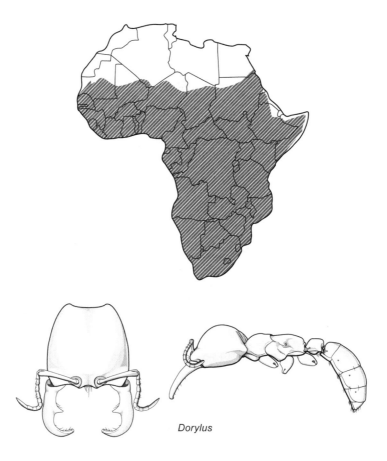

Dorylus

Worker caste strikingly polymorphic.

Prementum not visible when mouthparts closed.

Labrum with its distal (free) margin convex, not indented or cleft medially.

Palp formula 2,2 or 1,2.

Clypeus with an elongate, unpaired seta at the midpoint of its anterior margin.

Eyes absent; ocelli absent.

Frontal carinae present as a pair of raised ridges, sometimes weakly expanded into narrow frontal lobes, occasionally each with a short, posteriorly directed spine.

Antennal sockets fully exposed in full-face view.

Parafrontal ridges absent.

Antenna with 7–11 segments; funiculus filiform to gradually incrassate.

Antennal scrobes absent.

Promesonotal suture strongly present on the dorsum of the mesosoma but fused and inflexible.

Pronotal-mesopleural suture present on side of mesosoma but fully fused.

Mesonotum reduced to a very narrow strip across the dorsum of the mesosoma, between the pronotum and the long propodeum.

Metanotal groove absent.

Propodeal spiracle situated high on the side and in front of the midlength of the sclerite, subtended by a longitudinal impression and an endophragmal pit.

Propodeal lobes absent.

Tibial spurs: mesotibia 1, pectinate; metatibia 1, pectinate.

Metatibial gland present, located distally on the ventral surface.

Pretarsal claws simple, stout.

Waist of 1 segment (A2 = petiole).

Petiole (A2) sessile.

Helcium projects from the anterior surface of A3 (first gastral segment) at about its midheight or slightly lower.

Prora varies from a simple angle to a distinct prominence, or a short, longitudinal crest.

Abdominal segments 4, 5, and 6 all with differentiated presclerites.

Pygidium (the tergite of A7) large and distinct, flattened to concave dorsally and armed with a single, posteriorly directed short spine or tooth on each side.

Sting present but reduced and nonfunctional, usually not visible; used for trail-laying only.

AFROTROPICAL DORYLUS

acutus Santschi, 1937

aethiopicus Emery, 1895

affinis Shuckard, 1840
  = *planiceps* Haldeman, 1849
  = *oraniensis brevinodosus* (Mayr, 1862)
  affinis aegyptiacus Mayr, 1865
  = *brevinodosus abyssinicus* Emery, 1895
  affinis loewyi Forel, 1907
  affinis denudatus Santschi, 1910
  affinis exilis Santschi, 1914
  affinis ugandensis Santschi, 1914
  affinis parapsidalis Santschi, 1917
  affinis pulliceps Santschi, 1917
  affinis sudanicus Santschi, 1917
  affinis hirsutus Wheeler, W.M., 1922

agressor Santschi, 1923

alluaudi Santschi, 1914
  alluaudi lobatus Santschi, 1919

atratus Smith, F., 1859

atriceps Shuckard, 1840
  = *ritsemae* Dalla Torre, 1892
  = *shuckardi* Ritsema, 1874 (homonym)

attenuatus Shuckard, 1840
  = *attenuatus umbratipennis* Forel, 1909
  attenuatus acuminatus Emery, 1899
  attenuatus bondroiti Santschi, 1912
  attenuatus australis Santschi, 1919

attenuatus latinodis Forel, 1920

bequaerti Forel, 1913

bishyiganus (van Boven, 1972)

braunsi Emery, 1895
  = *helvolus impressus* Stitz, 1910
  braunsi anceps Forel, 1914

brevipennis Emery, 1895
  brevipennis marshalli Emery, 1901
  brevipennis zimmermanni Santschi, 1910

brevis Santschi, 1919

buyssoni Santschi, 1910
  buyssoni conjugens Santschi, 1910

congolensis Santschi, 1910

conradti Emery, 1895
  conradti berlandi Santschi, 1926

depilis Emery, 1895
  depilis clarior Santschi, 1917

diadema Gerstäcker, 1859
  diadema fusciceps Emery, 1899
  diadema arnoldi Forel, 1914

distinctus Santschi, 1910

ductor Santschi, 1939

emeryi Mayr, 1896
  emeryi pulsi Forel, 1904
  emeryi opacus Forel, 1909

erraticus (Smith, F., 1865)

faurei Arnold, 1946

fimbriatus (Shuckard, 1840)
  fimbriatus poweri Forel, 1914
  fimbriatus crampeli Santschi, 1919
  fimbriatus laevipodex Santschi, 1919
fulvus (Westwood, 1839)
  = *dahlbomii* (Westwood, 1840)
  = *shuckardi* (Westwood, 1840)
  = *kirbii* (Shuckard, 1840)
  = *spinolae* (Shuckard, 1840)
  = *thwaitsii* (Shuckard, 1840)
  = *clausii* (Joseph, 1882)
  fulvus glabratus Shuckard, 1840
  fulvus badius Gerstäcker, 1859
  = *fulvus rhodesiae* Forel, 1913
  fulvus dentifrons Wasmann, 1904
  = *fulvus stramineus* Stitz, 1910
  fulvus eurous Emery, 1915
  fulvus obscurior Wheeler, W.M., 1925
  fulvus saharensis Santschi, 1926
  fulvus mordax Santschi, 1931
  = *fulvus impressus* Santschi, 1928
    (homonym)
funereus Emery, 1895
  funereus acherontus Santschi, 1937
  funereus pardus Santschi, 1937
  funereus stygis Santschi, 1937
  funereus zumpti Santschi, 1937
furcatus (Gerstäcker, 1872)
fuscipennis (Emery, 1892)
  fuscipennis lugubris Santschi, 1919
  fuscipennis marginiventris Santschi,
    1919
gaudens Santschi, 1919
ghanensis van Boven, 1975
gribodoi Emery, 1892
  = *gerstaeckeri* Emery, 1895
  = *gribodoi confusa* Santschi, 1915
  = *gribodoi insularis* Santschi, 1937
  = *lamottei* Bernard, 1953
helvolus (Linnaeus, 1764)
  = *dorylus* (Lamarck, 1817)
  = *punctata* (Smith, F., 1858)
  = *planifrons* Mayr, 1865
  helvolus pretoriae Arnold, 1946
katanensis Stitz, 1911

kohli Wasmann, 1904
  = *gerstaeckeri quadratus* Santschi, 1914
  kohli minor Santschi, 1911
  kohli frenisyi Forel, 1916
  kohli victoriae Santschi, 1921
  kohli chapini Wheeler, W.M., 1922
  kohli langi Wheeler, W.M., 1922
  kohli militaris Santschi, 1923
  kohli indocilis Santschi, 1933
leo Santschi, 1919
mandibularis Mayr, 1896
  mandibularis pulchellus Santschi,
    1920
mayri Santschi, 1912
moestus Emery, 1895
  moestus schereri Forel, 1911
  moestus claripennis Santschi, 1919
  moestus morio Santschi, 1919
montanus Santschi, 1910
niarembensis (van Boven, 1972)
nigricans Illiger, 1802
  nigricans burmeisteri (Shuckard, 1840)
  nigricans arcens (Westwood, 1847)
  = *pubescens* (Roger, 1861)
  nigricans rubella (Savage, 1849)
  nigricans molesta (Gerstäcker, 1859)
  = *antinorii* (Emery, 1881)
  nigricans sjoestedti Emery, 1899
  nigricans sjostedtiwilverthi (Wasmann,
    1917)
  nigricans pallidus Santschi, 1921
  nigricans terrificus Santschi, 1923
ocellatus (Stitz, 1910)
politus Emery, 1901
rufescens Santschi, 1915
savagei Emery, 1895
  savagei mucronatus Emery, 1899
schoutedeni Santschi, 1923
spininodis Emery, 1901
  spininodis longiceps Viehmeyer, 1914
stadelmanni Emery, 1895
stanleyi Forel, 1909
staudingeri Emery, 1895
striatidens Santschi, 1910
termitarius Wasmann, 1911

titan Santschi, 1923
   titan vinalli Santschi, 1933
westwoodii (Shuckard, 1840)

wilverthi Emery, 1899
   = *nigritarsis* Strand, 1911
   = *nomadas* Santschi, 1935

EBUROPONE Borowiec (in preparation) gen. n.

Dorylinae

1 species, Afrotropical and Malagasy

PLATE 16

DIAGNOSTIC REMARKS: Among the dorylines, *Eburopone* lacks eyes, has the promesonotal suture present as an impressed groove, and the waist is a single segment. The propodeal spiracle is low on the side and situated behind the midlength of the sclerite, and propodeal lobes are present. The mesotibia has a single spur present, and the pretarsal claws lack a preapical tooth. Uniquely, the sternite of the second gastral segment (A4) has a pale patch that is probably glandular at the midpoint of its posterior margin.

DISTRIBUTION AND ECOLOGY: The single described species, *E. wroughtoni*, occurs in both regions under consideration here. It appears to be the only species in the Afrotropical region, confined to southern Africa, although other undescribed *Eburopone* occur in Madagascar.

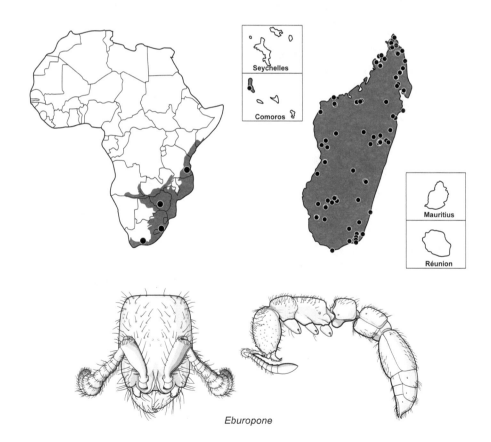

Eburopone

TAXONOMIC COMMENT: The genus *Eburopone* will be formally described by Marek Borowiec in a forthcoming genus-rank revision of Dorylinae currently in preparation. The name is not to be considered available here.

IDENTIFICATION OF SPECIES: Brown, 1975 (Afrotropical, as part of *Cerapachys*).

DEFINITION OF THE GENUS (WORKER): Characters of subfamily DORYLINAE, plus:

Prementum not visible when mouthparts closed.

Palp formula not known.

Eyes absent; ocelli absent.

Frontal carinae present as a pair of closely approximated raised ridges, abruptly truncated posteriorly, without horizontal frontal lobes.

Antennal sockets fully exposed in full-face view.

Parafrontal ridges present.

Antenna with 12 segments, gradually incrassate apically.

Antennal scrobes absent.

Promesonotal suture present as an incised line across the dorsum; suture fused and immobile.

Pronotal-mesopleural suture present on side of the mesosoma.

Metanotal groove absent.

Propodeal spiracle low on side and situated at or behind midlength of the sclerite, not subtended by a longitudinal impression.

Propodeal lobes present.

Tibial spurs: mesotibia 1, pectinate; metatibia 1, pectinate.

Metacoxa without a posterodorsal vertical cuticular flange or lamella.

Pretarsal claws simple.

Waist of 1 segment (petiole = A2).

Petiole (A2) sessile, not marginate dorsolaterally.

Helcium projects from the anterior surface of A3 (first gastral segment) at about its midheight.

Prora vestigial to absent.

Abdominal segment 4 with strongly developed presclerites.

Abdominal sternite 4 with a conspicuous, pale, oval or finger-like patch of cuticle present in the middle at the posterior margin; probably glandular.

Abdominal segments 5 and 6 without presclerites.

Pygidium (tergite of A7) large and distinct, flattened to concave dorsally and armed laterally, posteriorly, or both, with a series or row of denticles or peg-like spiniform setae.

Sting present, large and fully functional.

AFROTROPICAL EBUROPONE

wroughtoni (Forel, 1910)
  = *wroughtoni rhodesiana* (Forel, 1913)
  = *roberti* (Forel, 1914)

MALAGASY EBUROPONE

wroughtoni (Forel, 1910)

ECPHORELLA Forel, 1909

Dolichoderinae

1 species, Afrotropical

PLATE 16

DIAGNOSTIC REMARKS: Among the dolichoderines, the petiole (A2) of *Ecphorella* has a bluntly triangular scale that is distinct in profile; the petiole is not overhung from behind by the first gastral segment (A3). The mandibles have teeth but lack denticles (in contrast to all other regional dolichoderines, which have a combination of teeth and denticles). The fourth gastral sternite (sternite of A6) does not have a longitudinal keel posteriorly.

DISTRIBUTION AND ECOLOGY: This enigmatic monotypic genus is known only from a single collection, made in Angola prior to 1909.

IDENTIFICATION OF SPECIES: Shattuck, 1992.

DEFINITION OF THE GENUS (WORKER): Characters of subfamily DOLICHODERINAE, plus:

Masticatory margin of mandible with 7 teeth, without denticles.
Palp formula unknown.

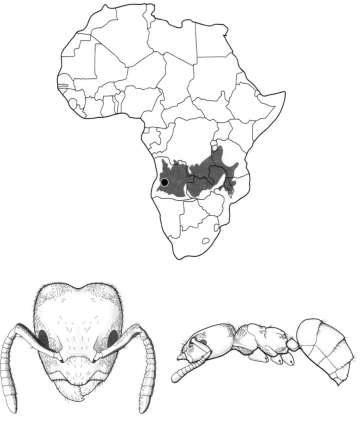

Ecphorella

Anterior clypeal margin shallowly convex, without a median notch.

Antenna with 12 segments.

Metathoracic spiracles dorsolateral.

Metanotal groove present, narrow.

Propodeum unarmed.

Propodeal spiracle dorsal and posterior.

Tibial spurs: mesotibia, unknown; metatibia 1, pectinate.

Abdominal segment 2 (petiole) a bluntly triangular scale.

Abdominal segment 3 (first gastral) does not project forward over the petiole; the petiole not concealed in dorsal view when the gaster and mesosoma are aligned.

Abdominal tergite 3 with a short longitudinal impression that accommodates the base of the petiole when the mesosoma and gaster are aligned.

Abdominal tergites 3–6 visible in dorsal view (i.e., 4 gastral tergites visible in dorsal view); tergite of A7 (pygidium) is reflexed onto the ventral surface of the gaster, where it abuts abdominal sternite 7 (hypopygium).

Abdominal sternite 6 (fourth gastral sternite) not keel-shaped posteriorly.

AFROTROPICAL ECPHORELLA

wellmani (Forel, 1909)

ERROMYRMA Bolton and Fisher gen. n.   Myrmicinae

Type-species: *Monomorium latinode* Mayr, 1872: 152, by present designation   PLATE 17

1 species, Afrotropical and Malagasy

DIAGNOSTIC REMARKS: Among the myrmicines, *Erromyrma* has 12-segmened antennae that terminate in a club of 3 segments. In terms of the Afrotropical and Malagasy regions, *Erromyrma* resembles *Monomorium* but is differentiated from it (and also from *Syllophopsis* and *Trichomyrmex*, former synonyms of *Monomorium*) by the following unique combination of characters: In *Erromyrma* the worker caste is polymorphic; the palp formula is 3,3; the mandible is unsculptured and has 5 teeth; the eye is far in front of the midlength of the head capsule; and the propodeal dorsum is transversely sculptured.

DISTRIBUTION AND ECOLOGY: The known distribution of the single species includes Bangladesh, Christmas Island (Indian Ocean), Comoro Islands, Hawaii, India, Japan, Madagascar, Malaysia (Sarawak), Mayotte Island (Indian Ocean), Mozambique, New Zealand, Sri Lanka, Taiwan, Tanzania, and Thailand. The Oriental region, particularly the Indian subcontinent, probably represents the native range of the species. As a casual introduction, *E. latinodis* has been recorded from the Netherlands and from Saint Helena Island in the Atlantic Ocean. If the Oriental region is in fact the region of origin, then its presence in the Afrotropical and Malagasy regions must both be regarded as introductions.

TAXONOMIC COMMENT: Apart from the original description by Mayr (1872), the single species included in this new genus, *E. latinodis* (Mayr, 1872) comb. n., has also been described in detail by Bolton (1987) and by Heterick (2006) under the name *Monomorium*

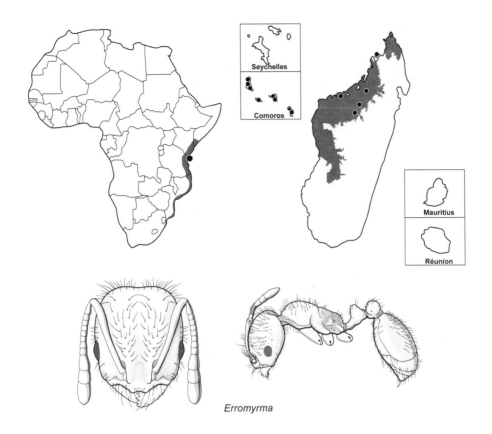

Erromyrma

latinode. From the time of its original description, this species has been regarded as a member of genus *Monomorium*. However, recent DNA analysis (Ward, et al. 2015) has shown it to be widely separated from *Monomorium* and sister to the monotypic Indo-Australian genus *Epelysidris* Bolton (1987). As is so often the case in Myrmicinae, no autapomorphy for *Erromyrma* can currently be designated. In consequence, the genus is defined by the combination of characters summarized above and listed below, and by the sequence of couplets that leads to its identity in the keys.

IDENTIFICATION OF SPECIES: Bolton, 1987 (Afrotropical, as *Monomorium latinode* group); Heterick, 2006 (Malagasy, as *Monomorium latinode* group).

DEFINITION OF THE GENUS (WORKER): Characters of subfamily MYRMICINAE, plus:

Worker caste polymorphic.

Mandible triangular, unsculptured except for setal pits; masticatory margin with 5 teeth.

Palp formula 3,3.

Stipes of the maxilla without a transverse crest.

Clypeus posteriorly narrowly inserted between the small frontal lobes; the width of the clypeus between the lobes greater than the width of one of the lobes. Median portion of the clypeus raised, projecting forward anteriorly, the raised section weakly longitudinally bicarinate dorsally; each carina terminates in a blunt angle at the anterior margin.

Clypeus with a distinct unpaired seta at the midpoint of the anterior margin; in profile the unpaired median seta arises from the anteriormost point of the clypeal projection.

Frontal carinae restricted to the small frontal lobes.

Antennal scrobes absent.

Antennal fossae with outer margins delineated by 1–3 very fine curved striolae.

Antenna with 12 segments, with an apical club of 3 segments.

Torulus with upper lobe visible in full-face view; maximum width of torulus lobe is posterior to the point of maximum width of the frontal lobes.

Eyes present and conspicuous, in full-face view located conspicuously in front of the midlength of the head capsule. Eye in profile roughly oval; never reniform or extended anteroventrally into a lobe.

Promesonotal suture absent from the dorsum of the mesosoma. Metanotal groove present; in dorsal view anteriorly arched medially, not transverse.

Propodeum unarmed; in profile the dorsum rounds evenly into the declivity.

Propodeal spiracle circular to subcircular, approximately at the midlength of the sclerite.

Propodeal dorsum and the declivity with distinct transverse sculpture.

Metasternal process absent.

Tibial spurs: mesotibia 0; metatibia 0.

Petiole (A2) with spiracle at the level of the anterior face of the node; peduncle ventrally with a shallow, crest-like process anteriorly.

Abdominal segment 4 (first gastral) tergite overlaps the sternite on the ventral gaster; gastral shoulders present.

Sting stout, strongly developed.

Main pilosity of dorsal head and body: fine and simple, present and numerous on all dorsal surfaces. Scapes, femora, and tibae all with numerous suberect to subdecumbent, fine projecting setae.

AFROTROPICAL SPECIES

latinodis (Mayr, 1872)
  = *voeltzkowi* (Forel, 1907)

MALAGASY SPECIES

latinodis (Mayr, 1872)

EUPONERA Forel, 1891 — Ponerinae

5 species (+ 1 subspecies), Afrotropical; 14 species, Malagasy — PLATE 17

DIAGNOSTIC REMARKS: Among the ponerines, *Euponera* combines the characters of presence of 2 spurs on the metatibia, absence of traction setae on the metabasitarsus, and location of the helcium at the base of the anterior face of the first gastral segment (A3), with the following: The mandible has a basal pit. The frontal lobes are moderate to large. The metanotal groove may be absent or present, and the propodeal spiracle is slit-shaped. The petiole (A2) is a node or scale.

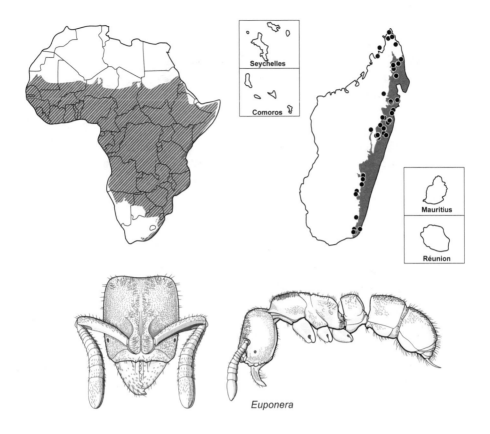

Euponera

DISTRIBUTION AND ECOLOGY: Most species of this genus are known from Madagascar but also occur in the Afrotropical, Oriental, and Malesian regions. Its species are ground nesters, either in rotten wood, or directly in the soil, or in the ground under surface objects, or in the walls of termitaries. Foraging takes place on the surface, in leaf litter, and in topsoil. The more common species appear to be generalist predators.

TAXONOMIC COMMENT: The species *E. suspecta* Santschi, 1914, is transferred from *Euponera* to *Parvaponera*; see the discussion under the latter name.

IDENTIFICATION OF SPECIES: Rakotonirina and Fisher, 2013b (Malagasy).

DEFINITION OF THE GENUS (WORKER): Characters of subfamily PONERINAE, plus:

Mandible stoutly triangular; with a basal pit but without a basal groove; masticatory margin with 6–12 teeth and basal angle angulate.

Palp formula 4,4, or 2,2.

Clypeus simple, without projecting lobes or teeth but the midpoint of the anterior margin may be distinctly indented or notched.

Frontal lobes moderate to large; in full-face view their anterior margins varying from close, to well behind, the anterior clypeal margin.

Eyes present, small to moderate in size, in front of the midlength of the head capsule in full-face view.

Metanotal groove varies from broad and deeply impressed to absent.

Epimeral sclerite present.

Mesopleuron divided into anepisternum and katepisternum by a weak transverse sulcus in some species; sulcus absent in others.

Metapleural gland orifice opens posteriorly, just above posteroventral angle.

Propodeal spiracle with orifice slit-shaped.

Tibial spurs: mesotibia 2; metatibia 2; on each tibia the anterior spur small and simple, the posterior spur larger and pectinate.

Metatibial posterior surface featureless or with a slightly depressed oval area of pale, very finely granular cuticle above the base of the spur that appears to be glandular.

Pretarsal claws simple.

Abdominal segment 2 (petiole) a thick node, a tall narrow node, or a thick scale.

Helcium located at the base of the anterior face of A3 (first gastral segment).

Prora present, conspicuous.

Abdominal segment 4 (second gastral segment) with presclerites strongly differentiated. Stridulitrum absent from the pretergite of A4.

AFROTROPICAL EUPONERA

aenigmatica (Arnold, 1949)	= *dentis* (Weber, 1942)
brunoi (Forel, 1913)	= *lamottei* (Bernard, 1953)
= *nigeriensis* (Santschi, 1914)	*fossigera* (Mayr, 1901)
= *nigeriensis katangana* (Santschi, 1933)	*sjostedti* (Mayr, 1896)
	wroughtonii (Forel, 1901)
= *bayoni* (Menozzi, 1933)	*wroughtonii crudelis* (Forel, 1901)

MALAGASY EUPONERA

agnivo (Rakotonirina and Fisher, 2013)	*mialy* (Rakotonirina and Fisher, 2013)
antsiraka (Rakotonirina and Fisher, 2013)	*nosy* (Rakotonirina and Fisher, 2013)
daraina (Rakotonirina and Fisher, 2013)	*rovana* (Rakotonirina and Fisher, 2013)
gorogota (Rakotonirina and Fisher, 2013)	*sikorae* (Forel, 1891)
haratsingy (Rakotonirina and Fisher, 2013)	*tahary* (Rakotonirina and Fisher, 2013)
	vohitravo (Rakotonirina and Fisher, 2013)
ivolo (Rakotonirina and Fisher, 2013)	
maeva (Rakotonirina and Fisher, 2013)	*zoro* (Rakotonirina and Fisher, 2013)

EURHOPALOTHRIX Brown and Kempf, 1961   Myrmicinae

1 species, Malagasy (Comoro Islands)   PLATE 17

DIAGNOSTIC REMARKS: Among the myrmicines, *Eurhopalothrix* is immediately diagnosed by its 7-segmented antennae that terminate in a 2-segmented club and deep antennal scrobes that pass below the eyes. No other genus shows these characters in combination.

DISTRIBUTION AND ECOLOGY: This genus currently contains 38 species, about half of which are Neotropical and the other half are widely distributed through the Malesian and Austral regions, with a number of species on Pacific Ocean islands. A single species,

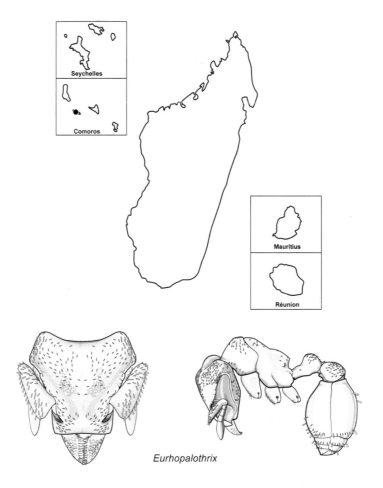

Eurhopalothrix

unidentified and apparently undescribed, is known from Comoro Islands, but otherwise the genus appears absent from the Afrotropical region and from Madagascar proper.

IDENTIFICATION OF SPECIES: Taylor, 1990 (Old World); Ketterl, Verhaagh, and Dietz, 2004 (New World), Longino, 2013 (New World).

DEFINITION OF THE GENUS (WORKER): Characters of subfamily MYRMICINAE, plus:

Mandible triangular to elongate-triangular; their masticatory margins oppose but do not overlap at full closure. Masticatory margin with 9–14 teeth or denticles in total.

Basal lamella absent on mandible.

Labrum not capable of reflexion over the anterior portion of the buccal cavity, not closing tightly over the apex of the labiomaxillary complex; labrum with trigger hairs present.

Palp formula 1,2.

Stipes of the maxilla without a transverse crest.

Clypeus large, broadly triangular, posteriorly broadly inserted between the frontal lobes.

Frontal carinae in full-face view, at level of antennal sockets, with their outer margins close to the apparent lateral margin of the head (actually the outer margin of the preocular carina).

Antennal scrobe present, deep and extensive, extending below the eye when eye present.

Antenna with 7 segments, with an apical club of 2 segments.

Torulus concealed by the frontal lobe in full-face view.

Eye absent or present; when present, located on the extreme dorsolateral rim of the scrobe.

Head capsule anteriorly strongly narrowed from side to side in full-face view.

Head anterolaterally with a preocular carina present that arises at the clypeus and extends posteriorly below the antennal socket.

Promesonotal suture absent from the dorsum.

Propodeum bidentate to bispinose.

Propodeal spiracle very close to or at the margin of the declivity.

Metasternal process absent.

Tibial spurs: mesotibia 0; metatibia 0.

Abdominal segments 2 and 3 (petiole and postpetiole) without spongiform tissue.

Abdominal tergite 4 (first gastral) does not broadly overlap the sternite on the ventral gaster; gastral shoulders absent.

Sting present.

Main pilosity of dorsal head and body: may be absent, frequently sparse; when present, short and stout, or bizarre (squamate, spatulate, remiform).

MALAGASY EURHOPALOTHRIX

1 undescribed species, Comoro Islands

EUTETRAMORIUM Emery, 1899   Myrmicinae

3 species, Malagasy   PLATE 17

DIAGNOSTIC REMARKS: Among the Malagasy myrmicines, *Eutetramorium* has 12-segmented antennae that terminate in a 3-segmented club. The lateral portions of the clypeus are raised into a narrow ridge on each side, in front of the antennal sockets, so that the sockets appear to be set in pits. A triangular point is present at the midpoint of the anterior clypeal margin. The sting is conspicuous and is simple, without a lamellate appendage near its dorsal apex.

DISTRIBUTION AND ECOLOGY: A small genus restricted to Madagascar, where it is found in both rainforest and tropical dry forest, nesting in rotten logs. Queens may be alate or extreme ergatoids.

IDENTIFICATION OF SPECIES: Bolton and Fisher, 2014.

DEFINITION OF THE GENUS (WORKER): Characters of subfamily MYRMICINAE, plus:

Mandible triangular; masticatory margin with 6–8 teeth.

Palp formula 4,3, or 4,2.

Stipes of the maxilla with a coarse transverse crest.

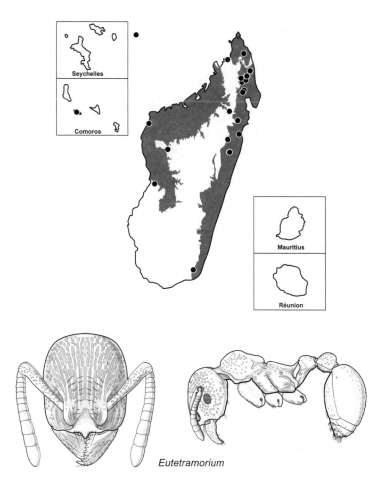

Eutetramorium

Clypeus posteriorly moderately broadly inserted between the frontal lobes; median portion of the clypeus broad, not bicarinate.

Clypeus with lateral portions raised into a shielding wall or sharp ridge in front of the antennal sockets.

Clypeus with a tooth or triangular point at the midpoint of the anterior margin; without an unpaired seta at the midpoint.

Frontal carinae restricted to the frontal lobes, never extending to the posterior margin of the head.

Antennal scrobes absent.

Antenna with 12 segments, with an apical club of 3 segments.

Antennal socket and the torulus subtended by a depressed antennal fossa; the anterior margin of the fossa is formed by the posterior surface of the narrow, raised, lateral portion of the clypeus.

Torulus with upper lobe concealed by the frontal lobe in full-face view. In profile the torulus lobe large, directed downward over the condyle of the scape.

Eyes present.

Head capsule without a median, longitudinal carina.

Metanotal groove impressed.

Propodeum bispinose or bidentate.

Metasternal process present.

Tibial spurs: mesotibia o or 1; metatibia o or 1; simple when present.

Abdominal segment 4 (first gastral) tergite does not broadly overlap the sternite on the ventral gaster; gastral shoulders absent.

Sting simple, without a spatulate to pennant-shaped lamellate appendage that projects from the dorsum of the shaft near or at its apex.

Main pilosity of dorsal head and body: simple.

MALAGASY EUTETRAMORIUM

mocquerysi Emery, 1899

monticellii Emery, 1899

parvum Bolton and Fisher, 2014

FEROPONERA Bolton and Fisher, 2008
Ponerinae

1 species, Afrotropical
PLATE 18

DIAGNOSTIC REMARKS: Among the ponerines, *Feroponera* combines the following characters: Stout, thickly spiniform or peg-like traction setae are present on the mesotibia, meso-basitarsus, and metabasitarsus (not on the metatibia). Eyes are absent. The subtriangular mandible has 5 teeth, and the anterior clypeal margin has a triangular tooth on each side that projects forward over the basal margin of the mandible. The helcium arises just below the midheight of the anterior face of the first gastral segment (A3), not at its base. The mesotibia and metatibia each have 2 spurs.

DISTRIBUTION AND ECOLOGY: Known only from the type-series, which was found in an empty termitary in Cameroon rainforest.

IDENTIFICATION OF SPECIES: Bolton and Fisher, 2008c; Schmidt and Shattuck, 2014.

DEFINITION OF THE GENUS (WORKER): Characters of subfamily PONERINAE, plus:

Mandible subtriangular, short and stout, with a basal groove. Masticatory margin with 5 teeth: apical tooth the largest by far, curved and acute, strongly crossing over the opposing mandible at full closure; also at full closure a space is present between the masticatory margins posterior to tooth 3; preapical tooth reduced; third tooth triangular, the largest after the apical; fourth tooth triangular, smaller than third; fifth tooth smaller still; proximal of the fifth tooth the basal angle is abruptly but bluntly rounded. Dorsum of mandible with a marked pale patch that is much lighter in color, adjacent to teeth 4 and 5.

Palp formula 2,3; apical labial palpomere globular.

Clypeal anterior margin on each side with a broadly triangular tooth that projects forward over the basal margins of the closed mandibles.

Frontal lobes moderate; in full-face view their anterior margins overhang the anterior

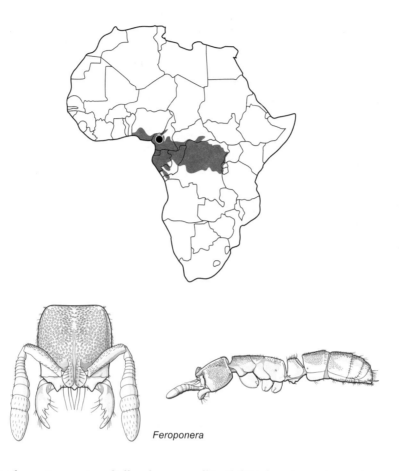

Feroponera

clypeal margin except medially, where a small tooth-like clypeal prominence is
 visible.

Eyes absent.

Antenna with a conspicuous 4-segmented club.

Metanotal groove absent.

Epimeral sclerite present.

Metapleural gland orifice posterior, near the posteroventral corner of the mesosoma.

Propodeal spiracle with orifice elliptical.

Procoxa much larger than mesocoxa and metacoxa.

Mesotibia, mesobasitarsus, and metabasitarsus with stout spiniform traction setae
 (none on metatibia).

Tibial spurs: mesotibia 2, both simple; metatibia 2, the anterior simple, the posterior
 larger and broadly pectinate.

Metatibial posterior surface with a slightly depressed broadly oval area of pale, very
 finely granular cuticle that extends proximally from the base of the spur and appears
 to be glandular.

Pretarsal claws simple.

Abdominal segment 2 (petiole) a node.

Helcium located just below the midheight of the anterior face of A3 (first gastral segment).

Prora a pair of insignificant ridges that arise on each side of the helcium base and
    extend weakly around the anteroventral corner of the first gastral sternite; the
    anterior face of the sternite between the ridges very feebly concave.

Abdominal segment 4 (second gastral segment) with distinctly differentiated preclerites.

Stridulitrum absent from the pretergite of A4.

AFROTROPICAL FEROPONERA

ferox Bolton and Fisher, 2008

FISHEROPONE Schmidt and Shattuck, 2014   Ponerinae

1 species, Afrotropical   PLATE 18

DIAGNOSTIC REMARKS: Among the ponerines, *Fisheropone* combines the characters of presence of a single spur on the metatibia and absence of traction setae on the metabasitarsus, with the following: Eyes are absent. The mandible does not have a basal pit and also lacks a basal groove. The sides of the propodeum strongly converge dorsally, so that the dorsum is very narrow, especially just behind the much broader mesonotum. The mesotibia does not have traction setae scattered among the normal setae, and the helcium arises from the base of the anterior face of the first gastral segment (A3).

DISTRIBUTION AND ECOLOGY: The single species of this genus has been found only a few times, in the forests of Gabon, South Sudan, Uganda, and northern Tanzania, in litter samples. Its presence in countries as far apart as Gabon and Uganda argues that it is probably also present in the intervening wide area.

IDENTIFICATION OF SPECIES: Schmidt and Shattuck, 2014.

DEFINITION OF THE GENUS (WORKER): Characters of subfamily PONERINAE, plus:

Mandible elongate triangular; without a basal groove, without a basal pit; masticatory
    margin with about 7 to 9 teeth. Basal angle angulate.

Palp formula apparently 3,2 (in situ count).

Clypeus simple, without projecting lobes or teeth.

Frontal lobes small and closely approximated; in full-face view their anterior margins
    well behind the anterior clypeal margin.

Eyes absent.

Metanotal groove present.

Epimeral sclerite usually absent.

Metapleural gland orifice opens posteriorly.

Propodeum with sides strongly convergent dorsally, so that the dorsum is very narrow,
    especially just behind the much broader mesonotum.

Propodeal spiracle with orifice roughly circular to short-elliptical.

Tibial spurs: mesotibia 1; metatibia 1, broadly pectinate.

Tibiae and basitarsi without traction setae.

Pretarsal claws simple.

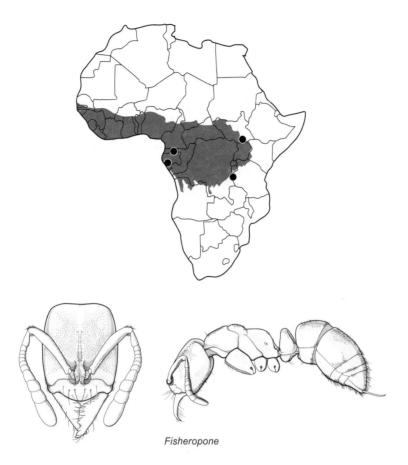

Fisheropone

Abdominal segment 2 (petiole) a scale.

Helcium located at the base of the anterior face of A3 (first gastral segment).

Prora present.

Abdominal segment 4 (second gastral segment) with presclerites differentiated.

Stridulitrum absent from the pretergite of A4.

AFROTROPICAL FISHEROPONE

ambigua (Weber, 1942)

  = *gulera* (Özdikmen, 2010) (redundant replacement name)

HAGENSIA Forel, 1901   Ponerinae

2 species (+ 4 subspecies), Afrotropical   PLATE 18

DIAGNOSTIC REMARKS: Among the ponerines, *Hagensia* combines the characters of presence of 2 spurs on the metatibia, absence of traction setae on the metabasitarsus, and location of the helcium at the base of the anterior face of the first gastral segment (A3), with the following: The mandible has an elongate groove or trench on its dorsal surface but does

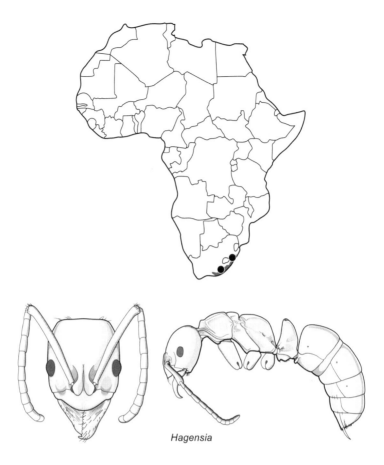

Hagensia

not have a basal groove. The pronotum is marginate dorsolaterally, the metanotal groove is present, and the propodeal spiracle is slit-shaped. The pretarsal claws of the metatarsus have a tooth on the inner curvature.

DISTRIBUTION AND ECOLOGY: This genus is known only from southern and eastern parts of South Africa. Its species nest directly in the ground and forage solitarily on the surface, where they are general scavengers and predators. The queen caste has been lost and reproduction is by a single mated gamergate in each colony.

IDENTIFICATION OF SPECIES: Arnold, 1951.

DEFINITION OF THE GENUS (WORKER): Characters of subfamily PONERINAE, plus:

Mandible triangular and multidentate; dorsal surface of the mandible with a longitudinal elongate groove or trench; basal groove absent.

Palp formula 4,4.

Clypeus simple, its anterior margin convex, without extended lobes or teeth on either the median or the lateral portions.

Frontal lobes moderate; in full-face view their anterior margins do not overhang the anterior clypeal margin.

Eyes present, of moderate size, slightly in front of the midlength of the head capsule.

Pronotum marginate dorsolaterally, sometimes only weakly so.

Metanotal groove present, shallowly to strongly impressed.

Epimeral sclerite present.

Mesopleuron with a transverse sulcus, may be weak or partial.

Metapleural gland orifice opens posteriorly.

Propodeal spiracle with orifice slit-shaped.

Tibial spurs: mesotibia 2; metatibia 2; on each tibia the anterior spur small, simple to barbulate, the posterior spur larger and pectinate.

Metatibial posterior surface with a slightly depressed, usually broadly oval, area of cuticle that extends proximally from the base of the spur and appears to be glandular.

Pretarsal claws each with a single tooth on its inner curvature.

Abdominal segment 2 (petiole) a scale.

Helcium located at the base of the anterior face of A3 (first gastral segment).

Prora present, conspicuous.

Abdominal segment 4 (second gastral segment) with presclerites weakly differentiated.

Stridulitrum absent from the pretergite of A4.

AFROTROPICAL HAGENSIA

havilandi (Forel, 1901)
  = *sulcigera* (Mayr, 1904)
  havilandi fochi (Forel, 1918)
  havilandi godfreyi Arnold, 1926
  havilandi marleyi Arnold, 1926
peringueyi (Emery, 1899)
  peringueyi saldanhae Arnold, 1951

HYPOPONERA Santschi, 1938 — Ponerinae

54 species, Afrotropical; 7 species (+ 4 subspecies), Malagasy — PLATE 18

DIAGNOSTIC REMARKS: Among the ponerines, *Hypoponera* combines the characters of presence of a single spur on the metatibia, absence of traction setae on the metabasitarsus, and location of the helcium at the base of the anterior face of the first gastral segment (A3), with the following: Eyes may be absent or present. The mandible is stoutly triangular, does not have a basal pit, and also lacks a basal groove. The sides of the propodeum do not converge strongly dorsally, so that its dorsum forms a distinct surface behind the mesonotum. The mesotibia does not have traction setae scattered among the normal setae. The petiole (A2) has a ventral process that usually lacks a translucent thin-spot or fenestra anteriorly (a fenestra occurs in a few species), and the posteroventral angles of the subpetiolar process are rounded to acute but never project posteriorly as a pair of small teeth.

DISTRIBUTION AND ECOLOGY: *Hypoponera* has an almost worldwide distribution and is especially species rich in the tropics. In the regions under consideration here it is usually the most common ponerine component encountered in samples of leaf litter or log mold.

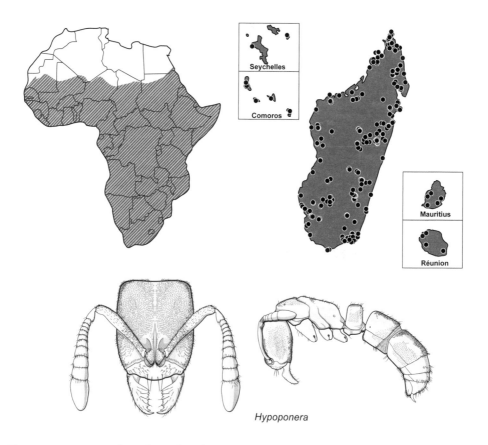

Hypoponera

Nests are constructed in the soil, either directly or under surface objects, between compressed leaves in the litter layer, in dead wood, ranging from small twigs to entire rotting branches, trunks, or stumps, or under flakes of bark on dead timber, or in the ground; no arboreal species are known. Foraging is carried out in all the habitats just noted, these ants being scavengers or general predators of small arthropods. Most species have normal alate queens and males, but some have ergatoid forms of one or both, and ergatoid intercastes between workers and queens occur in some species.

IDENTIFICATION OF SPECIES: Bernard, 1953 (West Africa); Bolton and Fisher, 2011 (Afrotropical).

DEFINITION OF THE GENUS (WORKER): Characters of subfamily PONERINAE, plus:

Mandible stoutly triangular, without a basal groove and without a basal pit; with 7 to about 18 teeth and denticles in total, and with a basal angle between masticatory and basal margins. When the mandibles are fully closed, there is no space between the masticatory margins nor between the basal margins of the mandibles and the clypeus.

Palp formula 1,1 or 1,2, the maxillary palp usually minute.

Clypeus simple, without extended lobes or teeth on either the median or the lateral portions and usually unarmed anteromedially.

Frontal lobes small, their anterior margins well behind the anterior clypeal margin.

Eyes absent or present; when present, always small to minute, lateral, and located well in front of the midlength of the head capsule.

Antenna with the apical 4–6 antennomeres gradually incrassate; only extremely rarely with a sharply differentiated club.

Mesopleuron fused to mesonotum in some species, without trace of a suture between the 2 sclerites.

Metanotal groove present or absent; when present, may be impressed.

Epimeral sclerite absent.

Metapleural gland orifice opens posteriorly.

Propodeal dorsum with its anterior half forming a distinctly defined transverse surface, never reduced to a narrow strip by extreme dorsal convergence of the sides.

Propodeal spiracle small, circular to slightly elliptical.

Tibial spurs: mesotibia 1; metatibia 1, pectinate.

Mesotibia without traction setae.

Pretarsal claws small and simple.

Abdominal segment 2 (petiole) a narrow node or a scale.

Abdominal sternite 2 (petiole sternite) with the ventral process usually lacking a translucent thin-spot or fenestra anteriorly, though fenestra are present in a few species. Posteroventral angles of subpetiolar process rounded to acute but not projecting into a pair of small teeth.

Helcium arises low down on the anterior face of A3 (first gastral segment).

Prora usually present (absent in 1 Afrotropical species): an arched tranverse crest that extends across the first gastral sternite below the helcium; usually the prora extends up the anterior face of the first sternite on each side, so that the entire prora is broadly U-shaped in the anterior view.

Abdominal segment 4 (second gastral segment) with distinctly differentiated presclerites.

Stridulitrum absent or present on the pretergite of A4.

AFROTROPICAL HYPOPONERA

angustata (Santschi, 1914)	*dulcis* (Forel, 1907)
aprora Bolton and Fisher, 2011	= *dulcis uncta* (Santschi, 1914)
austra Bolton and Fisher, 2011	= *rothkirchi* (Wasmann, 1918)
blanda Bolton and Fisher, 2011	= *lotti* (Weber, 1942)
boerorum (Forel, 1901)	= *muscicola* (Weber, 1942)
bulawayensis (Forel, 1913)	= *lamottei* (Bernard, 1953)
camerunensis (Santschi, 1914)	= *mandibularis* (Bernard, 1953)
coeca (Santschi, 1914)	= *villiersi* (Bernard, 1953)
= *myrmicariae* (Wasmann, 1918)	*eduardi* (Forel, 1894)
comis Bolton and Fisher, 2011	= *dideroti* (Forel, 1913)
defessa Bolton and Fisher, 2011	*exigua* Bolton and Fisher, 2011
dema Bolton and Fisher, 2011	*faex* Bolton and Fisher, 2011
dis Bolton and Fisher, 2011	*fatiga* Bolton and Fisher, 2011

hawkesi Bolton and Fisher, 2011

hebes Bolton and Fisher, 2011

ignavia Bolton and Fisher, 2011

importuna Bolton and Fisher, 2011

inaudax (Santschi, 1919)

jeanneli (Santschi, 1935)

   = *jeanneli abyssinica* (Santschi, 1938)

   = *coarctata imatongica* (Weber, 1942)

jocosa Bolton and Fisher, 2011

juxta Bolton and Fisher, 2011

lassa Bolton and Fisher, 2011

lea (Santschi, 1937) (unidentifiable)

lepida Bolton and Fisher, 2011

meridia Bolton and Fisher, 2011

mixta Bolton and Fisher, 2011

molesta Bolton and Fisher, 2011

natalensis (Santschi, 1914)

obtunsa Bolton and Fisher, 2011

occidentalis (Bernard, 1953)

   = *intermedia* (Bernard, 1953)

odiosa Bolton and Fisher, 2011

orba (Emery, 1915)

perparva Bolton and Fisher, 2011

petiolata (Bernard, 1953) (unidentifiable)

producta Bolton and Fisher, 2011

pulchra Bolton and Fisher, 2011

punctatissima (Roger, 1859)

   = *dulcis aemula* (Santschi, 1911)

= *ergatandria cognata* (Santschi, 1912)

= *sulcatinasis durbanensis* (Forel, 1914)

= *incisa* (Santschi, 1914)

= *ragusai sordida* (Santschi, 1914)

= *ergatandria petri* (Forel, 1916)

= *brevis* (Santschi, 1921)

= *mesoepinotalis* (Weber, 1942)

= *breviceps* (Bernard, 1953)

= *ursoidea* (Bernard, 1953)

quaestio Bolton and Fisher, 2011

ragusai (Emery, 1894)

   = *gleadowi aethiopica* (Forel, 1907)

regis Bolton and Fisher, 2011

rigida Bolton and Fisher, 2011

segnis Bolton and Fisher, 2011

sinuosa (Bernard, 1953)

spei (Forel, 1910)

   = *spei fidelis* (Santschi, 1926)

sulcatinasis (Santschi, 1914)

   = *spei devota* (Santschi, 1914)

surda Bolton and Fisher, 2011

tecta Bolton and Fisher, 2011

traegaordhi (Santschi, 1914)

transvaalensis (Arnold, 1947)

tristis Bolton and Fisher, 2011

ursa (Santschi, 1924)

venusta Bolton and Fisher, 2011

MALAGASY HYPOPONERA

grandidieri (Santschi, 1921)

indigens (Forel, 1895)

   indigens bellicosa (Forel, 1895)

johannae (Forel, 1891)

ludovicae (Forel, 1892)

madecassa (Santschi, 1938)

punctatissima (Roger, 1859)

   punctatissima jugata (Forel, 1892)

   punctatissima indifferens (Forel, 1895)

sakalava (Forel, 1891)

   sakalava excelsior (Forel, 1892)

LEPISIOTA Santschi, 1926

Formicinae

47 species (+ 29 subspecies), Afrotropical

PLATE 19

DIAGNOSTIC REMARKS: Among the formicines, *Lepisiota* species characteristically have 11-segmented antennae, and the petiole (A2) has a long posterior peduncle. Dorsally, the petiole is usually armed with a pair of spines or teeth but in some species is merely emarginate. The propodeum is usually armed in some way (only rarely rounded).

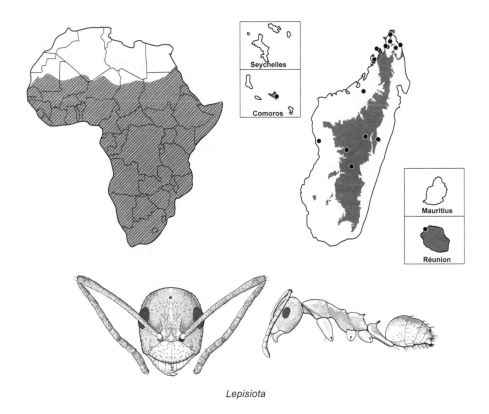

Lepisiota

DISTRIBUTION AND ECOLOGY: *Lepisiota* species occur throughout the Old World tropics, and the genus is also strongly represented in the southern Palaearctic. Species are common throughout the Afrotropical region, and undescribed species are present in Madagascar. Nests are constructed in a variety of locations: in hard-packed earth, including the walls of termitaries, in the ground under stones, in standing or fallen dead wood and tree stumps, in twigs on low vegetation, and in rotten holes in trees. Most species are avid tenders of homopterous insects.

TAXONOMIC COMMENT: The monophyly of this genus is in doubt. Its relationship with *Plagiolepis* in particular is in need of analysis.

IDENTIFICATION OF SPECIES: No review exists; the species-rank taxonomy is confused.

DEFINITION OF THE GENUS (WORKER): Characters of subfamily FORMICINAE, plus:

Mandible with 5–6 teeth; tooth 3 (counting from the apical) is smaller than tooth 4.
Palp formula 6,4.
Antenna with 11 segments.
Antennal sockets close to the posterior clypeal margin; the anterior arc of the torulus touches or slightly indents the clypeal suture.
Eyes present, located from slightly in front of, to slightly behind, the midlength of the sides of the head; ocelli present but may be indistinct in a few species.

Mesothorax usually constricted in its anterior half, sometimes very strongly constricted and distinctly elongate.

Metanotum forms a distinct isolated sclerite on the dorsum of the mesosoma in some species but not in others.

Metapleural gland present.

Propodeum only rarely unarmed; usually angulate, tuberculate, dentate, or short-spinose posteriorly.

Propodeal spiracle circular to subcircular, high and posterior; the spiracle may be immediately below the propodeal angle or may be incorporated into whatever posterior prominence or armament the propodeum possesses; in some species the spiracle may comprise most of the posterior propodeal prominence.

Propodeal foramen long: in ventral view the foramen extends anterior of a line that spans the anteriormost points of the metacoxal cavities.

Metacoxae widely separated: in ventral view, with the mesocoxae and metacoxae directed at right-angles to the long axis of the mesosoma, the distance between the bases of the mesocoxae is markedly less than the distance between the bases of the metacoxae.

Tibial spurs: mesotibia 0; metatibia 0.

Abdominal segment 2 (petiole) a scale or short node, its dorsal surface sometimes transverse but more commonly indented medially or armed with a pair of teeth or spines.

Petiole (A2) a scale or node with a shorter anterior face and a longer posterior face; petiole with a long posterior peduncle that is attached to A3 (first gastral segment) so low down that the peduncle appears to arise from the ventral surface of the gaster.

Petiole (A2) with its ventral margin U-shaped in section.

Abdominal segment 3 (first gastral) with a sharply margined longitudinal concavity or groove into which the posterior peduncle of the petiole fits when the mesosoma and gaster are aligned. In this position the anterior face of abdominal segment 3 overhangs and conceals the posterior peduncle of the petiole.

Abdominal segment 3 (first gastral) basally with complete tergosternal fusion for some distance on each side of the helcium; the line of fusion follows the edges of the impression in the anterior face of A3 and the unfused portions of the tergite and sternite commence some distance above the helcium, at the dorsolateral apices of the impression.

Abdominal sternite 3 without a transverse sulcus across the sclerite immediately behind the helcium sternite (i.e., the presternite and poststernite of A3 not separated by a sulcus).

Acidopore with a seta-fringed nozzle.

AFROTROPICAL LEPISIOTA

affinis (Santschi, 1937)	*angolensis* (Santschi, 1937)
albata (Santschi, 1935)	*arenaria* (Arnold, 1920)
alexis (Santschi, 1937)	*arnoldi* (Forel, 1913)
alexis dulcis (Santschi, 1937)	*arnoldi mota* (Santschi, 1937)
ambigua (Santschi, 1935)	*cacozela* (Stitz, 1916)

canescens (Emery, 1897)
  canescens latior (Santschi, 1935)
capensis (Mayr, 1862)
  capensis simplex (Forel, 1892)
  capensis guineensis (Mayr, 1902)
  capensis simplicoides (Forel, 1907)
  capensis laevis (Santschi, 1913)
  capensis anceps (Forel, 1916)
  capensis junodi (Forel, 1916)
  capensis minuta (Forel, 1916)
  capensis specularis (Santschi, 1935)
  capensis subopaciceps (Santschi, 1937)
  capensis acholli (Weber, 1943)
  capensis issore (Weber, 1943)
  capensis thoth (Weber, 1943)
capitata (Forel, 1913)
carbonaria (Emery, 1892)
  carbonaria baumi (Forel, 1910)
crinita (Mayr, 1895)
curta (Emery, 1897)
dendrophila (Arnold, 1949)
depilis (Emery, 1897)
deplanata (Stitz, 1911)
depressa (Santschi, 1914)
egregia (Forel, 1913)
  egregia santschii (Arnold, 1920)
erythraea (Forel, 1910)
foreli (Arnold, 1920)
  foreli convexa (Arnold, 1920)
  foreli impressa (Arnold, 1920)
gerardi (Santschi, 1915)
gracilicornis (Forel, 1892)
  gracilicornis abdominalis (Forel, 1894)

hirsuta (Santschi, 1914)
  hirsuta setosella (Santschi, 1935)
imperfecta (Santschi, 1926)
  imperfecta congolensis (Santschi, 1935)
incisa (Forel, 1913)
longinoda (Arnold, 1920)
megacephala (Weber, 1943)
mlanjiensis (Arnold, 1946)
monardi (Santschi, 1930)
  monardi australis (Santschi, 1930)
ngangela (Santschi, 1937)
nigrisetosa (Santschi, 1935)
nigriventris (Emery, 1899)
obtusa (Emery, 1901)
oculata (Santschi, 1935)
palpalis (Santschi, 1935)
piliscapa (Santschi, 1935)
  piliscapa longipilosa (Santschi, 1935)
  piliscapa punctifrons (Santschi, 1935)
quadraticeps (Arnold, 1944)
rubrovaria (Forel, 1910)
  rubrovaria pilosa (Forel, 1913)
rugithorax (Santschi, 1930)
schoutedeni (Santschi, 1935)
silvicola (Arnold, 1920)
somalica (Menozzi, 1927)
spinosior (Forel, 1913)
  spinosior ballaensis (Arnold, 1920)
  spinosior natalensis (Arnold, 1920)
submetallica (Arnold, 1920)
  submetallica aspera (Arnold, 1920)
tenuipilis (Santschi, 1935)
validiuscula (Emery, 1897)

LEPTANILLA Emery, 1870   Leptanillinae

3 species, Afrotropical   PLATE 19

DIAGNOSTIC REMARKS: *Leptanilla* species are minute ants with a waist consisting of 2 segments and with 12-segmented antennae. Eyes are absent. Frontal lobes are absent so that the antennal sockets are fully exposed and very close to the anterior margin of the head. The promesonotal suture is present, deeply impressed, and articulated. Propodeal lobes are absent, the pygidium is large and conspicuous, not armed with teeth or spines; a strong sting is present.

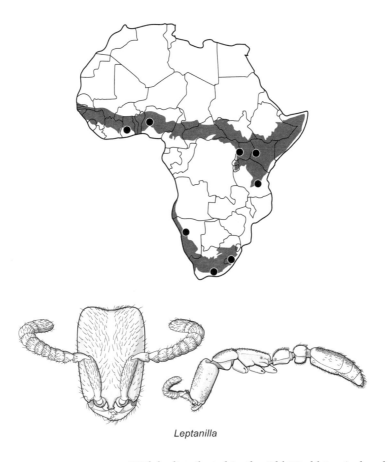

Leptanilla

DISTRIBUTION AND ECOLOGY: Widely distributed in the Old World tropical and temperate zones but absent from the Malagasy region and from the New World. *Leptanilla* species dwell and forage in the soil, sometimes at considerable depth. The only Afrotropical species in which the worker is known (*L. boltoni*) has been found a few times in soil samples from forested eastern Ghana. It is most probable that the genus is more widely represented in Africa, but the habitat of the species, deep in the soil, renders discovery of the workers more a matter of luck than judgment.

TAXONOMIC COMMENT: Like some of the doryline genera, *Leptanilla* has a dual taxonomy, with a large number of species named only from males. In the Afrotropics only a single species has been described from the worker (*L. boltoni*); the other 2 are male-based, from Nigeria and South Africa.

IDENTIFICATION OF SPECIES: Baroni Urbani, 1977.

DEFINITION OF THE GENUS (WORKER): Characters of subfamily LEPTANILLINAE, plus:

Mandible slender, with 1–4 teeth.
Palp formula 2,1, or 1,1.
Tibial spurs: mesotibia 1; metatibia 1.
Minute, slender ants; total body length usually less than 1.5 mm.

africana Baroni Urbani, 1977

australis Baroni Urbani, 1977

boltoni Baroni Urbani, 1977

LEPTOGENYS Roger, 1861 — Ponerinae

56 species, Afrotropical; 60 species, Malagasy — PLATE 19

DIAGNOSTIC REMARKS: Among the ponerines, *Leptogenys* combines the characters of presence of 2 spurs on the metatibia, absence of traction setae on the metabasitarsus, and location of the helcium at the base of the anterior face of the first gastral segment (A3), with the following: The mandible is triangular to linear and curved but never has more than 3 teeth (usually 2); there is no basal pit. The clypeus usually projects as a median lobe. The frontal lobes are small and only partially cover the antennal sockets. The pretarsal claws of the metatarsus are usually distinctly pectinate (reduced in a few minute species).

DISTRIBUTION AND ECOLOGY: *Leptogenys* is a pantropical genus and is the most species rich in subfamily Ponerinae. Its species can be found in an extremely wide variety of habitats, from semiarid to rainforest. Nests are usually constructed either in the ground or in and under rotten wood or in compressed leaf litter. A few species nest in dead branches or rotten holes in living trees, some distance above the ground, and some have been found in abandoned termitaries. The species appear to change nest sites frequently. Foraging takes place in the topsoil, on the surface of the ground, or in and under fallen timber. The species that nest in rotten pockets in standing trees forage on the trunks. At least in West Africa, *Leptogenys* species are not commonly retrieved from leaf-litter samples, and a number of species are known to be nocturnal. Many species of *Leptogenys* are known to prey on isopods, termites, or amphipods, so nocturnal foraging may be correlated with the maximum activity time of the prey. Queens of *Leptogenys* species are nearly always ergatoid, but 1 African species (*L. ergatogyna*) has a queen that is morphologically alate-like but always flightless.

IDENTIFICATION OF SPECIES: Bolton, 1975a (Afrotropical + Malagasy); Rakotonirina and Fisher, 2014 (Malagasy).

DEFINITION OF THE GENUS (WORKER): Characters of subfamily PONERINAE, plus:

Mandible subtriangular, short-linear, or elongate and curvilinear; in the last condition incapable of closing against the clypeus. Masticatory margin with 1–3 teeth (usually 2, apical and preapical). Basal groove variably developed: it may be a broad channel, or a narrow groove, or faint, vestigial, or absent.

Palp formula 4,4, or 4,3, or 3,3.

Clypeus usually projects as a distinct lobe and has a longitudinal sharp median carina dorsally (lobe reduced in *L. maxillosa* group).

Frontal lobes small and only partially cover the antennal sockets, the major portions of which are exposed in full-face view.

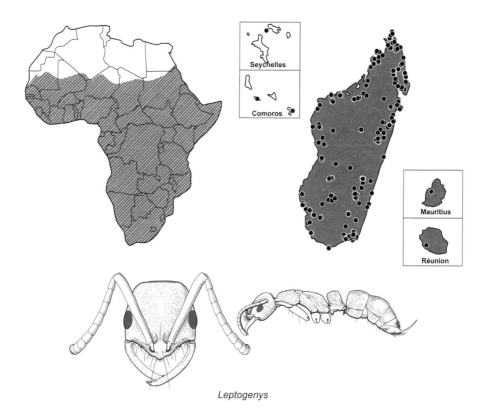

Leptogenys

Eyes present, minute to very large; located from slightly behind to distinctly in front of
the midlength of the head capsule.

Metanotal groove usually distinct and impressed; present but reduced in some minute
species.

Epimeral sclerite present.

Metapleural gland orifice opens posteriorly.

Propodeal spiracle with orifice usually circular to short-elliptical, sometimes an elongate
slit.

Tibial spurs: mesotibia 2; metatibia 2; on each tibia the anterior spur small, simple to
barbulate, the posterior spur larger and pectinate.

Pretarsal claws usually conspicuously pectinate, with 4–7 teeth; pectination reduced to
1–2 preapical teeth in some species and absent in one minute species.

Abdominal segment 2 (petiole) a node or a scale.

Helcium located at the base of the anterior face of A3 (first gastral segment).

Prora present.

Abdominal segment 4 (second gastral segment) usually with differentiated presclerites,
but demarcation may be very reduced (some species of *L. stuhlmanni* group);
presclerites are entirely absent in some species of the *L. stuhlmanni* group and
L. guineensis group.

Stridulitrum present on the pretergite of A4, frequently extensive and conspicuous.

amon Bolton, 1975

ankhesa Bolton, 1975

arnoldi Forel, 1913

attenuata (Smith, F., 1858)

   = *jaegerskjoeldi* Santschi, 1914

bellii Emery, 1901

bubastis Bolton, 1975

buyssoni Forel, 1907

camerunensis Stitz, 1910

castanea (Mayr, 1862)

   = *parva* Forel, 1901

   = *parva bellua* Forel, 1914

   = *parva dispar* Santschi, 1914

   = *hewitti* Santschi, 1932

conradti Forel, 1913

crassinoda Arnold, 1926

crustosa Santschi, 1914

   = *conradti rufipes* Santschi, 1937

   = *africanus* Weber, 1942

cryptica Bolton, 1975

diatra Bolton, 1975

elegans Bolton, 1975

ergatogyna Wheeler, W.M., 1922

   = *cursor* Arnold, 1954

erythraea Emery, 1902

excellens Bolton, 1975

falcigera Roger, 1861

ferrarii Forel, 1913

   = *ferrarii dentatula* Santschi, 1915

furtiva Arnold, 1926

guineensis Santschi, 1914

havilandi Forel, 1901

honoria Bolton, 1975

intermedia Emery, 1902

   = *nitida* (Smith, F., 1858) (homonym)

   = *tenuis* Stitz, 1911

   = *nitida gracilis* Santschi, 1914

   = *nitida insinuata* Santschi, 1914

   = *nitida adpressa* Forel, 1914

   = *nitida aena* Forel, 1914

   = *nitida grandior* Forel, 1915

   = *brevinodis deflocata* Santschi, 1926

   = *nitida speculans* Santschi, 1926

   = *nitida capensis* Baroni Urbani, 1971

   = *nitida brevinodis* Forel, 1915

   (homonym)

jeanneli Santschi, 1914

khaura Bolton, 1975

leiothorax Prins, 1965

longiceps Santschi, 1914

mactans Bolton, 1975

mastax Bolton, 1975

maxillosa (Smith, F., 1858)

   = *cribrata* Emery, 1895

microps Bolton, 1975

nebra Bolton, 1975

nuserra Bolton, 1975

occidentalis Bernard, 1953

   = *ferrarii bernardi* Baroni Urbani, 1971

   = *ferrarii sulcinodis* Bernard, 1953

   (homonym)

pavesii Emery, 1892

   = *maxillosa sericeus* Weber, 1942

peringueyi Forel, 1913

piroskae Forel, 1910

princeps Bolton, 1975

ravida Bolton, 1975

regis Bolton, 1975

schwabi Forel, 1913

spandax Bolton, 1975

sterops Bolton, 1975

strator Bolton, 1975

striatidens Bolton, 1975

stuhlmanni Mayr, 1893

stygia Bolton, 1975

sulcinoda (André, 1892)

terroni Bolton, 1975

testacea (Donisthorpe, 1948)

titan Bolton, 1975

trilobata Santschi, 1924

vindicis Bolton, 1975

zapyxis Bolton, 1975

acutirostris Emery, 1914

alamando Rakotonirina and Fisher, 2014

alatapia Rakotonirina and Fisher, 2014

alluaudi Emery, 1895

ambo Rakotonirina and Fisher, 2014

andritantely Rakotonirina and Fisher, 2014

angusta (Forel, 1892)

anjara Rakotonirina and Fisher, 2014

antongilensis Emery, 1899

arcirostris Santschi, 1926

avaratra Rakotonirina and Fisher, 2014

avo Rakotonirina and Fisher, 2014

barimaso Rakotonirina and Fisher, 2014

bezanozano Rakotonirina and Fisher, 2014

borivava Rakotonirina and Fisher, 2014

chrislaini Rakotonirina and Fisher, 2014

coerulescens Emery, 1895

comajojo Rakotonirina and Fisher, 2014

diana Rakotonirina and Fisher, 2014

edsoni Rakotonirina and Fisher, 2014

falcigera Roger, 1861

fasika Rakotonirina and Fisher, 2014

fiandry Rakotonirina and Fisher, 2014

fotsivava Rakotonirina and Fisher, 2014

gracilis Emery, 1899

grandidieri Forel, 1910

incisa Forel, 1891

imerinensis Forel, 1892

johary Rakotonirina and Fisher, 2014

lavavava Rakotonirina and Fisher, 2014

lohahela Rakotonirina and Fisher, 2014

lucida Rakotonirina and Fisher, 2014

malama Rakotonirina and Fisher, 2014

mangabe Rakotonirina and Fisher, 2014

manja Rakotonirina and Fisher, 2014

manongarivo Rakotonirina and Fisher, 2014

maxillosa (Smith, F., 1858)

= *vinsonnella* (Dufour, 1864)

mayotte Rakotonirina and Fisher, 2014

namana Rakotonirina and Fisher, 2014

namoroka Rakotonirina and Fisher, 2014

oswaldi Forel, 1891

pilaka Rakotonirina and Fisher, 2014

pavesii Emery, 1892

rabebe Rakotonirina and Fisher, 2014

rabesoni Rakotonirina and Fisher, 2014

ralipra Rakotonirina and Fisher, 2014

ridens Forel, 1910

sahamalaza Rakotonirina and Fisher, 2014

saussurei (Forel, 1891)

stuhlmanni Mayr, 1893

= *comorensis* Forel, 1907

suarensis Emery, 1895

tatsimo Rakotonirina and Fisher, 2014

toeraniva Rakotonirina and Fisher, 2014

truncatirostris Forel, 1897

tsingy Rakotonirina and Fisher, 2014

variabilis Rakotonirina and Fisher, 2014

vatovavy Rakotonirina and Fisher, 2014

vitsy Rakotonirina and Fisher, 2014

voeltzkowi Forel, 1897

zohy Rakotonirina and Fisher, 2014

LINEPITHEMA Mayr, 1866   Dolichoderinae

1 species, Afrotropical, introduced   PLATE 19

DIAGNOSTIC REMARKS: Among the dolichoderines, the petiole (A2) of *Linepithema* has a well-developed, narrow, erect scale that is not overhung from behind by the first gastral segment. The fourth gastral sternite (sternite of A6) is keel-shaped posteriorly.

DISTRIBUTION AND ECOLOGY: *Linepithema* is a Neotropical genus with 20 extant species, one of which, *L. humile*, is now widely distributed around the world in zones with a Medi-

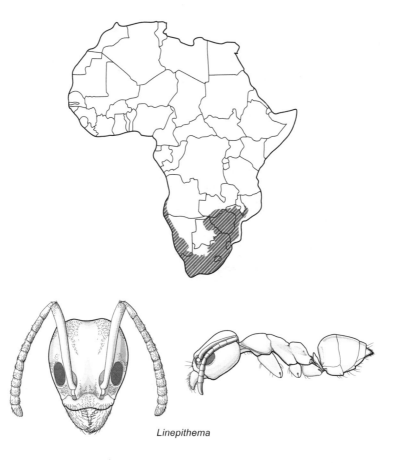

Linepithema

terranean climate, and is quite common in some parts of South Africa. The species nests in the ground, is invasive, and will enter buildings in search of food and water. The species has not been recorded from the Malagasy region.

IDENTIFICATION OF SPECIES: Wild, 2007.

DEFINITION OF THE GENUS (WORKER): Characters of DOLICHODERINAE, plus:

Masticatory margin of mandible with a combination of teeth and denticles (about 12–15 in total).

Palp formula usually 6,4.

Anterior clypeal margin with a broad, shallow concavity.

Antenna with 12 segments.

Metathoracic spiracles dorsal.

Metanotal groove present.

Propodeum unarmed.

Propodeal spiracle lateral and ventral of the dorsum.

Tibial spurs: mesotibia 1; metatibia 1, pectinate.

Abdominal segment 2 (petiole) a narrow, erect scale, slightly inclined anteriorly.

Abdominal segment 3 (first gastral) vertical, not projecting forward over the petiole, not concealing the petiole in dorsal view.

Abdominal tergites 3–6 visible in dorsal view (i.e., 4 gastral tergites visible in dorsal view); tergite of A7 (pygidium) is reflexed onto the ventral surface of the gaster where it abuts abdominal sternite 7 (hypopygium).

Abdominal sternite 6 (fourth gastral sternite) keel-shaped posteriorly.

AFROTROPICAL LINEPITHEMA

humile (Mayr, 1868)

LIOPONERA Mayr, 1879   Dorylinae

10 species, Afrotropical; 2 species, Malagasy   PLATE 20

DIAGNOSTIC REMARKS: Among the dorylines, *Lioponera* has eyes but lacks the promesonotal suture; the waist is of a single segment that is distinctly marginate dorsolaterally. The propodeal spiracle is low on the side and situated behind the midlength of the sclerite, and propodeal lobes are present. The mesotibia has a single spur present and the pretarsal claws lack a preapical tooth. Uniquely, the metacoxa has a posterodorsal cuticular flange that forms a vertical lamella.

DISTRIBUTION AND ECOLOGY: *Lioponera* species are present throughout the Old World tropics but are especially numerous in Australia. In the regions under consideration here, they are usually inhabitants of the leaf litter and topsoil, frequently nesting in or under rotten wood or under stones. Their prey is suspected to be other ants.

TAXONOMIC COMMENT: *Lioponera* is currently listed as a junior synonym of *Cerapachys*. It will be formally revived from synonymy by Marek Borowiec, in a genus-rank revision of Dorylinae that is in preparation.

IDENTIFICATION OF SPECIES: Brown, 1975 (Afrotropical, as part of *Cerapachys*).

DEFINITION OF THE GENUS (WORKER): Characters of subfamily DORYLINAE, plus:

Prementum not visible when mouthparts closed.

Palp formula 4,3.

Eyes present, moderate to large, located from slightly behind to well in front of the midlength of the head capsule in full-face view; ocelli absent.

Frontal carinae present as a pair of closely approximated raised ridges, abruptly truncated posteriorly, without horizontal frontal lobes.

Antennal sockets fully exposed in full-face view.

Parafrontal ridges present.

Antenna with 12 segments; funiculus gradually incrassate, the apical segment larger than the preapical but not strikingly enlarged.

Antennal scrobes absent.

Promesonotal suture absent on the dorsum of the mesosoma.

Pronotal-mesopleural suture present on the side of the mesosoma.

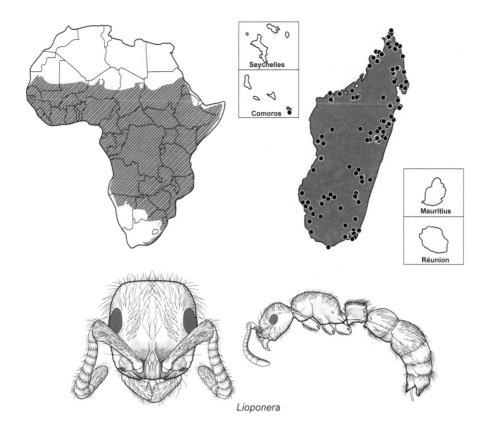

Lioponera

Metanotal groove absent.

Propodeal spiracle low on side and situated at or behind midlength of the sclerite, not subtended by a longitudinal impression.

Propodeal lobes present.

Tibial spurs: mesotibia 1, pectinate; metatibia 1, pectinate.

Metatibial gland usually present distally on ventral surface; its orifice usually a slit but sometimes a pore.

Metacoxae with a posterodorsal cuticular flange that forms a vertical lamella.

Pretarsal claws simple.

Waist of 1 segment (petiole = A2).

Petiole (A2) sessile; distinctly marginate dorsolaterally.

Helcium projects from anterior surface of A3 (first gastral segment) at about its midheight.

Prora present, usually conspicuous.

Abdominal segment 4 with strongly developed presclerites.

Abdominal segments 5 and 6 without presclerites.

Pygidium (tergite of A7) large and distinct, flattened to concave dorsally, and armed laterally, posteriorly, or both, with a series or row of denticles or peg-like spiniform setae.

Sting present, large and fully functional.

braunsi (Emery, 1902)	= *cooperi* Donisthorpe, 1939
braytoni (Weber, 1949)	(homonym)
coxalis (Arnold, 1926)	= *alfierii* Donisthorpe, 1939
decorsei Santschi, 1912	*nigra* Santschi, 1914
foreli (Santschi, 1914)	*nkomoensis* (Forel, 1916)
= *langi* (Wheeler, W.M., 1922)	= *cooperi congolensis* (Forel, 1916)
= *santschii* (Wheeler, W.M., 1922)	= *eidmanni* (Menozzi, 1942)
= *occipitalis* (Bernard, 1953)	*similis* Santschi, 1930
longitarsus Mayr, 1879	*vespula* (Weber, 1949)
= *aegyptiacus* (Brown, 1975)	= *cooperi* (Arnold, 1915) (homonym)

MALAGASY LIOPONERA

kraepelinii (Forel, 1895)
mayri (Forel, 1892)
  = *mayri brachynodus* (Forel, 1892)

LIVIDOPONE Bolton and Fisher gen. n.   Dorylinae

Type-species: *Cerapachys lividus* Brown, 1975: 64, by present designation

1 species, Malagasy   PLATE 20

DIAGNOSTIC REMARKS: Among the Old World dorylines, *Lividopone* is immediately diagnosed by its very high-set helcium, which in profile arises from very close to the dorsal margin of A3 (the helcium arises at about the midheight of A3 in all other genera). In addition, the promesonotal and pronotal-mesopleural sutures are absent, so that the pronotum is entirely fused to the mesonotum dorsally and the mesopleuron laterally.

DISTRIBUTION AND ECOLOGY: This uniquely Madagascan genus contains more than 20 species and displays a marked adaptive radiation on the island. *Lividopone* species occur in and under rotten wood and in the leaf litter of forested areas. All are undescribed except for the type-species.

TAXONOMIC COMMENT: The type-species, currently the only one formally described, is *Cerapachys lividus* Brown, 1975: 64, which was described in detail there. One other doryline genus, *Acanthostichus* Mayr, 1887, has the helcium in an elevated, amblyoponine-like position similar to *Lividopone*. However, this New World genus, with 22 extant species, retains the lateral suture between the pronotum and mesopleuron, has the palp formula 2,3, lacks parafrontal ridges, possesses narrow horizontal frontal lobes, and has a groove in the side of the head that extends posteriorly from the mandibular insertion.

IDENTIFICATION OF SPECIES: Brown, 1975 (as part of *Cerapachys*).

DEFINITION OF THE GENUS (WORKER): Characters of subfamily DORYLINAE, plus:

Prementum not visible when mouthparts closed.
Palp formula 3,2 (in situ count).

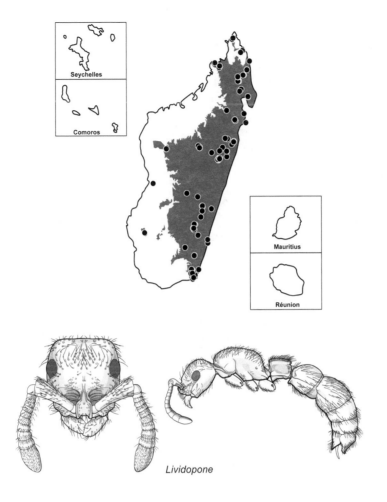

Lividopone

Eyes present, moderate to large, located from slightly in front of, to considerably behind, the midlength of the head capsule; ocelli absent.

Frontal carinae present as a pair of short elevated crests, usually truncated posteriorly but in a few species continued posteriorly as a pair of closely approximated low ridges.

Antennal sockets fully exposed in full-face view.

Parafrontal ridges present and conspicuous.

Antenna with 12 segments; apical antennomere enlarged, sometimes very large.

Antennal scrobes absent.

Head capsule, in ventral or ventrolateral view, usually with a carina that extends down the posterolateral margin and onto the ventral surface, where it passes through a near right-angle and extends to the ventral midline; in one species the carina continues along the ventrolateral margin, almost to the mandibular insertion.

Promesonotal suture absent from the dorsum of the mesosoma.

Pronotal-mesopleural suture absent on side of the mesosoma.

Mesopleuron with a transverse sulcus; the anepisternum above the sulcus without other visible boundaries, completely fused to the mesonotum.

Metanotal groove absent.

Propodeal spiracle low on side and situated at or behind the midlength of the sclerite, not subtended by a longitudinal impression.

Propodeal lobes present.

Tibial spurs: mesotibia 1, pectinate; metatibia 1, pectinate.

Metatibial gland absent, at least no orifice discernible.

Pretarsal claws simple.

Waist of 1 segment (petiole = A2).

Petiole (A2) sessile, not marginate dorsolaterally but bluntly angulate in some species.

Helcium projects from very high on the anterior surface of A3 (first gastral segment); in profile the dorsal surface of the helcium arises from immediately below the anterodorsal angle of A3.

Prora of A3 a U-shaped or V-shaped cuticular flange or sharp carina that separates the anterior face of the poststernite from the lateral and ventral surfaces; in a couple of species the prora reduced to a reinforced rim.

Abdominal segment 4 with strongly developed presclerites.

Abdominal segments 5 and 6 without presclerites.

Pygidium (tergite of A7) large, with marginal denticles or peg-like dentiform setae; sometimes with a bifid cuticular fork apically.

Sting present, large and fully functional.

MALAGASY LIVIDOPONE

livida (Brown, 1975)

LOBOPONERA Bolton and Brown, 2002　　　　　　　　　Ponerinae

9 species, Afrotropical　　　　　　　　　　　　　　　　　　PLATE 20

DIAGNOSTIC REMARKS: Among the ponerines, *Loboponera* combines the characters of presence of a single spur on the metatibia and absence of traction setae on the metabasitarsus with the following: The clypeus has a projecting median lobe, on each side of which there is also a small projecting lobe that overhangs the outer base of the mandible. The mandibles are triangular to subtriangular and have 2–6 teeth. The frontal lobes are very large, and the head capsule in profile has a prominent curved flange on its posterolateral curve. The propodeal dorsum lacks a median longitudinal groove, and both the mesofemur and metafemur have a mid-dorsal longitudinal groove, at least on the basal half. The second gastral segment (A4) has a long tergite that is strongly vaulted, arched, and downcurved posteriorly.

DISTRIBUTION AND ECOLOGY: *Loboponera* species are restricted to the rainforests of West and Central Africa. They nest in rotten wood on the ground, sometimes in deep shade, and forage both there and in the leaf litter. Specimens have also been found in abandoned termitaries.

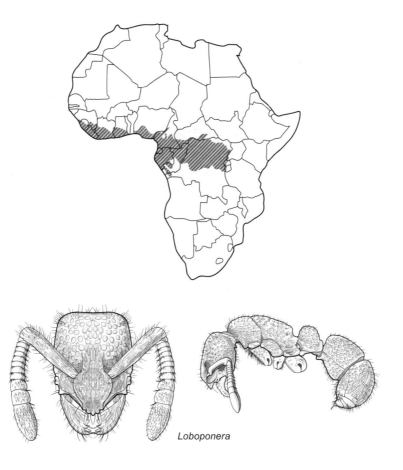

Loboponera

IDENTIFICATION OF SPECIES: Bolton and Brown, 2002; Fisher, 2006.

DEFINITION OF THE GENUS (WORKER): Characters of subfamily PONERINAE, plus:

Mandible triangular to subtriangular, with the basal angle rounded to bluntly angulate and the masticatory margin with 2–6 teeth; basal groove present, often weak and indistinct.

Palp formula 2,2.

Clypeus with a projecting median lobe; anterolaterally the clypeus also with a small projecting lobe that overhangs the outer base of the mandible.

Frontal lobes very large; in full-face view their anterior margins sometimes overlap the lateral portions of the anterior clypeal margin but not the median lobe.

Eyes present, small to minute, in front of midlength of the head capsule.

Head capsule in profile with a projecting curved flange on its posterolateral curve; the flange formed from the extended lateral portion of the occipital carina.

Mesopleural transverse sulcus present; the katepisternum below the sulcus usually well defined, but anepisternum fused to the mesonotum without trace of a suture between the 2 sclerites.

Metanotal groove absent.

Epimeral sclerite reduced to absent.

Metapleural gland orifice opens laterally.

Propodeal spiracle circular or very nearly so.

Propodeal dorsum without a median longitudinal groove or impression.

Mesofemur and metafemur with a mid-dorsal longitudinal groove, at least on the basal half.

Tibial spurs: mesotibia 1; metatibia 1; each spur pectinate.

Pretarsal claws simple.

Abdominal segment 2 (petiole) a node, strongly tapered apically in one species.

Abdominal sternite 2 (petiole sternite) with its anteroventral articulatory surface long and broad, the surface with a narrow, median, V-shaped longitudinal groove, or small, central, pore-like depression.

Helcium arises low down on the anterior face of A3 (first gastral segment).

Prora a conspicuous lobe.

Abdominal segment 4 (second gastral segment) with very strongly differentiated presclerites.

Abdominal segment 4 with the sternite very short; the tergite much longer, strongly vaulted, arched and downcurved posteriorly.

Stridulitrum absent from the pretergite of A4.

AFROTROPICAL LOBOPONERA

basalis Bolton and Brown, 2002	*politula* Bolton and Brown, 2002
edentula Bolton and Brown, 2002	*subatra* Bolton and Brown, 2002
nasica (Santschi, 1920)	*trica* Bolton and Brown, 2002
nobiliae Fisher, 2006	*vigilans* Bolton and Brown, 2002
obeliscata Bolton and Brown, 2002	

MALAGIDRIS Bolton and Fisher, 2014 — Myrmicinae

6 species, Malagasy — PLATE 20

DIAGNOSTIC REMARKS: Among the Malagasy myrmicines, *Malagidris* has 12-segmented antennae that terminate in a 3-segmented club and may be very slender. The mandibles are triangular and have at least 8 teeth. The clypeus has a stout unpaired seta at the midpoint of the convex anterior margin. The pronotum together with the anterior mesonotum is swollen and distinctly convex in profile, so that the dorsalmost point of the promesonotum is at a considerably higher level than the long, posteriorly bispinose, propodeal dorsum. The first gastral tergite (the tergite of A4) does not broadly overlap the sternite on the ventral gaster, and the sting is strongly developed and simple.

DISTRIBUTION AND ECOLOGY: A small genus of ants that is restricted to Madagascar. It inhabits both rainforest and dry forest, nesting in rotten wood or in the ground, either directly or under stones. Ground foraging is frequent, and workers have also been recovered from leaf litter and from low vegetation.

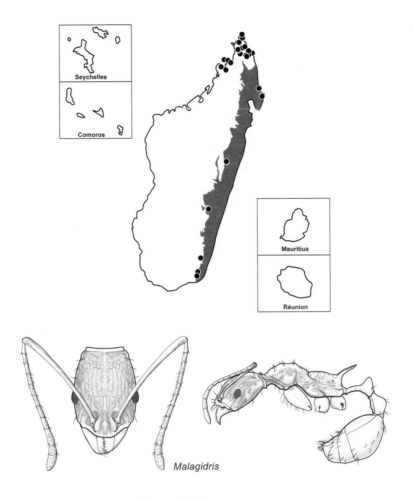

Malagidris

IDENTIFICATION OF SPECIES: Bolton and Fisher, 2014.

DEFINITION OF THE GENUS (WORKER): Characters of subfamily MYRMICINAE, plus:

Mandible triangular; masticatory margin with 8–13 teeth.

Palp formula 5,3.

Stipes of the maxilla with a weak transverse crest.

Clypeus posteriorly moderately broadly inserted between the frontal lobes; median portion of the clypeus longitudinally costulate, not bicarinate.

Clypeus with a stout unpaired seta at the midpoint of the convex anterior margin.

Frontal carinae short, restricted to well-defined but narrow frontal lobes.

Antennal scrobes absent.

Antenna with 12 segments, with an apical club of 3 segments that may be very slender.

Torulus with the upper lobe visible in full-face view; maximum width of torulus lobe is just posterior to the point of maximum width of the frontal lobes.

Eyes present, located in front of the midlength of the head capsule in full-face view.

Head capsule without a median, longitudinal carina; occipital carina conspicuous.

Pronotum plus anterior mesonotum swollen and distinctly convex in profile; dorsalmost point of the promesonotum at a considerably higher level than the long propodeal dorsum.

Propodeum bispinose; propodeal lobes rounded.

Propodeal spiracle behind midlength of the sclerite, at about midheight of the side of the propodeum and far in front of the margin of the declivity.

Metasternal process obsolete, at most a narrow crest on each side of the metasternal pit, each crest sometimes extended posteriorly as a narrow carina.

Tibial spurs: mesotibia 1; metatibia 1; both simple.

Abdominal segment 2 (petiole) with a long, slender anterior peduncle; the spiracle slightly in front of the midlength of the peduncle.

Abdominal segment 3 (postpetiole) elongate, not dorsoventrally flattened, not markedly broader than high; ventral surface in profile flat to shallowly convex.

Stridulitrum present on the pretergite of abdominal segment 4.

Abdominal tergite 4 (first gastral) does not broadly overlap the sternite on the ventral gaster; gastral shoulders absent.

Sting strongly developed, simple.

Main pilosity of dorsal head and body: simple, often sparse, usually absent from the propodeal dorsum.

MALAGASY MALAGIDRIS

alperti Bolton and Fisher, 2014
belti (Forel, 1895)
dulcis Bolton and Fisher, 2014
galokoa Bolton and Fisher, 2014
jugum Bolton and Fisher, 2014
sofina Bolton and Fisher, 2014

MEGAPONERA Mayr, 1862　　　　　　　　　　　　　　　　Ponerinae

1 species (+ 5 subspecies), Afrotropical　　　　　　　　　　　　　PLATE 21

DIAGNOSTIC REMARKS: Among the ponerines, *Megaponera* combines the characters of presence of 2 spurs on the metatibia, absence of traction setae on the metabasitarsus, and location of the helcium at the base of the anterior face of the first gastral segment (A3), with the following: The worker caste is polymorphic. The mandible does not have a basal pit or a basal groove. The head capsule has a longitudinal cuticular carina on each side that extends from the anterior margin of the eye to the anterior margin of the head. The metanotal groove is present, and the propodeal spiracle is slit-shaped. The pretarsal claws of the metatarsus have a stout tooth on the inner curvature.

DISTRIBUTION AND ECOLOGY: The single species of this purely Afrotropical genus is distributed over almost all of sub-Saharan Africa. Columns of its polymorphic, entirely termitophagous workers are instantly recognizable setting out on foraging expeditions from nests to local termitaries or returning from a raid loaded with dead termites. Nests are usually

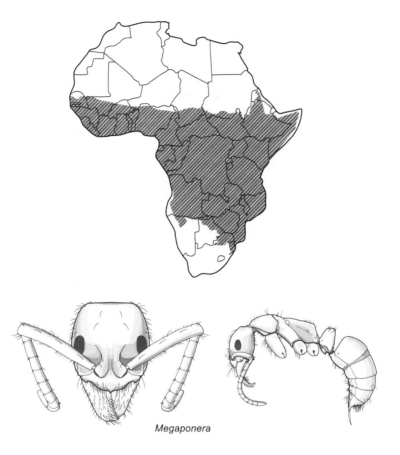

Megaponera

constructed directly in the ground or in deserted termitaries, and the queen of each colony is a single ergatoid. *M. analis* has an aromatic odor and stridulates audibly when disturbed.

TAXONOMIC COMMENT: Santschi (1930) produced a key to the subspecies known to that date. No more recent taxonomic work has been carried out to assess the real status of either these subspecies or the earlier synonyms noted below.

IDENTIFICATION OF SPECIES: No recent review exists.

DEFINITION OF THE GENUS (WORKER): Characters of subfamily PONERINAE, plus:

Worker caste polymorphic.

Mandible elongate-triangular; basal groove absent. Masticatory margin with more than 15 teeth.

Palp formula 4,4.

Clypeus simple, its anterior margin convex, without extended lobes or teeth on either the median or the lateral portions; the posterior portion of the clypeus broadly inserted between the frontal lobes.

Frontal lobes moderate; in full-face view their anterior margins do not overhang the anterior clypeal margin.

Eyes present, distinct, at about the midlength of the head capsule.

Head capsule with a longitudinal cuticular carina that extends from the anterior margin of the eye to the anterior margin of the head.

Antennal scapes conspicuously flattened.

Metanotal groove present, weakly impressed.

Epimeral sclerite present.

Metapleural gland orifice opens posteriorly; the orifice partially occluded by a lateral lobe.

Propodeal spiracle with orifice slit-shaped.

Tibial spurs: mesotibia 2; metatibia 2; on each tibia the anterior spur small, simple to barbulate, the posterior spur larger and pectinate.

Pretarsal claws each with a stout tooth on its inner curvature, the margin of the tooth adorned with 3–5 denticles.

Abdominal segment 2 (petiole) a node.

Helcium located at the base of the anterior face of A3 (first gastral segment).

Prora present, small.

Abdominal segment 4 (second gastral segment) with presclerites weakly differentiated.

Stridulitrum present on the pretergite of A4.

AFROTROPICAL MEGAPONERA

analis (Latreille, 1802)	*analis crassicornis* (Gerstäcker, 1859)
= *foetens* (Fabricius, 1793) (homonym)	*analis rapax* Santschi, 1914
= *abyssinica* (Guérin-Méneville, 1849)	*analis termitivora* Santschi, 1930
= *laeviuscula* (Gerstäcker, 1859)	*analis amazon* Santschi, 1935
= *dohrni* Emery, 1902	*analis subpilosa* Santschi, 1937

MELISSOTARSUS Emery, 1877 — Myrmicinae

3 species, Afrotropical; 1 species, Malagasy — PLATE 21

DIAGNOSTIC REMARKS: Among the myrmicines, *Melissotarsus* has only 6 antennal segments, of which the apical 2 form a distinct club; antennal scrobes are absent. The frontal lobes are confluent mid-dorsally, and the clypeus does not project back between them. The dorsal mesosoma has no trace of sutures or impressions. The lateral portion of the pronotum is reduced to a slender, down-pointing triangle, and the procoxa is much smaller than the mesocoxa and metacoxa, each of which is massive. The mesobasitarsus and metabasitarsus have apical circlets of traction spines. The mesonotum and petiole lack teeth or tubercles, and spongiform appendages are absent from the waist segments.

DISTRIBUTION AND ECOLOGY: The distribution of this very distinctive small genus is restricted to the Afrotropical and Malagasy regions. All species nest in tunnels and narrow galleries in heathy wood, under the bark of living trees. The tunnels are apparently excavated by the ants themselves and often contain live coccids as a food source. The gait of living workers is peculiar; when walking the fore and hind legs are in contact with the substrate, while the middle legs are elevated and in contact with the gallery roof. Strangely among ants,

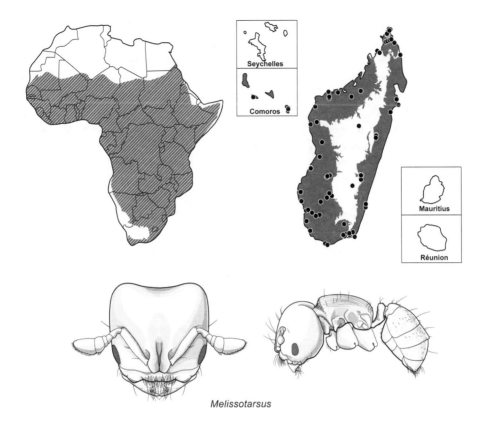

Melissotarsus

the adults, as well as the larvae, are able to produce silk. Nest samples of *M. weissi* often contain a few workers whose gasters are filled with a viscous, black, tarry substance. The composition and function of this material is unknown.

IDENTIFICATION OF SPECIES: Bolton, 1982 (Afrotropical).

DEFINITION OF THE GENUS (WORKER): Characters of subfamily MYRMICINAE, plus:

Mandible triangular, masticatory margin with 3–5 teeth, the apical tooth stout (margin often worn down to an edentate, or near-edentate, edge).

Palp formula 0,1, or 1,1.

Stipes of the maxilla without a transverse crest.

Clypeus posteriorly does not project back between the frontal lobes.

Frontal lobes confluent mid-dorsally; frontal lobes and antennal sockets extremely closely approximated.

Antennal scrobes absent.

Antenna with 6 segments, with an apical club of 2 segments; antenna very short.

Eyes well developed, distinctly longer than wide, located well in front of the midlength of the head capsule.

Promesonotal suture absent; metanotal groove absent.

Pronotum with its lateral portion reduced to a small, triangular sclerite, its apex ventral.

Propodeal lobes absent; propodeum unarmed, without trace of teeth or spines.

Propodeal spiracle with orifice circular, located close to the midlength of the sclerite.

Metasternal process present, small and bidentate.

Procoxa much smaller than the massive mesocoxa and metacoxa.

Tibial spurs: mesotibia o; metatibia o.

Basitarsi of middle and hind legs short, stout, and very broad, at least as broad as their respective tibiae; apex of each with a circlet of short traction spines; tarsomeres 2–4 very reduced. Probasitarsus greatly expanded.

Abdominal segment 3 (postpetiole) very broadly articulated to segment 4 (first gastral), the articulation high on the anterior face of A4.

Stridulitrum absent.

Abdominal segment 4 (first gastral) without presclerites.

Sting very reduced, apparently nonfunctional.

Main pilosity of dorsal head and body: simple, sparse, elongate, and fine.

AFROTROPICAL MELISSOTARSUS

beccarii Emery, 1877	*weissi* Santschi, 1910
emeryi Forel, 1907	= *major* Santschi, 1919
= *emeryi pilipes* Santschi, 1914	= *titubans* Delage-Darchen, 1972
= *compressus* Weber, 1952	

MALAGASY MELISSOTARSUS

insularis Santschi, 1911

MERANOPLUS Smith, F., 1853 — Myrmicinae

8 species, Afrotropical; 4 species, Malagasy — PLATE 21

DIAGNOSTIC REMARKS: Among the myrmicines, *Meranoplus* combines the characters of 9-segmented antennae that terminate in a 3-segmented club, with a uniquely constructed mesosoma. In dorsal view the promesonotum forms a broad shield that is always expanded laterally (overhangs the sides) and sometimes is also expanded posteriorly (over the propodeum). The lateral and/or the posterior margins of the promesonotal shield are commonly equipped with spines or lobes and sometimes with translucent fenestrae. In addition, the mandibles have only 4–5 teeth, deep antennal scrobes extend above the eye, the petiole (A2) is sessile, and the first gastral tergite (the tergite of A4) accounts for most or all of the visible gaster in dorsal view.

DISTRIBUTION AND ECOLOGY: Present throughout the Old World tropics, the genus has the vast majority of its species in Australia. In the regions currently under consideration, *Meranoplus* species are found in both forest and savanna habitats. All nest in the ground, either directly or among the roots of plants. Sometimes a small crater marks the entrance of a nest.

IDENTIFICATION OF SPECIES: Bolton, 1981a (Afrotropical); Boudinot and Fisher, 2013 (Malagasy).

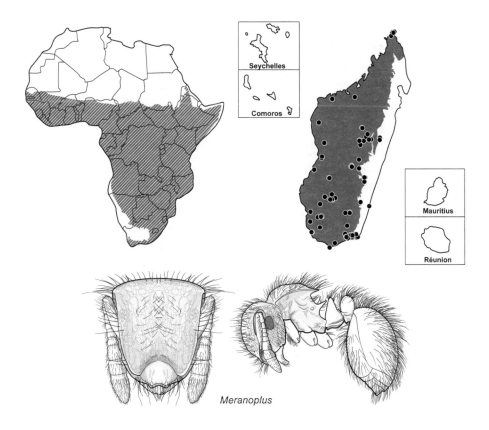

Meranoplus

DEFINITION OF THE GENUS (WORKER): Characters of subfamily MYRMICINAE, plus:

Mandible narrowly triangular; masticatory margin short and with 4–5 teeth.

Palp formula 5,3.

Stipes of the maxilla without a transverse crest.

Clypeus large, median portion usually carinate at each side; posteriorly broadly inserted between the frontal lobes.

Clypeus without an unpaired seta at the midpoint of the anterior margin.

Frontal carinae strongly developed, widely separated and extending back almost to the posterior margin of the head, forming the upper margins of the deep scrobes.

Antennal scrobe present, deep, extending above the eye.

Antenna with 9 segments, with an apical club of 3 segments.

Torulus concealed by the frontal lobe in full-face view.

Eyes present, large, located well behind the midlength of the head capsule, usually close to the ventral apex of the scrobe.

Head capsule with a lateroventral ridge or carina on each side.

Promesonotal suture vestigial to absent on dorsum. Promesonotum forms a broad shield that is always expanded laterally (overhangs the sides) and sometimes is also expanded posteriorly (over the propodeum). Lateral and/or posterior margins of the promesonotal shield commonly equipped with spines or lobes and sometimes with translucent fenestrae.

Propodeum usually bispinose but may be unarmed.

Metasternal process absent.

Tibial spurs: mesotibia 0 or 1; metatibia 0 or 1.

Abdominal segment 2 (petiole) sessile, the node usually cuneate in profile.

Abdominal tergite 4 (first gastral) does not broadly overlap the sternite on the ventral gaster; gastral shoulders absent. Tergite of A4 extensive, in dorsal view accounting for most or all of the visible gaster.

Stridulitrum absent.

Sting reduced, narrowly spatulate.

Main pilosity of dorsal head and body: simple, rarely absent.

AFROTROPICAL MERANOPLUS

clypeatus Bernard, 1953	= *simoni* Emery, 1895
glaber Arnold, 1926	= *simoni nitidiventris* Mayr, 1901
inermis Emery, 1895	= *simoni suturalis* Forel, 1910
= *nanus kiboshana* Forel, 1907	= *bondroiti* Santschi, 1915
= *nanus nanior* Forel, 1907	= *simoni springvalensis* Arnold, 1917
= *nanus soriculus* Wheeler, W.M.,	= *simoni diversipilosus* Santschi, 1932
1922	*nanus* André, 1892
= *nanus affinis* Baroni Urbani, 1971	*peringueyi* Emery, 1886
= *nanus similis* Karavaiev, 1931: 44	= *excisus* Arnold, 1914
(homonym)	*spininodis* Arnold, 1917
magrettii André, 1884	*sthenus* Bolton, 1981

MALAGASY MERANOPLUS

cryptomys Boudinot and Fisher, 2013

mayri Forel, 1910

radamae Forel, 1891

sylvarius Boudinot and Fisher, 2013

MESOPONERA Emery, 1900   Ponerinae

15 species (+ 5 subspecies), Afrotropical; 1 species (+ 1 subspecies), Malagasy   PLATE 21

DIAGNOSTIC REMARKS: Among the ponerines, *Mesoponera* combines the characters of presence of 2 spurs on the metatibia, absence of traction setae on the metabasitarsus, and location of the helcium at the base of the anterior face of the first gastral segment (A3), with the following: The mandible lacks a basal pit and usually lacks a basal groove. The frontal lobes are small. The metanotal groove is present and impressed, and the petiole (A2) is a scale. The pretarsal claws of the metatarsus are simple.

DISTRIBUTION AND ECOLOGY: Species of *Mesoponera* occur throughout the Old World tropics. The species nest in the ground, in active or abandoned termitaries, or in rotten wood in leaf litter and topsoil. Foraging takes place in these habitats and also directly on the surface of the ground. The few species that have been studied are predators of other

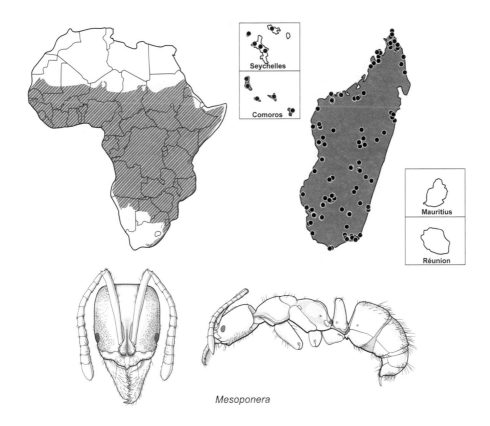

Mesoponera

insects, and at least one (*M. caffraria*) will also collect honeydew from homopterous insects. *M. melanaria macra* has been recorded only from the Seychelles and probably represents an introduction.

TAXONOMIC COMMENT: It is suspected that the Bernard (1953) names *novemdentata*, *picea*, *villiersi*, and *weberi*, listed below, represent different size variants of the very common West African species *M. ambigua* (André), which is not mentioned by Bernard.

IDENTIFICATION OF SPECIES: Bernard, 1953 (*Xiphopelta*). No recent review exists.

DEFINITION OF THE GENUS (WORKER): Characters of subfamily PONERINAE, plus:

Mandible triangular to elongate triangular; with or without a basal groove, without a
basal pit; masticatory margin with 8–13 or more teeth. Basal angle angulate.
Palp formula 4,4, or 4,3, or 3,3.
Clypeus simple, usually without projecting lobes or teeth but sometimes with a denticle
at the midpoint of the anterior clypeal margin.
Frontal lobes small; in full-face view their anterior margins well behind the anterior
clypeal margin.
Eyes present; small to moderate in size, well in front of the midlength of the head
capsule; always of more than one ommatidium.
Metanotal groove present and impressed.

Epimeral sclerite usually present, indistinct or absent only in minute species.

Metapleural gland orifice opens posteriorly.

Propodeal spiracle with orifice circular to slit-shaped.

Tibial spurs: mesotibia 2; metatibia 2; on each tibia the anterior spur small and simple, the posterior spur larger and pectinate.

Mesotibia without traction setae.

Metatibial posterior surface usually featureless, but sometimes a slightly depressed oval area of pale, very finely granular cuticle is present above the base of the spur that appears to be glandular.

Pretarsal claws simple.

Abdominal segment 2 (petiole) a scale.

Helcium located at the base of the anterior face of A3 (first gastral segment).

Prora present.

Abdominal segment 4 (second gastral segment) with presclerites differentiated.

Stridulitrum absent from the pretergite of A4.

AFROTROPICAL MESOPONERA

ambigua (André, 1890)	*ingesta* (Wheeler, W.M., 1922)
caffraria (Smith, F., 1858)	*nimba* (Bernard, 1953)
= *guineensis* (André, 1890)	= *neonimba* (Özdikmen, 2010)
caffraria affinis (Santschi, 1935)	(redundant replacement name)
caffraria caffra (Santschi, 1935)	*novemdentata* (Bernard, 1953)
elisae (Forel, 1891)	*picea* (Bernard, 1953)
elisae rotundata (Emery, 1895)	*scolopax* (Emery, 1899)
= *arnoldi* (Forel, 1913)	*senegalensis* (Santschi, 1914)
elisae redbankensis (Forel, 1913)	*subiridescens* (Wheeler, W.M., 1922)
elisae divaricata (Emery, 1915)	*testacea* (Bernard, 1953)
escherichi (Forel, 1910)	*villiersi* (Bernard, 1953)
flavopilosa (Weber, 1942)	*weberi* (Bernard, 1953)

MALAGASY MESOPONERA

elisae (Forel, 1891)

melanaria (Emery, 1893)

   melanaria macra (Emery, 1894)

MESSOR Forel, 1890   Myrmicinae

15 species, Afrotropical   PLATE 22

DIAGNOSTIC REMARKS: Among the myrmicines, *Messor* species are polymorphic and have 12-segmented antennae, either without an apical club or with the 3–4 apical segments slightly enlarged. The mandibles are large and powerful, especially in the largest workers. The clypeus is large and posteriorly is broadly inserted between the frontal lobes. A psammophore is present on the ventral surface of the head. The pronotum and anterior mesonotum in profile are swollen and convex, usually dome-like in profile; behind this the

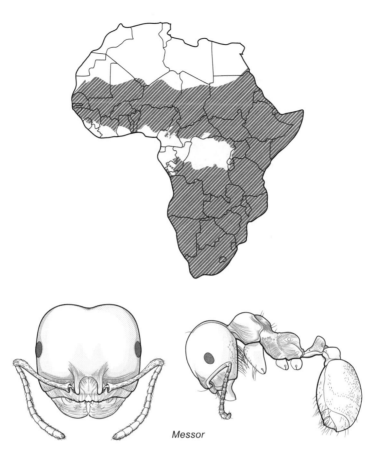

Messor

mesonotum slopes steeply posteriorly to the metanotal groove. The propodeal spiracle is large but never takes the form of an elongate slit. The sting is reduced.

DISTRIBUTION AND ECOLOGY: The vast majority of *Messor* species inhabit the Palaearctic region, with only a few species present in the Afrotropical region and the north of the Oriental region. The genus is absent from the Malesian, Austral, and Malagasy regions and from the New World. *Messor* species mainly occur in grassland and savanna but are also present in arid to desert conditions. Nests are constructed in the ground, often with a crater-like entrance, and all species are granivorous.

TAXONOMIC COMMENT: The subspecies *M. galla obscurus* Menozzi and Consani was overlooked in the revision of Bolton (1982). From its description, the subspecies *obscurus* was said to be separated from *M. galla* only by a trivial difference in shade of color, which in fact falls well within the range of color variation now known for *M. galla* across its wide range. We therefore synonymize *M. galla obscurus* under *M. galla*.

IDENTIFICATION OF SPECIES: Bolton, 1982.

DEFINITION OF THE GENUS (WORKER): Characters of subfamily MYRMICINAE, plus:

Worker caste polymorphic.

Mandible triangular, large and powerful; masticatory margin with up to 15 teeth in smaller workers, but this number decreases with increased body size; in largest workers only a few massive teeth or an edentate crushing edge remains. Masticatory margin longer than the basal margin.

Palp formula 5,3, or 4,3; within species often larger workers 5,3, smaller workers 4,3.

Stipes of the maxilla without a transverse crest.

Clypeus large, posteriorly broadly inserted between the frontal lobes.

Clypeus without an unpaired seta at the midpoint of the anterior margin.

Frontal carinae short, restricted to well-defined but narrow frontal lobes.

Antennal scrobes absent.

Antenna with 12 segments, filiform and without an apical club, or apical 3–4 segments slightly enlarged.

Torulus with the upper lobe visible in full-face view; maximum width of the torulus lobe is posterior to the point of maximum width of the frontal lobes.

Eyes well developed, located at or slightly behind the midlength of the head capsule.

Head massively constructed in larger workers; usually a psammophore present ventrally, but this is reduced or absent in a few species and may be more strongly developed in smaller workers.

Promesonotal suture usually represented by a weakly impressed vestige across the dorsum, less commonly obliterated.

Pronotum and anterior mesonotum in profile swollen and convex, usually dome-like; the mesonotum posteriorly sloping steeply to the metanotal groove.

Propodeum unarmed to strongly bidentate; in profile the propodeal dorsum on a much lower level than the apex of the convex promesonotum.

Propodeal spiracle large, its orifice circular or nearly so, located at or slightly behind the midlength of the sclerite.

Metasternal process present, large and conspicuous.

Tibial spurs: mesotibia 0 or 1; metatibia 0 or 1; in a few species the metatibial spur pectinate.

Abdominal segment 2 (petiole) with a long anterior peduncle.

Abdominal tergite 4 (first gastral) broadly overlaps the sternite on the ventral gaster; gastral shoulders absent.

Sting reduced.

Main pilosity of dorsal head and body: simple; sometimes sparse, may be absent in places.

AFROTROPICAL MESSOR

angularis Santschi, 1928

capensis (Mayr, 1862)
  = pseudoaegyptiaca (Emery, 1884)
  = braunsi Forel, 1913
  = capensis schencki Wheeler, W.M., 1922
  = donisthorpei Santschi, 1937

cephalotes (Emery, 1895)
  = plinii Santschi, 1912
collingwoodi Bolton, 1982
decipiens Santschi, 1917
  = capensis probus Wheeler, W.M., 1922
  = arcistriatus Santschi, 1928
denticornis Forel, 1910

= *denticornis brunni* Forel, 1910

= *denticornis parvidens* Forel, 1910

ferreri Collingwood, 1993

galla (Mayr, 1904)

    = *barbarus latinoda* Santschi, 1917

    = *galla nobilis* Santschi, 1928

    = *galla rufula* Finzi, 1939

    = *galla airensis* Bernard, 1950

    = *galla obscurus* Menozzi and

        Consani, 1952 syn. n.

incisus Stitz, 1923

luebberti Forel, 1910

piceus Stitz, 1923

regalis (Emery, 1892)

    = *regalis rubea* (Santschi, 1913)

    = *sculpturatus* (Stitz, 1916)

ruginodis Stitz, 1916

striatifrons Stitz, 1923

tropicorum Wheeler, W.M., 1922

    = *braunsi nigriventris* Stitz, 1923

    = *denticornis laevifrons* Stitz, 1923

METAPONE Forel, 1911 — Myrmicinae

1 species, Afrotropical; 3 species, Malagasy   PLATE 22

DIAGNOSTIC REMARKS: Among the myrmicines, *Metapone* is easily recognized. It combines the characters of 11-segmented antennae that terminate in a club of 3 segments; eyes that are small to absent; a large, anteriorly projecting clypeus that posteriorly is very broadly inserted between the frontal lobes; procoxae that are conspicuously smaller than the mesocoxae and metacoxae; and the presence of traction spines at the apices of the mesotibiae and metatibiae as well as on the basitarsi of all the legs.

DISTRIBUTION AND ECOLOGY: *Metapone* species are distributed throughout the Old World tropics but are only infrequently collected. All species collected have been retrieved from rotten wood that is infested with termites, which form their prey.

IDENTIFICATION OF SPECIES: Alpert, 2007 (Malagasy).

DEFINITION OF THE GENUS (WORKER): Characters of subfamily MYRMICINAE, plus:

Mandible short, with subparallel basal and outer margins; masticatory margin with 4–6 teeth.

Palp formula 2,3, or 1,3.

Stipes of the maxilla without a transverse crest.

Clypeus posteriorly very broadly inserted between the frontal lobes. Median portion of the clypeus raised and produced forward as a large, shield-like lobe that projects strongly over the mandibles.

Clypeus without an unpaired seta at the midpoint of the anterior margin.

Frontal carinae conspicuous, suddenly broadened behind the frontal lobes.

Antennal scrobes present, deep, extending above the eyes.

Antenna with 11 segments, with an apical club of 3 segments.

Torulus concealed by the frontal lobe in full-face view.

Eyes absent or present; when present, small and located behind the midlength of the head capsule; ocelli usually absent but sometimes present.

Promesonotal suture absent; metanotal groove present.

Propodeum unarmed.

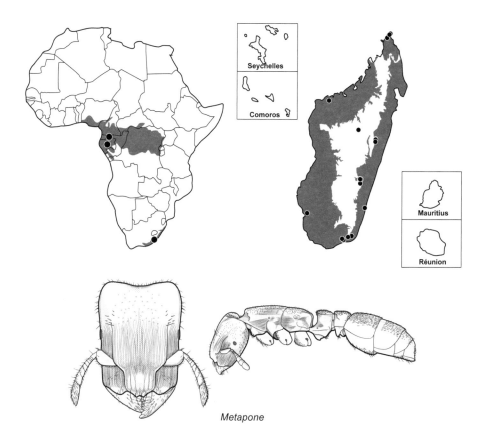

Metapone

Propodeal spiracle high on the side, located at about the midlength of the sclerite.

Procoxa smaller than mesocoxa and metacoxa.

Metafemur extremely anteroposteriorly compressed, very deep in anterior view.

Tibial spurs: mesotibia 1; metatibia 1; each spur pectinate.

Mesotibia and metatibia apically, and basitarsi of all legs, with traction spines.

Abdominal segment 2 (petiole) with a short anterior peduncle.

Abdominal segments 3 (postpetiole) and 4 (first gastral) with their articulation very broad.

Stridulitrum present on the pretergite of A4.

Abdominal segment 4 (first gastral) with very large presclerites.

Sting present.

Main pilosity of dorsal head and body: simple.

AFROTROPICAL METAPONE

1 undescribed species, Gabon

MALAGASY METAPONE

emersoni Gregg, 1958
madagascarica Gregg, 1958
vincimus Alpert, 2007

MICRODACETON Santschi, 1913   Myrmicinae

4 species, Afrotropical   PLATE 22

DIAGNOSTIC REMARKS: Among the Afrotropical myrmicines, *Microdaceton* has only 6 antennal segments, of which the apical 2 form a weak club; antennal scrobes are absent. The mandibles are linear, lack preapical teeth, and have an apical fork of 3 spiniform teeth in a vertical series. The frontal lobes are widely separated, and the large clypeus projects back between them. The procoxa is at least as large as the mesocoxa and metacoxa. The mesonotum and petiole always have a pair of teeth or tubercles, and spongiform appendages are absent from both waist segments. Pilosity is absent from the dorsal head and body.

DISTRIBUTION AND ECOLOGY: *Microdaceton* species are restricted to the Afrotropical region, where they occur throughout the forested zones but do not appear to be common anywhere. They are mainly denizens of the leaf-litter and topsoil layers but may also be found under the bark of rotting logs or occasionally in soil pockets on trees. Collembola and other small arthropods probably form their prey, and when hunting, *Microdaceton* workers advance with their mandibles locked open at about 180 degrees.

IDENTIFICATION OF SPECIES: Bolton, 1983; Bolton, 2000.

DEFINITION OF THE GENUS (WORKER): Characters of subfamily MYRMICINAE, plus:

Mandible linear and elongate; masticatory margins oppose but do not overlap at full closure. Mandible without preapical teeth and terminating in an apical fork of 3 teeth, arranged in a vertical series.

Basal process present on the mandible, projecting medially as a curved spur from the inner margin close to the base; the process is an extrusion of the mandible, not merely a modified tooth.

Labrum roughly T-shaped, not capable of reflexion over the anterior portion of the buccal cavity, not closing tightly over the apex of the labiomaxillary complex, which is permanently exposed; labrum anteriorly with 1, or rarely 2, trigger hairs present.

Palp formula 3,2.

Stipes of the maxilla without a transverse crest.

Clypeus large, broadly triangular, posteriorly very broadly inserted between the frontal lobes.

Frontal carinae in full-face view, at level of the antennal sockets, with their outer margins very close to, or slightly overhanging, the apparent lateral margin of the head (actually the outer margin of the preocular carina).

Antennal scrobe absent.

Antenna with 6 segments, with a weak apical club of 2 segments.

Torulus concealed by the frontal lobe in full-face view.

Eye not ventrolateral on side of the head.

Head capsule anteriorly strongly narrowed from side to side in full-face view; side of the head with an extensive gap between base of the mandible and margin of the head capsule.

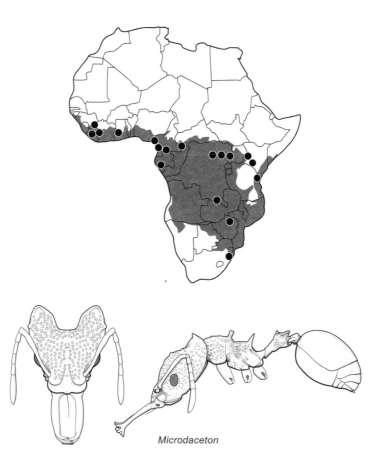

Microdaceton

Head anterolaterally with a preocular carina present that arises at the clypeus and extends posteriorly below the antennal socket.

Pronotum anteriorly, just before the occipital foramen, with an abruptly raised, thick, transverse rim or collar that is preceded by a broad transverse groove.

Promesonotal suture absent from dorsum; mesonotum with a pair of teeth or tubercles.

Propodeum bispinose.

Propodeal spiracle very close to the margin of the declivity.

Metasternal process absent.

Tibial spurs: mesotibia 0; metatibia 0.

Abdominal segments 2 and 3 (petiole and postpetiole) without spongiform tissue. Node of petiole bidentate. Postpetiole markedly dorsoventrally flattened and expanded into lateral wings.

Abdominal tergite 4 (first gastral) basally without a transverse cuticular crest or ridge (limbus absent).

Sting present.

Main pilosity of dorsal head and body: absent.

exornatum Santschi, 1913

  = *leakeyi* Patrizi, 1947

  = *exornatum laevior* Arnold, 1948

tanyspinosum Bolton, 2000

tibialis Weber, 1952

viriosum Bolton, 2000

MONOMORIUM Mayr, 1855   Myrmicinae

132 species, Afrotropical; 19 species, Malagasy   PLATE 22

DIAGNOSTIC REMARKS: Among the myrmicines, *Monomorium* has 11- or 12-segmented antennae, always with an apical club of 3 segments. The mandible has only 3–4 teeth, and eyes are present. Frontal carinae are absent behind the small frontal lobes, and scrobes are absent. The anterior clypeal margin has a single seta at the midpoint, and the median portion of the clypeus is usually longitudinally bicarinate. The lateral portions of the clypeus are not raised into a ridge or shield wall in front of the antennal sockets. The propodeum is usually unarmed, and the first gastral tergite (the tergite of A4) overlaps the sternite on the ventral surface of the gaster.

DISTRIBUTION AND ECOLOGY: This very large genus is almost worldwide in distribution, but the overwhelming majority of species are in the Old World, especially in the Afrotropical and Austral regions. Species of *Monomorium* occur in almost all locations, from harsh desert to rainforest, and from the soil to the canopy of trees, but only a few are entirely arboreal. Nests are constructed in the ground, either directly or under surface objects, in the compacted walls of termitaries, among the roots of plants, in leaf litter and fallen twigs, under flakes of bark, and in rotten wood, either on the ground or in standing timber. In a few species no nest material is used, as any appropriate cavity will suffice as a nest site. The species are mostly generalist scavengers and predators of a wide range of small arthropods, but some will also tend homopterous insects for honeydew. Queens of most species are normal alates, but some have ergatoid queens. Different species exhibit different degrees of reduction of the mesosomal morphology, from the usual alate form to a mesosoma that is almost entirely worker-like.

TAXONOMIC COMMENT: Ward, et al. (2015) have resurrected 2 genus-group names from synonymy under *Monomorium*, where they are found in the studies of Bolton (1987) and Heterick (2006). These are *Trichomyrmex* (the former *M. scabriceps* and *M. destructor* groups) and *Syllophopsis* (the former *M. fossulatum* or *M. hildebrandti* group). In addition, Bolton and Fisher (2014) have designated *Royidris* as a separate genus for Heterick's (2006) *M. shuckardi* group, and in this volume, above, *Erromyrma* is designated as a new genus for the former *M. latinode* group of both Bolton (1987) and Heterick (2006).

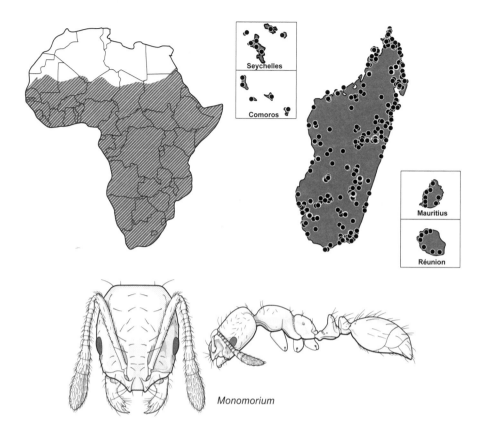

Monomorium

IDENTIFICATION OF SPECIES: Santschi, 1936 (Afrotropical *M. [Xeromyrmex]*); Bolton, 1987 (Afrotropical, as *Monomorium salomonis, M. setuliferum, M. monomorium,* and *M. hanneli* groups); Heterick, 2006 (Malagasy, as *M. salomonis, M. monomorium,* and *M. hanneli* groups).

DEFINITION OF THE GENUS (WORKER): Characters of subfamily MYRMICINAE, plus:

Worker caste monomorphic.

Mandible narrowly triangular, or with the basal and outer margins subparallel; masticatory margin with 3–4 teeth, most commonly with 4 (usually 4 distinct teeth, but the basalmost is sometimes reduced to an offset denticle).

Palp formula 2,2, or 1,2.

Stipes of the maxilla without a transverse crest.

Clypeus posteriorly narrowly inserted between the small frontal lobes; width of the clypeus between the lobes usually greater than the width of 1 of the lobes, uncommonly the 2 about equal. Median portion of the clypeus raised and usually narrowed, projecting forward anteriorly, the raised section longitudinally bicarinate dorsally (carinae reduced, or blunt and rounded, in some species); each carina usually terminates in an angle, denticle or tooth at the anterior margin (absent in a few species).

Clypeus with a distinct unpaired seta at the midpoint of the anterior margin; in profile the unpaired median seta arises from the anteriormost point of the clypeal projection.

Frontal carinae restricted to the small frontal lobes.

Antennal scrobes absent.

Antennal fossae not surrounded by fine, curved striolae or costulae.

Antenna with 11–12 segments (usually 12), with an apical club of 3 segments.

Torulus with the upper lobe visible in full-face view; maximum width of the torulus lobe is posterior to the point of maximum width of the frontal lobes.

Eyes present, small to very large; usually located at, or in front of, the midlength of the head capsule, only rarely slightly behind midlength. Eye in profile usually circular to roughly oval but may be reniform or may extend anteroventrally into a lobe.

Promesonotal suture absent from dorsum of mesosoma, or a slightly impressed remnant discernible. Metanotal groove usually present and impressed but sometimes vestigial; only rarely absent.

Propodeum usually unarmed and rounded but in a few species angulate, or with short angular lamellae, or weakly dentate.

Propodeal spiracle circular to subcircular, approximately at the midlength of the sclerite.

Propodeal dorsum only extremely rarely with transverse sculpture; usually reticulate-punctate to smooth.

Metasternal process absent.

Tibial spurs: mesotibia 0 or 1; metatibia 0 or 1.

Abdominal segment 4 (first gastral) tergite overlaps the sternite on the ventral gaster; gastral shoulders present.

Sting strongly developed to weak.

Main pilosity of dorsal head and body: simple, very variable in density and distribution; setae often numerous, sometimes sparse, sometimes absent on part or all of the body.

AFROTROPICAL MONOMORIUM

affabile Santschi, 1926

afrum André, 1884
  = *afrum asmarensis* Forel, 1910
  = *afrum fultor* Forel, 1913

alamarum Bolton, 1987

albopilosum Emery, 1895
  = *albopilosum thales* Forel, 1913
  = *albopilosum clarithorax* Santschi, 1919
  = *albopilosum paucipilosa* Santschi, 1919
  = *albopilosum fingo* Arnold, 1946

altinode Santschi, 1910
  = *altinode bondroiti* Santschi, 1920

anceps Emery, 1895

angustinode Forel, 1913

arboreum Weber, 1943

areniphilum Santschi, 1911
  = *salomonis pullula* Santschi, 1919
  = *salomonis lepineyi* Santschi, 1934

arnoldi Forel, 1913

australe Emery, 1886

balathir Bolton, 1987

bequaerti Forel, 1913

bevisi Arnold, 1944

bicolor Emery, 1877
  = *bicolor coerulescens* Santschi, 1912
  = *bicolor aequatoriale* Santschi, 1926
  = *bicolor tropicale* Santschi, 1926
  = *bicolor uelense* Santschi, 1926

boerorum Forel, 1910

borlei Santschi, 1937

braunsi Mayr, 1901

captator Santschi, 1936

carbo Forel, 1910

crawleyi Santschi, 1930

dakarense Santschi, 1914

damarense Forel, 1910
   = *salomonis unicolor* Stitz, 1923

delagoense Forel, 1894
   = *delagoense grahamstownense*
      Santschi, 1936
   = *delagoense lacrymans* Arnold, 1944

dictator Santschi, 1937

disertum Forel, 1913

disoriente Bolton, 1987

dolatu Bolton, 1987

drapenum Bolton, 1987

draxocum Bolton, 1987

ebangaense Santschi, 1937
   = *nyasae* Arnold, 1946

egens Forel, 1910
   = *jucundum* Santschi, 1926
   = *longiusculum* Santschi, 1926

esharre Bolton, 1987

excelsior Arnold, 1926
   = *speculiceps* Santschi, 1928

excensurae Forel, 1915

exiguum Forel, 1894
   = *exiguum bulawayensis* Forel, 1913
   = *faurei* Santschi, 1915
   = *exiguum flavescens* Forel, 1916
   = *minutissimum* Santschi, 1937

fasciatum Santschi, 1920

fastidium Bolton, 1987

firmum Santschi, 1926

floricola (Jerdon, 1851)

fridae Forel, 1905

fugelanum Bolton, 1987

gabrielense Forel, 1916

guillarmodi Arnold, 1946

guineense (Bernard, 1953)

hanneli Forel, 1907
   = *moestum* Santschi, 1914
   = *valtinum* Bolton, 1987

hannonis Santschi, 1910

havilandi Forel, 1910
   = *distinctum* Arnold, 1944
   = *distinctum leviceps* Arnold, 1958

hercules Viehmeyer, 1923

herero Forel, 1910

hirsutum Forel, 1910

holothir Bolton, 1987

ilgii Forel, 1894

inquietum Santschi, 1926

invidium Bolton, 1987

iyenasu Bolton, 1987

jacksoni Bolton, 1987

junodi Forel, 1910
   = *delagoense pretoriensis* Arnold, 1944

katir Bolton, 1987

kelapre Bolton, 1987

kineti Weber, 1943

kitectum Bolton, 1987

lene Santschi, 1920

lubricum Arnold, 1948

macrops Arnold, 1944

madecassum Forel, 1892
   = *minutum leopoldinum* Forel, 1905
   = *explorator* Santschi, 1920
   = *aequum* Santschi, 1928
   = *estherae* Weber, 1943

malatu Bolton, 1987
   = *altinode* (Santschi, 1935) (homonym)

manir Bolton, 1987

mantazenum Bolton, 1987

marshi Bolton, 1987

mavide Bolton, 1987

mediocre Santschi, 1920

micropacum Bolton, 1987

mictilis Forel, 1910

minor Stitz, 1923

mirandum Arnold, 1955

musicum Forel, 1910

nirvanum Bolton, 1987

notulum Forel, 1910
   = *setuliferum dolichops* Santschi, 1928
   = *setuliferum latior* Santschi, 1928

noxitum Bolton, 1987

nuptuale Forel, 1913

occidentale Bernard, 1953

ocelletum Arnold, in Santschi, 1920

opacior Forel, 1913

   = opacior Bolton, 1987

opacum Forel, 1913

ophthalmicum Forel, 1894

orangiae Arnold, 1956

osiridis Santschi, 1915

pacis Forel, 1915

pallidipes Forel, 1910

parvinode Forel, 1894

paternum Bolton, 1987

personatum Santschi, 1937

pharaonis (Linnaeus, 1758)

pulchrum Santschi, 1926

rabirium Bolton, 1987

rastractum Bolton, 1987

rhopalocerum Emery, 1895

   = minutum hottentota Emery, 1895

   = leimbachi Forel, 1914

rosae Santschi, 1920

   = cotterelli (Donisthorpe, 1942)

rotundatum Santschi, 1920

rufulum Stitz, 1923

   = monardi Santschi, 1937

schultzei Forel, 1910

senegalense Roger, 1862

setuliferum Forel, 1910

shilohense Forel, 1913

spectrum Bolton, 1987

speluncarum Santschi, 1914

springvalense Forel, 1913

sryetum Bolton, 1987

strangulatum Santschi, 1921

subdentatum Forel, 1913

subopacum (Smith, F., 1858)

   = subopacum liberta Santschi, 1927

sutu Bolton, 1987

symmotu Bolton, 1987

tablense Santschi, 1932

taedium Bolton, 1987

tanysum Bolton, 1987

tchelichofi Forel, 1914

termitarium Forel, 1910

termitobium Forel, 1892

   = exchao Santschi, 1926

   = binatu Bolton, 1987

torvicte Bolton, 1987

trake Bolton, 1987

tynsorum Bolton, 1987

vaguum Santschi, 1930

vatranum Bolton, 1987

vecte Bolton, 1987

viator Santschi, 1923

vonatu Bolton, 1987

westi Bolton, 1987

willowmorense Bolton, 1987

xanthognathum Arnold, 1944

zulu Santschi, 1914

MALAGASY MONOMORIUM

bifidoclypeatum Heterick, 2006

chnodes Heterick, 2006

denticulus Heterick, 2006

exiguum Forel, 1894

flavimembra Heterick, 2006

floricola (Jerdon, 1851)

hanneli Forel, 1907

lepidum Heterick, 2006

madecassum 1892

micrommaton Heterick, 2006

nigricans Heterick, 2006

pharaonis (Linnaeus, 1758)

platynode Heterick, 2006

sakalavum Santschi, 1928

subopacum (Smith, F., 1858)

termitobium Forel, 1892

   = minutum imerinense Forel, 1892

versicolor Heterick, 2006

willowmorense Bolton, 1987

xuthosoma Heterick, 2006

DIAGNOSTIC REMARKS: Among the myrmicines, *Myrmicaria* is recognized by its combined characters of 7-segmented antennae, lack of antennal scrobes, long petiolar peduncle, presence of gastral shoulders, and possession of a strongly developed spatulate sting. In addition, the articulation of the postpetiole (A3) to the first gastral segment (A4) is specialized. The posterior articulation of the postpetiole is shifted somewhat ventrally on the anterior face of A4. This allows the gaster (A4–A7), in life, to be carried flexed downward, almost at a right-angle to the mesosoma.

DISTRIBUTION AND ECOLOGY: Absent from the Malagasy region but represented by numerous species throughout the Afrotropical region and also well represented in the Oriental and the western Malesian regions. In sub-Saharan Africa most species of *Myrmicaria* are found primarily in open or savanna habitats, though some also occur in clearings and on paths within forested areas. Most species nest directly in the ground and are active in bright sunlight, foraging terrestrially, and sometimes make long, sunken runways on the surface of the ground. In contrast to this usual behavior, one small group of species is entirely arboreal. The larger terrestrial species are very efficient scavengers.

IDENTIFICATION OF SPECIES: Santschi, 1925 (out of date and containing many infraspecific and infrasubspecific names); no recent review exists.

DEFINITION OF THE GENUS (WORKER): Characters of subfamily MYRMICINAE, plus:

Mandible narrowly triangular; masticatory margin with 4–5 teeth, shorter than the basal margin.

Palp formula 3,3.

Stipes of the maxilla without a transverse crest.

Clypeus large, posteriorly broadly inserted between the frontal lobes.

Clypeus without an unpaired seta at the midpoint of the anterior margin.

Frontal carinae expand into well-defined frontal lobes that conceal the antennal sockets and continue for a short distance behind them.

Antennal scrobes absent.

Antenna with 7 segments; gradually incrassate apically or with a club of 3 segments.

Torulus with the upper lobe usually not visible in full-face view, but its outer margin just visible in very small species.

Eyes well developed, located well behind the midlength of the head capsule.

Side of head usually separated from ventral surface by a narrow longitudinal carina in larger species; carina absent in small species.

Promesonotal suture usually represented by a weak impression on the dorsum of the mesosoma, impression sometimes absent.

Metanotal groove present.

Propodeal lobes absent or at most represented by a very narrow carina.

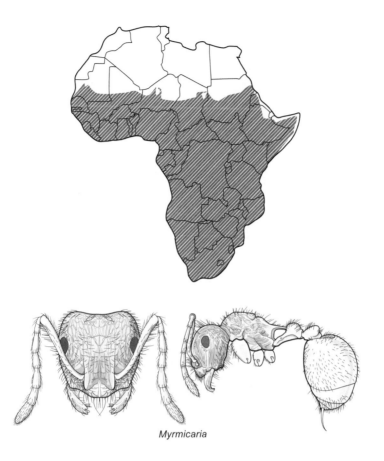

Myrmicaria

Propodeum bidentate to bispinose.

Propodeal spiracle circular, at about midlength of the sclerite.

Metasternal process usually present, absent in *M. exigua* group.

Legs long and slender.

Tibial spurs: mesotibia 0 or 1; metatibia 0 or 1.

Abdominal segment 2 (petiole) with a long, narrow, anterior peduncle.

Abdominal segment 3 (postpetiole) apparently with the tergite and sternite fused
    (requires more extensive investigation and confirmation).

Abdominal segment 3 with its posterior articulation shifted somewhat ventrally on the
    anterior face of A4; gaster (A4–A7) carried flexed downward, in life, almost at a
    right-angle to the mesosoma.

Abdominal segment 4 (first gastral) with presclerites tightly attached to each other,
    subfused; the tergite of A4 broadly overlaps the sternite on the ventral surface of the
    gaster; gastral shoulders present.

Stridulitrum absent.

Sting spatulate, strongly developed.

Main pilosity of dorsal head and body: simple.

anomala Arnold, 1960

arnoldi Santschi, 1925

basutorum Arnold, 1960

baumi Forel, 1901

   baumi occidentalis Santschi, 1920

distincta Santschi, 1925

   distincta abyssinica Santschi, 1925

   distincta vorax Santschi, 1933

exigua André, 1890

   exigua gracilis Stitz, 1910

   exigua rufiventris Forel, 1915

   exigua obscura Santschi, 1920

   exigua pulla Santschi, 1920

   exigua kisangani Wheeler, W.M., 1922

   exigua simplex Stitz, in Santschi, 1925

faurei Arnold, 1947

foreli Santschi, 1925

   foreli pallida Arnold, in Santschi, 1925

fumata Santschi, 1916

   = nitida brunnea Santschi, 1915
      (homonym)

   fumata linearis Santschi, 1925

fusca Stitz, 1911

   fusca nigerrima Arnold, 1916

   fusca consanguinea Santschi, 1925

irregularis Santschi, 1925

laevior Forel, 1910

natalensis (Smith, F., 1858)

   = sulcatus (Mayr, 1862)

natalensis eumenoides (Gerstäcker,
   1859)

natalensis verticalis Santschi, 1920

natalensis navicula Santschi, 1925

natalensis nigriventris Santschi, 1930

natalensis taeniata Santschi, 1930

natalensis obscuriceps Santschi, 1937

nigra (Mayr, 1862)

opaciventris Emery, 1893

   opaciventris congolensis Forel, 1909

   = nitida Stitz, 1910

   opaciventris crucheti Santschi, 1925

   opaciventris mesonotalis Santschi, 1925

   opaciventris obscuripes Santschi, 1937

reichenspergeri Santschi, 1925

rhodesiae Arnold, 1958

rustica Santschi, 1925

   rustica angustior Santschi, 1925

salambo Wheeler, W.M., 1922

striata Stitz, 1911

   striata buttgenbachi Forel, 1913

   striata insularis Santschi, 1920

   striata pilosa Arnold, 1926

striatula Santschi, 1925

tigreensis (Guérin-Méneville, 1849)

MYSTRIUM Roger, 1862

1 species, Afrotropical; 10 species, Malagasy

Amblyoponinae

PLATE 23

DIAGNOSTIC REMARKS: Among the amblyoponines, *Mystrium* is characterized by the possession of very long, apically blunt, linear mandibles that do not close tightly against the clypeus, the absence of an anterior peduncle on the petiole, and the presence of spatulate or otherwise bizarre setae on the head and body.

DISTRIBUTION AND ECOLOGY: As well as the numerous Malagasy species, and 1 in the Afrotropical region, members of this genus are also found in the wet forest zones of the Oriental region and in Malaysia and Indonesia to New Guinea and northern Australia. They usually inhabit rainforest zones, where they nest in the ground or in and under rotten wood. Foraging takes place mostly in the leaf litter and topsoil. Queens include both alate and

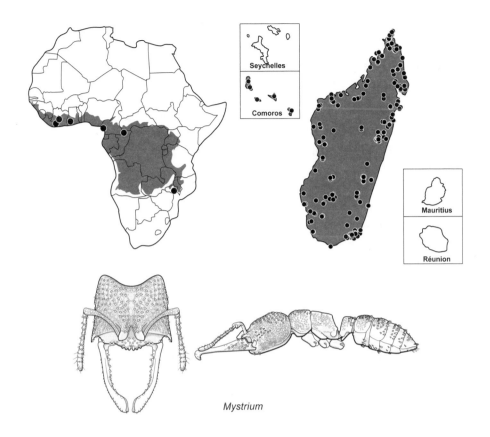

Mystrium

ergatoid forms, sometimes also with intermediates between them. Gamergates have been recorded in some species.

IDENTIFICATION OF SPECIES: Menozzi, 1929; Yoshimura and Fisher, 2014 (Malagasy).

DEFINITION OF THE GENUS (WORKER): Characters of subfamily AMBLYOPONINAE, plus:

Mandible linear and very long, extensively crossing over when fully closed. Apex of mandible blunt or truncated, its apical tooth located ventrally; inner margin of the mandible with 10–30 small teeth or denticles arranged in 2 rows. When mandibles are fully closed, there is a marked gap between the masticatory margins and between the basal margins of the mandibles and the clypeus.

Palp formula 4,3.

Gena on each side, immediately behind the clypeus, often with a prominent angle or an anteriorly directed tooth.

Frontal lobes conspicuous and widely separated, their anterior margins behind the anterior clypeal margin.

Eyes present, located slightly to conspicuously behind the midlength of the head capsule.

Antenna with 12 segments, with the apical 4 antennomeres usually forming a weak club.

Metanotal groove absent to weakly present.

Tibial spurs: mesotibia 1 or 2; metatibia 2, main spur pectinate.

Abdominal segment 2 (petiole) sessile; in profile with a steep anterior face and weakly convex dorsum that may slope weakly downward posteriorly but without a distinctly differentiated posterior face; the petiole broadly attached to abdominal segment 3 (first gastral segment).

Helcium with its dorsal surface arising very high on the anterior face of the first gastral segment (A3), so that in profile A3 above the helcium has no free anterior face (at most a short, rounded angle present between the anterior and dorsal surfaces).

Prora absent or present.

Abdominal segment 4 with distinctly differentiated presclerites.

Hypopygium (sternite of A7) with 1 or 2 stout spines present posteriorly.

Main pilosity of head and body usually bizarre; squamate, clavate, or spatulate, and with spatulate setae present on the clypeus.

AFROTROPICAL MYSTRIUM

silvestrii Santschi, 1914

MALAGASY MYSTRIUM

barrybressleri Yoshimura and Fisher, 2014	= *stadelmanni* Forel, 1895
eques Yoshimura and Fisher, 2014	*oberthueri* Forel, 1897
janovitzi Yoshimura and Fisher, 2014	*rogeri* Forel, 1899
labyrinth Yoshimura and Fisher, 2014	*shadow* Yoshimura and Fisher, 2014
mirror Yoshimura and Fisher, 2014	*voeltzkowi* Forel, 1897
mysticum Roger, 1862	= *fallax* Forel, 1897

NESOMYRMEX Wheeler, W.M., 1910

Myrmicinae

25 species, Afrotropical; 4 species, Malagasy

PLATE 23

DIAGNOSTIC REMARKS: Among the myrmicines, *Nesomyrmex* has 11- or 12-segmented antennae that terminate in an apical club of 3 segments. The mandible has only 3–5 teeth. The clypeus posteriorly is broadly inserted between the frontal lobes, and its anterior margin never has an unpaired median seta. Eyes are present at about the midlength of the head capsule. The first gastral tergite (A4) broadly overlaps the sternite on the ventral surface of the gaster. In habitus, species of *Nesomyrmex* are only likely to be confused with those of *Temnothorax*, but the structure of the clypeus is different. In *Nesomyrmex* the median portion of the clypeus forms an anteriorly projecting lobe or apron that fits very tightly over the dorsal surfaces of the mandibles and distinctly overlaps the basal portion of the mandible; when seen in profile the anterior clypeal margin is closely adherent to the dorsal surface of the mandible, not elevated away from it.

DISTRIBUTION AND ECOLOGY: *Nesomyrmex* species are present in both the Afrotropical and the Malagasy region, with the greatest number of species in South Africa. The genus is also strongly represented in the Neotropics, and a couple of species occur in the southern Palaearctic and the southern Nearctic regions, but these are northerly outliers of a basically

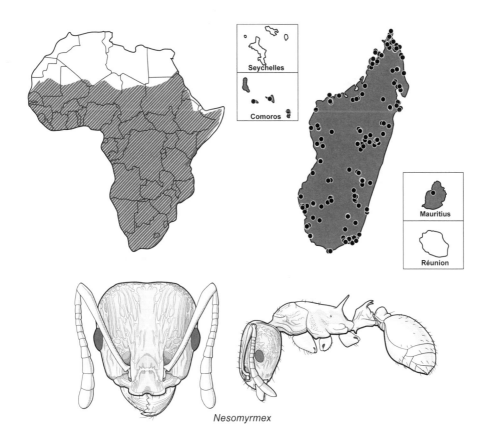

Nesomyrmex

southerly genus. It is absent elsewhere in the Old World. Most species nest and forage in the soil, leaf litter, and log mold, but a few are arboreal and nest in trees.

IDENTIFICATION OF SPECIES: Bolton, 1982 (Afrotropical, as part of *Leptothorax*); Mbanyana and Robertson, 2008 (Southern Africa).

DEFINITION OF THE GENUS (WORKER): Characters of subfamily MYRMICINAE, plus:

Mandible short triangular and stout; masticatory margin with 3–5 teeth.

Palp formula 5,3.

Stipes of the maxilla without a transverse crest.

Clypeus posteriorly broadly inserted between the frontal lobes. Anterior margin of the median portion of the clypeus forms a prominent lobe that fits tightly over the dorsa of the closed mandibles. Clypeal lobe conceals the basal 1–2 mandibular teeth unless the lobe is notched medially. In profile the anterior clypeus overlaps and is closely adherent to the dorsal surface of the mandible.

Clypeus without an unpaired seta at the midpoint of the anterior margin.

Frontal carinae usually consist only of the frontal lobes; uncommonly extended posteriorly.

Antennal scrobes absent.

Antenna with 11 or 12 segments, with an apical club of 3 segments.

Torulus concealed by the frontal lobe in full-face view.

Eyes well developed, located at the midlength of the head capsule.

Promesonotal suture vestigial to absent on the dorsum; the pronotum not domed, the mesonotum not elongate.

Metasternal process absent.

Propodeum unarmed or bidentate to bispinose.

Propodeal spiracle approximately at midlength of the sclerite.

Tibial spurs: mesotibia 0 or 1; metatibia 0 or 1.

Abdominal segment 2 (petiole) frequently (but not universally) equipped with tubercles, teeth, or spines.

Abdominal tergite 4 (first gastral) broadly overlaps the sternite on the ventral gaster; gastral shoulders present.

Stridulitrum present on pretergite of A4.

Sting simple.

Main pilosity of dorsal head and body: usually short and stubbly to moderately long, usually truncated apically; sometimes very sparse or absent.

AFROTROPICAL NESOMYRMEX

angulatus (Mayr, 1862)
= *angulatus ilgii* (Forel, 1894)
= *latinodis* (Mayr, 1895)
= *angulatus concolor* (Santschi, 1914)
antoinetteae Mbanyana and Robertson, 2008
braunsi (Forel, 1912)
cataulacoides (Snelling, 1992)
cederbergensis Mbanyana and Robertson, 2008
denticulatus (Mayr, 1901)
entabeni Mbanyana and Robertson, 2008
evelynae (Forel, 1916)
ezantsi Mbanyana and Robertson, 2008
grisoni (Forel, 1916)
humerosus (Emery, 1896)
innocens (Forel, 1913)
inye Mbanyana and Robertson, 2008

karooensis Mbanyana and Robertson, 2008
koebergensis Mbanyana and Robertson, 2008
larsenae Mbanyana and Robertson, 2008
mcgregori Mbanyana and Robertson, 2008
nanniae Mbanyana and Robertson, 2008
njengelanga Mbanyana and Robertson, 2008
ruani Mbanyana and Robertson, 2008
saasveldensis Mbanyana and Robertson, 2008
simoni (Emery, 1895)
stramineus (Arnold, 1948)
tshiguvhoae Mbanyana and Robertson, 2008
vannoorti Mbanyana and Robertson, 2008

MALAGASY NESOMYRMEX

gibber (Donisthorpe, 1946)
madecassus (Forel, 1892)
retusispinosus (Forel, 1892)
sikorai (Emery, 1896)

NYLANDERIA Emery, 1906

Formicinae

17 species, Afrotropical; 10 species (+ 6 subspecies), Malagasy

PLATE 23

DIAGNOSTIC REMARKS: Among the formicines, *Nylanderia* species have 12-segmented antennae, the sockets of which are close to the clypeal margin (separated by a distance less than the basal width of the scape), and the scapes always have projecting setae present. The mandible is unsculptured and has 6–7 teeth, the third of which is reduced in size, smaller than the fourth. The metapleural gland orifice is conspicuous, and the propodeal dorsum always lacks setae.

DISTRIBUTION AND ECOLOGY: Distributed throughout the tropics, with strong representation in subtropical and some temperate zones. In the regions under consideration here, most species are found in rainforest, but some widespread species occur in drier habitats. Nests are constructed in the ground or among the leaf litter, with some species nesting in rotten wood.

IDENTIFICATION OF SPECIES: LaPolla, Hawkes, and Fisher, 2011 (Afrotropical).

DEFINITION OF THE GENUS (WORKER): Characters of subfamily FORMICINAE, plus:

Mandible usually with 6 teeth, rarely with 7; tooth 3 (counting from the apical) is smaller than tooth 4.

Palp formula usually 6,4 but 5,3 in 1 species.

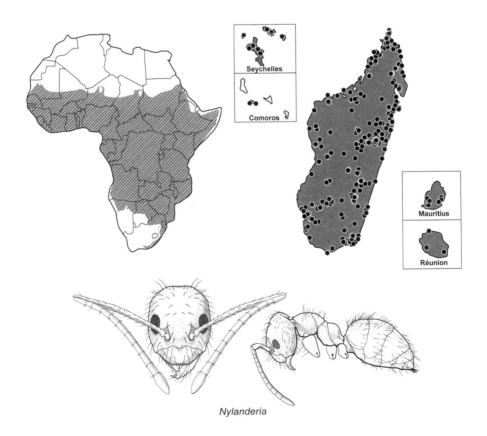

Nylanderia

Antenna with 12 segments.

Antennal scapes with projecting setae present.

Antennal sockets close to the posterior clypeal margin; anterior arc of the torulus may be separated from the clypeal suture by a narrow strip of cuticle, but often the torulus just touches the clypeal suture.

Eyes present, located at or slightly in front of the midlength of the sides of the head; ocelli usually absent but sometimes present, small and indistinct.

Mesothorax not constricted behind the pronotum.

Mesonotum with stout setae that are arranged in pairs.

Metapleural gland present.

Propodeum unarmed, without setae on its dorsal surface.

Propodeal spiracle circular.

Propodeal foramen long: in ventral view the foramen extends anterior of a line that spans the anteriormost points of the metacoxal cavities.

Metacoxae widely separated: in ventral view, with the mesocoxae and metacoxae directed at right-angles to the long axis of the mesosoma, the distance between the bases of the mesocoxae is markedly less than the distance between the bases of the metacoxae.

Tibial spurs: mesotibia 0 or 1; metatibia 0 or 1.

Abdominal segment 2 (petiole) an unarmed scale.

Petiole (A2) scale with a shorter anterior face and a longer posterior face; with a posterior peduncle.

Petiole (A2) with its ventral margin U-shaped in section.

Abdominal segment 3 (first gastral) with a sharply margined longitudinal concavity or groove, into which the posterior face of the petiole fits when the mesosoma and gaster are aligned.

Abdominal segment 3 basally with complete tergosternal fusion for some distance on each side of the helcium; the line of fusion follows the edges of the impression in the anterior face of A3 and the unfused portions of the tergite and sternite commence some distance above the helcium, at the dorsolateral apices of the impression.

Abdominal sternite 3 without a transverse sulcus across the sclerite immediately behind the helcium sternite (i.e., presternite and poststernite of A3 not separated by a sulcus).

Acidopore with a seta-fringed nozzle.

AFROTROPICAL NYLANDERIA

boltoni LaPolla and Fisher, 2011	= zelotypa (Santschi, 1915)
brevisetula LaPolla and Fisher, 2011	= weissi nimba (Bernard, 1953)
bourbonica (Forel, 1886)	lepida (Santschi, 1915)
impolita LaPolla and Fisher, 2011	= grisoni (Forel, 1916)
incallida (Santschi, 1915)	= grisoni fuscula (Menozzi, 1942)
= arlesi (Bernard, 1953)	luteafra LaPolla and Fisher, 2011
jaegerskioeldi (Mayr, 1904)	mendica (Menozzi, 1942)
= traegaordhi (Forel, 1904)	natalensis (Forel, 1915)
= weissi (Santschi, 1911)	scintilla LaPolla and Fisher, 2011

silvula LaPolla and Fisher, 2011

umbella LaPolla and Fisher, 2011

usambarica LaPolla and Fisher, 2011

vaga (Forel, 1901)

vividula (Nylander, 1846)

waelbroecki (Emery, 1899)

MALAGASY NYLANDERIA

amblyops (Forel, 1892)
  amblyops rubescens (Forel, 1892)
bourbonica (Forel, 1886)
  bourbonica farquharensis (Forel, 1907)
  bourbonica ngasiyana (Forel, 1907)
comorensis (Forel, 1907)
dodo (Donisthorpe, 1946)
madagascarensis (Forel, 1886)
  madagascarensis ellisii (Forel, 1891)

madagascarensis sechellensis (Emery, 1894)
madagascarensis rufescens (Wheeler, W.M., 1922)
gracilis (Forel, 1892)
humbloti (Forel, 1891)
jaegerskioeldi (Mayr, 1904)
mixta (Forel, 1897)
sikorae (Forel, 1892)

OCHETELLUS Shattuck, 1992

Dolichoderinae

1 species Malagasy, probably introduced

PLATE 24

DIAGNOSTIC REMARKS: Among the dolichoderines, the petiole (A2) of *Ochetellus* has a well-developed, narrow, tall, erect scale that is not overhung from behind by the first gastral segment. In profile the outline of the propodeal declivity is distinctly concave. The fourth gastral sternite (sternite of A6) is not keel-shaped posteriorly.

DISTRIBUTION AND ECOLOGY: *Ochetellus* contains 6 species and 3 subspecies, distributed from Japan to Australia. A seventh species, *O. vinsoni*, was described from Mauritius and is also known from Réunion Island. This probably represents an introduction and is possibly a synonym of one of the more easterly species. The genus is absent from the Afrotropical region and from Madagascar itself.

IDENTIFICATION OF SPECIES: No review exists.

DEFINITION OF THE GENUS (WORKER): Characters of DOLICHODERINAE, plus:

Masticatory margin of the mandible with a combination of teeth and denticles (about 7–12 in total).

Palp formula 6,4.

Anterior clypeal margin evenly concave.

Antenna with 12 segments.

Metathoracic spiracles dorsolateral.

Metanotal groove present.

Propodeum unarmed, the declivity concave in profile.

Propodeal spiracle lateral and ventral of the dorsum.

Tibial spurs: mesotibia 1; metatibia 1, pectinate.

Abdominal segment 2 (petiole) a narrow, erect scale, not inclined anteriorly.

Abdominal segment 3 (first gastral) with anterior face vertical, not projecting forward over the petiole, not concealing the petiole in dorsal view.

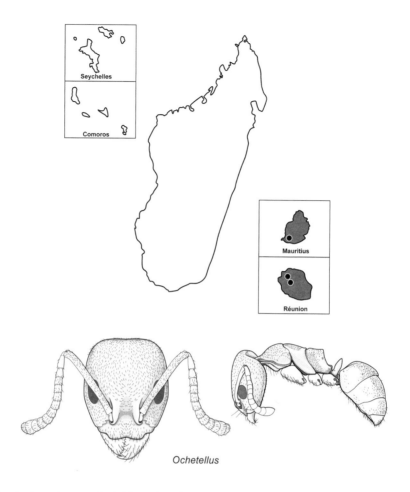

Ochetellus

Abdominal tergites 3 to 6 visible in dorsal view (i.e., 4 gastral tergites visible in dorsal view); the tergite of A7 (pygidium) is reflexed onto the ventral surface of the gaster, where it abuts abdominal sternite 7 (hypopygium).

Abdominal sternite 6 (fourth gastral sternite) not keel-shaped posteriorly.

MALAGASY OCHETELLUS

vinsoni (Donisthorpe, 1946)

OCYMYRMEX Emery, 1886

Myrmicinae

34 species, Afrotropical

PLATE 24

DIAGNOSTIC REMARKS: Among the Afrotropical myrmicines, *Ocymyrmex* has 12-segmented antennae that are filiform and do not have an apical club. The eyes are well behind the midlength of the head capsule in full-face view, and the ventral surface of the head always has a large psammophore. The mesothoracic spiracles are open, located dorsally or high on the side and clearly visible in dorsal view; the orifices of the spiracles are slit-like to

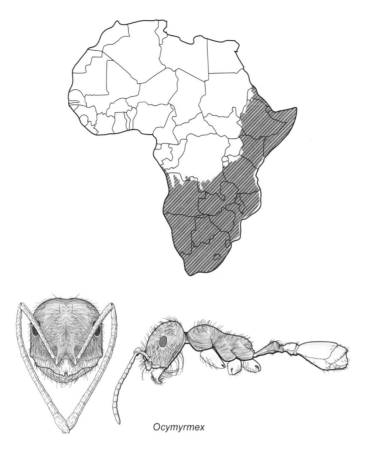

Ocymyrmex

crescent-shaped. The propodeum is unarmed, and its spiracles are conspicuous, very long, and slit shaped. All the legs are extremely long and slender, with large, powerful coxae.

DISTRIBUTION AND ECOLOGY: The unmistakably long-legged, fast-moving species of this genus are all Afrotropical. They are strikingly thermophilic and found from the harsh desert conditions of the Namib across all of the grassland, savanna, and scrub zones of southern and eastern Africa but are absent from the rainforest zone and do not penetrate the sub-Saharan Sahelian zone. Nests are constructed directly in the earth, sometimes in the open, sometimes in very rocky soil, and may extend to a considerable depth. In very unstructured loose sand, such as in dry riverbeds in the Namib Desert, the tunnels and chambers of the nest run alongside the roots of shrubs and trees. Nest entrances in loose or sandy soil frequently have a crater-like entrance, sometimes adorned with small stones. The prey of *Ocymyrmex* consists mostly of scavenged dead or heat-stressed insects, but some species will attack termites, and others have been credited with taking seeds and other plant material. The ergatoid queens of this genus are among the most worker-like of all ants; the 2 castes cannot be distinguished without microscopic examination. Morphological queens often make up a good proportion of a colony, but only 1 is mated and reproductive and is physogastric; the rest behave as supplementary workers.

IDENTIFICATION OF SPECIES: Bolton, 1981b; Bolton and Marsh, 1989.

DEFINITION OF THE GENUS (WORKER): Characters of subfamily MYRMICINAE, plus:

Mandible short and powerful; masticatory margin with 5 (usually) or 4 (rarely) teeth. Counting from the apical, the third and fourth teeth (when 5 present in total), or only the third tooth (when 4 present in total), paired, having a flanking tooth internally on the margin that is only visible when the mandible is open. Masticatory margin at most as long as basal margin, often shorter.

Palp formula 5,3, or 4,3, or 4-3,3 (incomplete fusion of the maxillary palpomeres), or 3,3, or 2,3.

Stipes of the maxilla without a transverse crest.

Clypeus large, posteriorly broadly inserted between the frontal lobes.

Clypeus without an unpaired seta at the midpoint of the anterior margin.

Frontal carinae short, restricted to well-defined but narrow frontal lobes.

Antennal scrobes absent.

Antenna with 12 segments, filiform, without an apical club.

Torulus with the upper lobe visible in full-face view; maximum width of the torulus lobe is posterior to the point of maximum width of the frontal lobes.

Eyes well developed, located behind the midlength of the head capsule.

Head capsule ventrally with a strongly developed psammophore.

Promesonotal suture absent from the dorsum of the mesosoma.

Pronotum and anterior mesonotum in profile convex to dome-like; the mesonotum posteriorly sloping to the propodeum.

Mesothoracic spiracles open, located dorsally or high on the side, clearly visible in dorsal view; orifice slit-like to crescent-shaped.

Metanotal groove absent.

Propodeum unarmed, without a trace of spines or teeth; in profile the propodeal dorsum is on a much lower level than the apex of the convex promesonotum.

Propodeal spiracle extremely elongate, slit-shaped and very conspicuous.

Metasternal process absent.

Legs extremely long and slender; coxae large and powerful.

Tibial spurs: mesotibia 0 or 1; metatibia 0 or 1.

Abdominal segment 2 (petiole) with a long, narrow, anterior peduncle.

Abdominal tergite 4 (first gastral) broadly overlaps the sternite on the ventral gaster; gastral shoulders absent.

Stridulitrum present on the pretergite of A4.

Sting very reduced in size, apparently not functional.

Main pilosity of dorsal head and body: simple.

AFROTROPICAL OCYMYRMEX

afradu Bolton and Marsh, 1989	*celer* Weber, 1943
alacer Bolton and Marsh, 1989	*cursor* Bolton, 1981
ankhu Bolton, 1981	*dekerus* Bolton and Marsh, 1989
barbiger Emery, 1886	*engytachys* Bolton and Marsh, 1989
cavatodorsatus Prins, 1965	*flavescens* Stitz, 1923

flaviventris Santschi, 1914

foreli Arnold, 1916

fortior Santschi, 1911

   = *weitzeckeri transversus* Santschi, 1911

   = *arnoldi* Forel, 1913

   = *weitzeckeri abdominalis* Santschi, 1914

   = *weitzeckeri usakosensis* Stitz, 1923

hirsutus Forel, 1910

ignotus Bolton and Marsh, 1989

kahas Bolton and Marsh, 1989

laticeps Forel, 1901

micans Forel, 1910

monardi Santschi, 1930

nitidulus Emery, 1892

okys Bolton and Marsh, 1989

phraxus Bolton, 1981

picardi Forel, 1901

   = *carpenteri* Donisthorpe, 1933

resekhes Bolton and Marsh, 1989

robecchii Emery, 1892

robustior Stitz, 1923

shushan Bolton, 1981

sobek Bolton, 1981

sphinx Bolton, 1981

tachys Bolton and Marsh, 1989

turneri Donisthorpe, 1931

velox Santschi, 1932

weitzeckeri Emery, 1892

   = *weitzeckeri wroughtoni* Forel, 1910

zekhem Bolton, 1981

ODONTOMACHUS Latreille, 1804   Ponerinae

2 species, Afrotropical; 3 species, Malagasy   PLATE 24

DIAGNOSTIC REMARKS: Only 2 genera of the ponerines have linear mandibles articulated very close together near the midpoint of the anterior margin of the head: *Odontomachus* and *Anochetus*. They are easily separated because in *Odontomachus* the petiole (A2) always terminates in a single dorsal spine. In addition, the carina that separates the dorsal from the posterior surface of the head (the nuchal carina) forms a V-shape at the midline and also receives a pair of prominent, dark, posterior apophyseal lines that converge to form a longitudinal sharp median groove on the vertex.

DISTRIBUTION AND ECOLOGY: *Odontomachus* is predominantly a pantropical genus, with only a few species found outside the tropics. Some are restricted to wet forest zones, but others are much more widespread, in open or loosely wooded areas. Nests are usually in rotten wood or in the ground, either under surface objects or at the bases of trees. Foraging takes place mainly on the surface of the ground and in the leaf-litter layer, but workers may also be found on trees. The species are opportunist predators that take a wide range of prey but on occasion will also gather drops of honeydew. Most species have normal alate queens, but *O. coquereli* has ergatoids. The Malagasy collections of *O. simillimus*, a species otherwise distributed through the entire Indo-Pacific area, are from the Seychelles, where it is certainly an introduction.

IDENTIFICATION OF SPECIES: Brown, 1976 (Afrotropical + Malagasy); Fisher and Smith, 2008 (Malagasy).

DEFINITION OF THE GENUS (WORKER): Characters of subfamily PONERINAE, plus:

Mandibles linear and nearly parallel when closed, articulated close to the midline of the anterior margin of the head. Mandible equipped apically with a vertical series

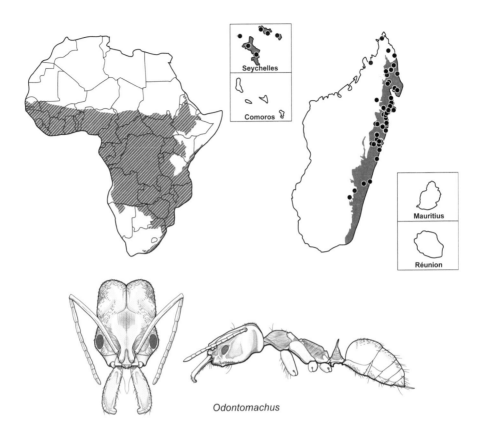

Odontomachus

of 3 teeth, the uppermost (apicodorsal) tooth usually truncated (acute only in the Madagascan *O. coquereli*). A pair of trigger setae present below the mandibles.

Palp formula 4,4, or 4,3.

Clypeus with anterior margin projecting on each side of the mandibles.

Frontal lobes small; in full-face view their anterior margins do not overhang the anterior clypeal margin.

Eyes present, in front of the midlength of the head capsule; each eye located on a broad, rounded cuticular prominence.

Head capsule with eye separated from frons by a broad impression in the dorsum.

Vertex with a median furrow that extends to the posterior margin and is continuous posteriorly with a nuchal carina and also with a V-shaped pair of dark apophyseal lines that extend down to the occipital foramen.

Metanotum present on the dorsal mesosoma as a narrow, sometimes poorly delimited, transverse arc between the epimeral sclerites.

Epimeral sclerite present.

Metapleural gland orifice opens posteriorly.

Propodeal spiracle with orifice subcircular, or a very small ellipse.

Tibial spurs: mesotibia 2; metatibia 2; on each tibia the anterior spur small, simple to barbulate, the posterior spur larger and pectinate.

Pretarsal claws simple.

Abdominal segment 2 (petiole) surmounted by a single spine.

Helcium located at the base of the anterior face of A3 (first gastral segment).

Prora present, small.

Abdominal segment 4 (second gastral segment) with very weakly differentiated presclerites.

Stridulitrum present or absent on the pretergite of A4.

OECOPHYLLA Smith, F., 1860 — Formicinae

1 species (+ 7 subspecies), Afrotropical — PLATE 24

DIAGNOSTIC REMARKS: Among the formicines, *Oecophylla* is dimorphic and has 12-segmented antennae, the sockets of which are located well behind the posterior clypeal margin (a distance greater than the basal width of the scape). Eyes are conspicuous, and the clypeus is large, curved, and shield-like. On the powerfully constructed mandibles, the teeth do not decrease in size from apical to basal; instead, tooth 4 is enlarged, noticeably larger than tooth 3. The mesosoma is elongate and unarmed, with a conspicuously strangulated mesothorax. The metapleural gland orifice is absent. The petiole (A2) is elongate and low, and in life the gaster can be reflexed over the mesosoma. The helcium arises at about the midheight of the first gastral segment (A3).

DISTRIBUTION AND ECOLOGY: A dimorphic arboreal species, the minor workers of which are only rarely seen outside of nests. Uniquely, the species constructs nests by using larval silk to stitch leaves together in a living tree. Silk is also used to build protective tents over aggregations of large arboreal coccids, which are avidly tended for the honeydew they produce. Foraging is mainly arboreal, but major workers will descend trunks and sometimes cross the ground between trees, seizing arthropod prey anywhere it is encountered. At present, only 2 extant *Oecophylla* species are recognized, together with a number of fossil forms. One living species is extremely widespread in Afrotropical forests, and 1 is very widely dis-

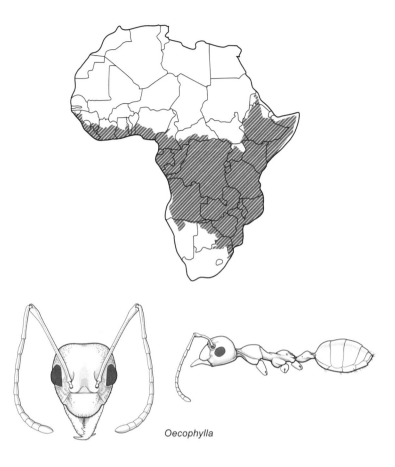

Oecophylla

tributed in tropical forests from Sri Lanka and India to northern Australia. The genus does not occur in the Malagasy region.

TAXONOMIC COMMENT: The 7 currently recognized subspecies of *O. longinoda* have never been subjected to a proper taxonomic analysis. One or more of these subspecies may represent valid species.

IDENTIFICATION OF SPECIES: No recent review exists.

DEFINITION OF THE GENUS (WORKER): Characters of subfamily FORMICINAE, plus:

Worker caste dimorphic but minors usually confined to nests.

Mandible with 9–16 teeth; tooth 3 (counting from the apical) is distinctly smaller than tooth 4.

Palp formula 5,4.

Antenna with 12 segments; first funicular segment very long.

Antennal sockets located well behind the posterior clypeal margin (a distance greater than the basal width of the scape).

Eyes present, located at the midlength of the sides of the head; ocelli absent.

Mesosoma with the anterior portion of the mesothorax strongly constricted.

Metapleural gland absent.

Propodeal spiracle elliptical.

Propodeal lobes present.

Propodeal foramen short: in ventral view the foramen does not extend anterior of a line that spans the anteriormost points of the metacoxal cavities.

Metacoxae closely approximated: in ventral view, with the mesocoxae and metacoxae directed at right-angles to the long axis of the mesosoma, the distance between the bases of the mesocoxae is about equal to the distance between the bases of the metacoxae.

Tibial spurs: mesotibial spur vestigial to absent; metatibial spur vestigial to absent.

Pretarsal claws large and strongly developed.

Abdominal segment 2 (petiole) an elongate low node, without teeth or spines.

Petiole (A2) with its ventral margin V-shaped in section anteriorly.

Helcium in profile at about midheight of abdominal segment 3 (first gastral).

Abdominal segment 3 without tergosternal fusion on each side of the helcium.

Abdominal sternite 3 without a transverse sulcus across the sclerite immediately behind the helcium sternite (i.e., presternite and poststernite of A3 not separated by a sulcus).

Gaster (A3–A7) capable of reflexion over the mesosoma in life.

Acidopore with a seta-fringed nozzle.

AFROTROPICAL OECOPHYLLA

longinoda (Latreille, 1802)	*longinoda rubriceps* Wheeler, W.M., 1922
= *brevinodis* André, 1890	
longinoda fusca Emery, 1899	*longinoda textor* Wheeler, W.M., 1922
longinoda annectens Wheeler, W.M., 1922	*longinoda claridens* Santschi, 1928
	longinoda rufescens Santschi, 1928
	longinoda taeniata Santschi, 1928

OOCERAEA Roger, 1862 — Dorylinae

1 species, Malagasy — PLATE 25

DIAGNOSTIC REMARKS: Among the dorylines, *Ooceraea* usually lacks eyes (sometimes 1 or 2 minute ommatidia are present) and always lacks the promesonotal suture; the waist is of 2 nodiform segments, and the antennae have only 9 segments. The propodeal spiracle is low on the side and situated behind the midlength of the sclerite, and propodeal lobes are present. The mesotibia has a single spur present, and the pretarsal claws lack a preapical tooth.

DISTRIBUTION AND ECOLOGY: The single species represented, *O. biroi*, is a tramp species originally described from Singapore but now widely distributed and well established in parts of Madagascar. Species of *Ooceraea* occur from India to Australia, and there are also a couple of endemic species in the Fiji Islands, but the genus has never been found in the Afrotropical region.

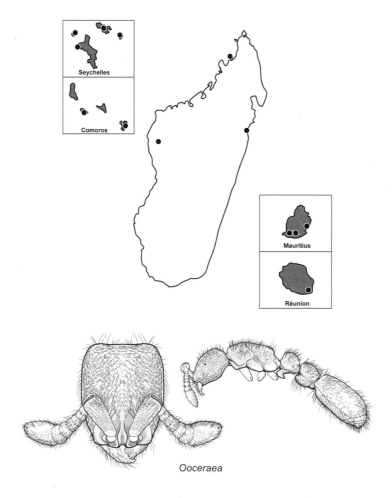

Ooceraea

TAXONOMIC COMMENT: *Ooceraea* is currently a junior synonym of *Cerapachys*. It will be formally revived from synonymy by Marek Borowiec in a genus-rank revision of Dorylinae that is in preparation.

IDENTIFICATION OF SPECIES: No review exists.

DEFINITION OF THE GENUS (WORKER): Characters of subfamily DORYLINAE, plus:

Prementum not visible when mouthparts closed.

Palp formula not confirmed.

Eyes usually absent, sometimes minutely present (of 1–2 ommatidia only); ocelli absent.

Frontal carinae present as a pair of closely approximated, raised ridges, abruptly truncated posteriorly, without horizontal frontal lobes.

Antennal sockets fully exposed in full-face view.

Parafrontal ridges present.

Antenna with 9 segments; the funiculus gradually incrassate apically.

Antennal scrobes absent.

Promesonotal suture absent on the dorsum of the mesosoma.

Pronotal-mesopleural suture present but fused on the side of the mesosoma.

Metanotal groove absent.

Propodeal spiracle low on the side and situated at or behind the midlength of the sclerite, not subtended by a longitudinal impression.

Propodeal lobes present.

Tibial spurs: mesotibia 1, pectinate; metatibia 1, pectinate.

Metatibial gland present.

Metacoxae without a posterodorsal vertical lamella.

Pretarsal claws simple.

Waist of 2 segments (petiole and postpetiole = A2 and A3).

Petiole (A2) sessile, not marginate dorsolaterally.

Helcium projects from the anterior surface of A3 (first gastral segment) at about its midheight.

Prora present.

Abdominal segment 4 with strongly developed presclerites; in profile, the tergite of A4 not folding over the sternite, so that the anterior portion of the sternite is visible equally to the tergite.

Abdominal segments 5 and 6 without presclerites.

Pygidium (tergite of A7) large and distinct, flattened to concave dorsally and armed laterally with a series or row of denticles or peg-like spiniform setae.

Sting present, large and fully functional.

MALAGASY OOCERAEA

biroi (Forel, 1907)

OPHTHALMOPONE Forel, 1890 — Ponerinae

5 species (+ 1 subspecies), Afrotropical — PLATE 25

DIAGNOSTIC REMARKS: Among the ponerines, *Ophthalmopone* combines the characters of presence of 2 spurs on the metatibia, absence of traction setae on the metabasitarsus, and location of the helcium at the base of the anterior face of the first gastral segment (A3), with the following: Large to very large eyes are present, at or behind the midlength of the head. The mandible lacks both a basal pit and a basal groove. The metanotal groove is present and impressed, and the propodeal spiracle is slit-shaped. The pretarsal claws of the metatarsus are usually simple, but a basal tooth sometimes occurs. The hypopygium (sternite of A7) has a row of teeth or spiniform setae on the posterior part of its dorsal margin, which project upward, outside the pygidium (the tergite of A7).

DISTRIBUTION AND ECOLOGY: *Ophthalmopone* is restricted to the Afrotropical region, where its species are most commonly found in the drier parts of the south and east of the continent, although, strangely, 2 species have been described from São Tomé Island, as discussed below. As far as is known, all species are usually ground nesters, either directly or under stones, but abandoned termitaries may sometimes be utilized. Foraging takes place

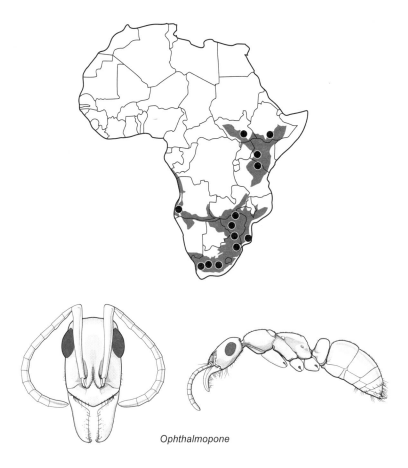

Ophthalmopone

on the ground, and the prey consists entirely of termites. *Ophthalmopone* species have lost the queen caste; reproduction is entirely by gamergates.

TAXONOMIC COMMENT: There is a remarkable record by Emery (1902) of 2 species, *O. depilis* and *O. mocquerysi*, on the island of São Tomé in the Gulf of Guinea. These species have never been recorded from the island again, and there are no records of *Ophthalmopone* from any other locality in West or Central Africa, despite quite extensive recent collecting. The 2 supposed species are very similar and bear a close resemblance to the long-headed *O. berthoudi* and *O. ilgii*. The suspicion must be entertained that both the São Tomé names represent a casual introduction of 1 of these species or perhaps are mislabeled as to locality.

IDENTIFICATION OF SPECIES: No review exists.

DEFINITION OF THE GENUS (WORKER): Characters of subfamily PONERINAE, plus:

Mandible stoutly elongate-triangular, multidentate; basal groove absent.
Palp formula 4,4.
Clypeus simple, its anterior margin convex, without extended lobes or teeth on either
  the median or the lateral portions.

Frontal lobes small; in full-face view their anterior margins do not overhang the anterior clypeal margin.

Eyes present, large to very large, located at or distinctly behind the midlength of the head capsule.

Metanotal groove present, impressed.

Epimeral sclerite present.

Metapleural gland orifice opens posteriorly.

Propodeal spiracle with orifice slit-shaped.

Tibial spurs: mesotibia 2; metatibia 2; on each tibia the anterior spur small, simple to barbulate, the posterior spur larger and pectinate.

Pretarsal claws usually unarmed but a single tooth may occur basally on its inner curvature.

Abdominal segment 2 (petiole) a node.

Helcium located at the base of the anterior face of A3 (first gastral segment).

Prora present, small.

Abdominal segment 4 (second gastral segment) with presclerites weakly differentiated.

Stridulitrum present on the pretergite of A4.

Hypopygium (sternite of A7) with a row of teeth or spiniform setae on its dorsal margin posteriorly; when viewed in profile, these teeth project upward, outside the pygidium (tergite of A7).

AFROTROPICAL OPHTHALMOPONE

berthoudi Forel, 1890	= *lanceolata* Mayr, 1895
berthoudi pubescens Weber, 1942	*ilgii* Forel, 1894
depilis Emery, 1902	*mocquerysi* Emery, 1902
hottentota (Emery, 1886)	

PALTOTHYREUS Mayr, 1862 — Ponerinae

1 species (+ 6 subspecies), Afrotropical — PLATE 25

DIAGNOSTIC REMARKS: Among the ponerines, *Paltothyreus* combines the characters of presence of 2 spurs on the metatibia, absence of traction setae on the metabasitarsus, and location of the helcium at the base of the anterior face of the first gastral segment (A3), with the following: The clypeus has a prominent median lobe that is truncated apically. The propodeal spiracle is slit-shaped. The petiole (A2) is a thick scale, and the first gastral tergite (tergite of A3) has conspicuous anterolateral angles in dorsal view. The pretarsal claws of the metatarsus have a stout tooth on the inner curvature.

DISTRIBUTION AND ECOLOGY: Restricted to the Afrotropical region, the large, distinct, foul-smelling workers of *Paltothyreus* are commonly encountered in forested zones. Nests of the species are constructed directly in the ground and may be extensive. The workers forage singly or in small processions on the surface and prey on a wide range of arthropods, though termites make up a very large part of the diet.

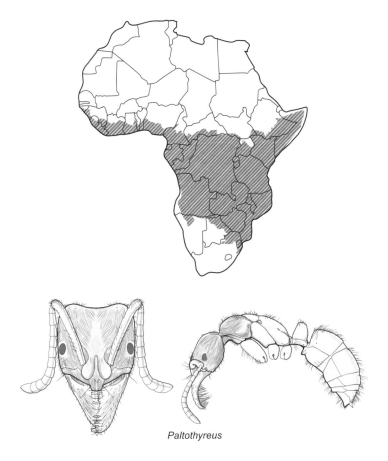

Paltothyreus

TAXONOMIC COMMENT: Apart from a very old key to subspecies produced by Santschi (1919), there has been no attempt at a serious analysis of the infraspecific taxa attached to *P. tarsatus*.

IDENTIFICATION OF SPECIES: Schmidt and Shattuck, 2014.

DEFINITION OF THE GENUS (WORKER): Characters of subfamily PONERINAE, plus:

Mandible stoutly elongate-triangular; basal groove present. Masticatory margin with more than 15 teeth, the apical 2–3 of which are the largest.

Palp formula 4,4.

Clypeus with a projecting rectangular median lobe that is truncated anteriorly and concave dorsally.

Frontal lobes moderate; in full-face view their anterior margins do not overhang the anterior clypeal margin.

Eyes present, distinct, in front of the midlength of the head capsule.

Metanotal groove present but only as a weakly developed faint line, not impressed.

Epimeral sclerite present.

Metapleural gland orifice opens laterally; immediately behind the orifice is a vertical cuticular ridge that shields the opening.

Propodeal spiracle with orifice slit-shaped.

Tibial spurs: mesotibia 2; metatibia 2; on each tibia the anterior spur small, simple to barbulate, the posterior spur larger and pectinate.

Pretarsal claws each with a single stout tooth on its inner curvature, apex of the tooth subtruncate.

Abdominal segment 2 (petiole) a thick scale.

Helcium located at the base of the anterior face of A3 (first gastral segment).

Prora present, conspicuous, subtended by a crest along the midline of the anterior half of abdominal sternite 3 (first gastral sternite).

Abdominal tergite 3 in dorsal view with blunt but pronounced anterolateral angles.

Abdominal segment 4 (second gastral segment) with distinct presclerites.

Pygidium (tergite of A7) with short spines on its dorsolateral margins, among the regular long setae.

Hypopygium (sternite of A7) with a series of short spines or spiniform setae posteriorly on its dorsal margin; these spines directed dorsally, concealed among the dense, regular setae of the hypopygial apex.

Stridulitrum absent from the pretergite of A4.

AFROTROPICAL PALTOTHYREUS

tarsatus (Fabricius, 1798)	*tarsatus robustus* Santschi, 1919
= *gagates* (Guérin-Méneville, 1844)	*tarsatus striatidens* Santschi, 1919
= *pestilentia* (Smith, F., 1858)	*tarsatus subopacus* Santschi, 1919
= *spiniventris* (Smith, F., 1858)	*tarsatus striatus* Santschi, 1930
= *simillima* (Smith, F., 1858)	= *tarsata kaya* (Özdikmen, 2010)
tarsatus delagoensis Emery, 1899	(redundant replacement
tarsatus medianus Santschi, 1919	name)

PARAPARATRECHINA Donisthorpe, 1947   Formicinae

11 species, Afrotropical; 5 species, Malagasy   PLATE 25

DIAGNOSTIC REMARKS: Among the formicines, *Paraparatrechina* species have 12-segmented antennae, the sockets of which are close to the clypeal margin (separated by a distance less than the basal width of the scape), and the scapes usually lack projecting setae (present in a single species). The mandible has 5–6 teeth, the third of which is reduced in size, smaller than the fourth. The metapleural gland orifice is conspicuous, and the propodeal dorsum always possesses projecting setae (usually just 1 pair). Eyes are usually present, but in 1 species-group eyes are very reduced or absent, and the workers of this group are polymorphic.

DISTRIBUTION AND ECOLOGY: The genus is distributed throughout the Old World tropics but is absent from the New World. Most species have large eyes and are found in leaf litter in rainforest or in forest clearings, but the polymorphic species of the *P. weissi* group have eyes that are very reduced or absent and tend to be subterranean. Nests of large-eyed species are usually constructed directly in the litter layer, or under stones, or in and under twigs or larger pieces of rotten wood. The workers may ascend trees to forage and have been

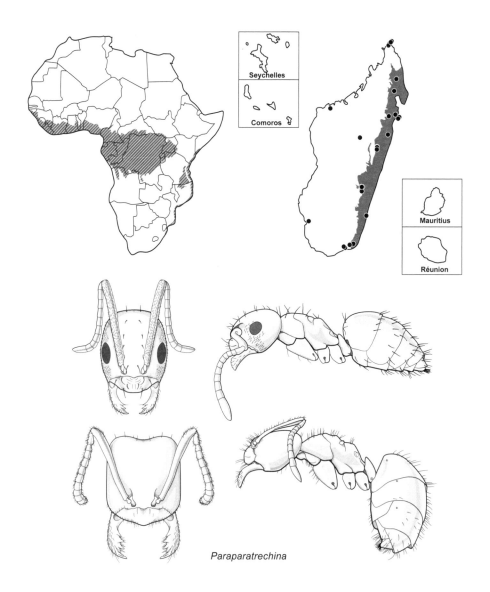

Paraparatrechina

retrieved from the forest canopy as well as from low vegetation. The subterranean species of the *P. weissi* group can be found in and under very rotten wood as well as in the soil around the roots of plants. Species of this group actively avoid light but may be found on the surface of the ground at night.

IDENTIFICATION OF SPECIES: LaPolla, 2004b (Afrotropical, part); LaPolla, Cheng, and Fisher, 2010 (Afrotropical + Malagasy); LaPolla and Fisher, 2014a (Afrotropical + Malagasy).

DEFINITION OF THE GENUS (WORKER): Characters of subfamily FORMICINAE, plus:

Worker caste polymorphic in *P. bufona*, *P. sordida*, and *P. weissi*; all others monomorphic.
Mandible with 5–6 teeth; tooth 3 (counting from the apical) is smaller than tooth 4.
Palp formula usually 6,4, but 3,3 at least in *P. weissi*.

Antenna with 12 segments.

Antennal scapes without projecting setae except in *P. bufona*.

Antennal sockets close to the posterior clypeal margin; anterior arc of the torulus may be separated from the clypeal suture by a narrrow strip of cuticle or may just touch the clypeal suture.

Eyes present or absent; when present, located at or slightly in front of the midlength of the sides of the head (in the polymorphic *P. bufona*, *P. sordida*, and *P. weissi*, the eyes very reduced in majors; absent in minors); ocelli present or absent.

Mesothorax not constricted behind the pronotum.

Mesosoma dorsally with setae that are arranged in pairs: 2 pairs on the pronotum, 1 pair on the mesonotum, and 1 pair on the propodeum (except in *P. bufona*, where more are present).

Metapleural gland present.

Propodeum unarmed.

Propodeal spiracle circular.

Propodeal foramen long: in ventral view the foramen extends anterior of a line that spans the anteriormost points of the metacoxal cavities.

Metacoxae widely separated: in ventral view, with the mesocoxae and metacoxae directed at right-angles to the long axis of the mesosoma, the distance between the bases of the mesocoxae is markedly less than the distance between the bases of the metacoxae.

Tibial spurs: mesotibia 0; metatibia 0.

Abdominal segment 2 (petiole) an unarmed scale.

Petiole (A2) scale with a shorter anterior face and a longer posterior face, with at least a short posterior peduncle.

Petiole (A2) with its ventral margin U-shaped in section.

Abdominal segment 3 (first gastral) with a sharply margined longitudinal concavity or groove, into which the posterior face of the petiole fits when the mesosoma and gaster are aligned.

Abdominal segment 3 basally with complete tergosternal fusion for some distance on each side of the helcium; the line of fusion follows the edges of the impression in the anterior face of A3 and the unfused portions of the tergite and sternite commence some distance above the helcium, at the dorsolateral apices of the impression.

Abdominal sternite 3 without a transverse sulcus across the sclerite immediately behind the helcium sternite (i.e., presternite and poststernite of A3 not separated by a sulcus).

Acidopore with a seta-fringed nozzle.

AFROTROPICAL PARAPARATRECHINA

albipes (Emery, 1899)	= *bucculentus* (Wheeler, W.M., 1922)
brunnella LaPolla and Cheng, 2010	= *gowdeyi* (Wheeler, W.M., 1922)
bufona (Wheeler, W.M., 1922)	*splendida* LaPolla and Cheng, 2010
concinnata LaPolla and Cheng, 2010	*subtilis* (Santschi, 1920)
gnoma LaPolla and Cheng, 2010	= *subtilis termitophila* (Santschi, 1921)
oreias LaPolla and Cheng, 2010	*umbranatis* LaPolla and Cheng, 2010
sordida (Santschi, 1914)	*weissi* (Santschi, 1910)

= *bayonii* (Menozzi, 1924)

= *myersi* (Weber, 1943)

= *myersi occipitalis* (Weber and Anderson, 1950)

MALAGASY PARAPARATRECHINA

glabra (Forel, 1891)

illusio LaPolla and Fisher, 2014

luminella LaPolla and Fisher, 2014

myops LaPolla and Fisher, 2010

ocellatula LaPolla and Fisher, 2010

PARASYSCIA Emery, 1882 Dorylinae

14 species, Afrotropical; 1 species, Malagasy PLATE 26

DIAGNOSTIC REMARKS: Among the dorylines, *Parasyscia* has small to moderate eyes, and the apical antennomere is swollen and bulbous. The promesonotal suture is absent; the waist is of a single segment, the dorsal surface of which rounds into the sides. The propodeal spiracle is low on the side and situated behind the midlength of the sclerite, and propodeal lobes are present. The mesotibia has a single spur present, and the pretarsal claws lack a preapical tooth. The metacoxa does not have a posterodorsal cuticular flange that forms a vertical lamella.

DISTRIBUTION AND ECOLOGY: Species of *Parasyscia* occur throughout the Old World tropics, and a few species are present in the subtropics. The habitat of most species is in and under rotten wood or in the earth under stones.

TAXONOMIC COMMENT: *Parasyscia* is currently recorded as a junior synonym of *Cerapachys*. It will be formally revived from synonymy by Marek Borowiec, in a genus-rank revision of Dorylinae that is in preparation.

IDENTIFICATION OF SPECIES: Brown, 1975 (Afrotropical, as part of *Cerapachys*).

DEFINITION OF THE GENUS (WORKER): Characters of subfamily DORYLINAE, plus:

Prementum not visible when mouthparts closed.

Palp formula 3,2, or 2,2.

Eyes present, small to moderate, at or in front of the midlength of the head capsule in full-face view; ocelli absent.

Frontal carinae present as a pair of closely approximated raised ridges, abruptly truncated posteriorly, without horizontal frontal lobes.

Antennal sockets fully exposed in full-face view.

Parafrontal ridges present.

Antenna with 11 or 12 segments, the apical segment very strongly enlarged with respect to the preapical segment.

Antennal scrobes absent.

Promesonotal suture absent on the dorsum of the mesosoma.

Pronotal-mesopleural suture absent, or a weakly defined fused line on the side of the mesosoma.

Metanotal groove absent.

Propodeal spiracle low on the side and situated at or behind the midlength of the sclerite, not subtended by a longitudinal impression.

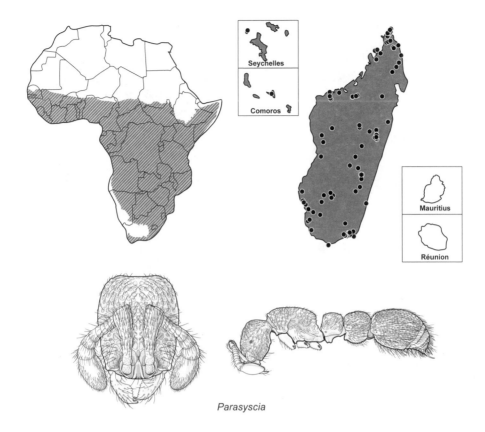

Parasyscia

Propodeal lobes present.

Tibial spurs: mesotibia 1, pectinate; metatibia 1, pectinate.

Metacoxae without a posterodorsal vertical cuticular lamella.

Pretarsal claws simple.

Waist usually of 1 segment (petiole = A2), but A3 sometimes reduced and appearing subpostpetiolate.

Petiole (A2) sessile, not dorsolaterally marginate, the dorsum rounds into the sides.

Helcium projects from the anterior surface of A3 (first gastral segment) at about its midheight.

Prora present, varying in appearance from a simple cuticular rim to strongly prominent.

Abdominal segment 4 with strongly developed presclerites.

Abdominal segments 5 and 6 without presclerites.

Pygidium (tergite of A7) large and distinct, flattened to concave dorsally and armed laterally, posteriorly, or both, with a series or row of denticles or peg-like spiniform setae.

Sting present, large and fully functional.

AFROTROPICAL PARASYSCIA

afer (Forel, 1907)	*centurio* (Brown, 1975)
= *cooperi* (Arnold, 1915)	*cribrinodis* Emery, 1899
arnoldi (Forel, 1914)	*faurei* (Arnold, 1949)
= *arnoldi hewitti* (Arnold, 1926)	*kenyensis* (Consani, 1951)

lamborni (Crawley, 1923)

  = pigra (Weber, 1942)

natalensis (Forel, 1901)

nitidula (Brown, 1975)

  = nitida (Weber, 1949) (homonym)

peringueyi Emery, 1886

  = peringueyi latiuscula Emery, 1895

sudanensis (Weber, 1942)

  = variolosus (Arnold, 1949)

sylvicola (Arnold, 1955)

valida (Arnold, 1960)

villiersi (Bernard, 1953)

MALAGASY PARASYSCIA

imerinensis Forel, 1891

PARATRECHINA Motschoulsky, 1863   Formicinae

3 species, Afrotropical; 3 species, Malagasy   PLATE 26

DIAGNOSTIC REMARKS: Among the formicines, *Paratrechina* species have 12-segmented antennae, the sockets of which are close to the clypeal margin (separated by a distance less than the basal width of the scape), and the scapes may lack or possess projecting setae. The mandible usually has 5 teeth (1 species has 8), the third of which is reduced in size, smaller than the fourth. The metapleural gland orifice is conspicuous, and the propodeal dorsum always lacks projecting setae. Eyes are conspicuous.

DISTRIBUTION AND ECOLOGY: Of the 5 species in this genus, 4 are restricted to the Afrotropical or the Malagasy region, but the fifth, *P. longicornis*, is one of the world's most successful tramp species. It may be found almost anywhere and is frequently encountered in plantations of tree crops, in botanical gardens, in houses, or anywhere that conditions are suitable and food is available. It seems equally at home in cocoa plantations in forested West Africa and quite arid coconut groves on the coast.

IDENTIFICATION OF SPECIES: LaPolla, Brady, and Shattuck, 2010; LaPolla, Hawkes, and Fisher, 2013; LaPolla and Fisher, 2014b (Afrotropical + Malagasy).

DEFINITION OF THE GENUS (WORKER): Characters of subfamily FORMICINAE, plus:

Mandible usually with 5 teeth, 1 species with 8 teeth; tooth 3 (counting from the apical) is smaller than tooth 4.

Palp formula 6,4.

Antenna with 12 segments.

Antennal scape long; when laid straight back from its insertion, in full-face view, one-half to nearly two-thirds of the scape length projects beyond the posterior margin of the head. Scapes with or without projecting setae.

Antennal sockets very close to the posterior clypeal margin; anterior arc of the torulus separated from the clypeal suture by a narrow space just touching the clypeal suture.

Eyes conspicuous, located at or behind the midlength of the sides of the head; ocelli present or absent.

Mesothorax not strongly constricted immediately behind the pronotum.

Mesonotum with stout setae present that are arranged in distinct pairs.

Metapleural gland present.

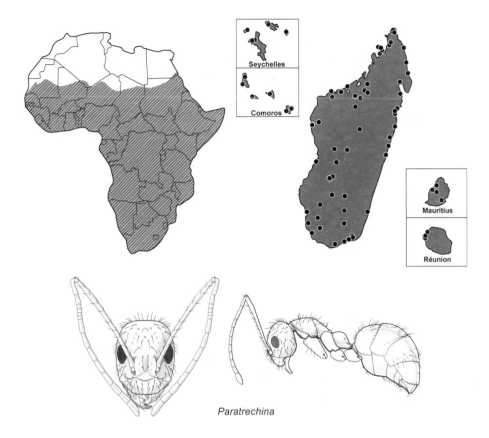

Paratrechina

Propodeum unarmed, without setae on its dorsal surface.

Propodeal spiracle circular.

Propodeal foramen long: in ventral view the foramen extends anterior of a line that spans the anteriormost points of the metacoxal cavities.

Metacoxae widely separated: in ventral view, with the mesocoxae and metacoxae directed at right-angles to the long axis of the mesosoma, the distance between the bases of the mesocoxae is markedly less than the distance between the bases of the metacoxae.

Tibial spurs: mesotibia 1; metatibia 1.

Abdominal segment 2 (petiole) a low, unarmed scale.

Petiole (A2) scale with a shorter anterior face and a slightly longer posterior face; without a long posterior peduncle.

Petiole (A2) with its ventral margin U-shaped in section.

Abdominal segment 3 (first gastral) with a sharply margined longitudinal concavity or groove, into which the posterior face of the petiole fits when the mesosoma and gaster are aligned.

Abdominal segment 3 basally with complete tergosternal fusion for some distance on each side of the helcium; the line of fusion follows the edges of the impression in the anterior face of A3 and the unfused portions of the tergite and sternite commence some distance above the helcium, at the dorsolateral apices of the impression.

Abdominal sternite 3 without a transverse sulcus across the sclerite immediately behind the helcium sternite (i.e., presternite and poststernite of A3 not separated by a sulcus).

Acidopore with a seta-fringed nozzle.

AFROTROPICAL PARATRECHINA

kohli (Forel, 1916)

longicornis (Latreille, 1802)

  = *longicornis hagemanni* (Forel, 1901)

zanjensis LaPolla, Hawkes, and Fisher, 2013

MALAGASY PARATRECHINA

ankarana LaPolla and Fisher, 2014

antsingy LaPolla and Fisher, 2014

longicornis (Latreille, 1802)

PARVAPONERA Schmidt and Shattuck, 2014   Ponerinae

1 species (+ 1 subspecies), Afrotropical; 1 subspecies, Malagasy   PLATE 26

DIAGNOSTIC REMARKS: Among the ponerines, *Parvaponera* combines the characters of presence of 2 spurs on the metatibia, absence of traction setae on the metabasitarsus, and location of the helcium at the base of the anterior face of the first gastral segment (A3), with the following: The eyes are usually absent (rarely a single ommatidium is present). The mandible has no basal pit and no basal groove. The frontal lobes are small. The metanotal groove is present, the propodeum in dorsal view is not strongly bilaterally constricted behind the mesonotum, and the propodeal spiracle is circular. The petiole (A2) is a scale or a high, narrow node. The pretarsal claws of the metatarsus are simple.

DISTRIBUTION AND ECOLOGY: Members of this little understood genus are distributed across the Old World tropics. None appears to be common, and the widely scattered collections known from the Afrotropical and Malagasy regions have all been from samples of leaf litter and log mold. Several previously described taxa in the genus are currently listed as subspecies of *darwinii* and are in need of taxonomic revision; a number of undescribed Afrotropical species are known.

TAXONOMIC COMMENT: *Parvaponera suspecta* (Santschi, 1914) is a new combination here for *Euponera suspecta* Santschi, 1914. The species was retained in *Euponera* by Schmidt and Shattuck (2014), but the type-specimen shows no trace of a mandibular pit, and its propodeal spiracle is circular—characters that exclude it from *Euponera*.

IDENTIFICATION OF SPECIES: No review exists.

DEFINITION OF THE GENUS (WORKER): Characters of subfamily PONERINAE, plus:

Mandible triangular; without a basal groove, without a basal pit; masticatory margin with 5 to 10 teeth. Basal angle angulate.

Palp formula not known.

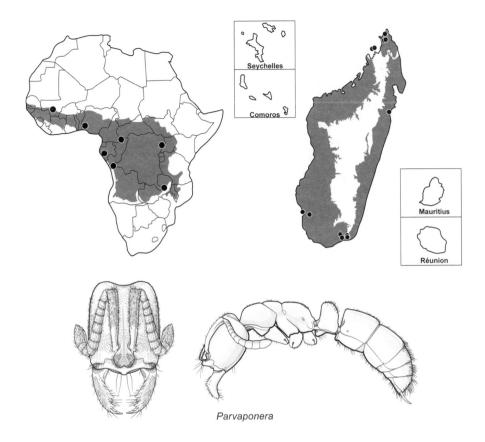

Parvaponera

Clypeus simple, usually without projecting lobes or teeth but sometimes with a minute denticle just above the midpoint of the anterior clypeal margin.

Frontal lobes small; in full-face view their anterior margins well behind the anterior clypeal margin.

Eyes usually absent; extremely rarely a single, minute ommatidium is present on each side.

Metanotal groove present, weakly impressed.

Epimeral sclerite present.

Metapleural gland orifice opens posteriorly.

Propodeum in dorsal view not strongly bilaterally constricted anteriorly.

Propodeal spiracle with orifice circular.

Tibial spurs: mesotibia 2; metatibia 2; on each tibia the anterior spur small and simple, the posterior spur larger and pectinate.

Mesobasitarsus with short traction setae present (not confirmed for all species); such setae absent from mesotibia.

Pretarsal claws simple.

Abdominal segment 2 (petiole) a scale or a tall, narrow node; subpetiolar process often with a weak anterior fenestra and usually terminating posteroventrally in a pair of short, acute teeth.

Helcium located at the base of the anterior face of A3 (first gastral segment).

Prora present.

Abdominal segment 4 (second gastral segment) with presclerites differentiated.

Stridulitrum absent on the pretergite of A4.

AFROTROPICAL PARVAPONERA

darwinii (Forel, 1893)

  = *lamarki* (Santschi, 1913)

  darwinii africana (Forel, 1909)

suspecta (Santschi, 1914) comb. n.

MALAGASY PARVAPONERA

darwinii (Forel, 1893)

  darwinii madecassa (Emery, 1899)

PETALOMYRMEX Snelling, 1979 — Formicinae

1 species, Afrotropical — PLATE 26

DIAGNOSTIC REMARKS: Among the formicines, *Petalomyrmex* has only 9 antennal segments, has a tooth on the basal margin of the mandible, and has a tall, narrow petiolar scale. Only 1 other genus shares this suite of characters, *Aphomomyrmex*. But *Petalomyrmex* is monomorphic and has eyes that are located lower on the head capsule, so that in full-face view they appear quite close to the lateral margins of the head. In addition, the single species of *Petalomyrmex* is brownish yellow and has only 3 maxillary palp segments, while that of the polymorphic *Aphomomyrmex* is black and has 5 maxillary palp segments.

DISTRIBUTION AND ECOLOGY: This monotypic genus has been collected only in the Central African rainforests of Cameroon and Gabon. The single species inhabits myrmecophytes, in an association that appears to be obligatory.

IDENTIFICATION OF SPECIES: Snelling, 1979.

DEFINITION OF THE GENUS (WORKER): Characters of subfamily FORMICINAE, plus:

Mandible with 6 teeth; tooth 3 (counting from the apical) is smaller than tooth 4; basal tooth slightly offset onto basal margin.

Palp formula 3,3.

Frontal triangle broad and shallow, not depressed and only weakly differentiated from the clypeus and inner margins of the frontal carinae, the latter very weakly represented and obtuse.

Antenna with 9 segments.

Antennal sockets close to the posterior clypeal margin; anterior arc of the torulus touches or slightly indents the clypeal suture.

Eyes present, about at the midlength of the sides of the head; ocelli present, located very close to the posterior margin.

Mesothorax not constricted in its anterior half.

Metapleural gland present.

Propodeal spiracle circular, large.

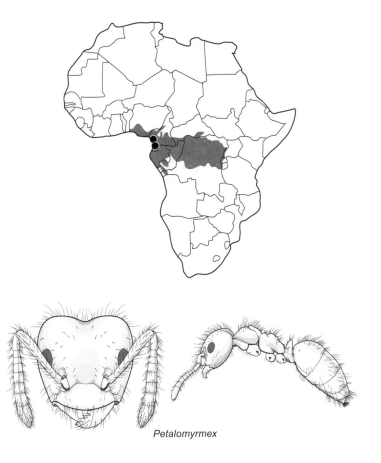

Petalomyrmex

Propodeal foramen long: in ventral view the foramen extends anterior of a line that spans the anteriormost points of the metacoxal cavities.

Metacoxae widely separated: in ventral view, with the mesocoxae and metacoxae directed at right-angles to the long axis of the mesosoma, the distance between the bases of the mesocoxae is markedly less than the distance between the bases of the metacoxae.

Tibial spurs: mesotibia 0; metatibia 0.

Abdominal segment 2 (petiole) a tall scale, its dorsal surface without teeth or spines.

Petiole (A2) with a short posterior peduncle that is attached low down on A3 (first gastral segment).

Petiole (A2) with its ventral margin U-shaped in section.

Abdominal segment 3 (first gastral) with a longitudinal concavity into which the posterior peduncle of the petiole fits when the mesosoma and gaster are aligned.

Abdominal segment 3 basally with complete tergosternal fusion for some distance on each side of the helcium; the line of fusion follows the edges of the impression in the anterior face of A3 and the unfused portions of the tergite and sternite commence some distance above the helcium, at the dorsolateral apices of the impression.

Abdominal sternite 3 without a transverse sulcus across the sclerite immediately behind the helcium sternite (i.e., presternite and poststernite of A3 not separated by a sulcus).

Acidopore with a seta-fringed nozzle.

AFROTROPICAL PETALOMYRMEX

phylax Snelling, 1979

PHASMOMYRMEX Stitz, 1910 — Formicinae

4 species (+ 2 subspecies), Afrotropical — PLATE 27

DIAGNOSTIC REMARKS: Among the formicines, *Phasmomyrmex* species have 12-segmented antennae, the sockets of which are located well behind the posterior clypeal margin (a distance greater than the basal width of the scape). Eyes are present, located behind the midlength of the head. The metapleural gland orifice is absent. Mandibular teeth decrease in size from apical to basal; tooth 3 is not strikingly reduced compared to tooth 4. The pronotum is sometimes dentate; the propodeum is unarmed to bidentate and may be marginate laterally. The petiole (A2) bears a pair of teeth, spines, or laterally projecting lobes. Abdominal tergite 3 (first gastral) is short, at most only slightly longer than the tergite of A4, often shorter; A3 tergite, in dorsal view or in profile, accounts for distinctly less than half the length of the gaster.

DISTRIBUTION AND ECOLOGY: All taxa in this genus are confined to the Afrotropical region, and all are arboreal, nesting and foraging in trees and very rarely descending to ground level. The 4 described species are confined to the rainforest zone of West and Central Africa, but at least 1 undescribed species is known from the forests of Tanzania and South Africa.

TAXONOMIC COMMENT: This genus is artificial and curently contains the few Afrotropical species that do not conform fully to the definitions of *Camponotus* or *Polyrhachis*. Its 4 species are currently distributed through 3 different subgenera; see the synopsis in Bolton, 2003.

IDENTIFICATION OF SPECIES: No review exists.

DEFINITION OF THE GENUS (WORKER): Characters of subfamily FORMICINAE, plus:

Mandible with 5 teeth that decrease in size from apical to basal; tooth 3 (counting from the apical) is larger than tooth 4.

Palp formula 6,4.

Frontal carinae strongly developed.

Antenna with 12 segments.

Antennal sockets located a considerable distance behind the posterior clypeal margin (a distance greater than the basal width of the scape).

Eyes present, located behind the midlength of the sides of the head; ocelli absent.

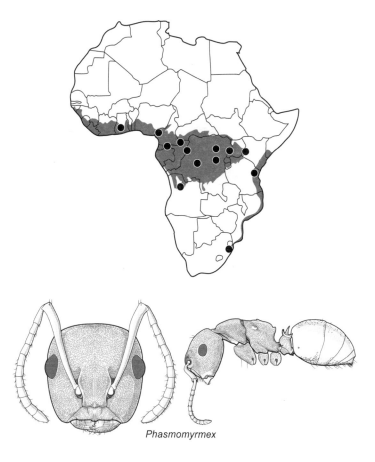

Phasmomyrmex

Mesosoma with pronotal humeri rounded to dentate; propodeum unarmed to bidentate and may be marginate laterally.

Metapleural gland absent.

Propodeal spiracle elliptical to slit-shaped, located posteriorly on the side and low down.

Propodeal foramen short: in ventral view the foramen does not extend anterior of a line that spans the anteriormost points of the metacoxal cavities.

Metacoxae closely approximated: in ventral view, with the mesocoxae and metacoxae directed at right-angles to the long axis of the mesosoma, the distance between the bases of the mesocoxae is about equal to the distance between the bases of the metacoxae.

Tibial spurs: mesotibia 1; metatibia 1.

Abdominal segment 2 (petiole) with a pair of teeth, spines, or laterally projecting lobes.

Petiole (A2) with its ventral margin V-shaped in section.

Helcium tergite dorsally with a U-shaped emargination in its anterior margin.

Abdominal tergite 3 (first gastral) short, at most only slightly longer than the tergite of A4; A3 tergite, in dorsal view or in profile, accounts for less than half of the length of the gaster.

Abdominal segment 3 without tergosternal fusion on each side of the helcium.

Abdominal sternite 3 with a transverse sulcus or mobile suture across the sclerite, immediately behind the helcium sternite (i.e., presternite and poststernite of A3 separated by a sulcus).

Acidopore always open, usually with a seta-fringed nozzle.

AFROTROPICAL PHASMOMYRMEX

aberrans (Mayr, 1895)	*paradoxus* (André, 1892)
buchneri (Forel, 1886)	= *polyrhachioides* (Emery, 1898)
= *sericeus* Stitz, 1910	*paradoxus cupreus* Santschi, 1923
buchneri griseus Santschi, 1937	*wolfi* (Emery, 1920)

PHEIDOLE Westwood, 1839　　　　　　　　　　　　　　Myrmicinae

72 species (+ 69 subspecies), Afrotropical; 25 species (+ 6 subspecies), Malagasy　　　PLATE 27

DIAGNOSTIC REMARKS: Among the myrmicines, *Pheidole* has 12-segmented antennae that terminate in a strongly defined club of 3 segments. The worker caste is strikingly dimorphic, the major worker with a very large head. The clypeus is large and is broadly inserted posteriorly between the frontal lobes. One or more hypostomal teeth are usually present on the posterior margin of the buccal cavity (occasionally absent). The pronotum, and usually also the anterior mesonotum, is swollen and convex in profile, usually dome-like; behind this the mesonotum slopes steeply to the metanotal groove, or the mesonotum may form a second, separate convexity before sloping to the metanotal groove. The first gastral tergite (the tergite of A4) broadly overlaps the sternite on the ventral gaster.

DISTRIBUTION AND ECOLOGY: This genus is distributed nearly worldwide, with a vast number of species in the New World. In the regions under consideration here, its species occur in all vegetation zones, from semidesert to rainforest. Many nest in the ground, either directly or beneath stones and other surface objects or among the roots of plants. Some nest in rotten wood on the ground, and small species may be found beneath the bark of standing trees or in rotten holes in trunks. A number of small species are characteristic of the leaf litter and topsoil, where they may be numerous, and some are commonly found in termitaries. Foraging usually takes place on or in the ground, but some ascend trees from ground-based nests, sometimes in large numbers, to tend homopterous insects in the canopy.

IDENTIFICATION OF SPECIES: Fischer, Hita Garcia, and Peters, 2012 (Afrotropical *P. pulchella* group); Fischer and Fisher, 2013 (Southwest Indian Ocean Islands); no complete review exists and the species-rank taxonomy is otherwise confused. The taxonomy of the very common and widely distributed *P. megacephala* group is particularly impenetrable.

DEFINITION OF THE GENUS (WORKER): Characters of subfamily MYRMICINAE, plus:

Worker caste dimorphic; major worker with disproportionately enlarged head.

Mandible triangular, large and powerful in majors, more delicate in minors. In minors, masticatory margin of the mandible with 8–18 teeth; the third tooth (counting from the apex) smaller than the fourth, or the reduced third tooth followed by a minute

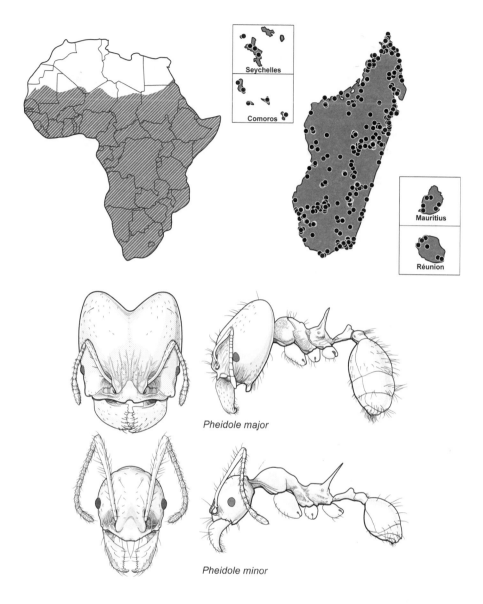

Pheidole major

Pheidole minor

denticle before the larger fourth tooth. In majors, with 2 large apical teeth followed by a long diastema and then 1 or 2 (rarely 3) basal teeth. Zero to 4 hypostomal teeth present on the posterior margin of the buccal cavity.

Palp formula 3,2, or 2,2.

Stipes of the maxilla without a transverse crest.

Clypeus large, posteriorly broadly inserted between the frontal lobes.

Clypeus without an unpaired seta at the midpoint of the anterior margin.

Frontal carinae short in all minors and most majors, restricted to well-defined but narrow frontal lobes; in majors of some species the carinae extend almost to the posterior margin of the head.

Antennal scrobes absent in minors but variably developed in majors; in majors of some species, distinct scrobes are present that extend above the eyes.

Antenna with 12 segments, with a strongly defined apical club of 3 segments.

Torulus with upper lobe visible in full-face view in minors; maximum width of the torulus lobe is posterior to the point of maximum width of the frontal lobes. Majors may have the torulus lobe concealed by the frontal lobe.

Eyes present, located at, or more usually in front of, the midlength of the head capsule.

Head massively constructed in major workers.

Promesonotal suture represented across the dorsum in some species by a weak impression or by a narrow line; in others, vestigial or entirely absent.

Pronotum, or pronotum plus anterior mesonotum, swollen and convex in profile, usually dome-like; posteriorly the mesonotum slopes steeply to the metanotal groove, or the mesonotum may form a second, separate convexity before sloping to the metanotal groove.

Propodeum unarmed to bispinose; in profile the propodeal dorsum is usually on a much lower level than the apex of the convex promesonotum.

Propodeal spiracle large, its orifice circular or nearly so, located at or slightly behind the midlength of the sclerite.

Metasternal process minute or absent.

Tibial spurs: mesotibia 0; metatibia 0.

Abdominal segment 2 (petiole) with a distinct anterior peduncle.

Abdominal tergite 4 (first gastral) broadly overlaps the sternite on the ventral gaster; gastral shoulders present.

Main pilosity of dorsal head and body: usually simple, rarely stubbly, sometimes bifid at extreme apex; may be sparse.

AFROTROPICAL PHEIDOLE

aeberlii Forel, 1894
  aeberlii erythraea Emery, 1901
akermani Arnold, 1920
albidula Santschi, 1928
andrieui Santschi, 1930
areniphila Forel, 1910
  = *arenicola* Forel, 1910 (homonym)
  areniphila aurora Santschi, 1928
arnoldi Forel, 1913
  arnoldi ballaensis Arnold, 1920
  arnoldi rufescens Arnold, 1920
aspera Mayr, 1862
aurivillii Mayr, 1896
  aurivillii attenuata Santschi, 1910
  aurivillii kasaiensis Forel, 1911
  aurivillii rubricalva Forel, 1915
batrachorum Wheeler, W.M., 1922

bequaerti Forel, 1913
buchholzi Mayr, 1901
caffra Emery, 1895
  caffra abyssinica Forel, 1910
  caffra amoena Forel, 1911
  caffra bayeri Forel, 1916
  caffra senilifrons Wheeler, W.M., 1922
  caffra montivaga Santschi, 1939
capensis Mayr, 1862
  capensis dregei Emery, 1895
  capensis reddenburgensis Forel, 1913
  capensis modestior Santschi, 1919
christinae Fischer, Hita Garcia, and
  Peters, 2012
clavata (Emery, 1877)
concinna Santschi, 1910
corticicola Santschi, 1910

crassinoda Emery, 1895
   crassinoda ruspolii Emery, 1897
   crassinoda pluto Arnold, 1920
   crassinoda sordidula Santschi, 1937
cuitensis Forel, 1910
darwini Fischer, Hita Garcia, and
   Peters, 2012
dea Santschi, 1921
decarinata Santschi, 1929
escherichii Forel, 1910
excellens Mayr, 1862
   excellens weissi Santschi, 1910
   excellens fulvobasalis Santschi, 1921
foreli Mayr, 1901
   foreli pubens Forel, 1910
glabrella Fischer, Hita Garcia, and
   Peters, 2012
guineensis (Fabricius, 1793)
heliosa Fischer, Hita Garcia, and
   Peters, 2012
hewitti Santschi, 1932
   = capensis hewitti Arnold, 1944
irritans (Smith, F., 1858)
katonae Forel, 1907
kitschneri Forel, 1910
kohli Mayr, 1901
liengmei Forel, 1894
   liengmei malindana Forel, 1913
   liengmei micrartifex Forel, 1913
   liengmei shinsendensis Forel, 1913
maufei Arnold, 1920
mayri Forel, 1894
megacephala (Fabricius, 1793)
   = perniciosa (Gerstäcker, 1859)
   megacephala talpa Gerstäcker, 1871
   megacephala rotundata Forel, 1894
   megacephala impressifrons Wasmann,
     1905
   = megacephala impressiceps
     Wasmann, 1904 (homonym)
   megacephala ilgi Forel, 1907
   megacephala speculifrons Stitz, 1911
   megacephala melancholica Santschi,
     1912

megacephala costauriensis Santschi,
   1914
megacephala nkomoana Forel, 1916
megacephala duplex Santschi, 1937
mentita Santschi, 1914
   mentita pullata Santschi, 1914
minima Mayr, 1901
   minima catella Santschi, 1914
   minima faurei Santschi, 1920
   minima malelana Wheeler, W.M.,
     1922
minuscula Bernard, 1953
mylognatha Wheeler, W.M., 1922
neokohli Wilson, 1984
   = kohli (Wasmann, 1915) (homonym)
nigeriensis Santschi, 1914
nigritella Bernard, 1953
nimba Bernard, 1953
njassae Viehmeyer, 1914
   njassae legitima Santschi, 1916
   = njassae sculptior Viehmeyer, 1914
     (homonym)
occipitalis André, 1890
   occipitalis neutralis Santschi, 1914
   occipitalis adami Santschi, 1939
philippi Emery, 1915.
prelli Forel, 1911
   prelli redbankensis Forel, 1913
   prelli ingenita Santschi, 1928
pulchella Santschi, 1910
   = niapuana Wheeler, W.M., 1922
   = pulchella achantella Santschi, 1939
punctulata Mayr, 1866
   punctulata atrox Forel, 1913
   = inquilina Forel, 1914
   punctulata subatrox Santschi, 1937
rebeccae Fischer, Hita Garcia, and
   Peters, 2012
retronitens Santschi, 1930
rohani Santschi, 1925
   rohani monardi Santschi, 1930
   rohani pellax Santschi, 1930
rugaticeps Emery, 1877
saxicola Wheeler, W.M., 1922

scabriuscula Gerstäcker, 1871

schoutedeni Forel, 1913

   schoutedeni platycephala Stitz, 1916

schultzei Forel, 1910

   schultzei gwaaiensis Forel, 1913

   schultzei ebangana Santschi, 1937

   schultzei woodvalensis Arnold, 1960

sculpturata Mayr, 1866

   sculpturata berthoudi Forel, 1894

   sculpturata areolata Forel, 1911

   sculpturata rhodesiana Forel, 1913

   sculpturata welgelegenensis Forel, 1913

   sculpturata zambesiana Forel, 1913

   sculpturata dignata Santschi, 1915

   sculpturata particeps Santschi, 1921

semidea Fischer, Hita Garcia, and
   Peters, 2012

setosa Fischer, Hita Garcia, and Peters,
   2012

speculifera Emery, 1877

   speculifera ascara Emery, 1901

speculifera cubangensis Forel, 1901

speculifera bispecula Santschi, 1930

spinulosa Forel, 1910

   spinulosa conigera Forel, 1910

   spinulosa messalina Forel, 1910

squalida Santschi, 1910

strator Forel, 1910

   strator fugax Arnold, 1920

   strator tabida Menozzi, 1933

tenuinodis Mayr, 1901

   tenuinodis bothae Forel, 1901

   tenuinodis sipapomae Arnold, 1920

   tenuinodis robusta Stitz, 1923

termitophila Forel, 1904

   termitophila liberiensis Forel, 1911

tricarinata Santschi, 1914

vanderveldi Forel, 1913

variolosa Emery, 1892

xocensis Forel, 1913

   xocensis bulawayensis Forel, 1913

MALAGASY PHEIDOLE

annemariae Forel, 1918

bessonii Forel, 1891

braueri Forel, 1897

decepticon Fischer and Fisher, 2013

dodo Fischer and Fisher, 2013

ensifera Forel, 1897

fervens Smith, F., 1858

grallatrix Emery, 1899

jonas Forel, 1907

komori Fischer and Fisher, 2013

loki Fischer and Fisher, 2013

longispinosa Forel, 1891

   longispinosa scabrata Forel, 1895

lucida Forel, 1895

madecassa Forel, 1892

megacephala (Fabricius, 1793)

   = *megacephala scabrior* Forel, 1891

   = *picata* Forel, 1891

   = *picata gietleni* Forel, 1905

   = *picata bernhardae* Emery, 1915

megatron Fischer and Fisher, 2013

nemoralis Forel, 1892

   nemoralis petax Forel, 1895

oculata (Emery, 1899)

oswaldi Forel, 1891

   oswaldi decollata Forel, 1892

parva Mayr, 1865

   = *flavens farquharensis* Forel, 1907

   = *tardus* Donisthorpe, 1947

punctulata Mayr, 1866.

   punctulata spinosa Forel, 1891

ragnax Fischer and Fisher, 2013

sikorae Forel, 1891

   sikorae litigiosa Forel, 1892

teneriffana Forel, 1893

   = *voeltzkowii* Forel, 1894

veteratrix Forel, 1891

   veteratrix angustinoda Forel, 1892

PHRYNOPONERA Wheeler, W.M., 1920   Ponerinae

5 species, Afrotropical   PLATE 27

DIAGNOSTIC REMARKS: Among the ponerines, *Phrynoponera* combines the characters of presence of 2 spurs on the metatibia, absence of traction setae on the metabasitarsus, and location of the helcium at the base of the anterior face of the first gastral segment (A3), with the following: The propodeum has a pair of spines. The petiole (A2) is a tall, stout scale that curves posteriorly over the base of the first gastral segment (A3) and is armed dorsally with 5 long teeth or spines. There is no visible prora.

DISTRIBUTION AND ECOLOGY: *Phrynoponera* is characteristic of the wet forests of the Afrotropical region, especially the West and Central African rainforest belt, but more easterly species occur in shrubland. The species nest in and under rotten wood and sometimes directly in compacted soil; a couple of species have been found in abandoned termitaries, but this is uncommon. Foragers are sometimes encountered in the leaf-litter layer, but most are found in rotten timber on the ground, where they appear to be generalist predators.

IDENTIFICATION OF SPECIES: Bolton and Fisher, 2008b.

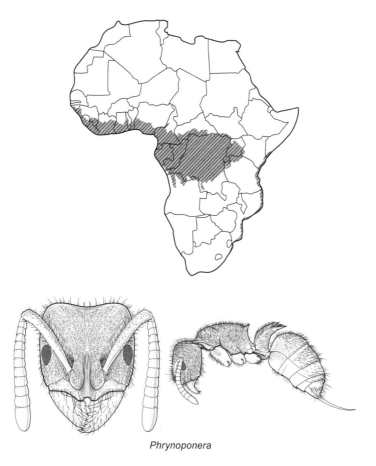

Phrynoponera

DEFINITION OF THE GENUS (WORKER): Characters of subfamily PONERINAE, plus:

Mandible at most subtriangular, usually the basal and external margins roughly parallel, with a weak basal groove. Masticatory margin somewhat oblique, with 3–8 teeth; basalmost tooth at the rounded basal angle.

Palp formula 4,4.

Clypeus with anterior margin rounded medially or with a pair of blunt teeth anteromedially.

Frontal lobes large but not hypertrophied; in full-face view their anterior margins do not overhang the anterior clypeal margin.

Eyes present, distinct, at or slightly in front of the midlength of the head capsule.

Metanotal groove vestigial to absent, not impressed.

Epimeral sclerite present.

Metapleural gland orifice opens posterolaterally.

Propodeum stoutly bispinose.

Propodeal spiracle with orifice slit-shaped.

Tibial spurs: mesotibia 2; metatibia 2; on each tibia the anterior spur small, simple to barbulate, the posterior spur larger and pectinate.

Pretarsal claws simple.

Abdominal segment 2 (petiole) a high, stout scale that curves posteriorly over the base of the first gastral segment (A3) and is armed dorsally with 5 long teeth or spines.

Abdominal sternite 2 (petiole sternite), when detached and in posterior view, complex: in the posterior third of its length, the sternite bifurcates into an externally visible broad and concave ventral plate and a slightly shorter, internally projecting sclerite that is completely concealed by the external plate in normal view.

Helcium located at the base of the anterior face of A3 (first gastral segment).

Prora apparently absent (not visible in whole specimens) but actually unusually modified and concealed. When the helcium is disarticulated from the petiole, the reduced prora can be seen, inserted between, and fused to the ventral apices of the helcium tergite. This makes the helcium double-chambered, with the upper chamber floored by the helcium sternite and the lower chamber floored by the modified prora.

Abdominal segment 4 (second gastral segment) with differentiated presclerites that are usually concealed by the tergite and sternite of A3, so that the gaster usually appears small, roughly globular, and compact.

Stridulitrum usually absent from the pretergite of A4, vestigially present in 1 species.

AFROTROPICAL PHRYNOPONERA

bequaerti Wheeler, W.M., 1922

gabonensis (André, 1892)

  = *gabonensis striatidens* (Santschi, 1914)

  = *armata* (Santschi, 1919)

  = *gabonensis robustior* (Santschi, 1919)

  = *gabonensis esta* Wheeler, W.M., 1922

  = *gabonensis fecunda* Wheeler, W.M., 1922

  = *gabonensis umbrosa* Wheeler, W.M., 1922

  = *heterodus* Wheeler, W.M., 1922

pulchella Bolton and Fisher, 2008

sveni (Forel, 1916)

transversa Bolton and Fisher, 2008

PILOTROCHUS Brown, 1978

Myrmicinae

1 species, Malagasy

PLATE 28

DIAGNOSTIC REMARKS: Among the Malagasy myrmicines, *Pilotrochus* combines the characters of an 8-segmented antenna that terminates in a strong 2-segmented club and deep antennal scrobes. There is also a large mesopleural gland that takes the form of a roughly circular impression or pit that is filled with centrally directed fine, pale, hairs. The waist segments lack spongiform tissue, and the propodeum is unarmed.

DISTRIBUTION AND ECOLOGY: This genus is represented by a single leaf-litter species from Madagascar.

TAXONOMIC COMMENT: *Pilotrochus* may initially be confused with some species of *Strumigenys*, but the combination of 8-segmented antennae and a lack of spongiform tissue on the waist segments, together with its unique dentition (described below), are immediately diagnostic of *Pilotrochus*.

IDENTIFICATION OF SPECIES: Brown, 1978.

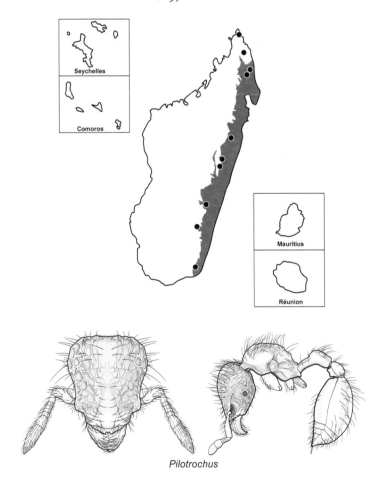

Pilotrochus

DEFINITION OF THE GENUS (WORKER): Characters of subfamily MYRMICINAE, plus:

Mandible triangular; masticatory margins oppose but do not overlap at full closure. Mandible with 7 main teeth that increase in size from apex to base, so that the basal tooth is the largest; a minute denticle is present between each of the main teeth.

Labrum linguiform, not capable of reflexion over the anterior portion of buccal cavity, not closing tightly over the apex of the labiomaxillary complex.

Palp formula 3,2.

Stipes of the maxilla without a transverse crest.

Clypeus large, broadly triangular, posteriorly broadly inserted between the frontal lobes.

Frontal carinae in full-face view, at level of the antennal sockets, with their outer margins overhanging and concealing the apparent lateral margin of the head (actually the outer margin of the preocular carina).

Antennal scrobe present, deep, extending above the eye.

Antenna with 8 segments, with an apical club of 2 segments.

Torulus concealed by the frontal lobe in full-face view.

Eye ventrolateral on side of head, at the ventral margin of the scrobe.

Head capsule anteriorly narrowed from side to side in full-face view.

Promesonotal suture vestigial across the dorsum.

Mesopleuron with a conspicuously developed gland dorsally that takes the form of a roughly circular impression or pit that is filled with fine pale hairs.

Propodeum unarmed.

Propodeal spiracle very close to the margin of the declivity.

Metasternal process absent.

Tibial spurs: mesotibia 0; metatibia 0.

Abdominal segments 2 and 3 (petiole and postpetiole) without spongiform tissue.

Abdominal segment 4 (first gastral) basally without a raised, transverse cuticular crest or ridge across the tergite.

Sting present.

Main pilosity of dorsal head and body: numerous fine, simple setae on all surfaces.

MALAGASY PILOTROCHUS

besmerus Brown, 1978

PLAGIOLEPIS Mayr, 1861   Formicinae

20 species (+ 7 subspecies), Afrotropical; 2 species, Malagasy   PLATE 28

DIAGNOSTIC REMARKS: Among the formicines, *Plagiolepis* is characterized by its 11-segmented antennae, presence of the metanotum as a separate sclerite on the dorsal mesosoma (demarcated by conspicuous sutures in front of it and behind it), and the propodeum and petiole lacking teeth or spines.

DISTRIBUTION AND ECOLOGY: Species of *Plagiolepis* occur throughout the Old World tropics, and the genus is also strongly represented in the Palaearctic region; there is also at least 1 successful tropical tramp species (*P. alluaudi*). Some species are arboreal, nesting

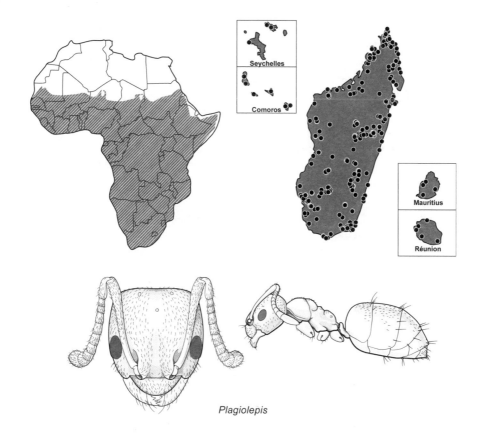

Plagiolepis

and foraging in trees and on their trunks. Others tend to nest in hard-packed earth or in rotten twigs on the ground, but some of these also ascend trees or low vegetation to forage.

TAXONOMIC COMMENT: The relationship between the various species-groups of *Plagiolepis* and *Lepisiota* is in need of review; the monophyly of the latter is in doubt.

IDENTIFICATION OF SPECIES: No review exists; the species-rank taxonomy is confused.

DEFINITION OF THE GENUS (WORKER): Characters of subfamily FORMICINAE, plus:

Mandible with 5–6 teeth; tooth 3 (counting from the apical) is smaller than tooth 4.

Palp formula 6,4.

Antenna with 11 segments.

Antennal sockets close to the posterior clypeal margin; the anterior arc of the torulus touches or slightly indents the clypeal suture.

Eyes present, usually distinctly in front of the midlength of the sides of the head, uncommonly at about the midlength; ocelli absent or present.

Mesothorax not constricted in its anterior half.

Metanotum forms a distinct isolated sclerite on the dorsum of the mesosoma.

Metapleural gland present.

Propodeum without teeth or spines.

Propodeal spiracle circular, usually high on the side.

Propodeal foramen long: in ventral view the foramen extends anterior of a line that spans the anteriormost points of the metacoxal cavities.

Metacoxae widely separated: in ventral view, with the mesocoxae and metacoxae directed at right-angles to the long axis of the mesosoma, the distance between the bases of the mesocoxae is markedly less than the distance between the bases of the metacoxae.

Tibial spurs: mesotibia 0; metatibia 0.

Abdominal segment 2 (petiole) a scale, its dorsal surface without teeth or spines.

Petiole (A2) scale with a shorter anterior face and a longer posterior face; petiole with a short posterior peduncle that is attached low down on A3 (first gastral segment).

Petiole (A2) with its ventral margin U-shaped in section.

Abdominal segment 3 (first gastral) with a sharply margined longitudinal concavity or groove, into which the posterior peduncle of the petiole fits when the mesosoma and gaster are aligned. In this position the anterior face of abdominal segment 3 overhangs and conceals the posterior peduncle of the petiole.

Abdominal segment 3 basally with complete tergosternal fusion for some distance on each side of the helcium; the line of fusion follows the edges of the impression in the anterior face of A3 and the unfused portions of the tergite and sternite commence some distance above the helcium, at the dorsolateral apices of the impression.

Abdominal sternite 3 without a transverse sulcus across the sclerite immediately behind the helcium sternite (i.e., presternite and poststernite of A3 not separated by a sulcus).

Acidopore with a seta-fringed nozzle.

AFROTROPICAL PLAGIOLEPIS

abyssinica Forel, 1894	*livingstonei* Santschi, 1926
alluaudi Emery, 1894	*mediorufa* Forel, 1916
brunni Mayr, 1895	*montivaga* Arnold, 1958
brunni nilotica Mayr, 1904	*pictipes* Santschi, 1914
brunni pubescens Forel, 1913	*puncta* Forel, 1910
capensis Mayr, 1865	*pygmaea* (Latreille, 1798)
chirindensis Arnold, 1949	*pygmaea bulawayensis* Arnold, 1922
decora Santschi, 1914	*pygmaea mima* Arnold, 1922
deweti Forel, 1904	*simoni* Emery, 1921
funicularis Santschi, 1919	*sudanica* Weber, 1943
fuscula Emery, 1895	*vanderkelleni* Forel, 1901
intermedia Emery, 1895	*vanderkelleni tricolor* Forel, 1910
intermedia minutula Emery, 1921	*vanderkeleni polita* Santschi, 1914
jouberti Forel, 1910	

MALAGASY PLAGIOLEPIS

alluaudi Emery, 1894
madecassa Forel, 1892

DIAGNOSTIC REMARKS: Among the ponerines, *Platythyrea* combines the characters of presence of 2 spurs on the metatibia (both of which are pectinate) and absence of traction setae on the metabasitarsus with the following: The frontal lobes are widely separated. The helcium is at about midheight on the anterior face of the first gastral segment (A3). The pretarsal claws of the metatarsus have a tooth on the inner curvature. There is no prora at the base of the first gastral segment (A3). Pruinose sculpture is present on all surfaces of the head and body.

DISTRIBUTION AND ECOLOGY: A pantropical genus which contains a number of arboreal species that nest in rotten holes in trunks and branches or in hollow twigs or beetle burrows. Other species are terrestrial nesters, being found under stones, in and under fallen timber, at the bases of termitaria, or in the leaf litter. Some of the terrestrial nesters also forage on tree trunks and low vegetation. Most species are general predators, but a few specialists are known to prey on termites or beetles. Alate queens are often present, but some species have ergatoids, or gamergates, or more than 1 reproductive form.

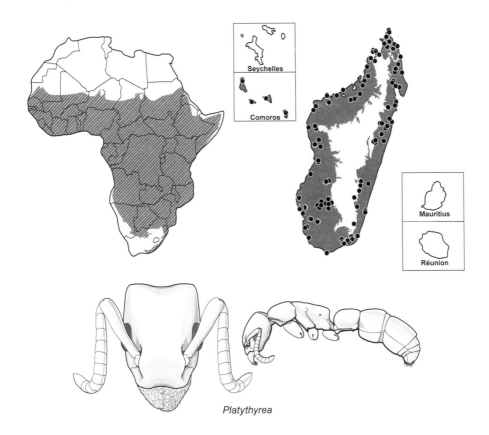

Platythyrea

TAXONOMIC COMMENT: Among the Afrotropical and Malagasy ponerines, the pruinose sculpture characteristic of *Platythyrea* is duplicated only in a few *Leptogenys* species of the *L. maxillosa* group; it also occurs in the proceratiine genus *Probolomyrmex* and in some *Discothyrea*.

IDENTIFICATION OF SPECIES: Brown, 1975 (Afrotropical + Malagasy).

DEFINITION OF THE GENUS (WORKER): Characters of subfamily PONERINAE, plus:

Mandible triangular, usually multidentate but sometimes edentate; with or without a basal groove, without a basal pit.

Palp formula very variable, 6,4, or 4,4, or 4,3, or 3,3, or 3,2.

Clypeus simple, without projecting lobes or teeth, posteriorly broadly inserted between the horizontal frontal lobes.

Frontal lobes large and widely separated; in full-face view their anterior margins well behind the anterior clypeal margin.

Eyes present, moderate to large, located in front of the midlength of the head capsule.

Metanotal groove usually vestigial to absent, uncommonly distinct.

Epimeral sclerite present.

Metapleural gland orifice opens laterally.

Metacoxal cavities open in some species, closed but with a suture in the annulus in others.

Propodeal spiracle with orifice circular to slit-shaped.

Tibial spurs: mesotibia 2; metatibia 2; on each tibia both spurs are pectinate, the posterior spur is larger than the anterior.

Pretarsal claws each with a single preapical tooth.

Abdominal segment 2 (petiole) a node.

Abdominal sternite 2 (petiole sternite) varies from simple to complex; in complex forms, when detached and in posterior view, the posterior portion of the sternite length bifurcates to produce a broad ventral flange that considerably overlaps, and partially conceals, the articulatory apex of the sclerite.

Helcium located at about the midheight of the anterior face of A3 (first gastral segment).

Prora absent.

Abdominal segment 4 (second gastral segment) with presclerites distinctly differentiated.

Stridulitrum present on the pretergite of A4.

Sculpture of head and body entirely pruinose.

AFROTROPICAL PLATYTHYREA

arnoldi Forel, 1913
  = *lamellosa apicalis* Santschi, 1937
conradti Emery, 1899
  = *monodi* Bernard, 1953
cooperi Arnold, 1915
cribrinodis (Gerstäcker, 1859)
  = *cribrinodis brevinodis* Santschi, 1914

  = *cribrinodis brevidentata* Wheeler, W.M., 1922
  = *cribrinodis punctata* Arnold, 1915 (homonym)
crucheti Santschi, 1911
frontalis Emery, 1899
gracillima Wheeler, W.M., 1922

lamellosa (Roger, 1860)
  = lamellosa longinoda Forel, 1894
  = lamellosa suturalis Forel, 1909
  = aethiops grisea (Forel, 1910)
  = lamellosa rhodesiana Forel, 1913
matopoensis Arnold, 1915
modesta Emery, 1899

occidentalis André, 1890
schultzei Forel, 1910
  = schultzei bequaerti Forel, 1913
  = cyriluli Forel, 1922
  = schultzei lata Santschi, 1930
tenuis Emery, 1899
viehmeyeri Santschi, 1914

MALAGASY PLATYTHYREA

arthuri Forel, 1910
bicuspis Emery, 1899
mocquerysi Emery, 1899
  = mocquerysi debilior Forel, 1907
parallela (Smith, F., 1859)
  = wroughtoni sechellensis Forel, 1912

PLECTROCTENA Smith, F., 1858

Ponerinae

16 species, Afrotropical

PLATE 28

DIAGNOSTIC REMARKS: Among the ponerines, *Plectroctena* combines the characters of presence of a single spur on the metatibia and absence of traction setae on the metabasitarsus with the following: The mandibles are linear, have 0–2 teeth on the inner margin, and have a longitudinal groove or trench on the dorsal surface. To the side of the base of each mandible, the head capsule has a distinct, roughly semicircular excavation. The frontal lobes are large. The mesofemur and metafemur both have a longitudinal groove mid-dorsally, at least on the basal half.

DISTRIBUTION AND ECOLOGY: *Plectroctena* is restricted to the Afrotropical region but is very widely distributed within it, being present in habitats that range from rainforest to tropical dry forest, to savanna. Nests are usually constructed directly in the ground, or in and under rotten timber, or occasionally abandoned termitaries. Foraging takes place in these locations as well as in the topsoil and leaf-litter layers and under the bark of rotting fallen trees. Most species appear to be specialist predators of millipedes or their eggs, but some have also been reported as feeding on termites and other arthropods, including other ants. The majority of species have normal alate queens, but several have ergatoids.

IDENTIFICATION OF SPECIES: Bolton, 1974b; Bolton and Brown, 2002; Fisher, 2006.

DEFINITION OF THE GENUS (WORKER): Characters of subfamily PONERINAE, plus:

Mandible stoutly linear, blunt or obliquely truncated apically, with 0–2 teeth on the inner margin; with a basal groove and also with a longitudinal groove or trench on the dorsal surface. Mandibles strongly cross over when at rest.

Labrum minutely but extremely densely, strongly sculptured.

Palp formula 3,4, or 2,3, or 2,2.

Clypeus simple; median portion of anterior margin shallowly concave.

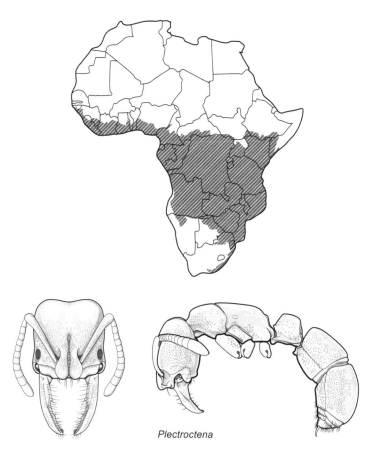

Plectroctena

Head capsule anterolaterally with an extensive, roughly semicircular excavation around the mandibular articulation.

Frontal lobes large, in full-face view their anterior margins immediately above, or somewhat overhanging, the anterior clypeal margin.

Eyes usually present, rarely absent; when present, small to moderate in size and located far forward on the head capsule, usually close to the posterior clypeal margin.

Metanotal groove absent to vestigially present, not impressed.

Epimeral sclerite present.

Metapleural gland orifice opens laterally.

Propodeal dorsum usually without a mid-dorsal longitudinal groove; present in a few species.

Propodeal spiracle circular or very nearly so.

Mesofemur and metafemur with a longitudinal groove mid-dorsally, at least on the basal half.

Tibial spurs: mesotibia 1; metatibia 1; each spur pectinate.

Pretarsal claws simple.

Abdominal segment 2 (petiole) a node.

Abdominal sternite 2 (petiole sternite) with its anterior articulatory surface long and broad, the surface with a narrow, median, V-shaped longitudinal groove, or small central, pore-like depression.

Helcium arises low down on the anterior face of A3 (first gastral segment).

Prora a conspicuous lobe or flange.

Abdominal segment 4 (second gastral segment) with very strongly differentiated presclerites.

Stridulitrum absent from the pretergite of A4.

AFROTROPICAL PLECTROCTENA

anops Bolton, 1974	= *caffra major* Forel, 1894
cristata Emery, 1899	= *minor conjugata* Santschi, 1914
= *cristata semileavis* Santschi, 1924	= *mandibularis integra* Santschi, 1924
cryptica Bolton, 1974	*minor* Emery, 1892
dentata Santschi, 1912	= *gabonensis* Santschi, 1919
= *emeryi* Santschi, 1924	= *minor insularis* Santschi, 1924
gestroi Menozzi, 1922	= *minor liberiana* Santschi, 1924
hastifera (Santschi, 1914)	= *minor perusta* Santschi, 1924
laevior Stitz, in Santschi, 1924	*strigosa* Emery, 1899
latinodis Santschi, 1924	*subterranea* Arnold, 1915
lygaria Bolton, Gotwald, and Leroux,	= *punctatus* Santschi, 1924
1979	*thaui* Fisher, 2006
macgeei Bolton, 1974	*ugandensis* Menozzi, 1933
mandibularis Smith, F., 1858	

POLYRHACHIS Smith, F., 1857   Formicinae

48 species, Afrotropical   PLATE 29

DIAGNOSTIC REMARKS: Among the formicines, *Polyrhachis* species have 12-segmented antennae, the sockets of which are located well behind the posterior clypeal margin (a distance greater than the basal width of the scape). Eyes are present, located behind the mid-length of the head. The metapleural gland orifice is absent. Mandibular teeth decrease in size from apical to basal; tooth 3 is not strikingly reduced compared to tooth 4. The pronotum is usually bidentate to bispinose, only rarely unarmed. The propodeum is unarmed to bidentate, and part or all of the mesosoma is usually marginate laterally. The petiole (A2) is always armed, bearing 2–6 spines or teeth. Abdominal tergite 3 (the first gastral) is long, always distinctly longer than the tergite of A4; A3 tergite, in dorsal view or in profile, accounts for at least half the length of the gaster, often more.

DISTRIBUTION AND ECOLOGY: Absent from Madagascar and the New World, this genus has a few southern Palaearctic species, but hundreds of species in the Austral, Malesian, and Oriental regions. By comparison with these, the Afrotropical region is relatively depauperate. The vast majority of Afrotropical species are associated with trees and are entirely arboreal in terms of nesting and foraging. Only a few species—mostly inhabitants of rather arid regions where large trees are scarce, or in savanna, scrubland, and grasslands—have

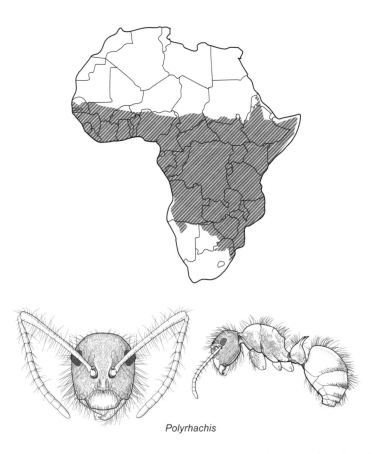

Polyrhachis

abandoned their arboreal habitat and taken to nesting in the ground, either directly, or under stones, or among the roots of grasses. When foraging, these ground-nesting species frequently ascend grasses and bushes. The arboreal species nest in hollow stems and branches, or in rotten holes in branches and trunks of standing trees, or uncommonly in tree stumps. Some specialized species occupy hollow galls or thorns. A number of species build nests of silk mixed with vegetable fragments that adhere to the undersides of leaves or are between 2 leaves that are glued together with silk. At least 2 species that construct nests in rotten holes in standing trees conceal the nest entrance behind a mat of vegetable fiber and silk.

TAXONOMIC COMMENT: *Polyrhachis* currently contains 13 subgenera worldwide. All the Afrotropical species within subgenus *P. (Myrma)*; see the synopsis in Bolton, 2003.

IDENTIFICATION OF SPECIES: Bolton, 1973a.

DEFINITION OF THE GENUS (WORKER): Characters of subfamily FORMICINAE, plus:

Mandible with 5–6 teeth that decrease in size from apical to basal; tooth 3 (counting from the apical) is larger than tooth 4.
Palp formula 6,4.
Frontal carinae strongly developed.

Antenna with 12 segments.

Antennal sockets located a considerable distance behind the posterior clypeal margin (a distance greater than the basal width of the scape).

Eyes present, located behind the midlength of the sides of the head; ocelli absent.

Mesosoma usually with spines or teeth at the pronotal humeri (rarely absent); propodeum usually with spines, teeth, or transverse ridges posteriorly (sometimes absent). Dorsolateral margins of the mesosoma often marginate in part or entirely, but some emarginate.

Metapleural gland absent.

Propodeal spiracle elliptical to slit-shaped, located posteriorly on the side and low down.

Propodeal foramen short: in ventral view the foramen does not extend anterior of a line that spans the anteriormost points of the metacoxal cavities.

Metacoxae closely approximated: in ventral view, with the mesocoxae and metacoxae directed at right-angles to the long axis of the mesosoma, the distance between the bases of the mesocoxae is about equal to the distance between the bases of the metacoxae.

Tibial spurs: mesotibia 0 or 1; metatibia 0 or 1.

Abdominal segment 2 (petiole) a node or a scale, with 2–6 spines or teeth.

Petiole (A2) with its ventral margin V-shaped in section.

Helcium tergite dorsally usually without a U-shaped emargination in its anterior margin, present but shallow and weakly developed in a few species.

Abdominal tergite 3 (first gastral) long, always distinctly longer than the tergite of A4; A3 tergite, in dorsal view or in profile, accounts for at least half of the length of the gaster.

Abdominal segment 3 without tergosternal fusion on each side of the helcium.

Abdominal sternite 3 with a transverse sulcus or mobile suture across the sclerite, immediately behind the helcium sternite (i.e., presternite and poststernite of A3 separated by a sulcus).

Acidopore concealed by the pygidium when not in use, without a seta-fringed nozzle.

AFROTROPICAL POLYRHACHIS

aenescens Stitz, 1910

aerope Wheeler, W.M.,1922

alexisi Forel, 1916

alluaudi Emery, 1892

   = *alluaudi anteplana* Forel, 1916

andrei Emery, 1921

arnoldi Forel, 1914

asomaningi Bolton, 1973

braxa Bolton, 1973

concava André, 1889

cornuta Stitz, 1910

cubaensis Mayr, 1862

   = *gerstaeckeri* Forel, 1886

   = *cubaensis striolatorugosa* Mayr, 1893

   = *cubaensis wilmsi* Forel, 1910

curta André, 1890

   = *maynei* Forel, 1911

   = *lyrifera* Stitz, 1933

decellei Bolton, 1973

decemdentata André, 1889

   = *decemdentata fernandensis* Forel, 1901

   = *decemdentata flavipes* Stitz, 1910

   = *decemdentata gustavi* Emery, 1921

   = *decemdentata tenuistriata* Menozzi, 1933

durbanensis Forel, 1914

epinotalis Santschi, 1924

esarata Bolton, 1973

fissa Mayr, 1902

   = *bequaerti* Wheeler, W.M., 1922

   = *fissa ugandensis* Arnold, 1954

gagates Smith, F., 1858

   = *gagates congolensis* Santschi, 1910

   = *nigriseta* Santschi, 1910

   = *nigriseta clariseta* Santschi, 1910

   = *gagates indefinita* Forel, 1913

   = *gagates obsidiana* Emery, 1921

gamaii Santschi, 1917

khepra Bolton, 1973

laboriosa Smith, F., 1858

   = *laboriosa architecta* Santschi, 1924

   = *hortulana* Arnold, 1955

lanuginosa Santschi, 1910

   = *lanuginosa conradti* Santschi, 1923

   = *lanuginosa santschii* Emery, 1921

     (homonym)

   = *lanuginosa felici* Emery, 1925

latharis Bolton, 1973

latispina Emery, 1925

   = *atalanta* Wheeler, W.M., 1922

     (homonym)

   = *iperpunctata* Menozzi, 1942

lauta Santschi, 1910

   = *lauta localis* Forel, 1913

   = *lauta laeta* Emery, 1921

lestoni Bolton, 1973

limitis Santschi, 1939

medusa Forel, 1897

militaris (Fabricius, 1782)

   = *militaris cupreopubescens* Forel,
     1879

   = *militaris striativentris* Emery, 1892

   = *militaris calabarica* Forel, 1907

   = *militaris ssibangensis* Forel, 1907

   = *militaris bruta* Santschi, 1912

   = *militaris nkomoensis* Santschi, 1924

monista Santschi, 1910

nigrita Mayr, 1895

   = *schoutedeni* Santschi, 1919

otleti Forel, 1916

phidias Forel, 1910

platyomma Emery, 1921

regesa Bolton, 1973

revoili André, 1887

   = *natalensis* Santschi, 1914

   = *revoili donisthorpei* Forel, 1916

rufipalpis Santschi, 1910

   = *rufipalpis mayumbensis* Forel, 1913

schistacea (Gerstäcker, 1859)

   = *rugulosus* Mayr, 1862

   = *carinatus* Smith, F., 1858

     (homonym)

   = *militaris cafrorum* Forel, 1879

   = *schistacea atrociliata* Santschi, 1913

   = *schistacea gagatoides* Santschi, 1913

   = *schistacea divina* Forel, 1913

   = *schistacea fracta* Santschi, 1914

   = *schistacea divinoides* Arnold, 1924

schlueteri Forel, 1886

   = *schlueteri plebeia* Santschi, 1914

   = *schlueteri indigens* Forel, 1914

spinicola Forel, 1894

   = *cubaensis gallicola* Forel, 1894

spitelleri Forel, 1916

sulcata André, 1895

transiens Bolton, 1973

viscosa Smith, F., 1858

   = *antinorii* Emery, 1877

   = *viscosa spretula* Santschi, 1923

   = *cubaensis imatongica* Weber, 1943

volkarti Forel, 1916

   = *kohli* Forel, 1916

weissi Santschi, 1910

   = *revoili conduensis* Forel, 1915

   = *revoili crassa* Emery, 1921

   = *revoili balli* Santschi, 1939

wellmani Forel, 1909

PONERA Latreille, 1804   Ponerinae

1 species, Afrotropical, introduced; 3 species, Malagasy, introduced   PLATE 29

DIAGNOSTIC REMARKS: Among the ponerines, *Ponera* combines the characters of presence of a single spur on the metatibia, absence of traction setae on the metabasitarsus, and location of the helcium at the base of the anterior face of the first gastral segment (A3), with the following: The mandible does not have a basal pit and also lacks a basal groove. The sides of the propodeum do not converge strongly dorsally, so that its dorsum forms a distinct surface behind the mesonotum. The mesotibia does not have traction setae scattered among the normal setae. The petiole (A2) has a ventral process that has a translucent thin-spot or fenestra anteriorly, and the posteroventral angles of the subpetiolar process always project posteriorly as a pair of small teeth.

DISTRIBUTION AND ECOLOGY: The genus currently contains 55 extant species, mostly distributed in the Oriental and Malesian regions, but with outlying species in the holarctic. No endemic species of *Ponera* are known from the Afrotropical or Malagasy regions. The Afrotropical record consists of a single discovered introduction in a plantation on Zanzibar Island, Tanzania. The Malagasy records are from litter samples from the Indian Ocean

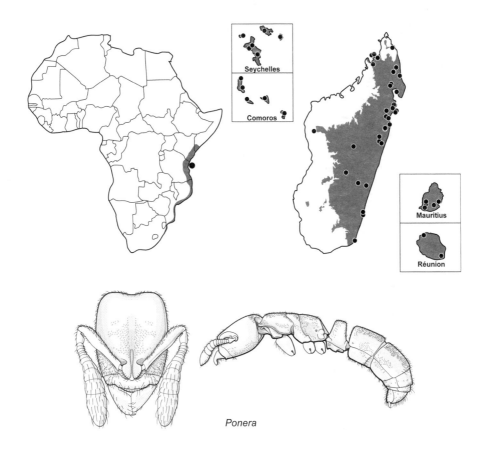

Ponera

islands of the region and, in the case of *P. swezeyi*, also from Madagascar itself. The species-rank identifications are tentative and await confirmation.

IDENTIFICATION OF SPECIES: Taylor, 1967.

DEFINITION OF THE GENUS (WORKER): Characters of subfamily PONERINAE, plus:

Mandible stoutly triangular, without a basal groove and without a basal pit; with 7 to about 13 teeth and denticles in total, with a basal angle between the masticatory and basal margins. When the mandibles are fully closed, there is no space between the masticatory margins nor between the basal margins of the mandibles and the clypeus.

Palp formula 2,2.

Clypeus simple, without extended lobes or teeth on either the median or the lateral portions and usually unarmed anteromedially.

Frontal lobes small, their anterior margins well behind the anterior clypeal margin.

Eyes absent or present; when present, always small to minute, lateral, and located well in front of the midlength of the head capsule.

Antenna gradually incrassate or with an apical club of 4 or 5 segments.

Metanotal groove present or absent; when present, may be impressed.

Propodeal dorsum with its anterior half forming a distinct defined transverse surface, never reduced to a narrow strip by extreme dorsal convergence of the sides.

Epimeral sclerite absent.

Metapleural gland orifice opens posteriorly.

Propodeal spiracle small, circular to slightly elliptical.

Tibial spurs: mesotibia 1; metatibia 1, pectinate.

Mesotibia without traction setae.

Pretarsal claws small and simple.

Abdominal segment 2 (petiole) a scale.

Abdominal sternite 2 (petiole sternite) with a ventral process that has a translucent thin-spot or fenestra anteriorly and has the posteroventral angles projecting as a pair of small teeth; sometimes the petiolar sternite itself is bifurcated posteroventrally.

Helcium arises low down on the anterior face of A3 (first gastral segment).

Prora present.

Abdominal segment 4 (second gastral segment) with distinctly differentiated presclerites.

Stridulitrum absent or present on the pretergite of A4.

AFROTROPICAL PONERA
swezeyi (Wheeler, W.M., 1933)

MALAGASY PONERA
exotica Smith, M.R., 1962
incerta (Wheeler, W.M., 1933)
swezeyi (Wheeler, W.M., 1933)

PRIONOPELTA Mayr, 1866 Amblyoponinae

3 species, Afrotropical; 6 species, Malagasy PLATE 29

DIAGNOSTIC REMARKS: Among the amblyoponines, *Prionopelta* is characterized by the possession of narrowly triangular mandibles that have only 3 teeth (of which the median tooth is the smallest) and which close tightly against the clypeus.

DISTRIBUTION AND ECOLOGY: This genus occurs throughout the world's tropics. It is most often retrieved from leaf litter and topsoil samples but is also found in rotten wood.

IDENTIFICATION OF SPECIES: No Afrotropical review exists, but species can be diagnosed from Terron, 1974; Overson and Fisher, 2015 (Malagasy).

DEFINITION OF THE GENUS (WORKER): Characters of subfamily AMBLYOPONINAE, plus:

Mandible narrowly triangular, with 3 teeth, of which the second is the smallest. When mandibles are fully closed, there is no space between the masticatory margins nor between basal margins of the mandibles and the clypeus.

Palp formula 2,2.

Clypeus simple, without extended lobes on either the median or the lateral portions.

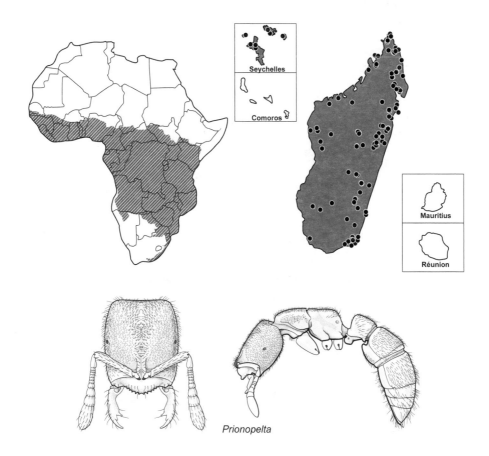

Prionopelta

Frontal lobes small and closely approximated, their anterior margins well behind the anterior clypeal margin.

Eyes absent or minutely present; when present, located at or slightly behind the midlength of the head capsule.

Antenna with 8–12 segments, with the apical 3–4 antennomeres forming a club.

Pronotum not marginate anteriorly.

Metanotal groove present or absent.

Tibial spurs: mesotibia 0 or 1; metatibia 1, pectinate.

Abdominal segment 2 (petiole) sessile; in profile with a steep anterior face and weakly convex dorsum that may slope weakly downward posteriorly but without a distinctly differentiated posterior face; the petiole broadly attached to abdominal segment 3 (first gastral segment).

Helcium with its dorsal surface arising very high on the anterior face of the first gastral segment (A3), so that in profile A3 above the helcium has no free anterior face (at most a short, rounded angle present between anterior and dorsal surfaces).

Prora present.

Abdominal segment 4 with distinctly differentiated presclerites.

AFROTROPICAL PRIONOPELTA

aethiopica Arnold, 1949

amieti Terron, 1974

humicola Terron, 1974

MALAGASY PRIONOPELTA

descarpentriesi Santschi, 1924

laurae Overson and Fisher, 2015

seychelles Overson and Fisher, 2015

subtilis Overson and Fisher, 2015

talos Overson and Fisher, 2015

vampira Overson and Fisher, 2015

xerosilva Overson and Fisher, 2015

PRISTOMYRMEX Mayr, 1866 — Myrmicinae

5 species, Afrotropical; 3 species, Malagasy — PLATE 29

DIAGNOSTIC REMARKS: Among the myrmicines, *Pristomyrmex* has 11-segmented antennae that terminate apically in a 3-segmented club. Uniquely, the frontal lobes in this genus are vestigial to absent so that in full-face view the antennal sockets are exposed and the depressed area that contains the antennal sockets is clearly visible; the sockets themselves are close to the anterior margin of the head. The anterior clypeal margin is usually armed with a few blunt denticles or crenulae.

DISTRIBUTION AND ECOLOGY: *Pristomyrmex* is distributed throughout the Old World tropics, usually in forested areas, where the species are components of the topsoil and leaf-litter fauna and nest around plant roots, in rotten wood, or in rotten parts of standing trees.

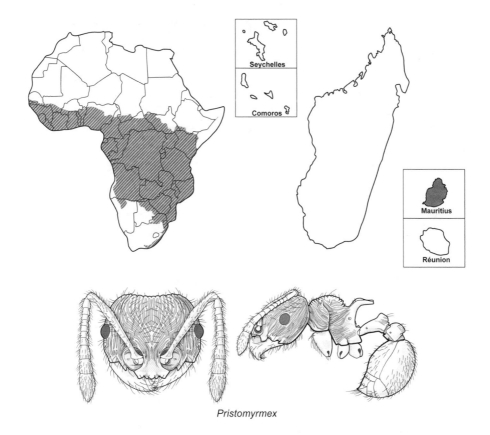

Pristomyrmex

The 3 Malagasy species are endemics restricted to Mauritius, with 1 of them also present on Réunion Island, but are absent from the rest of the region. *Pristomyrmex* species are predators and scavengers. Most species have normal alate queens, but in a few only ergatoids are known.

IDENTIFICATION OF SPECIES: Bolton, 1981b (Afrotropical); Wang, 2003 (Afrotropical + Malagasy).

DEFINITION OF THE GENUS (WORKER): Characters of subfamily MYRMICINAE, plus:

Mandible with distal portion suddenly broadened relative to proximal portion; masticatory margin with 3–5 teeth, sometimes also with a tooth on the basal margin.

Palp formula 5,3, or 4,3, or 4,2, or 2,3, or 2,2, or 1,3, or 1,2.

Stipes of the maxilla without a transverse crest.

Labrum with anterior margin abruptly angled downward, so that the margin appears thick in anterior view; at the point where the labrum curves down, there is a transverse raised ridge, or 2–3 small teeth, or both of these. Labrum large; when fully retracted, it covers most of the labiomaxillary complex.

Clypeus posteriorly broadly inserted between the frontal carinae.

Clypeus with lateral portions, in front of the antennal sockets, reduced to a thin plate or ridge; the anterior margin of the median portion of the clypeus crenulate or dentate.

Clypeus without an unpaired seta at the midpoint of the anterior margin.

Frontal carinae present but frontal lobes vestigial to absent, so that the antennal sockets are exposed.

Antennal sockets close to the anterior margin of head.

Antennal scrobes absent to very weakly present, extending above the eyes when present.

Antenna with 11 segments, with an apical club of 3 segments.

Torulus exposed in full-face view because of absence of frontal lobes.

Eyes present, variable in size, located approximately at the midlength of the head capsule or slightly behind the midlength.

Promesonotal suture absent from the dorsum of the mesosoma; pronotum with a pair of teeth or tubercles in some species.

Propodeum bidentate to bispinose.

Propodeal spiracle position variable, at or behind the midlength of the sclerite.

Metasternal process absent.

Tibial spurs: mesotibia 0 or 1; metatibia 0 or 1.

Stridulitrum present on the pretergite of A4.

Abdominal segment 4 (first gastral) tergite does not broadly overlap the sternite on the ventral gaster; gastral shoulders absent.

Sting long and slender, its apical portion very thin and hair-like.

Main pilosity of dorsal head and body: simple, may be very sparse or absent in places.

AFROTROPICAL PRISTOMYRMEX

africanus Karavaiev, 1931	*cribrarius* Arnold, 1926
= *myersi* (Weber, 1941)	*fossulatus* (Forel, 1910)
= *myersi beni* (Weber, 1952)	*orbiceps* (Santschi, 1914)
= *myersi mbomu* (Weber, 1952)	= *laevigatus* (Weber, 1952)
= *myersi primus* (Weber, 1952)	*trogor* Bolton, 1981

MALAGASY PRISTOMYRMEX

bispinosus (Donisthorpe, 1949)
browni Wang, 2003
trispinosus (Donisthorpe, 1946)

PROBOLOMYRMEX Mayr, 1901 — Proceratiinae

3 species, Afrotropical; 3 species, Malagasy · PLATE 30

DIAGNOSTIC REMARKS: Among the proceratiines, *Probolomyrmex* species do not have the tergite of A4 (the second gastral tergite) vaulted; the segments behind A4 are directed posteriorly. The clypeus and anterior portion of the frons form a shelf that projects out over the mandibles, and the latter have 6–8 small teeth. The antennal sockets are located very close to the anterior margin of the frontoclypeal shelf, and the frontal carinae are fused into a single median lamina between the sockets. Eyes are absent, and the antennae have 12 segments.

DISTRIBUTION AND ECOLOGY: Members of this small genus are cryptic, hypogaeic ants that occur throughout the tropical regions of the world; its species do not seem common anywhere. Specimens are most commonly retrieved from samples of leaf litter, but nests

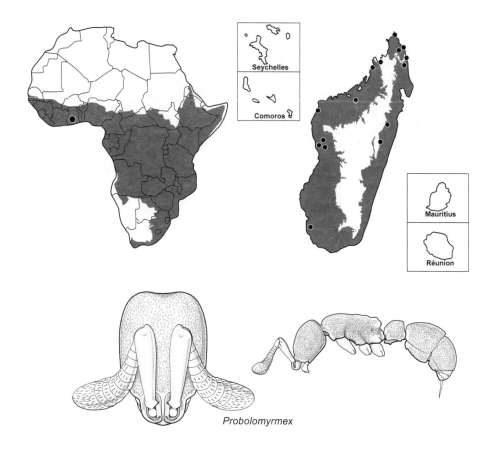

Probolomyrmex

are known from pieces of rotten wood in the litter layer and also from the compacted walls of termitaries.

IDENTIFICATION OF SPECIES: Taylor, 1965 (Afrotropical); Hita Garcia and Fisher, 2014c (Malagasy).

DEFINITION OF THE GENUS (WORKER): Characters of subfamily PROCERATIINAE, plus:

Mandible small and narrowly triangular, masticatory margin with 6–8 small teeth.

Palp formula 4,2.

Clypeus and anterior portion of the frons form a strongly projecting frontoclypeal shelf that overhangs the mandibles.

Frontal carinae fused to form a single median lamina between the antennal sockets.

Antennal sockets located very close to the anterior margin of the frontoclypeal shelf, well in front of the level of the bases of the mandibles.

Eyes absent.

Antenna with 12 segments.

Metanotal groove absent.

Tibial spurs: mesotibia 1, pectinate; metatibia 1, pectinate.

Abdominal segment 2 (petiole) with tergite and sternite not fused.

Helcium sternite not visible in profile. With helcium detached and in anterior view, the tergite overlaps the sternite on each side; the sternite is transverse and retracted, and does not bulge ventrally below the level of the tergal apices.

Prora present.

Abdominal segment 4 (second gastral segment) with strongly developed presclerites.

Abdominal segment 4 with tergite not vaulted, not hypertrophied with respect to the sternite.

Sculpture of head and body entirely pruinose.

AFROTROPICAL PROBOLOMYRMEX

brevirostris (Forel, 1910)
  = *parvus* Weber, 1949
filiformis Mayr, 1901
guineensis Taylor, 1965

MALAGASY PROBOLOMYRMEX

curciliformis Hita Garcia and Fisher, 2014
tani Fisher, 2007
zahamena Hita Garcia and Fisher, 2014

PROCERATIUM Roger, 1863   Proceratiinae

9 species, Afrotropical; 3 species, Malagasy   PLATE 30

DIAGNOSTIC REMARKS: Among the proceratiines, *Proceratium* species have the tergite of A4 (the second gastral tergite) enlarged, arched, and strongly vaulted so that the remaining segments point forward. The antennal sockets may be very close to the anterior margin of the head, but no frontoclypeal shelf is developed. The mandibles behind the apical tooth have 2–13 other teeth or denticles. Eyes may be absent or present. The antennae always have 12 segments, and the apical segment is moderately enlarged but never bulbous.

DISTRIBUTION AND ECOLOGY: A mainly tropical genus with species distributed worldwide but also with a number of species that occur outside the tropics. Workers are usually encountered in samples of leaf litter, where some species may also nest, but most nest sites are in the ground, either directly or under stones or in fallen timber. Nests may be constructed in the soil around the bases of trees, from which the workers emerge to forage on the surface or in the litter layer. Perhaps uniquely, the Mauritius endemic species *P. avium* nests in rot pockets in the branches of standing trees. The recorded prey is arthropod eggs, especially those of spiders.

IDENTIFICATION OF SPECIES: Baroni Urbani and De Andrade, 2003 (Afrotropical + Malagasy); Hita Garcia, Hawkes, and Alpert, 2014 (Afrotropical).

DEFINITION OF THE GENUS (WORKER): Characters of subfamily PROCERATIINAE, plus:

Mandible subtriangular, masticatory margin with an apical tooth plus 2–13 teeth or denticles.

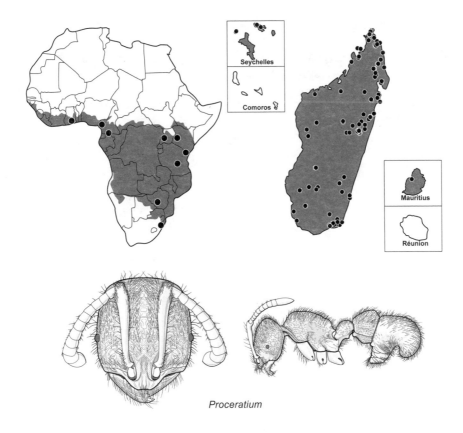

Proceratium

Palp formula 4,3, or 3,3, or 3,2, or 2,2. Second segment of the maxillary palp hammer-shaped.

Clypeus narrow, sometimes with a median lobe that overhangs the mandibles, sometimes so reduced that the antennal sockets are at the anterior margin of the head and overhang the mandibles.

Frontal carinae usually present and elevated, divergent posteriorly, not or only partially covering the antennal sockets; rarely reduced and fused to form a single median lamina.

Antennal sockets not located on a shelf-like frontoclypeal structure that projects forward above the mandibles.

Eyes absent or present; when present, small and located at or just in front of the midlength of the head capsule.

Antenna with 12 segments, the apical antennomere not enormously swollen.

Metanotal groove usually absent, only rarely weakly represented.

Tibial spurs: mesotibia 0 or 1; metatibia 1, pectinate.

Abdominal segment 2 (petiole) with tergite and sternite fused.

Helcium sternite not visible in profile. With helcium detached and in anterior view, the tergite overlaps the sternite on each side; the sternite is transverse and retracted, and does not bulge ventrally below the level of the tergal apices.

Prora present, often small and sometimes very reduced.

Abdominal segment 4 (second gastral segment) with distinctly differentiated presclerites.

Abdominal segment 4 with tergite enlarged and strongly vaulted, the tergite hypertrophied with respect to the sternite, which is reduced.

AFROTROPICAL PROCERATIUM

arnoldi Forel, 1913

boltoni Leston, 1971

burundense De Andrade, 2003

carri Hita Garcia, Hawkes, and Alpert, 2014

lunatum Terron, 1981

nilo Hita Garcia, Hawkes, and Alpert, 2014

sali Hita Garcia, Hawkes, and Alpert, 2014

sokoke Hita Garcia, Hawkes, and Alpert, 2014

terroni Bolton, 1995

= *coecum* Terron, 1981 (homonym)

MALAGASY PROCERATIUM

avium Brown, 1974

= *avioide* De Andrade, 2003

diplopyx Brown, 1980

google Fisher, 2005

PROMYOPIAS Santschi, 1914 Ponerinae

1 species, Afrotropical PLATE 30

DIAGNOSTIC REMARKS: Among the ponerines, *Promyopias* combines the following characters. Stout, thickly spiniform or peg-like traction setae are present on the mesotibia, mesobasitarsus, and metabasitarsus (not on the metatibia). Eyes are absent. The helcium arises close to the midheight of the anterior face of A3 (the first gastral segment), not at its base. The mandible is elongate, sublinear and narrow, with a small tooth at about the midlength of the inner margin and armed apically with a vertical series of 3–4 small teeth. The anterior clypeal margin does not have teeth. The mesotibia and metatibia each have 2 spurs. The prora is a longitudinal, thick, bluntly convex crest that extends from just below the helcium almost to the apex of the first gastral (A3) sternite.

DISTRIBUTION AND ECOLOGY: An uncommon but very widely distributed monotypic genus in the forested zones of Africa, where it inhabits leaf litter and rotten wood.

IDENTIFICATION OF SPECIES: Bolton and Fisher, 2008c; Schmidt and Shattuck, 2014.

DEFINITION OF THE GENUS (WORKER): Characters of subfamily PONERINAE, plus:

Mandible elongate, sublinear and narrow, with weak basal groove but without a dorsal longitudinal groove. Apex of mandible armed with a short vertical series of 3–4 small teeth (may be worn and indistinct). Apical half of inner margin of mandible concave. Basal angle of mandible at about the midlength of the inner margin (where a small tooth is present) and proximal of this the long basal margin is shallowly convex.

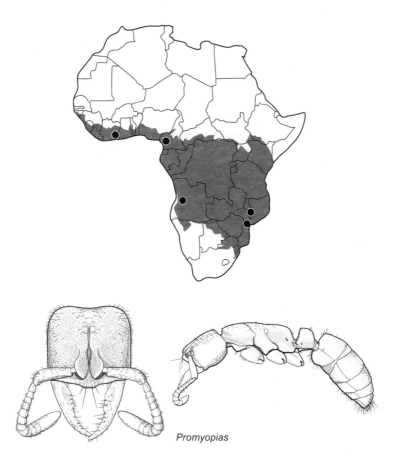

Promyopias

Palp formula 4,4.

Clypeus with median portion slightly projecting anteriorly; anterior margin of projecting portion transverse.

Frontal lobes with their anterior margins close to, but not overhanging, the anterior
clypeal margin; in full-face view the distance from the most anterior point of a
frontal lobe to the anterior clypeal margin is about equal to the basal width of the
scape.

Eyes absent.

Antenna with scape somewhat dorsoventrally flattened.

Metanotal groove vestigial to moderately developed in dorsal view; in profile the
propodeum continues the line of the mesonotum.

Epimeral sclerite present.

Metapleural gland orifice opens posteriorly as a curved slit that is shielded from lateral
view by a small lobe of cuticle, the orifice about level with the upper portion of the
propodeal lobe.

Propodeal spiracle broadly elliptical, almost round.

Mesotibia, mesobasitarsus, and metabasitarsus with strongly sclerotized spiniform traction setae (none on metatibia).

Tibial spurs: mesotibia 2; metatibia 2; on each tibia the anterior spur small and simple, the posterior larger and pectinate.

Pretarsal claws simple.

Abdominal segment 2 (petiole) a node.

Helcium located close to midheight of the anterior face of A3 (first gastral segment).

Prora a longitudinal, thick, bluntly convex crest that extends from just below the helcium almost to the apex of the first gastral (A3) sternite.

Abdominal segment 4 (second gastral segment) with differentiated presclerites.

Stridulitrum absent.

AFROTROPICAL PROMYOPIAS

silvestrii (Santschi, 1914)
  = *asili* Crawley, 1916

PSALIDOMYRMEX André, 1890

Ponerinae

6 species, Afrotropical

PLATE 30

DIAGNOSTIC REMARKS: Among the ponerines, *Psalidomyrmex* combines the characters of presence of a single spur on the metatibia and absence of traction setae on the metabasitarsus with the following: The labrum projects as a narrow lobe in front of the anterior clypeal margin (clearly visible in full-face view and usually transversely striate). The apical tooth of the mandible is attenuated, and the masticatory margin is edentate or crenulate. The frontal lobes are large. The mesofemur and metafemur do not have a longitudinal groove mid-dorsally, but the propodeal dorsum always has a median longitudinal groove or impression.

DISTRIBUTION AND ECOLOGY: *Psalidomyrmex* is mostly restricted to the rainforests of West and Central Africa but extends its range to Ugandan wet forest in the east. Nests are constructed in rotten wood that is usually in an advanced state of decay or directly in the soil beneath rotten logs. Individual foragers have also been found in termitaries, deep leaf litter and log mold. Very little is known about the biology of the species, but their principal prey is earthworms.

IDENTIFICATION OF SPECIES: Bolton, 1975b; Bolton and Brown, 2002.

DEFINITION OF THE GENUS (WORKER): Characters of subfamily PONERINAE, plus:

Mandible triangular to falcate, the apical tooth attenuated and the basal angle rounded; masticatory margin edentate or with a number of small, blunt teeth near the basal angle; basal groove present.

Labrum with upper portion directed anteriorly, projecting as a narrow but conspicuous lobe in front of the clypeal margin; this lobe clearly visible in full-face view and usually transversely striate.

Palp formula 3,4.

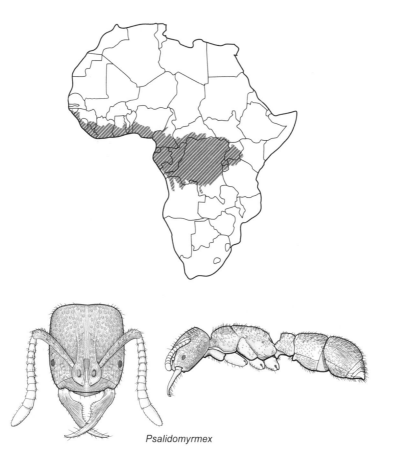

Psalidomyrmex

Clypeus simple, without projecting lobes or teeth.

Frontal lobes large, in full-face view their anterior margins behind the anterior clypeal margin.

Eyes present, in front of the midlength of the head capsule.

Metanotal groove vestigial to absent, not impressed.

Epimeral sclerite present.

Metapleural gland orifice opens laterally.

Propodeal spiracle circular or very nearly so.

Propodeal dorsum with a median longitudinal groove or impression.

Mesofemur and metafemur without a mid-dorsal longitudinal groove.

Tibial spurs: mesotibia 1; metatibia 1; each spur pectinate.

Pretarsal claws simple.

Abdominal segment 2 (petiole) a node.

Abdominal sternite 2 (petiole sternite) with its anterior articulatory surface long and broad, the surface with a narrow, median, V-shaped longitudinal groove, or small, central, pore-like depression.

Helcium arises low down on the anterior face of A3 (first gastral segment).

Prora a conspicuous lobe.

Abdominal segment 4 (second gastral segment) with very strongly differentiated presclerites.

Stridulitrum absent from the pretergite of A4.

AFROTROPICAL PSALIDOMYRMEX

feae Menozzi, 1922
　= *feae impressa* Menozzi, 1922
foveolatus André, 1890
procerus Emery, 1901
　= *longiscapus* Santschi, 1920
　= *obesus* Wheeler, W.M., 1922

　= *procerus collarti* Santschi, 1937
reichenspergeri Santschi, 1913
　= *mandibularis mabirensis* (Arnold, 1954)
sallyae Bolton, 1975
wheeleri Santschi, 1923

RAVAVY Fisher, 2009　　　　　　　　　　　　　　　　Dolichoderinae

1 species, Malagasy　　　　　　　　　　　　　　　　　　　　　PLATE 31

DIAGNOSTIC REMARKS: Among the dolichoderines, the worker of *Ravavy* closely resembles that of *Tapinoma* but is separated from other Malagasy taxa by its dentition and the presence of a transverse crest on the propodeum.

DISTRIBUTION AND ECOLOGY: Currently restricted to Madagascar, but Afrotropical species are possible, as mentioned below. Nesting and foraging habits have not been recorded.

TAXONOMIC COMMENT: The single species of this genus was originally described from the male sex; the worker caste is noted and defined here for the first time. In the Afrotropical region there are a couple of species, currently included in *Tapinoma*, that also have a propodeal crest. It is possible that these properly belong to *Ravavy*, but confirmation must await a taxonomic revision of the *Tapinoma* fauna of both regions. A second *Ravavy* species from Madagascar is known but remains undescribed.

IDENTIFICATION OF SPECIES: Fisher, 2009.

DEFINITION OF THE GENUS (WORKER): (previously undescribed) Characters of DOLICHODERINAE, plus:

Masticatory margin of mandible with 4 teeth followed by 5 denticles; apical tooth two times longer and two times broader than preapical; tooth 3 from the apical end smaller than tooth 4. Basal margin either unarmed or at most with 2–3 microscopic serrations following the basal denticle.

Palp formula 6,4.

Anterior clypeal margin concave medially, the apices of the concavity form a marked angle on the margin on each side.

Antenna with 12 segments.

Metathoracic spiracles dorsolateral.

Metanotal groove absent, merely a suture across the dorsum.

Propodeum with a short, transverse cuticular crest between the dorsum and declivity; crest appears as a small triangular tooth in profile.

Propodeal spiracle very close to the margin of the declivity.

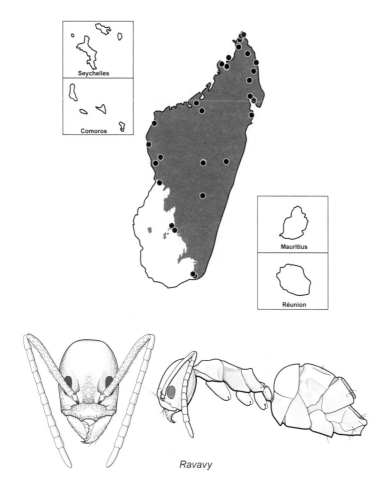

Ravavy

Tibial spurs: mesotibia 1; metatibia 1, pectinate and very long.

Abdominal segment 2 (petiole) extremely reduced, merely a low, narrow segment without a dorsal node or scale.

Abdominal segment 3 (first gastral) projects forward over the reduced petiole and entirely conceals the petiole in dorsal view when the gaster and mesosoma are aligned.

Abdominal segment 3, on ventral surface of projecting portion, with a longitudinal impression or groove that accommodates the entire petiole when the mesosoma and gaster are aligned.

Abdominal tergites 3–6 visible in dorsal view (i.e., 4 gastral tergites visible in dorsal view); tergite of A7 (pygidium) is reflexed onto the ventral surface of the gaster, where it abuts abdominal sternite 7 (hypopygium).

Abdominal sternite 6 (fourth gastral sternite) not keel-shaped posteriorly.

MALAGASY RAVAVY

miafina Fisher, 2009

DIAGNOSTIC REMARKS: Among the Malagasy myrmicines, *Royidris* has 12-segmented antennae that terminate in a club of 3 or 4 segments. Frontal carinae are restricted to the frontal lobes, and antennal scrobes are absent. The mandible has 5 distinct teeth, and the midpoint of the anterior clypeal margin has a single, unpaired, stout seta. The propodeum is unarmed, the petiolar spiracle is close to the midlength of the peduncle, and the first gastral tergite (the tergite of A4) does not overlap the sternite on the ventral surface of the gaster.

DISTRIBUTION AND ECOLOGY: This genus is restricted to Madagascar, where it inhabits tropical dry forest, spiny forest, shrubland, and savanna. Nests are constructed under stones, in dead twigs on the ground, or in rotten logs. Foraging takes place on the ground and in leaf litter, but some have been found in burned savanna or running on barren rock.

IDENTIFICATION OF SPECIES: Heterick, 2006 (as *Monomorium shuckardi* group); Bolton and Fisher, 2014.

DEFINITION OF THE GENUS (WORKER): Characters of subfamily MYRMICINAE, plus:

Mandible triangular, masticatory margin with 5 teeth.

Palp formula 5,3.

Stipes of the maxilla with a vestigial transverse crest or without a crest.

Clypeus posteriorly moderately broadly inserted between the frontal lobes; median portion of the clypeus with a fine, weak, longitudinal rugula on each side; the median longitudinal strip unsculptured.

Clypeus with a stout, unpaired seta at the midpoint of the anterior margin.

Clypeus with lateral portions not raised into a shielding wall or sharp ridge in front of the antennal sockets.

Frontal carinae short, restricted to well-defined but narrow frontal lobes.

Antennal scrobes absent.

Antenna with 12 segments, with an apical club of 3 or 4 segments.

Torulus with upper lobe visible in full-face view.

Eyes present, relatively large, located at about the midlength of the head capsule, or slightly in front of the midlength.

Head capsule without a median longitudinal carina.

Pronotum plus anterior mesonotum often swollen and distinctly convex in profile, so that the dorsalmost point of the promesonotum is on a considerably higher level than the propodeal dorsum (not in *admixta* group).

Propodeum unarmed, the dorsum and declivity separated by a blunt angle; propodeal lobes small and rounded.

Propodeal spiracle at about the midlength of the sclerite and close to the dorsal mar-

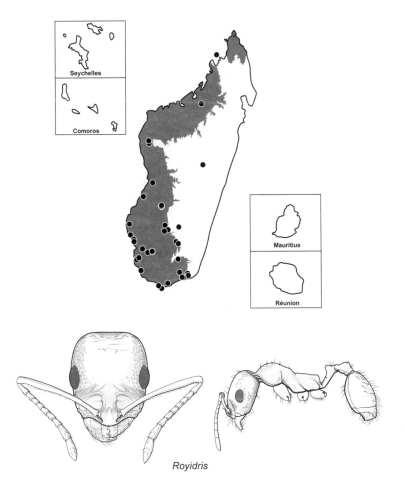

Royidris

gin, far in front of the margin of the declivity and separated from the apex of the metapleural gland bulla by more than the spiracle's diameter.

Metasternal process absent.

Tibial spurs: mesotibia 1; metatibia 1; spurs simple.

Abdominal segment 2 (petiole) with an anterior peduncle; spiracle slightly behind the midlength of the peduncle.

Subpetiolar process a minute crest.

Abdominal segment 3 (postpetiole) not dorsoventrally flattened in profile, about as high as broad.

Stridulitrum present on the pretergite of abdominal segment 4.

Abdominal tergite 4 (first gastral) does not overlap the sternite on the ventral surface of the gaster; gastral shoulders absent to weakly present.

Sting present, usually weakly developed.

Main pilosity of dorsal head and body: simple, usually absent from the propodeal dorsum.

admixta Bolton and Fisher, 2014	*notorthotenes* (Heterick, 2006)
anxietas Bolton and Fisher, 2014	*pallida* Bolton and Fisher, 2014
clarinodis (Heterick, 2006)	*peregrina* Bolton and Fisher, 2014
depilosa Bolton and Fisher, 2014	*pulchra* Bolton and Fisher, 2014
diminuta Bolton and Fisher, 2014	*robertsoni* (Heterick, 2006)
etiolata Bolton and Fisher, 2014	*shuckardi* (Forel, 1895)
gravipuncta Bolton and Fisher, 2014	*singularis* Bolton and Fisher, 2014
longiseta Bolton and Fisher, 2014	

SANTSCHIELLA Forel, 1916 — Formicinae

1 species, Afrotropical — PLATE 31

DIAGNOSTIC REMARKS: Among the Afrotropical formicines, *Santschiella* is unique, as it is the only species in which the antennal scape, when laid back in its natural resting position, passes below the eye rather than above it. In addition, the eyes are enormous—far larger than in any other Afrotropical ant. They extend from close behind the antennal sockets to the posterior margin of the head. The long axes of the eyes are convergent anteriorly, and the antennal sockets are in front of the eyes, on the lines of their long axes. Ocelli are also present.

DISTRIBUTION AND ECOLOGY: A single species of this exclusively Afrotropical genus is known, represented by only a couple of collections from the Central African rainforests of Gabon and the Democratic Republic of Congo. A related genus is present in the Oriental region and the Malesian territories of Malaysia, Indonesia, and the Philippines, and 2 related fossil genera have been described. *Santschiella* is arboreal, morphologically highly specialized, and apparently very uncommon.

IDENTIFICATION OF SPECIES: No review exists, but the single species is unlikely to be confused with any other Afrotropical ant.

DEFINITION OF THE GENUS (WORKER): Characters of subfamily FORMICINAE, plus:

Mandible with 5–6 teeth.

Palp formula 6,4.

Frontal carinae absent; antennal sockets concealed dorsally by an extension of the torulus.

Antenna with 12 segments.

Antennal scape, when laid back in its normal resting position, passes below the eyes.

Antennal sockets located at the posterior clypeal margin; the median portion of the clypeus projects back between the sockets.

Eyes enormous, extending from close behind the antennal sockets to the posterior margin of the head; in full-face view the eyes occupy the posterior angles and the lateral parts of the posterior margin; ocelli present.

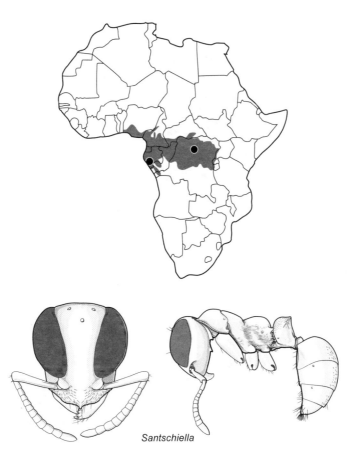

Santschiella

Eyes with their long axes convergent anteriorly; the antennal sockets are in line with the long axes of the eyes and directly in front of the anterior margins of the eyes.

Head capsule ventrolaterally with a translucent cuticular lamina along its posterior half.

Metapleural gland present.

Propodeal spiracle circular.

Propodeum armed posteriorly with a pair of short spines.

Propodeal foramen short: in ventral view the foramen does not extend anterior of a line that spans the anteriormost points of the metacoxal cavities (not confirmed in *Santschiella* but present in related genera).

Procoxa with a short spine on its anterior surface.

Metacoxae closely approximated: in ventral view, with the mesocoxae and metacoxae directed at right-angles to the long axis of the mesosoma, the distance between the bases of the mesocoxae is about equal to the distance between the bases of the metacoxae (not confirmed in *Santschiella* but present in related genera).

Tibial spurs: mesotibia 1; metatibia 1.

Abdominal segment 2 (petiole) with a short, tall node that has a transverse ridge across its width posteriorly; the ridge terminates in an angle or tooth at each side.

Petiole (A2) with its ventral margin V-shaped in section.

Abdominal segment 3 (first gastral) without tergosternal fusion on each side of the helcium.

Abdominal sternite 3 presumably with a transverse sulcus across the sclerite immediately behind the helcium sternite (i.e., presternite and poststernite of A3 presumably separated by a sulcus; not confirmed in *Santschiella* but present in related genera).

Acidopore with a seta-fringed nozzle.

AFROTROPICAL SANTSCHIELLA

kohli Forel, 1916

SIMOPONE Forel, 1891 — Dorylinae

18 species, Afrotropical; 16 species, Malagasy — PLATE 31

DIAGNOSTIC REMARKS: Among the dorylines, *Simopone* species have large eyes and also have ocelli. The antennae are 11-segmented, and the antennal sockets are not fully exposed. The head is unique as there is no differentiated vertical posterior surface above the occipital foramen. The promesonotal suture is an incised groove to vestigial; the waist is of a single segment, which is angulate to marginate dorsolaterally. The propodeal spiracle is low on the side and situated behind the midlength of the sclerite, and propodeal lobes are present. The mesotibia lacks spurs and the pretarsal claws have a preapical tooth.

DISTRIBUTION AND ECOLOGY: Primarily a genus of the Afrotropical and Malagasy regions. Only 4 extralimital species are known, 1 from the Oriental region and 3 from the Malesian. All species nest and forage arboreally, though workers may sometimes descend to the ground to forage in leaf litter or in rotten wood. Nests are made in hollow twigs, rotten stems and branches, or rotten holes in trees. The prey of a couple of *Simopone* species is known and consists of the brood of other ants. Known queens of the Afrotropical species are all alate, but among the Malagasy species no queen recognizable by external morphology has ever been seen, even within nest samples. Excluding the possibility that alates have somehow been overlooked, it is suspected that they are remarkably ergatoid or have even been replaced by gamergates

IDENTIFICATION OF SPECIES: Brown, 1975 (Afrotropical); Bolton and Fisher, 2012 (Afrotropical + Malagasy).

DEFINITION OF THE GENUS (WORKER): Characters of subfamily DORYLINAE, plus:

Worker caste of most species with considerable size variation.

Prementum not visible when mouthparts closed.

Palp formula usually 6,4, rarely 5,3.

Eyes present and usually large, in full-face view located from slightly in front of, to distinctly behind, the midlength of the head capsule; ocelli present.

Frontal carinae present as a pair of ridges, expanded anteriorly into narrow frontal lobes.

Antennal sockets partially concealed in full-face view.

Parafrontal ridges variably developed, usually present but sometimes absent.

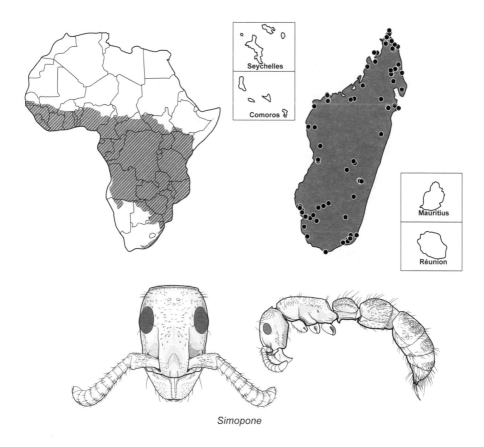

Simopone

Antenna with 11 segments; apical antennomere large but subcylindrical, not swollen and
 bulbous.

Antennal scrobes usually absent, present in only 1 species.

Head capsule without a differentiated vertical posterior surface above the occipital
 foramen; instead, the vertex slopes evenly down to the upper margin of the occipital
 foramen, which is visible in full-face view.

Head capsule, in ventral or ventrolateral view, with a carina that extends down the
 posterolateral margin and onto the ventral surface, where it terminates or fades out
 well before meeting the ventral midline.

Promesonotal suture at least a transverse vestige on the dorsum of the mesosoma.

Pronotal-mesopleural suture present on the side of the mesosoma.

Metanotal groove usually absent, although sometimes vestigially present to obvious.

Propodeal spiracle low on the side and situated at or behind the midlength of the
 sclerite, not subtended by a longitudinal impression.

Propodeal lobes present.

Tibial spurs: mesotibia 0; metatibia 1, pectinate.

Metatibial gland absent.

Metabasitarsus ventrally with a longitudinal glandular groove that occupies at least the
 basal half of the tarsomere length.

Pretarsal claws of all legs with a single preapical tooth on the inner surface of each claw.

Waist of 1 segment (petiole = A2), but A3 (first gastral segment) always deeply separated from A4 and subpostpetiolate.

Petiole (A2) sessile, flattened or shallowly convex dorsally; angulate to marginate dorsolaterally in dorsal view.

Helcium projects from the anterior surface of A3 at about its midheight.

Prora of A3 a simple curved cuticular rim or carina that separates the anterior face of the poststernite from its lateral and ventral surfaces.

Abdominal segment 4 with strongly developed presclerites.

Abdominal segments 5 and 6 without presclerites.

Pygidium (tergite of A7) large and distinct, flattened to concave dorsally and armed laterally, posteriorly, or both, with a series or row of denticles or peg-like spiniform setae; sometimes with a bifid cuticular fork apically.

Sting present, large and fully functional.

AFROTROPICAL SIMOPONE

amana Bolton and Fisher, 2012	*marleyi* Arnold, 1915
annettae Kutter, 1976	*matthiasi* Kutter, 1977
brunnea Bolton and Fisher, 2012	*miniflava* Bolton and Fisher, 2012
conradti Emery, 1899	*occulta* Bolton and Fisher, 2012
dryas Bolton and Fisher, 2012	*persculpta* Bolton and Fisher, 2012
fulvinodis Santschi, 1923	*rabula* Bolton and Fisher, 2012
grandis Santschi, 1923	*schoutedeni* Santschi, 1923
laevissima Arnold, 1954	*vepres* Bolton and Fisher, 2012
latiscapa Bolton and Fisher, 2012	*wilburi* Weber, 1949

MALAGASY SIMOPONE

consimilis Bolton and Fisher, 2012	= *satagia* Bolton, 1995
dignita Bolton and Fisher, 2012	*merita* Bolton and Fisher, 2012
dux Bolton and Fisher, 2012	*nonnihil* Bolton and Fisher, 2012
elegans Bolton and Fisher, 2012	*rex* Bolton and Fisher, 2012
emeryi Forel, 1892	*sicaria* Bolton and Fisher, 2012
fera Bolton and Fisher, 2012	*silens* Bolton and Fisher, 2012
grandidieri Forel, 1891	*trita* Bolton and Fisher, 2012
inculta Bolton and Fisher, 2012	*victrix* Bolton and Fisher, 2012
mayri Emery, 1911	

SOLENOPSIS Westwood, 1840 — Myrmicinae

12 species (+ 9 subspecies), Afrotropical; 2 species, Malagasy — PLATE 32

DIAGNOSTIC REMARKS: Among the myrmicines, *Solenopsis* species can usually be recognized by the presence of 10-segmented antennae that terminate in a strongly developed 2-segmented club (1 undescribed African species has smaller workers with 10 segments, and larger workers with 11), coupled with a single, unpaired long seta at the midpoint of the

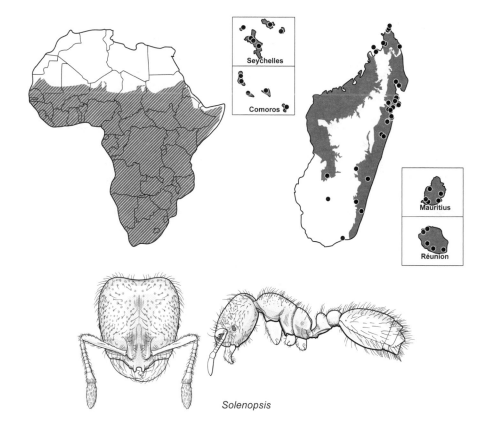

Solenopsis

anterior clypeal margin, and a propodeum that is always unarmed and rounded. The head lacks frontal carinae behind the frontal lobes and lacks antennal scrobes; the median portion of the clypeus is narrowly inserted between the frontal lobes and bicarinate.

DISTRIBUTION AND ECOLOGY: This genus is distributed throughout the world, but the majority of its species are in the Neotropical region. The few species known from the regions under consideration here are taxonomically confused and in need of revision. Most taxa appear to be endemic, but *S. geminata* is an introduction from the Neotropics. It is a polymorphic species with relatively large major workers that has established itself in some parts of West Africa. The endemic species all tend to be small and are mainly found under stones, in the soil at the base of trees, or in leaf litter.

IDENTIFICATION OF SPECIES: No review exists; the taxonomy is confused.

DEFINITION OF THE GENUS (WORKER): Characters of subfamily MYRMICINAE, plus:

Worker caste usually with size variation, some species polymorphic.

Mandible narrowly triangular; masticatory margin with 3–4 teeth (margin edentate in largest worker of 1 polymorphic species).

Palp formula 2,2, or 1,2; the maxillary palp often geniculate.

Stipes of the maxilla without a transverse crest.

Clypeus posteriorly narrowly inserted between the small frontal lobes. Median portion

of the clypeus usually raised, narrow, and longitudinally bicarinate dorsally; each
   carina usually terminates in a denticle or tooth at the anterior margin.

Clypeus with an unpaired seta at the midpoint of the anterior margin.

Frontal carinae restricted to the small frontal lobes.

Antennal scrobes absent.

Antenna with 10–11 segments, with a strong apical club of 2 segments (usually
   10-segmented, majors of 1 currently undescribed species have 11 antennomeres).

Torulus with upper lobe visible in full-face view; maximum width of the torulus lobe is
   posterior to the point of maximum width of the frontal lobes.

Eyes present but may be very small, located at, or in front of, the midlength of the side
   of the head capsule.

Promesonotal suture absent from the dorsum of the mesosoma, or a slightly impressed
   remnant discernible. Metanotal groove impressed.

Propodeum unarmed.

Propodeal spiracle approximately at the midlength of the sclerite.

Metasternal process absent.

Tibial spurs: mesotibia 0 or 1; metatibia 0 or 1.

Abdominal segment 4 (first gastral) tergite overlaps the sternite on the ventral gaster;
   gastral shoulders present.

Sting strongly developed.

Main pilosity of dorsal head and body: simple, sometimes sparse.

AFROTROPICAL SOLENOPSIS

africana Santschi, 1914
capensis Mayr, 1866
geminata (Fabricius, 1804)
   = *geminata innota* Santschi, 1915
georgica Menozzi, 1942
gnomula Emery, 1915
insinuans Santschi, 1933
maligna Santschi, 1910
orbuloides André, 1890
punctaticeps Mayr, 1865
   punctaticeps caffra Forel, 1894
   = *punctaticeps diversipilosa* Mayr, 1901
   = *punctaticeps cyclops* Santschi, 1914

punctaticeps cleptomana Santschi,
   1914
punctaticeps indocilis Santschi, 1914
punctaticeps erythraea Emery, 1915
punctaticeps kibaliensis Wheeler,
   W.M., 1922
punctaticeps fur Santschi, 1926
punctaticeps juba Weber, 1943
saevissima (Smith, F., 1855)
   saevissima itinerans Forel, 1911
ugandensis Santschi, 1933
   ugandensis congolensis Santschi, 1935
zambesiae Arnold, 1926

MALAGASY SOLENOPSIS

mameti Donisthorpe, 1946
seychellensis Forel, 1909

STIGMATOMMA Roger, 1859 — Amblyoponinae

2 species, Afrotropical; 1 species, Malagasy

PLATE 32

DIAGNOSTIC REMARKS: Among the amblyoponines, *Stigmatomma* is characterized by the possession of apically pointed linear mandibles that do not close tightly against the clypeus, the presence of dentiform setae that resemble teeth on the anterior clypeal margin, the absence of an anterior peduncle on the petiole, and the absence of spatulate setae on the head and body.

DISTRIBUTION AND ECOLOGY: Members of this genus occur worldwide. They are predominantly found in leaf litter and rotten wood but sometimes penetrate deep into the soil. A few have been found in the compacted earth walls of termite nests. Their prey appears to be geophilomorph centipedes. Numerous undescribed species are known from both the Afrotropical and the Malagasy regions.

IDENTIFICATION OF SPECIES: No review exists.

DEFINITION OF THE GENUS (WORKER): Characters of subfamily AMBLYOPONINAE, plus:

Mandibles linear, short to long, always acute apically and crossing over near the apex when fully closed. Inner margin of the mandible with 5–10 teeth; apical tooth large,

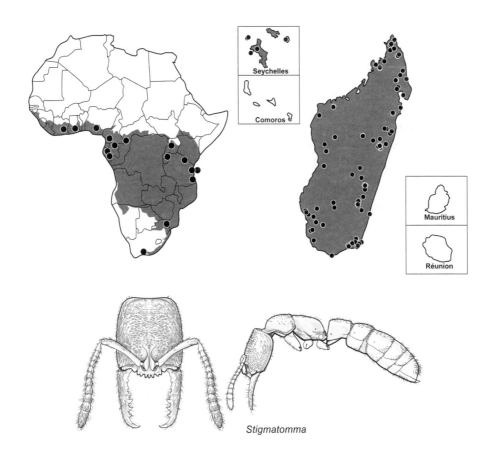

Stigmatomma

second tooth small and sometimes absent; teeth 3–6 are frequently double-ranked; basal tooth is sometimes enlarged into a subtriangular lamella. When the mandibles are fully closed, there is usually a gap between the masticatory margins and between the basal margins of the mandibles and the clypeus.

Palp formula 4,2 (*S. pluto*).

Gena on each side, immediately behind the clypeus, often with an anteriorly directed tooth.

Clypeus variably developed; antennal sockets may be considerably behind the anterior margin of the head or may overhang the anterior margin.

Clypeal anterior margin with dentiform setae that usually appear as teeth under low magnification.

Frontal lobes present, widely separated to closely approximated, partially to entirely concealing the antennal sockets.

Eyes absent or present; when present, located behind the midlength of the head capsule.

Antenna with 9–12 segments, the funiculus gradually incrassate apically or with a weakly defined club of 3 or 4 segments.

Metanotal groove absent to weakly present.

Tibial spurs: mesotibia 1 or 2; metatibia 2, main spur pectinate.

Abdominal segment 2 (petiole) sessile; in profile petiole with a steep anterior face and a weakly convex dorsum but without a differentiated posterior face; petiole broadly attached to abdominal segment 3 (first gastral segment).

Petiole (A2) sternite not reduced to a small medioventral sclerite.

Helcium with its dorsal surface arising very high on the anterior face of the first gastral segment (A3), so that in profile A3 above the helcium has no free anterior face (at most a short, rounded angle present between anterior and dorsal surfaces).

Prora usually present but may be only weakly developed.

Abdominal segment 4 with distinctly differentiated presclerites.

AFROTROPICAL STIGMATOMMA

pluto (Gotwald and Lévieux, 1972)
santschii Menozzi, 1922

MALAGASY STIGMATOMMA

besucheti (Baroni Urbani, 1978)

STREBLOGNATHUS Mayr, 1862   Ponerinae

2 species, Afrotropical   PLATE 32

DIAGNOSTIC REMARKS: Among the ponerines, *Streblognathus* combines the characters of presence of 2 spurs on the metatibia, absence of traction setae on the metabasitarsus, and location of the helcium at the base of the anterior face of the first gastral segment (A3) with the following: The clypeus has a broad, shallow median lobe, and the anterior margin of this lobe terminates on each side in an angle or a tooth. The propodeum is armed with a pair of short teeth. The petiole (A2) in profile is slender and very tall, reaching above the

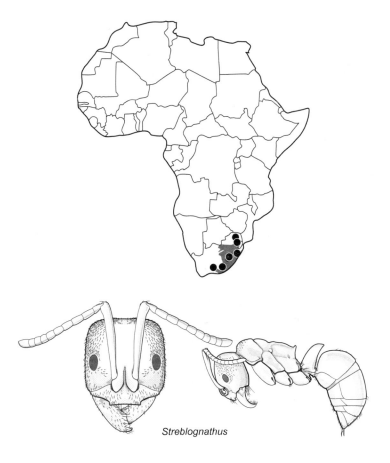

Streblognathus

level of the propodeal dorsum. The upper portion of the anterior petiolar face, together with its dorsum, form a single acute, sharp-edged, longitudinal ridge.

DISTRIBUTION AND ECOLOGY: The 2 species of this Afrotropical genus are restricted to grassland and karoo areas in South Africa and Lesotho. Both species nest in the ground and forage on the surface and are very conspicuous because of their size (up to 25 mm); they stridulate audibly when disturbed. The queen caste has been lost and reproduction depends upon a single gamergate per colony. The recorded prey of *Streblognathus* consists of tenebrionid beetles.

IDENTIFICATION OF SPECIES: Robertson, 2002.

DEFINITION OF THE GENUS (WORKER): Characters of subfamily PONERINAE, plus:

Mandibles with basal and external margins subparallel, weakly divergent anteriorly, with a triangular gap between basal margins and the clypeus when closed; a weak basal groove present. Masticatory margin with 5–7 teeth; basal angle angulate.
Palp formula 4,4.
Clypeus with a broad but shallow median lobe; anterior margin of the lobe terminates on each side in an angle or a tooth.

Frontal lobes small; in full-face view their anterior margins do not overhang the anterior clypeal margin.

Eyes present, distinct, at or slightly in front of the midlength of the head capsule and well in from the outer margins of the head in full-face view.

Metanotal groove present, weakly impressed.

Epimeral sclerite present.

Metapleural gland orifice followed posteriorly by a cuticular ridge.

Propodeal spiracle with orifice slit-shaped.

Propodeal dorsum posteriorly with a pair of short teeth.

Tibial spurs: mesotibia 2; metatibia 2; on each tibia the anterior spur small, simple to barbulate, the posterior spur larger and pectinate.

Pretarsal claws small, stout and simple.

Abdominal segment 2 (petiole) in profile slender and very tall, far surpassing the level of the propodeal dorsum. Petiole with the upper portion of the anterior face and dorsum forming an acute, sharp-edged, longitudinal ridge.

Helcium located at the base of the anterior face of A3 (first gastral segment).

Prora almost absent; merely a low transverse ridge immediately behind the helcium sternite (may be difficult to see if gaster is flexed down).

Abdominal segment 4 (second gastral segment) with presclerites weakly differentiated.

Stridulitrum present on the pretergite of A4.

AFROTROPICAL STREBLOGNATHUS

aethiopicus (Smith, F., 1858)

peetersi Robertson, 2002

STRUMIGENYS Smith, F., 1860   Myrmicinae

135 species, Afrotropical; 90 species, Malagasy   PLATE 32

DIAGNOSTIC REMARKS: Among the myrmicines, *Strumigenys* is easily identified. It has only 4–6 antennal segments, of which the apical 2 form a distinct club; antennal scrobes are usually conspicuous. The frontal lobes are widely separated, and the large clypeus projects back between them. The procoxa is at least as large as the mesocoxa and metacoxa. The mesobasitarsus and metabasitarsus lack apical circlets of traction spines. The mesonotum and petiole always lack teeth or tubercles, and spongiform appendages are usually present on 1 or both waist segments. The mandibles are triangular to linear. Bizarre pilosity is usually present.

DISTRIBUTION AND ECOLOGY: A very large genus (about 800 species) of almost worldwide distribution. They are usually small to minute cryptic ants, predominantly found in the soil, leaf litter, rotten twigs, or larger pieces of rotten wood, where they nest and forage. Only a very few Afrotropical species are known to nest in rot holes in trees and forage arboreally. A few species apparently always nest with, or very close to, the nests of much larger ponerine ants, and a few have been found in termitaries. Many ground-nesting species have

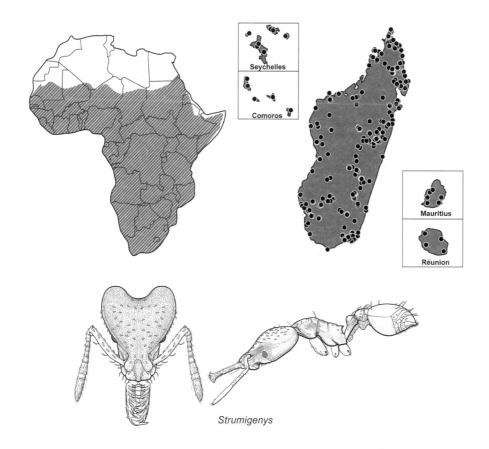

Strumigenys

a waxy film on some surfaces of the body, and some retain a coat of fine particles of soil, held in place by specialized setae, apparently for camouflage. The prey of *Strumigenys* is predominantly Collembola, but a wide range of other small arthropods is also taken. Long-mandibulate species are capable of an enormous gape, locking the mandibles open at more than 180 degrees when hunting.

TAXONOMIC COMMENT: *Strumigenys* has a large number of genus-group names as junior synonyms. Most of the references that follow refer to various groups of *Strumigenys* as they were understood at the time of publication. A full list of current synonyms is given in the introduction.

IDENTIFICATION OF SPECIES: Santschi, 1913a (Afrotropical *Strumigenys*); Brown, 1952 (*Serrastruma*); Brown, 1953 (Afrotropical *Smithistruma*); Brown, 1954 (Afrotropical + Malagasy *Strumigenys*); Bolton, 1972 (*Epitritus*); Bolton, 1983 (Afrotropical *Cladarogenys, Epitritus, Glamyromyrmex, Quadristruma, Serrastruma, Smithistruma, Strumigenys, Trichoscapa*); Bolton, 2000 (Afrotropical *Pyramica, Strumigenys*; Malagasy *Pyramica*); Fisher, 2000 (Malagasy *Strumigenys*).

DEFINITION OF THE GENUS (WORKER): Characters of subfamily MYRMICINAE, plus:

Mandible triangular to elongate-triangular, blade-like, sublinear, or linear; their masticatory margins oppose but do not overlap at full closure. If mandible triangular to blade-like, then masticatory margin with 7 to more than 25 teeth or denticles in total; if linear, mandible with 0–6 preapical teeth or denticles and terminating either in an apical fork of 2–3 teeth arranged in a vertical series (apicodorsal tooth the longest) or in a vertical series of denticles, the apicodorsal of which may be much enlarged.

Basal lamella present on mandible, projecting medially from the inner margin close to the base (proximal to the basal tooth); the lamella is an extrusion of the mandible, not merely a modified tooth.

Labrum not capable of reflexion over anterior portion of buccal cavity, not closing tightly over the apex of the labiomaxillary complex; labrum with trigger hairs present.

Palp formula 0,1 or 1,1.

Stipes of the maxilla without a transverse crest.

Clypeus large, broadly triangular, posteriorly broadly inserted between the frontal lobes.

Frontal carinae in full-face view, at level of the antennal sockets, with their outer margins varying from very close to, to overhanging and concealing, the apparent lateral margin of the head (actually the outer margin of the preocular carina).

Antennal scrobe usually present, extending above the eye (scrobes very rarely reduced).

Antenna with 4–6 segments, with an apical club of 2 segments.

Torulus concealed by the frontal lobe in full-face view.

Eye ventrolateral on side of head, at or close to the ventrolateral margin.

Head capsule anteriorly strongly narrowed from side to side in full-face view.

Head anterolaterally with a preocular carina present that arises at the clypeus and extends posteriorly below the antennal socket.

Promesonotal suture absent from the dorsum; the mesonotum without a pair of teeth or tubercles.

Propodeum usually bidentate to bispinose, only rarely unarmed.

Propodeal spiracle high on the side, very close to or at the margin of the declivity.

Metasternal process absent.

Femoral and tibial dorsal gland orifices frequently visible; may occur on first, second, or third legs, or any 1 or 2 out of the 3, or on all 3; occurrence and distribution is quite variable and gland orifices are frequently absent.

Tibial spurs: mesotibia 0; metatibia 0.

Abdominal segments 2 and 3 (petiole and postpetiole) usually with discrete lobes or patches of spongiform tissue on both tergites and sternites; these are reduced or absent in some species.

Abdominal tergite 4 (first gastral) basally with a raised, transverse cuticular crest or ridge (the limbus) across the sclerite posterior to the presclerite and above it. Sternite of A4 usually truncated basally.

Sting present.

Main pilosity of dorsal head and body: bizarre setae frequently present (flagellate, remiform, clavate, spatulate, squamate, orbicular, spoon-shaped); sometimes simple, sometimes absent.

adrasora Bolton, 1983

africana (Bolton, 1983)

agnosta (Bolton, 2000)

alessandrae Rigato, 2006

anarta (Bolton, 1983)

anorbicula (Bolton, 2000)

arahana (Bolton, 1983)

arnoldi Forel, 1913

bartolozzii Rigato, 2006

behasyla (Bolton, 1983)

belial (Bolton, 2000)

bellatrix (Bolton, 2000)

bequaerti Santschi, 1923

bernardi Brown, 1960

bitheria Bolton, 1983

cacaoensis Bolton, 1971

cavinasis (Brown, 1950)

cenagra Bolton, 2000

chyatha (Bolton, 1983)

concolor Santschi, 1914

cryptura (Bolton, 1983)

dagon (Bolton, 1983)

datissa (Bolton, 1983)

depilosa (Bolton, 2000)

dextra Brown, 1954

dictynna (Bolton, 2000)

didyma Bolton, 2000

dotaja (Bolton, 1983)

dromoshaula Bolton, 1983

dyshaula Bolton, 1983

emarginata Mayr, 1901

emmae (Emery, 1890)

enkara (Bolton, 1983)

ettillax Bolton, 1983

exunca (Bolton, 2000)

faurei Arnold, 1948

fenkara (Bolton, 1983)

fisheri (Bolton, 2000)

fulda (Bolton, 1983)

gatuda (Bolton, 1983)

geoterra (Bolton, 1983)

hastyla Bolton, 1983

havilandi Forel, 1905

helytruga Bolton, 1983

hensekta (Bolton, 1983)

impidora (Bolton, 1983)

inquilina (Bolton, 1983)

irrorata Santschi, 1913

katapelta Bolton, 1983

kerasma (Bolton, 1983)

korahyla Bolton, 1983

lasia (Brown, 1976)

laticeps (Brown, 1962)

londianensis (Patrizi, 1946)

loveridgei (Brown, 1953)

lucifuga (Bolton, 2000)

ludovici Forel, 1904

    = *alluaudi* Santschi, 1911

    = *alluaudi nigeriensis* Santschi, 1914

    = *rothkirchi* Wasmann, 1918

    = *escherichi lotti* Weber, 1943

lujae Forel, 1902

    = *reticulata* Stitz, 1910

    = *glanduscula* Santschi, 1919

    = *calypso* Santschi, 1923

    = *gerardi* Santschi, 1923

    = *aequalis* Menozzi, 1942

malaplax (Bolton, 1983)

marchosias Bolton, 2000

marginata (Santschi, 1914)

marleyi Arnold, 1914

maxillaris Baroni Urbani, 2007

    = *mandibularis* (Szabó, 1909)

    (homonym)

maynei Forel, 1916

    = *maynei latiuscula* Forel, 1916

mekaha (Bolton, 1983)

membranifera Emery, 1869

mesahyla Bolton, 1983

miccata (Bolton, 1983)

minima (Bolton, 1972)

minkara (Bolton, 1983)

mira (Bolton, 2000)

mormo (Bolton, 2000)

murshila Bolton, 1983

nimbrata Bolton, 1983

nimravida (Bolton, 2000)

ninda (Bolton, 1983)

noara (Bolton, 2000)

nykara (Bolton, 1983)

ogyga (Bolton, 2000)

omalyx Bolton, 1983

oxysma (Bolton, 1983)

pallestes Bolton, 1971

paranax Bolton, 1983

percrypta Bolton, 2000

petiolata Bernard, 1953

piliversa (Bolton, 2000)

placora (Bolton, 1983)

pretoriae Arnold, 1949

rakkota Bolton, 2000

ravidura (Bolton, 1983)

relahyla Bolton, 1983

robertsoni (Bolton, 2000)

rogeri Emery, 1890

   = *sulfurea* Santschi, 1915

roomi (Bolton, 1972)

rufobrunea Santschi, 1914

rusta (Bolton, 1983)

rukha Bolton, 1983

sahura (Bolton, 1983)

sardonica (Bolton, 2000)

sarissa Bolton, 1983

semicrypta Bolton, 2000

serrula Santschi, 1910

   = *uelensis* Santschi, 1923

sharra (Bolton, 1983)

shaula Bolton, 1983

sibyna Bolton, 2000

simoni Emery, 1895

   = *escherichi* Forel, 1910

   = *cognata* Santschi, 1911

   = *biconvexa* Santschi, 1913

   = *cognata boerorum* Santschi, 1913

   = *escherichi cliens* Forel, 1913

   = *escherichi limbata* Forel, 1913

   = *escherichi fusciventris* Emery, 1924

sistrura (Bolton, 1983)

spathoda Bolton, 1983

stygia Santschi, 1913

subsessa (Bolton, 2000)

sulumana (Bolton, 1983)

synkara (Bolton, 1983)

syntacta Bolton, 2000

tacta (Bolton, 1983)

terroni (Bolton, 1983)

tethepa (Bolton, 2000)

tetragnatha (Taylor, 1966)

tetraphanes Brown, 1954

thuvida (Bolton, 1983)

tiglath (Bolton, 1983)

tigrilla (Brown, 1973)

tolomyla (Bolton, 1983)

totyla Bolton, 1983

traegaordhi Santschi, 1913

transenna Bolton, 2000

transversa Santschi, 1913

truncatidens (Brown, 1950)

   = *dendexa* (Bolton, 1983)

trymala (Bolton, 1983)

tukulta (Bolton, 1983)

ultromalyx Bolton, 2000

vodensa (Bolton, 1983)

vazerka Bolton, 1983

weberi (Brown, 1959)

xenohyla Bolton, 1983

zandala Bolton, 1983

MALAGASY STRUMIGENYS

abdera Fisher, 2000

actis Fisher, 2000

admixta Fisher, 2000

adsita Fisher, 2000

agetos Fisher, 2000

agra Fisher, 2000

alapa Fisher, 2000

alperti Fisher, 2000

ambatrix (Bolton, 2000)

ampyx Fisher, 2000

apios Fisher, 2000

balux Fisher, 2000

bathron Fisher, 2000

bibiolana Fisher, 2000

bola Fisher, 2000

cabira Fisher, 2000

carisa Fisher, 2000

carolinae Fisher, 2000

charino Fisher, 2000

chilo Fisher, 2000

chroa Fisher, 2000

coveri Fisher, 2000

covina Fisher, 2000

deverra Fisher, 2000

dexis Fisher, 2000

dicomas Fisher, 2000

diota Fisher, 2000

diux Fisher, 2000

dolabra Fisher, 2000

dora Fisher, 2000

doxa Fisher, 2000

ection Fisher, 2000

emmae (Emery, 1890)

epulo Fisher, 2000

erynnes (Bolton, 2000)

europs Fisher, 2000

exiguaevitae Baroni Urbani, 2007
  = *hoplites* (Bolton, 2000) (homonym)

fanano Fisher, 2000

fautrix (Bolton, 2000)

finator Fisher, 2000

fronto Fisher, 2000

glycon Fisher, 2000

gorgon Fisher, 2000

grandidieri Forel, 1892

hathor (Bolton, 2000)

heliani Fisher, 2000

hilaris Fisher, 2000

inatos Fisher, 2000

ipsea Fisher, 2000

khakaura (Bolton, 2000)

labaris Fisher, 2000

langrandi Fisher, 2000

levana Fisher, 2000

lexex Fisher, 2000

livens Fisher, 2000

luca Fisher, 2000

lucomo Fisher, 2000

ludovicae Forel, 1904

lura Fisher, 2000

lutron Fisher, 2000

lysis Fisher, 2000

manga Fisher, 2000

maxillaris Baroni Urbani, 2007
  = *mandibularis* (Szabó, 1909)
    (homonym)

micrans Fisher, 2000

milae Fisher, 2000

mola Fisher, 2000

nambao Fisher, 2000

norax Fisher, 2000

odacon Fisher, 2000

olsoni (Bolton, 2000)

origo Fisher, 2000

peyrierasi Fisher, 2000

rabesoni Fisher, 2000

ravola Fisher, 2000

rogeri Emery, 1890

rubigus Fisher, 2000

schuetzi Fisher, 2000

scotti Forel, 1912

serket (Bolton, 2000)

seti (Bolton, 2000)

simoni Emery, 1895
  = *raymondi* Donisthorpe, 1946

sphera Fisher, 2000

sylvaini Fisher, 2000

symmetrix (Bolton, 2000)

tathula (Bolton, 2000)

tegar Fisher, 2000

toma Fisher, 2000

vazimba Fisher, 2000

victrix (Bolton, 2000)

wardi Fisher, 2000

DIAGNOSTIC REMARKS: Among the myrmicines, *Syllophopsis* has 12-segmented antennae that terminate in a large, well-defined, apical club of 3 segments. Eyes are usually minute, of only 1–2 ommatidia (rarely slightly larger). With the head in profile, the unpaired median clypeal seta arises from a vertical, or concave-vertical, surface that is well below the anteriormost point of the clypeal projection. The clypeus posteriorly is very narrow, so that the antennal sockets are very close together, and elevated. The propodeal dorsum is angulate to denticulate between the dorsum and declivity.

DISTRIBUTION AND ECOLOGY: Widespread in the Old World tropics, and with 1 species described from the Caribbean, which may possibly be an introduction from the Old World. All species are cryptic, nesting and foraging in leaf litter and rotten wood.

TAXONOMIC COMMENT: Until recently, members of this genus were included in *Monomorium*, but *Syllophopsis* species have the combination of characters given above, which is not seen in *Monomorium*. The taxonomy of the species in this genus is not well understood and in need of further study.

IDENTIFICATION OF SPECIES: Santschi, 1921 (*Syllophopsis*); Bolton, 1987 (Afrotropical, as *Monomorium fossulatum* group); Heterick, 2006 (Malagasy, as *M. hildebrandti* group); Sharaf and Aldawood, 2013 (Afrotropical, as *M. hildebrandti* group).

DEFINITION OF THE GENUS (WORKER): Characters of subfamily MYRMICINAE, plus:

Mandible narrowly triangular, smooth; masticatory margin usually oblique; usually with 4 teeth, uncommonly with 5.

Palp formula usually 2,2, rarely 3,2.

Stipes of the maxilla without a transverse crest.

Clypeus posteriorly very narrowly inserted between the frontal lobes; width of the clypeus between the lobes narrower than one of the lobes; antennal insertions consequently very closely approximated. Median portion of the clypeus raised and narrowed, weakly longitudinally bicarinate at least posteriorly. Anterior clypeal margin without a pair of projecting angles or teeth.

Clypeus with an unpaired median seta, projecting from the midpoint of the anterior margin in full-face view; in profile the unpaired median seta is seen to arise from a vertical or concave-vertical surface below and behind the anteriormost point of the clypeal projection.

Frontal carinae restricted to the small frontal lobes.

Antennal scrobes absent.

Antennal fossae not surrounded by fine, curved striolate or costulate sculpture.

Antenna with 12 segments, with a large, strongly defined apical club of 3 segments.

Torulus with upper lobe visible in full-face view; maximum width of the torulus lobe is posterior to the point of maximum width of the frontal lobes.

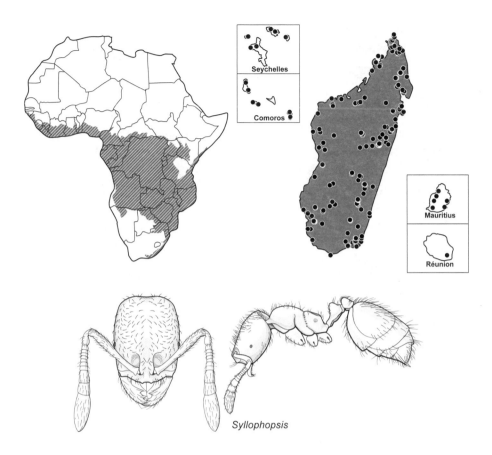

Syllophopsis

Eyes present, usually minute (of only 1–2 ommatidia) but larger in some Malagasy
species; located at the midlength of the head capsule.

Promesonotal suture absent from the dorsum of the mesosoma. Metanotal groove
usually present and impressed but sometimes vestigial.

Propodeum angulate to denticulate between the dorsum and declivity; without transverse sculpture dorsally.

Propodeal spiracle at, or slightly behind, the midlength of the sclerite.

Metasternal process absent.

Tibial spurs: mesotibia 0; metatibia 0.

Petiole (A2) with a long anterior peduncle; the spiracle at the node, or at the anterior
face of the node. Ventral surface of peduncle with fine transverse rugulae in larger
Malagasy species.

Abdominal segment 4 (first gastral) tergite overlaps the sternite on the ventral gaster;
gastral shoulders present.

Sting present.

Main pilosity of dorsal head and body: simple; sparse to absent on propodeum.

arnoldi Santschi, 1921
  = *jonesi* (Arnold, 1952) (redundant
    replacement name)
cryptobia Santschi, 1921
elgonensis Santschi, 1935
malamixta (Bolton, 1987)
modesta (Santschi, 1914)

= *modestum boerorum* (Santschi, 1915)
    (homonym)
= *modestum transwaalensis* (Emery,
    1922) (replacement name)
= *modestum smutsi* (Wheeler,
    W.M.,1922) (replacement name)
sersalata (Bolton, 1987)
thrascolepta (Bolton, 1987)

MALAGASY SYLLOPHOPSIS

adiastolon (Heterick, 2006)
aureorugosa (Heterick, 2006)
cryptobia Santschi, 1921
ferodens (Heterick, 2006)
fisheri (Heterick, 2006)
gongromos (Heterick, 2006)

hildebrandti (Forel, 1892)
infusca (Heterick, 2006)
modesta (Santschi, 1914)
sechellensis (Emery, 1894)
  = *fossulatum* (Emery, 1895)

TANIPONE Bolton and Fisher, 2012

Dorylinae

10 species, Malagasy

PLATE 33

DIAGNOSTIC REMARKS: Among the dorylines, *Tanipone* has large eyes that are slightly behind the midlength of the head and also has ocelli present. The maxillary palps are extremely long. The antennae are 12-segmented, and the antennal sockets are fully exposed. The promesonotal suture is absent; the waist is of a single segment and is not marginate dorsolaterally. The propodeal spiracle is low on the side and situated behind the midlength of the sclerite, and propodeal lobes are present. The mesotibia lacks spurs, and the pretarsal claws of the metatarsus have a preapical tooth. Uniquely, the first gastral tergite (tergite of A3) has a pair of subovate glandular patches on the posterior half (absent in 1 species), and the posterior margin of A3 tergite always has a transverse band of pale cuticle or a pair of pale patches.

DISTRIBUTION AND ECOLOGY: This genus is restricted to Madagascar, where it is characteristically an inhabitant of tropical dry forest, spiny forest, or desert scrub forest. It is predominantly terrestrial, being found under stones, in and under rotten logs, and in tree stumps, and foraging on the ground and on low vegetation. Some species are more arboreal, dwelling in live or dead stems and branches, or in rotten holes in trees. Suspected queens are extreme ergatoids; no trace of an alate form has been discovered.

IDENTIFICATION OF SPECIES: Bolton and Fisher, 2012.

DEFINITION OF THE GENUS (WORKER): Characters of subfamily DORYLINAE, plus:

Prementum not visible when mouthparts closed.

Palp formula 6,4. Maxillary palps extremely long: with mouthparts retracted the apices of the maxillary palps, when extended back on underside of the head, extend beyond

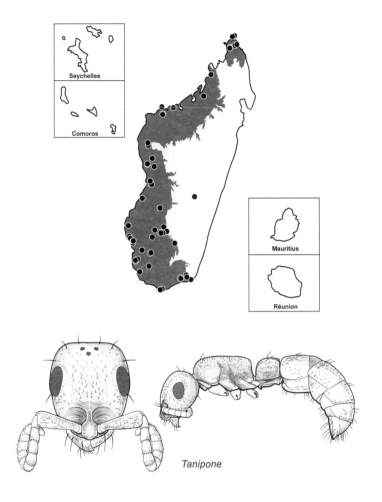

Tanipone

the level of the posterior margin of the eye and usually approach the occipital foramen.

Eyes present, large, in full-face view located slightly behind the midlength of the head capsule; ocelli present.

Frontal carinae present as a pair of short ridges, constricted and abruptly terminated posteriorly.

Antennal sockets fully exposed in full-face view.

Parafrontal ridges short and indistinct, weakly developed.

Antenna with 12 segments; funiculus tapers apically, not incrassate; apical antennomere not swollen and bulbous.

Antennal scrobes usually absent, present in only 1 species.

Head capsule, in ventral or ventrolateral view, with a carina that extends down the posterolateral margin and onto the ventral surface, where it passes through a near right-angle and extends to the ventral midline.

Promesonotal suture absent on the dorsum of the mesosoma.

Pronotal-mesopleural suture present on the side of the mesosoma.

Metanotal groove vestigial to absent.

Propodeal spiracle low on the side and situated at or behind the midlength of the sclerite, not subtended by a longitudinal impression.

Propodeal lobes present.

Tibial spurs: mesotibia 0; metatibia 1, pectinate.

Metatibial gland present, its orifice usually small and indistinct.

Pretarsal claws of at least the metatarsus with a single preapical tooth on the inner surface of each claw.

Waist of 1 segment (petiole = A2).

Petiole (A2) sessile, not marginate dorsolaterally.

Helcium projects from the anterior surface of A3 at about its midheight.

Abdominal tergite 3 (first gastral tergite) with a pair of subovate glandular patches on the posterior half (absent in 1 species). Posterior margin of A3 tergite with a transverse band of pale cuticle, or a pair of pale patches, that subtend the glands (this feature retained in the one species where glands are not apparent).

Prora of A3 a small, simple, convex cuticular boss or prominence, not delimited by a sharp, curved carina that separates the anterior face of the poststernite from the lateral and ventral surfaces.

Abdominal segment 4 with strongly developed presclerites.

Abdominal segments 5 and 6 without presclerites.

Pygidium (tergite of A7) large, flattened dorsally; pygidial denticles sparse, restricted to a short apical arc along a narrow prominence at the extreme apex of the sclerite.

Sting present, large and fully functional.

MALAGASY TANIPONE

aglandula Bolton and Fisher, 2012	*pilosa* Bolton and Fisher, 2012
aversa Bolton and Fisher, 2012	*scelesta* Bolton and Fisher, 2012
cognata Bolton and Fisher, 2012	*subpilosa* Bolton and Fisher, 2012
hirsuta Bolton and Fisher, 2012	*varia* Bolton and Fisher, 2012
maculata Bolton and Fisher, 2012	*zona* Bolton and Fisher, 2012

TAPINOLEPIS Emery, 1925   Formicinae

11 species (+ 3 subspecies), Afrotropical   PLATE 33

DIAGNOSTIC REMARKS: Among the formicines, *Tapinolepis* can be identified by its combination of 11-segmented antennae, presence of ocelli, lack of a differentiated metanotum on the mesosoma, lack of spines or teeth on the propodeum and petiole, and the absence of a stout spur on the metatibia. In addition, the eyes are never strikingly behind the midlength of the head, and standing setae are absent from the head in Afrotropical species but may be sparsely present in Malagasy forms.

DISTRIBUTION AND ECOLOGY: *Tapinolepis* species are widely distributed in Africa outside the rainforest zones, and a number of species occur in Africa north of the Sahara desert.

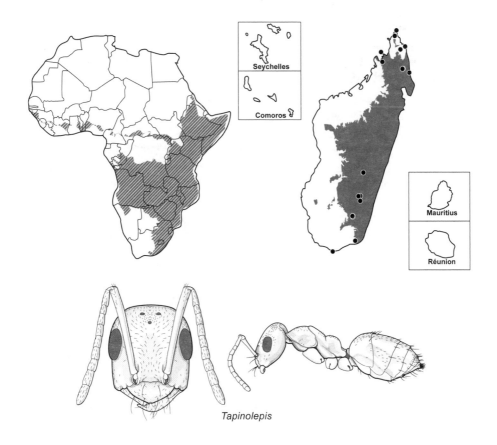

Tapinolepis

Undescribed species are present in Madagascar, but otherwise the genus is absent from other Old World regions and absent from the New World. The species are characteristically ground nesters and foragers. Replete workers have been found in the nests of some species, although whether the production of these specialized food-storing workers is universal remains to be seen.

IDENTIFICATION OF SPECIES: No review exists.

DEFINITION OF THE GENUS (WORKER): Characters of subfamily FORMICINAE, plus:

Some (perhaps all) species with replete workers within the nest.

Mandible with 5 teeth; tooth 3 (counting from the apical) is smaller than tooth 4.

Mandibles at full closure mostly concealed by the clypeus.

Palp formula 6,4.

Frontal carinae scarcely exceed posteriorly apices of the toruli.

Antenna with 11 segments.

Antennal sockets close to the posterior clypeal margin; the anterior arc of the torulus slightly indents the clypeal suture.

Eyes present, located from slightly in front of, to slightly behind, the midlength of the sides of the head; ocelli present.

Cephalic dorsum behind the clypeus usually without erect setae (1 pair may be present in Malagasy species, close to posterior margin).

Metanotum does not form a distinct isolated sclerite on the dorsum of the mesosoma.

Metapleural gland present.

Propodeal spiracle circular.

Propodeal foramen long: in ventral view the foramen extends anterior of a line that spans the anteriormost points of the metacoxal cavities.

Metacoxae widely separated: in ventral view, with the mesocoxae and metacoxae directed at right-angles to the long axis of the mesosoma, the distance between the bases of the mesocoxae is markedly less than the distance between the bases of the metacoxae.

Tibial spurs: mesotibia 0; metatibia 0.

Abdominal segment 2 (petiole) a scale, without spines or teeth.

Petiole (A2) scale inclined anteriorly, with a short anterior face and a longer posterior face; petiole usually with a posterior peduncle.

Petiole (A2) with its ventral margin U-shaped in section.

Abdominal segment 3 (first gastral) with a sharply margined longitudinal concavity into which the petiole fits when the mesosoma and gaster are aligned. In this position the anterior face of abdominal segment 3 overhangs at least the posterior peduncle of the petiole.

Abdominal segment 3 basally with complete tergosternal fusion for some distance on each side of the helcium; the line of fusion follows the edges of the impression in the anterior face of A3, and the unfused portions of the tergite and sternite commence some distance above the helcium, at the dorsolateral apices of the impression.

Abdominal sternite 3 without a transverse sulcus across the sclerite immediately behind the helcium sternite (i.e., presternite and poststernite of A3 not separated by a sulcus).

Acidopore with a seta-fringed nozzle.

AFROTROPICAL TAPINOLEPIS

bothae (Forel, 1907)	*melanaria ochraceotincta* (Arnold,
candida (Santschi, 1928)	1949)
deceptor (Arnold, 1922)	*pernix* (Viehmeyer, 1923)
decolor (Emery, 1895)	*trimenii* (Forel, 1895)
litoralis (Arnold, 1958)	*trimenii karrooensis* (Arnold, 1922)
macgregori (Arnold, 1922)	*trimenii angolensis* (Santschi, 1930)
macrophthalma (Arnold, 1962)	*tumidula* (Emery, 1915)
melanaria (Arnold, 1922)	

TAPINOMA Foerster, 1850   Dolichoderinae

14 species (+ 6 subspecies), Afrotropical; 5 species, Malagasy   PLATE 33

DIAGNOSTIC REMARKS: Among the dolichoderines, the petiole (A2) of *Tapinoma* is extremely reduced, merely a low, slender segment, without a node or scale. It is overhung from behind by the first gastral segment (A3), the anterior surface of which bears a groove that accommodates and conceals the petiole when the 2 segments are aligned. This

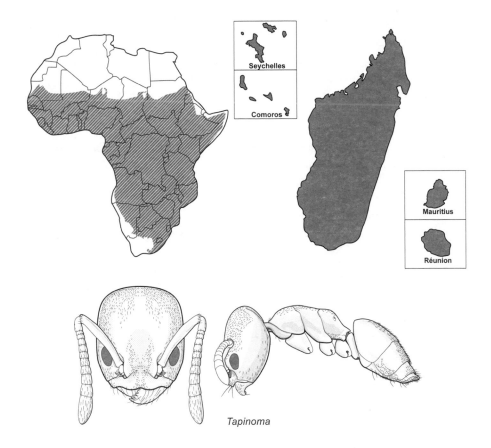

Tapinoma

development is also shared by *Technomyrmex*, but in this genus the pygidium is visible in dorsal view, whereas in *Tapinoma* it is reflexed onto the ventral surface. Because of this, only 4 gastral tergites (A3–6) are visible in dorsal view in *Tapinoma*, whereas 5 (A3–7) can be seen in *Technomyrmex*.

DISTRIBUTION AND ECOLOGY: *Tapinoma* has an almost worldwide distribution, with endemic species in all zoogeographical regions. Many *Tapinoma* species are terrestrial, nesting directly in the ground or under stones or rotten wood, or in compressed leaf litter, but some species are arboreal. The very common *T. melanocephalum* is mainly a pantropical tramp species but can survive in temperate climates in heated buildings and greenhouses. In West Africa it is often encountered indoors.

TAXONOMIC COMMENT: It is possible that a few Afrotropical species may correctly belong in the otherwise Malagasy genus *Ravavy*, as noted there, but the taxonomy of these species has not yet been investigated.

IDENTIFICATION OF SPECIES: No review exists.

DEFINITION OF THE GENUS (WORKER): Characters of DOLICHODERINAE, plus:

Masticatory margin of mandible with 3–6 teeth followed by about 7 denticles; apical tooth subequal to, or at most only slightly larger than, the preapical tooth. Basal margin denticulate, at least in part.

Palp formula usually 6,4; reduced to 6,3 in one Madagascan species.

Anterior clypeal margin transverse to notched medially.

Antenna with 11–12 segments.

Metathoracic spiracles dorsolateral.

Metanotal groove present, may be reduced to a mere line across the dorsum.

Propodeum unarmed, or with a raised rim around the sides and dorsum of the declivity, or with a small, median, dentiform process posterodorsally (may be a projection of the rim).

Propodeal spiracle usually close to the margin of the declivity; rarely otherwise.

Tibial spurs: mesotibia 1; metatibia 1.

Abdominal segment 2 (petiole) extremely reduced, merely a low, narrow segment without a dorsal node or scale.

Abdominal segment 3 (first gastral) projects forward over the reduced petiole and entirely conceals the petiole in dorsal view when the gaster and mesosoma are aligned.

Abdominal segment 3, on ventral surface of projecting portion, with a longitudinal impression or groove that accommodates the entire petiole when the mesosoma and gaster are aligned.

Abdominal tergites 3–6 visible in dorsal view (i.e., 4 gastral tergites visible in dorsal view); tergite of A7 (pygidium) is reflexed onto the ventral surface of the gaster, where it abuts abdominal sternite 7 (hypopygium).

Abdominal sternite 6 (fourth gastral sternite) not keel-shaped posteriorly.

AFROTROPICAL TAPINOMA

acuminatum Forel, 1907	*luridum connexum* Santschi, 1914
albinase (Forel, 1910)	*luridum longiceps* Wheeler, W.M.,
arnoldi Forel, 1913	1922
arnoldi tectum Santschi, 1917	*luridum sokolovi* Karavaiev, 1931
carininotum Weber, 1943	*luteum* (Emery, 1895)
chiaromontei Menozzi, 1930	*luteum natalicum* Özdikmen, 2010
danitschi Forel, 1915	= *luteum emeryi* (Forel, 1910)
danitschi bevisi Forel, 1915	(homonym)
demissum Bolton, 1995	*melanocephalum* (Fabricius, 1793)
= *gracile* Forel, 1913 (homonym)	*minimum* Mayr, 1895
lugubre Santschi, 1917	*modestum* Santschi, 1932
luridum Emery, 1908	*schultzei* (Forel, 1910)

MALAGASY TAPINOMA

aberrans (Santschi, 1911)

fragile (Smith, F., 1876)

melanocephalum (Fabricius, 1793)

pomonae Donisthorpe, 1947

subtile Santschi, 1911

DIAGNOSTIC REMARKS: Among the dolichoderines, the petiole (A2) of *Technomyrmex* is extremely reduced, merely a low, slender segment, without a node or scale. It is overhung from behind by the first gastral segment (A3), the anterior surface of which bears a groove that accommodates and conceals the petiole when the 2 segments are aligned. This development is also shared by *Tapinoma*, but in this genus the pygidium is not visible in dorsal view because it is reflexed onto the ventral surface. In consequence, 5 gastral tergites (those of A3–7) can be seen in *Technomyrmex*, whereas only 4 gastral tergites (those of A3–6) are visible in dorsal view in *Tapinoma*.

DISTRIBUTION AND ECOLOGY: *Technomyrmex* has endemic species throughout the Old World tropics, and the genus also contains some of the subfamily's most accomplished tramp species. Oddly, there are also a couple of isolated endemic species in the Neotropical region. Some species occur almost entirely in the forest leaf-litter layer, where they may be locally abundant. Many species nest in the ground, either directly or under stones, or in and under rotten wood on the ground, or in tree stumps. These species forage on the ground and in the leaf litter, but many also ascend trees in search of food. Some species are entirely arboreal, with nests under bark, in twigs, or in rotten parts of standing trees. One species builds nests of carton upon foliage (*T. anterops*), and one species appears to have an obligatory relationship with myrmecophytes (*T. laurenti*). The tramp species are particularly versatile, able to establish colonies in almost any suitable location, and may invade houses, where they sometimes nest in such bizarre locations as electricity transformers or light fixtures. Castes in the *T. albipes* species-group are complex in both sexes. Among females as well as

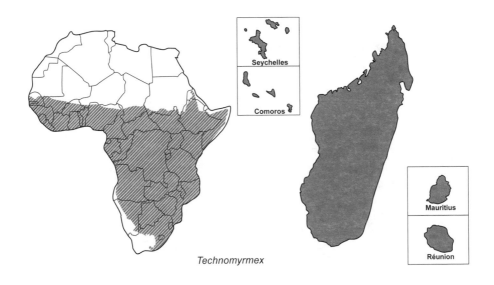

Technomyrmex

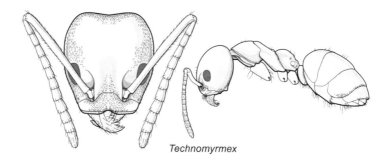

Technomyrmex

normal workers and alate queens, there is usually also a series of worker-queen intercastes; in males both alate and ergatoid forms are produced.

IDENTIFICATION OF SPECIES: Bolton, 2007 (Afrotropical + Malagasy).

DEFINITION OF THE GENUS (WORKER): Characters of subfamily DOLICHODERINAE, plus:

Masticatory margin of the mandible with a combination of teeth and denticles (about 9–25 in total).

Palp formula usually 6,4; reduced to 5,3 in one species (*T. lasiops*), to 4,3 in one species (*T. lujae*).

Anterior clypeal margin transverse to very deeply notched or impressed.

Antenna with 12 segments.

Metathoracic spiracles dorsolateral to dorsal.

Metanotal groove present.

Propodeum unarmed.

Propodeal spiracle usually at about midheight and close to the margin of the declivity; rarely otherwise.

Tibial spurs: mesotibia 1; metatibia 1, pectinate.

Abdominal segment 2 (petiole) extremely reduced, merely a low, narrow segment without a dorsal node or scale.

Abdominal segment 3 (first gastral) projects forward over the reduced petiole and entirely conceals the petiole in dorsal view when the gaster and mesosoma are aligned.

Abdominal segment 3, on ventral surface of projecting portion, with a longitudinal impression or groove that accommodates the entire petiole when the mesosoma and gaster are aligned.

Abdominal tergites 3–7 aligned and visible in dorsal view (i.e., 5 gastral tergites visible in dorsal view); tergite of A7 (pygidium) small but not reflexed onto the ventral surface of the gaster.

Abdominal sternite 6 (fourth gastral sternite) not keel-shaped posteriorly.

AFROTROPICAL TECHNOMYRMEX

albipes (Smith, F., 1861)	= *andrei schereri* Forel, 1911
andrei Emery, 1899	= *wolfi* (Forel, 1916)

= *allecta* (Stitz, 1916)

= *zumpti* Santschi, 1937

arnoldinus Forel, 1913

camerunensis Emery, 1899

hostilis Bolton, 2007

ilgi (Forel, 1910)

= *stygium* (Santschi, 1911)

= *gowdeyi* (Wheeler, W.M., 1922)

lasiops Bolton, 2007

laurenti (Emery, 1899)

= *kohli* (Forel, 1916)

= *laurenti congolensis* (Forel, 1916)

lujae (Forel, 1905)

= *lujae wasmanni* (Forel, 1916)

= *griseopubens* (Wheeler, W.M., 1922)

= *lujae pulliceps* (Santschi, 1926)

menozzii (Donisthorpe, 1936)

metandrei Bolton, 2007

moerens Santschi, 1913

= *albipes congolensis* Karavaiev, 1926

= *moerens nigricans* Santschi, 1930

= *longiscapus* Weber, 1943

= *nequitus* Bolton, 1995

= *incisus* Weber, 1943 (homonym)

nigriventris Santschi, in Forel, 1910

pallipes (Smith, F., 1876)

= *atrichosus* Viehmeyer, 1922

= *brevicornis* Santschi, 1930

= *foreli affinis* Santschi, 1930

= *albipes truncicolus* Weber, 1943

parandrei Bolton, 2007

parviflavus Bolton, 2007

pilipes Emery, 1899

rusticus Santschi, 1930

schoedli Bolton, 2007

schoutedeni Forel, 1910

= *zimmeri okiavoensis* (Forel, 1916)

semiruber Emery, 1899

senex Bolton, 2007

sycorax Bolton, 2007

taylori (Santschi, 1930)

vapidus Bolton, 2007

voeltzkowi (Forel, 1907)

= *voeltzkowi rhodesiae* (Forel, 1913)

zimmeri (Forel, 1911)

MALAGASY TECHNOMYRMEX

albipes (Smith, F., 1861)

anterops Fisher and Bolton, in Bolton, 2007

curiosus Fisher and Bolton, in Bolton, 2007

difficilis Forel, 1892

= *mayri nitidulans* Santschi, 1930

docens Fisher and Bolton, in Bolton, 2007

fisheri Bolton, 2007

innocens Fisher and Bolton, in Bolton, 2007

madecassus Forel, 1897

= *madecassus fusciventris* Forel, 1907

mayri Forel, 1891

pallipes (Smith, F., 1876)

= *albipes foreli* Emery, 1893

= *primroseae* Donisthorpe, 1949

vitiensis Mann, 1921

voeltzkowi (Forel, 1907)

TEMNOTHORAX Mayr, 1861   Myrmicinae

6 species, Afrotropical   PLATE 34

DIAGNOSTIC REMARKS: Among the myrmicines, *Temnothorax* has 12-segmented antennae that terminate in an apical club of 3 segments. The mandible has only 5 teeth. The clypeus posteriorly is broadly inserted between the frontal lobes, and its anterior margin never has an unpaired median seta. Eyes are present at about the midlength of the head capsule. The first gastral tergite (A4) broadly overlaps the sternite on the ventral surface of the gas-

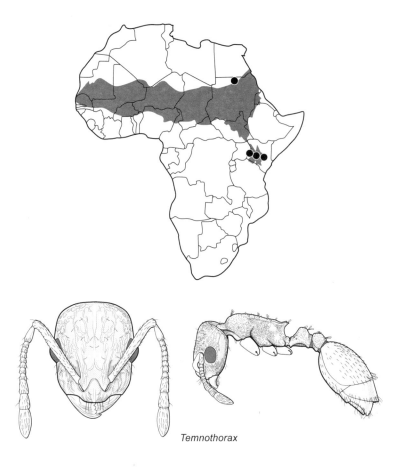

Temnothorax

ter. In habitus, species of *Temnothorax* are only likely to be confused with those of *Nesomyrmex*, but the structure of the clypeus is different. In *Temnothorax* the anterior margin of the median portion of the clypeus is convex in full-face view, but it does not form an apron that fits very tightly against the dorsal surfaces of the mandibles; when seen in profile, the anterior clypeal margin is elevated away from the dorsal surface of the mandible.

DISTRIBUTION AND ECOLOGY: A large, predominantly Holarctic genus, *Temnothorax* is absent from the Malagasy region and has only 6 poorly known species in the Afrotropical region, 1 from Sudan and 5 from Kenya. Workers have been found on the ground and in leaf litter.

IDENTIFICATION OF SPECIES: Bolton, 1982 (Afrotropical, as part of *Leptothorax*); Prebus, 2015 (Afrotropical).

DEFINITION OF THE GENUS (WORKER): Characters of subfamily MYRMICINAE, plus:

Mandible short triangular and stout; masticatory margin with 5 teeth.
Palp formula 5,3.
Stipes of the maxilla without a transverse crest.

Clypeus posteriorly broadly inserted between the frontal lobes. Anterior margin of the median portion of the clypeus does not form a prominent lobe that fits tightly over the dorsa of the closed mandibles. In profile the anterior clypeal margin is elevated slightly away from the dorsal surface of the mandible.

Clypeus without an unpaired seta at the midpoint of the anterior margin.

Frontal carinae consist only of the frontal lobes or are very feebly extended posteriorly.

Antennal scrobes absent.

Antenna with 12 segments, with an apical club of 3 segments.

Torulus concealed by frontal lobe in full-face view.

Eyes well developed, located at the midlength of the head capsule.

Promesonotal suture vestigial to absent on dorsum.

Metasternal process absent.

Propodeum bidentate.

Propodeal spiracle approximately at midlength of the sclerite.

Tibial spurs: mesotibia 0; metatibia 0.

Abdominal segment 2 (petiole) without tubercles, teeth, or spines.

Abdominal tergite 4 (first gastral) broadly overlaps the sternite on the ventral gaster; gastral shoulders present.

Stridulitrum present on the pretergite of A4.

Sting simple.

Main pilosity of dorsal head and body: short and stubbly, usually truncated apically.

AFROTROPICAL TEMNOTHORAX

brevidentis Prebus, 2015

cenatus (Bolton, 1982)

megalops (Hamann and Klemm, 1967)

mpala Prebus, 2015

rufus Prebus, 2015

solidinodus Prebus, 2015

TERATANER Emery, 1912   Myrmicinae

6 species, Afrotropical; 6 species, Malagasy   PLATE 34

DIAGNOSTIC REMARKS: Among the myrmicines, *Terataner* has 12-segmented antennae that terminate in a club of 3 segments. The posterior corners of head, in full-face view, are angulate to sharply denticulate, and frontal carinae are present. The pronotum is usually marginate both anteriorly and laterally, and the humeri are angulate to dentate. The ventral margin of the side of the metapleuron has a broad groove that runs forward to the mesopleuron. The propodeum is unarmed to bidentate but never has long sharp spines. The petiole (A2) node dorsally has either a transverse crest, or is indented medially, or has a pair of teeth or spines. In contrast to the equally arboreal *Atopomyrmex*, all *Terataner* species are monomorphic.

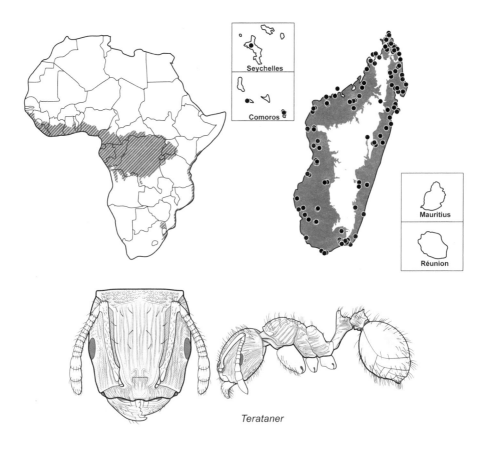

Terataner

DISTRIBUTION AND ECOLOGY: *Terataner* species are present in both Africa and Madagascar but do not occur anywhere else. All species are arboreal, with nests constructed in twigs, hollow stems, or rotten parts of standing trees, often a considerable distance above the ground. Madagascar has many undescribed species.

IDENTIFICATION OF SPECIES: Bolton, 1981b (Afrotropical + Malagasy).

DEFINITION OF THE GENUS (WORKER): Characters of subfamily MYRMICINAE, plus:

Mandible short, triangular and stout; masticatory margin with 5–6 teeth.

Palp formula 5,3, or 4,3.

Stipes of the maxilla with a transverse crest.

Clypeus large, median portion shield-like, posteriorly broadly inserted between the frontal lobes. Anterior clypeal margin fits tightly against the bases of the closed mandibles, without a prominent lobe that overlaps the mandibles.

Clypeus without an unpaired seta at the midpoint of the anterior margin.

Frontal carinae extend back from the narrow frontal lobes, varying in expression from weak to distinct; carinae are frequently roughly parallel and extend back well beyond the level of the posterior margins of the eyes, fading out posteriorly or apically angled toward the sides of the head.

Antennal scrobes absent or broadly but very shallowly present.

Antenna with 12 segments, with an apical club of 3 segments.

Torulus concealed by the frontal lobe in full-face view.

Eyes well developed, located at or in front of the midlength of the head capsule.

Head capsule with posterior corners angulate, tuberculate, or denticulate in full-face view.

Pronotum marginate laterally; pronotal humeri angulate to dentate in dorsal view.

Promesonotal suture usually vestigial to absent on the dorsum; metanotal groove present.

Metapleuron in profile with a broad, strong groove that extends forward from the orifice of the metapleural gland.

Metathorax ventrally without a deep pit between the metacoxae; metasternal process absent.

Propodeum unarmed, bituberculate, or bidentate.

Propodeal spiracle in front of the margin of the declivity.

Tibial spurs: mesotibia 0 or 1; metatibia 0 or 1.

Metafemur markedly incrassate medially.

Abdominal segment 2 (petiole) usually appears cuneate, conical, or spiniform in profile; the dorsum of the node is narrow and forms either a transverse crest, or is indented medially, or is bidentate or bispinose.

Abdominal tergite 4 (first gastral) does not broadly overlap the sternite on the ventral gaster; gastral shoulders absent.

Stridulitrum present on the pretergite of A4.

Sting simple.

Main pilosity of dorsal head and body: simple or short and stout; may be very sparse; rarely absent.

AFROTROPICAL TERATANER

bottegoi (Emery, 1896)

elegans Bernard, 1953

luteus (Emery, 1899)

piceus Menozzi, 1942

transvaalensis Arnold, 1952

velatus Bolton, 1981

MALAGASY TERATANER

alluaudi (Emery, 1895)

foreli (Emery, 1899)

rufipes Emery, 1912

scotti (Forel, 1912)

steinheili (Forel, 1895)

xaltus Bolton, 1981

DIAGNOSTIC REMARKS: Among the myrmicines, *Tetramorium* has 10-, 11-, or 12-segmented antennae that terminate in a distinct club of 3 segments, and eyes are always present. The clypeus is broadly inserted between widely separated frontal lobes, and the head, or head plus clypeus, usually has a longitudinal median carina. The lateral portions of the clypeus are usually raised into a shielding wall or sharp ridge in front of the antennal sockets (but this is secondarily lost in a few species). The pronotum and anterior mesonotum never form a raised dome in profile view. The first gastral tergite (the tergite of A4) does not broadly overlap the sternite on the ventral gaster, and the sting has a translucent, spatulate to pennant-shaped lamellate appendage that projects from the dorsum of the shaft near or at its apex (lost in only 1 species).

DISTRIBUTION AND ECOLOGY: The distribution of *Tetramorium* is worldwide, but only a few endemic species are present in the New World. It is easily the most morphologically diverse and species-rich genus in the Afrotropical and Malagasy regions. Its species can be found in all localities, from harsh desert to rainforest and swampland, and from deep in the soil to the tops of trees. Almost any imaginable nest site can be used by its various species: in the ground, either directly or under objects; in termitaries; in the leaf litter, either directly or in twigs, stems, and small pieces of rotting wood; among the roots of plants; in tree stumps and rotten logs, either under the bark or deep in the tissues; in rotten holes anywhere in standing trees, in hollow twigs, or even in curled leaves. A few species apparently always nest close to colonies of large ponerine ants, and some arboreal species construct nests of silk and vegetable fragments that are attached to the ventral surfaces of leaves, or in the axils of larger leaves, in the canopy. Many species are carnivorous. A few are obligate termitophages, but most other arthropod groups feature as prey, including other ants or their brood. All members of the *T. solidum* group, and a scattering of other species, are granivorous, but some species are strongly attracted to the honeydew produced by homopterous insects.

TAXONOMIC COMMENT: *Tetramorium* has a large number of genus-group names as junior synonyms. Some of the references that follow refer to various groups of *Tetramorium* as they were understood at the time of publication. A full list of current synonyms is given in the introduction.

IDENTIFICATION OF SPECIES: Brown, 1964 (*Rhoptromyrmex*); Bolton, 1976 (Afrotropical *Decamorium, Rhoptromyrmex, Triglyphothrix*); Bolton, 1979 (Malagasy *Tetramorium*, in old restricted sense); Bolton, 1980 (Afrotropical *Tetramorium*, in old restricted sense); Bolton, 1986 (*Rhoptromyrmex*), Hita Garcia, Fischer, and Peters, 2010 (Afrotropical + Malagasy *T. weitzeckeri* group); Hita Garcia and Fisher, 2011 (Malagasy *T. bicarinatum, T. obesum, T. sericeiventre, T. tosii* groups); Hita Garcia and Fisher, 2012a (Malagasy *T. bessonii, T. bonibony, T. dysalum, T. marginatum, T. tsingy, T. weitzeckeri* groups); Hita Garcia and Fisher, 2012b (Malagasy *T. kelleri, T. tortuosum* groups); Hita Garcia and Fisher, 2013 (Afrotropical

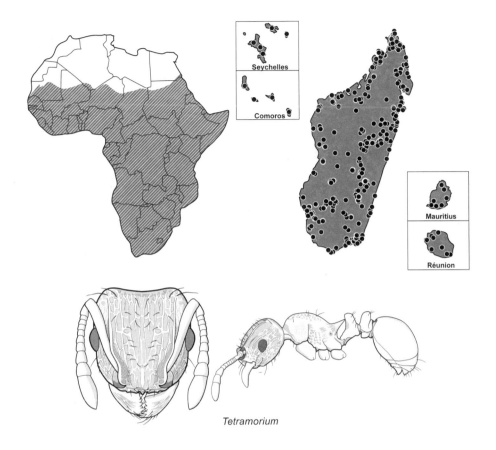

Tetramorium

T. capillosum complex); Hita Garcia and Fisher, 2014a (*T. decem* group); Hita Garcia and Fisher, 2014c (Malagasy *T. naganum, T. plesiarum, T. schaufussii, T. severini* groups); Hita Garcia and Fischer, 2014 (*T. weitzeckeri* complex); Hita Garcia and Fisher, 2015 (key to Malagasy species-groups).

DEFINITION OF THE GENUS (WORKER): Characters of subfamily MYRMICINAE, plus:

Mandible triangular; masticatory margin with 6–11 teeth (only very rarely 6, usually 7 or more), arranged with 2–3 larger teeth apically, followed by at least 4 smaller teeth or denticles; length of the masticatory margin equal to or greater than the length of the basal margin.

Palp formula 4,3, or 4,2, or 3,3, or 3,2, or 2,2.

Stipes of the maxilla usually with a distinct transverse crest (weak in a few minute species).

Clypeus posteriorly broadly inserted between the widely separated frontal lobes; median portion of the clypeus broad, flat to convex but not elevated and not bicarinate; usually a median longitudinal carina present.

Clypeus with lateral portions usually raised into a shielding wall or sharp ridge in front of the antennal sockets (secondarily lost in 7–8 species).

Clypeus without an unpaired seta at the midpoint of the anterior margin.

Frontal carinae may be restricted to frontal lobes or may extend to the posterior margin of the head.

Antennal scrobes usually present, extending above the eyes; scrobes reduced or lost in some species and some species-groups.

Antenna with 10–12 segments, with an apical club of 3 segments.

Antennal socket and torulus within, and surrounded by, a depressed antennal fossa; the anterior margin of the fossa is formed by the posterior surface of the narrow, raised, lateral portion of the clypeus.

Torulus with the upper lobe usually concealed by the frontal lobe in full-face view (some exceptions).

Eyes present, minute to large.

Head capsule usually with a median, longitudinal carina, uncommonly absent; carina frequently extends anteriorly onto the clypeus.

Promesonotal suture absent from the dorsum of the mesosoma; pronotum not domed, mesonotum not elongate.

Propodeum usually bispinose or bidentate, only very rarely unarmed; propodeal lobes conspicuous.

Propodeal spiracle usually low on side and behind the midlength of the sclerite, uncommonly otherwise.

Metasternal process present, usually conspicuous; posterior to the process is a pair of carinae that extend to the propodeal foramen or the foramen itself that extends anteriorly almost to the metasternal pit.

Tibial spurs: mesotibia 0 or 1; metatibia 0 or 1.

Abdominal segment 4 (first gastral) tergite does not broadly overlap the sternite on the ventral gaster; gastral shoulders absent.

Sting with a spatulate to pennant-shaped lamellate appendage that projects from the dorsum of the shaft near or at its apex (absent only in *T. africanum*).

Main pilosity of dorsal head and body: usually simple but may be bizarre (squamate, spatulate, bifid, trifid, quadrifid, flabellate); may be short and stubbly; sometimes absent from part or all of body.

AFROTROPICAL TETRAMORIUM

aculeatum (Mayr, 1866)
  = *wasmanni* (Forel, 1901)
  = *aculeatum andricum* Emery, 1908
  = *aculeatum major* Forel, 1915
  = *aculeatum rubroflava* Forel, 1916
  = *aculeatus melanogyne* (Santschi, 1923)
  = *aculeatus pulchellus* (Santschi, 1924)
  = *aculeatus militaris* (Santschi, 1924)
  = *zumpti* (Santschi, 1937)
  = *viridis* (Weber, 1943)
  = *aculeatus inermis* (Bernard, 1953)

africanum (Mayr, 1866)
  = *tessmanni* (Forel, 1910)
  = *lamottei* Bernard, 1953
agile Arnold, 1960
agnum (Santschi, 1935)
akengense (Wheeler, W.M., 1922)
akermani Arnold, 1926
  = *akermani myersi* Arnold, 1958
altivagans Santschi, 1914
  = *simillimum isis* Weber, 1943
amatongae Bolton, 1980

amaurum Bolton, 1980

amentete Bolton, 1980

amissum Bolton, 1980

angulinode Santschi, 1910

  = *angulinode daphnis* (Santschi, 1920)

  = *papyri* (Weber, 1943)

  = *humerosum* Bernard, 1953

  = *nullispinum* Bolton, 1980

antrema (Bolton, 1976)

anxium Santschi, 1914

argenteopilosum Arnold, 1926

arnoldi (Forel, 1913)

asetyum Bolton, 1980

ataxium Bolton, 1980

avium Bolton, 1980

barbigerum Bolton, 1980

baufra (Bolton, 1976)

bellicosum Bolton, 1980

bendai Hita Garcia, Fischer, and Peters,
  2010

bequaerti Forel, 1913

  = *humile* Santschi, 1914

berbiculum Bolton, 1980

bevisi Arnold, 1958

bicarinatum (Nylander, 1846)

boehmei Hita Garcia and Fisher, 2010

boltoni Hita Garcia, Fischer, and Peters,
  2010

bothae Forel, 1910

  = *guillarmodi* Arnold, 1960

brevispinosum (Stitz, 1910)

  = *nion* (Bernard, 1953)

browni Bolton, 1980

bulawayense Forel, 1913

  = *bequaerti bruni* Santschi, 1917

buthrum Bolton, 1980

caldarium (Roger, 1857)

  = *pauper transformans* Santschi, 1914

  = *pusillum hemisi* Wheeler, W.M., 1922

calinum Bolton, 1980

camerunense Mayr, 1895

candidum Bolton, 1980

capense Mayr, 1865

  = *braunsi* Forel, 1913

  = *popovici* Forel, 1914

capillosum Bolton, 1980

caritum (Bolton, 1986)

chloe (Santschi, 1920)

clunum Forel, 1913

coloreum Mayr, 1901

  = *humerosum muscicola* Bernard, 1953

constanciae Arnold, 1917

  = *longispinosa* (Arnold, 1956)

convexum Bolton, 1980

cristatum Stitz, 1910

  = *guineense medje* Wheeler, W.M., 1922

critchleyi (Bolton, 1976)

crypticum (Bolton, 1976)

decem Forel, 1913

dedefra (Bolton, 1976)

delagoense Forel, 1894

desertorum (Forel, 1910)

dichroum Santschi, 1932

diomandei Bolton, 1980

distinctum (Bolton, 1976)

do Forel, 1914

dogieli Karavaiev, 1931

dolichosum Bolton, 1980

dominum Bolton, 1980

doriae Emery, 1881

dumezi Menozzi, 1942

  = *dumezi* Bolton, 1980

dysderke Bolton, 1980

edouardi Forel, 1894

  = *kivuense* Stitz, 1911

  = *tersum* Santschi, 1911

  = *kivuense atrinodis* (Santschi, 1928)

elidisum Bolton, 1980

emeryi Mayr, 1901

  = *emeryi cristulatum* Forel, 1913

eminii (Forel, 1894)

  = *marthae* (Forel, 1911)

  = *marleyi* (Forel, 1914)

  = *marleyi akermani* Arnold, 1926

  = *guillodi* (Santschi, 1937)

  = *guillodi mus* (Santschi, 1937)

  = *cinereus* (Weber, 1943)

erectum Emery, 1895

= *bacchus* Forel, 1910

ericae Arnold, 1917

  = *pauper* (Santschi, 1917) (homonym)

flabellum Bolton, 1980

flaviceps Arnold, 1960

  = *do mus* Arnold, 1960

flavithorax (Santschi, 1914)

frenchi Forel, 1914

frigidum Arnold, 1926

  = *akermani drakensbergensis* Arnold, 1926

furtivum (Arnold, 1956)

gabonense (André, 1892)

  = *gabonensis soyauxi* (Forel, 1901)

  = *mucidus* (Forel, 1909)

  = *areolatus* (Stitz, 1910)

  = *gabonensis boulognei* (Forel, 1916)

  = *areolata burgeoni* (Santschi, 1935)

  = *gabonensis kamerunensis* (Santschi, 1937)

galoasanum Santschi, 1910

gazense Arnold, 1958

gegaimi Forel, 1916

geminatum Bolton, 1980

gestroi (Menozzi, 1933)

ghindanum Forel, 1910

glabratum Stitz, 1923

  = *rutilum* Prins, 1973

gladstonei Forel, 1913

globulinode (Mayr, 1901)

  = *globulinodis alberti* (Forel, 1916)

  = *globulinodis obscurus* (Santschi, 1932)

gracile Forel, 1894

grandinode Santschi, 1913

  = *grandinode hopensis* Forel, 1914

granulatum Bolton, 1980

grassii Emery, 1895

  = *grassii laevigatum* Mayr, 1901

  = *grassii simulans* Santschi, 1914

  = *joffrei* Forel, 1914

  = *joffrei algoa* Arnold, 1917

  = *grassii mayri* Emery, 1924

guineense (Bernard, 1953)

hapale Bolton, 1980

hecate Hita Garcia and Fisher, 2013

hortorum Arnold, 1958

humbloti Forel, 1891

  = *humbloti pembensis* Forel, 1907

  = *humblotii victoriensis* Forel, 1913

ictidum Bolton, 1980

imbelle (Emery, 1915)

incruentatum Arnold, 1926

  = *arnoldi* (Santschi, 1916) (homonym)

inezulae (Forel, 1914)

  = *hepburni* Arnold, 1917

  = *hepburni mashonana* (Arnold, 1949)

intermedium Hita Garcia, Fischer, and Peters, 2010

intextum Santschi, 1914

  = *intextum cataractae* Santschi, 1916

intonsum Bolton, 1980

invictum Bolton, 1980

isipingense Forel, 1914

jauresi Forel, 1914

  = *latens* Arnold, 1948

jejunum Arnold, 1926

jordani Santschi, 1937

  = *aspinatum* Prins, 1973

jugatum Bolton, 1980

kakamega Hita Garcia, Fischer, and Peters, 2010

kestrum Bolton, 1980

khyarum Bolton, 1980

krynitum Bolton, 1980

laevithorax Emery, 1895

  = *jeanae* Weber, 1943

legone Bolton, 1980

lobulicorne Santschi, 1916

longicorne Forel, 1907

longoi Forel, 1915

lucayanum Wheeler, W.M., 1905

  = *camerunense waelbroeki* Forel, 1909

  = *rectinodis* Menozzi, 1942

luteipes Santschi, 1910

luteolum Arnold, 1926

magnificum Bolton, 1980

matopoense Arnold, 1926

menkaura (Bolton, 1976)

meressei Forel, 1916

metactum Bolton, 1980

microgyna Santschi, 1918

microps (Mayr, 1901)

   = *auropunctatus* (Forel, 1910)

   = *auropunctatus pallens* (Forel, 1910)

   = *auropunctatus fusciventris* (Forel, 1913)

   = *auropunctata rhodesiana* (Forel, 1913)

   = *auropunctatus bulawayensis* Arnold,
     1917

minimum (Bolton, 1976)

minusculum (Santschi, 1914)

   = *minusculus amen* (Weber, 1943)

miserabile Santschi, 1918

mkomazi Hita Garcia, Fischer, and
   Peters, 2010

monardi (Santschi, 1937)

mossamedense Forel, 1901

   = *caespitum schultzei* Forel, 1910

mpala Hita Garcia and Fischer, 2014

muralti Forel, 1910

   = *muralti trilineata* (Santschi, 1919)

muscorum Arnold, 1926

nautarum Santschi, 1918

nefassitense Forel, 1910

nigrum Forel, 1907

   = *brevis* Weber, 1943

nodiferum (Emery, 1901)

notiale Bolton, 1980

   = *guineense striatum* Arnold, 1917
     (homonym)

nube Weber, 1943

occidentale (Santschi, 1916)

   = *insularis* (Menozzi, 1924)

oculatum Forel, 1913

opacum (Emery, in Forel, 1909)

   = *opacus esta* (Forel, 1909)

   = *opacus laeviceps* (Santschi, 1916)

   = *opacus monodi* (Bernard, 1953)

osiris (Bolton, 1976)

parasiticum Bolton, 1980

pauper Forel, 1907

peringueyi Arnold, 1926

perlongum Santschi, 1923

petersi Forel, 1910

peutli Forel, 1916

phasias Forel, 1914

   = *guineense hertigi* Santschi, 1937

philippwagneri Hita Garcia, Fischer,
   and Peters, 2010

pialtum Bolton, 1980

pinnipilum Bolton, 1980

platynode Bolton, 1980

plumosum Bolton, 1980

pogonion Bolton, 1980

postpetiolatum Santschi, 1919

poweri Forel, 1914

praetextum Bolton, 1980

psymanum Bolton, 1980

pulcherrimum (Donisthorpe, 1945)

pullulum Santschi, 1924

   = *uelensis* (Santschi, 1935)

   = *fernandensis* Menozzi, 1942

pusillum Emery, 1895

   = *caespitum ladismithensis* Forel, 1913

   = *pusillum tablensis* Forel, 1914

pylacum Bolton, 1980

quadridentatum Stitz, 1910

   = *commodum* Santschi, 1924

qualarum Bolton, 1980

raptor Hita Garcia 2014

regulare Bolton, 1980

renae Hita Garcia, Fischer, and Peters,
   2010

repentinum Arnold, 1926

reptana (Bolton, 1976)

rhetidum Bolton, 1980

rimytyum Bolton, 1980

robertsoni Hita Garcia, Fischer, and
   Peters, 2010

rogatum Bolton, 1980

rothschildi (Forel, 1907)

rotundatum (Santschi, 1924)

rubrum Hita Garcia, Fischer, and
   Peters, 2010

rufescens Stitz, 1923

saginatum Bolton, 1980

schoutedeni Santschi, 1924

semireticulatum Arnold, 1917
   = *semireticulatum politum* Arnold, 1948
sepositum Santschi, 1918
sepultum Bolton, 1980
sericeiventre Emery, 1877
   = *quadrispinosum* Emery, 1886
   = *sericeiventre debile* Forel, 1894
   = *sericeiventre femoratum* Emery, 1895
   = *neuvillei* Forel, 1907
   = *blochmanni nigriventre* Stitz, 1910
   = *sericeiventre inversa* Santschi, 1910
   = *blochmanni continentis* Forel, 1910
   = *quadrispinosum elegans* Santschi, 1918
   = *quadrispinosum eudoxia* Santschi,
     1918
   = *sericeiventre bipartita* Santschi, 1918
   = *sericeiventre cinnamomeum*
     Santschi, 1918
   = *sericeiventre gamaii* Santschi, 1918
   = *sericeiventre hori* Santschi, 1918
   = *sericeiventre jasonis* Santschi, 1918
   = *sericeiventre munda* Santschi, 1918
   = *sericeiventre vascoi* Santschi, 1918
   = *blochmanni calvum* Stitz, 1923
   = *quadrispinosum beirae* Arnold, 1926
   = *quadrispinosum otaviensis* Arnold,
     1926
   = *sericeiventre repertum* Santschi, 1926
   = *sericeiventre vividum* Santschi, 1926
   = *quadrispinosum angolense* Santschi,
     1930
   = *hortensis* (Bernard, 1948)
sericeum Arnold, 1926
setigerum Mayr, 1901
   = *setigerum quaerens* Forel, 1914
   = *setigerum anteversa* Santschi, 1921
setuliferum Emery, 1895
   = *setuliferum cucalense* Santschi, 1911
   = *setuliferum triptolemus* Arnold, 1917
shilohense Forel, 1913
sigillum Bolton, 1980
signatum Emery, 1895
   = *solidum lugubre* Forel, 1910
   = *solidum grootensis* Forel, 1913

   = *solidum tuckeri* Arnold, 1926
simillimum (Smith, F., 1851)
   = *pygmaeum* Emery, 1877
   = *pusillum bantouana* Santschi, 1910
   = *pusillum exoleta* Santschi, 1914
simulator Arnold, 1917
sitefrum Bolton, 1980
snellingi Hita Garcia, Fischer, and
   Peters, 2010
solidum Emery, 1886
somniculosum Arnold, 1926
squaminode Santschi, 1911
subcoecum Forel, 1907
   = *subcoecum inscia* Forel, 1913
sudanense (Weber, 1943)
surrogatum Bolton, 1985
   = *silvestrii* (Emery, 1915) (homonym)
susannae Hita Garcia, Fischer, and
   Peters, 2010
tabarum Bolton, 1980
talpa (Bolton, 1976)
tanaense Hita Garcia, Fischer, and
   Peters, 2010
tenebrosum Arnold, 1926
termitobium Emery, 1908
thoth (Bolton, 1976)
titus Forel, 1910
traegaordhi Santschi, 1914
transversinode (Mayr, 1901)
   = *steini* (Forel, 1913)
   = *transversinodis pretoriae* (Arnold,
     1926)
trimeni (Emery, 1895)
trirugosum Hita Garcia, Fischer and
   Peters, 2010
tychadion Bolton, 1980
typhlops Bolton, 1980
ubangense Santschi, 1937
umtaliense Arnold, 1926
uelense Santschi, 1923
   = *decem nimba* (Bernard, 1953)
ultor Forel, 1913
unicum Bolton, 1980
venator Hita Garcia, 2014

versiculum Bolton, 1980

vexator Arnold, 1926

viticola Weber, 1943

wadje Bolton, 1980

warreni Arnold, 1926

weitzeckeri Emery, 1895

   = *escherichi* Forel, 1910

   = *ebeninum* Arnold, 1926

   = *weitzeckeri nigellus* (Santschi, 1932)

   = *weitzeckeri edithae* (Weber, 1943)

xuthum Bolton, 1980

yarthiellum (Bolton, 1976)

youngi Bolton, 1980

zambezium Santschi, 1939

zapyrum Bolton, 1980

zonacaciae Weber, 1943

MALAGASY TETRAMORIUM

adamsi Hita Garcia and Fisher, 2012

aherni Hita Garcia and Fisher, 2012

ala Hita Garcia and Fisher, 2012

alperti Hita Garcia and Fisher, 2014

ambanizana Hita Garcia and Fisher, 2012

ambatovy Hita Garcia and Fisher, 2012

andohahela Hita Garcia and Fisher, 2012

andrei Forel, 1892

   = *andrei robustior* Forel, 1892

ankarana Hita Garcia and Fisher, 2012

anodontion Bolton, 1979

artemis Hita Garcia and Fisher, 2012

aspis Hita Garcia and Fisher, 2014

avaratra Hita Garcia and Fisher, 2012

bessonii Forel, 1891

   = *bessonii orientale* Forel, 1895

bicarinatum (Nylander, 1846)

bonibony Hita Garcia and Fisher, 2012

bressleri Hita Garcia and Fisher, 2014

caldarium (Roger, 1857)

camelliae Hita Garcia and Fisher, 2014

cavernicola Hita Garcia and Fisher, 2015

cognatum Bolton, 1979

coillum Bolton, 1979

dalek Hita Garcia and Fisher, 2014

degener Santschi, 1911

delagoense Forel, 1894

   = *simillimum madecassum* Forel, 1895

dysalum Bolton, 1979

electrum Bolton, 1979

elf Hita Garcia and Fisher, 2012

enkidu Hita Garcia and Fisher, 2014

freya Hita Garcia and Fisher, 2014

gilgamesh Hita Garcia and Fisher, 2014

gladius Hita Garcia and Fisher, 2014

gollum Hita Garcia and Fisher, 2014

hector Hita Garcia and Fisher, 2012

hobbit Hita Garcia and Fisher, 2014

humbloti Forel, 1891

ibycterum Bolton, 1979

insolens (Smith, F., 1861)

isectum Bolton, 1979

isoelectrum Hita Garcia and Fisher, 2012

jedi Hita Garcia and Fisher, 2012

kali Hita Garcia and Fisher, 2012

karthala Hita Garcia and Fisher, 2014

kelleri Forel, 1887

lanuginosum Mayr, 1870

   = *striatidens felix* (Forel, 1912)

   = *mauricei* (Donisthorpe, 1946)

latreillei Forel, 1895

mackae Hita Garcia and Fisher, 2012

mahafaly Hita Garcia and Fisher, 2011

malagasy Hita Garcia and Fisher, 2012

mallenseana Hita Garcia and Fisher, 2012

marginatum Forel, 1895

marojejy Hita Garcia and Fisher, 2012

mars Hita Garcia and Fisher, 2014

merina Hita Garcia and Fisher, 2014

monticola Hita Garcia and Fisher, 2014

myrmidon Hita Garcia and Fisher, 2014

naganum Bolton, 1979

nassonowii Forel, 1892

nazgul Hita Garcia and Fisher, 2012

nify Hita Garcia and Fisher, 2012

noeli Hita Garcia and Fisher, 2012

norvigi Hita Garcia and Fisher, 2012

nosybe Hita Garcia and Fisher, 2012

obiwan Hita Garcia and Fisher, 2014

olana Hita Garcia and Fisher, 2012

orc Hita Garcia and Fisher, 2012

orientale Forel, 1895

pacificum Mayr, 1870

pleganon Bolton, 1979

plesiarum Bolton, 1979

popell Hita Garcia and Fisher, 2012

proximum Bolton, 1979

pseudogladius Hita Garcia and Fisher, 2014

quasirum Bolton, 1979

rala Hita Garcia and Fisher, 2014

ranarum Forel, 1895

robitika Hita Garcia and Fisher, 2012

rumo Hita Garcia and Fisher, 2014

ryanphelanae Hita Garcia and Fisher, 2012

sabatra Hita Garcia and Fisher, 2012

sada Hita Garcia and Fisher, 2012

sargina Hita Garcia and Fisher, 2012

schaufussii Forel, 1891

scutum Hita Garcia and Fisher, 2014

scytalum Bolton, 1979

sericeiventre Emery, 1877

   = blochmannii Forel, 1887

= blochmannii montanum Forel, 1891

severini (Emery, 1895)

shamshir Hita Garcia and Fisher, 2012

sikorae Forel, 1892

   = latior (Santschi, 1926)

silvicola Hita Garcia and Fisher, 2012

simillimum (Smith, F., 1851)

singletonae Hita Garcia and Fisher, 2012

smaug Hita Garcia and Fisher, 2012

steinheili Forel, 1892

tantillum Bolton, 1979

tenuinode Hita Garcia and Fisher, 2014

tosii Emery, 1899

trafo Hita Garcia and Fisher, 2012

tsingy Hita Garcia and Fisher, 2012

tyrion Hita Garcia and Fisher, 2012

xanthogaster Santschi, 1911

valky Hita Garcia and Fisher, 2012

voasary Hita Garcia and Fisher, 2012

vohitra Hita Garcia and Fisher, 2012

vony Hita Garcia and Fisher, 2012

wardi Hita Garcia and Fisher, 2012

yammer Hita Garcia and Fisher, 2012

zenatum Bolton, 1979

TETRAPONERA Smith, F., 1852   Pseudomyrmecinae

30 species (+ 14 subspecies), Afrotropical; 21 species (+ 5 subspecies), Malagasy   PLATE 35

DIAGNOSTIC REMARKS: In *Tetraponera* eyes are always present, as are slender frontal carinae. The promesonotal suture is conspicuous, fully articulated and flexible in fresh specimens, and the metapleural gland orifice is a simple hole. The waist consists of 2 segments. The metatibia has 2 spurs, and the pretarsal claw of the metatarsus has a tooth present in its inner curvature. A strong sting is always apparent. With disarticulation, the tergite of the helcium, in anterior view, overlaps the sternite on each side.

DISTRIBUTION AND ECOLOGY: Pseudomyrmecinae contains only 3 genera, 2 of which are restricted to the New World, primarily to the Neotropical region. The third, *Tetraponera*, is distributed throughout the Old World tropics, with numerous species in the Oriental, Malesian, and Austral regions, as well as in the 2 regions considered here. All species are arboreal; most nest in twigs or hollow branches, but some have very specialized associations with thorns, vines, or particular species of trees. The larvae of pseudomyrmecines are unique in their possession of a trophothylax, a pocket in the ventral surface of the thorax in which workers deposit pellets of pulped insect prey for the larvae to consume.

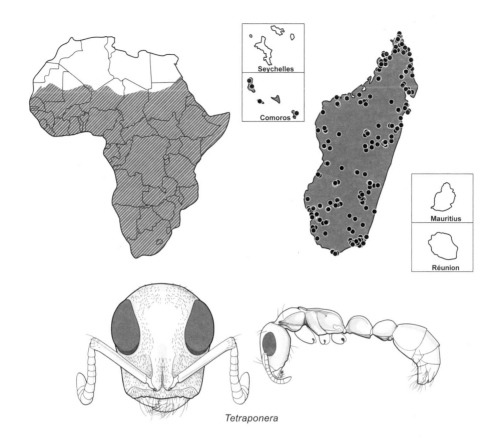

Tetraponera

IDENTIFICATION OF SPECIES: Wheeler, W.M., 1922 (*Pachysima*); Ward, 2006 (Afrotropical *T. ambigua* group); Ward, 2009 (Malagasy *T. grandidieri* group).

DEFINITION OF THE GENUS (WORKER): Characters of subfamily PSEUDOMYRMECINAE, plus:

Mandible with 3–6 teeth; sometimes also with 1 or 2 teeth on the basal border.

Palp formula usually 6,4, very rarely reduced to 4,3 or 3,3.

Metanotum varying from a distinctly differentiated dorsal sclerite to absent.

Metanotal groove usually present and impressed.

Tibial spurs: mesotibia 1 or 2; metatibia 2, the main spur pectinate.

Metabasitarsus with a longitudinal sulcus on its anterior face.

AFROTROPICAL TETRAPONERA

aethiops Smith, F., 1877
  = *spininoda* (André, 1892)
ambigua (Emery, 1895)
  = *bifoveolata* (Mayr, 1895)
  = *ambigua erythraea* (Emery, 1895)

= *bifoveolata maculifrons* (Santschi, 1912)
= *ambigua rhodesiana* (Forel, 1913)
= *bifoveolata syriaca* (Wheeler and Mann, 1916)
= *encephala* (Santschi, 1919)

= *ophthalmica angolensis* Santschi, 1930

= *ambigua occidentalis* Menozzi, 1934

andrei (Mayr, 1895)

angusta (Arnold, 1949)

anthracina (Santschi, 1910)

braunsi (Forel, 1913)

   braunsi durbanensis (Forel, 1914)

capensis (Smith, F., 1858)

claveaui (Santschi, 1913)

clypeata (Emery, 1886)

emacerata (Santschi, 1911)

   emacerata oberbecki (Forel, 1911)

   emacerata odiosa (Forel, 1916)

emeryi (Forel, 1911)

gerdae (Stitz, 1911)

latifrons (Emery, 1912)

ledouxi Terron, 1969

lemoulti (Santschi, 1920)

liengmei (Forel, 1894)

mayri (Forel, 1901)

mocquerysi (André, 1890)

   mocquerysi elongata (Stitz, 1911)

   mocquerysi lutea (Stitz, 1911)

   mocquerysi lepida Wheeler, W.M., 1922

mocquerysi biozellata (Karavaiev, 1931)

monardi (Santschi, 1937)

natalensis (Smith, F., 1858)

   natalensis obscurata (Emery, 1895)

   natalensis cuitensis (Forel, 1911)

   natalensis usambarensis (Forel, 1911)

   natalensis caffra (Santschi, 1914)

ophthalmica (Emery, 1912)

= *ophthalmica tenebrosa* Santschi, 1928

= *ophthalmica unidens* Santschi, 1928

= *nasuta* Bernard, 1953

parops Ward, 2006

penzigi (Mayr, 1907)

   penzigi continua (Forel, 1907)

   penzigi praestigiatrix Santschi, 1937

poultoni Donisthorpe, 1931

prelli (Forel, 1911)

schulthessi (Santschi, 1915)

scotti Donisthorpe, 1931

tessmanni (Stitz, 1910)

= *tessmanni castanea* (Wheeler, W.M., 1922)

triangularis (Stitz, 1910)

   triangularis illota (Santschi, 1914)

zavattarii (Menozzi, 1939)

MALAGASY TETRAPONERA

arrogans (Santschi, 1911)

demens (Santschi, 1911)

diana (Santschi, 1911)

exasciata (Forel, 1892)

fictrix (Forel, 1897)

flexuosa (Santschi, 1911)

grandidieri (Forel, 1891)

= *grandidieri hildebrandti* (Forel, 1891)

hespera Ward, 2009

hirsuta Ward, 2009

hysterica (Forel, 1892)

   hysterica dimidiata (Forel, 1895)

   hysterica inflata (Emery, 1899)

inermis Ward, 2009

manangotra Ward, 2009

mandibularis (Emery, 1895)

merita Ward, 2009

morondaviensis (Forel, 1891)

perlonga (Santschi, 1928)

phragmotica Ward, 2006

plicatidens (Santschi, 1926)

rakotonis (Forel, 1891)

sahlbergii (Forel, 1887)

   sahlbergii longula (Emery, 1895)

   sahlbergii spuria (Forel, 1897)

   sahlbergii deplanata (Forel, 1904)

variegata (Forel, 1895)

DIAGNOSTIC REMARKS: Among the myrmicines, *Trichomyrmex* has 12-segmented antennae that either lack an apical club, or terminate in a club of 3 or 4 segments. The median portion of the clypeus is short and does not project sharply forward anteromedially, and the anterior margin of the median clypeus is only shallowly convex to concave. The antennal fossae are surrounded by a series of fine, curved striolae or costulae. The propodeal dorsum has transverse sculpture, and the worker caste is usually polymorphic.

DISTRIBUTION AND ECOLOGY: The species present in the regions under consideration here fall into 2 species-groups. The first contains only *T. abyssinicus*, which belongs to a group that is otherwise mainly Oriental (*T. scabriceps* group). It ranges across the entire Sahelian zone, nests in the ground, and is granivorous. In the second group, 2 species (*T. destructor* and *T. mayri*) are tramps that are also extremely widespread; the remaining species are endemics. They occur in both forest and savanna zones, nesting in the ground, either directly or under stones, or in rotten wood, or sometimes in more specialized habitats such as termitaries. The species of this group are general predators and scavengers.

TAXONOMIC COMMENT: Until recently, members of this genus were included in *Monomorium*, but *Trichomyrmex* species have the combination of characters given above, which are not seen in *Monomorium*.

IDENTIFICATION OF SPECIES: Bolton, 1987 (Afrotropical, as *Monomorium scabriceps* and *M. destructor* groups); Heterick, 2006 (Malagasy, as *Monomorium destructor* group).

DEFINITION OF THE GENUS (WORKER): Characters of subfamily MYRMICINAE, plus:

Worker caste usually polymorphic; 1 species monomorphic.

Mandible short-triangular or with basal and outer margins subparallel; masticatory margin with 3–4 teeth, when 4 the basalmost tooth is reduced to an offset denticle or blunt angle.

Palp formula 2,2.

Stipes of the maxilla without a transverse crest.

Clypeus posteriorly narrowly inserted between the frontal lobes; width of the clypeus between the lobes broader than one of the lobes. Median portion of the clypeus longitudinally bicarinate to rounded. Median portion of the clypeus short, not sharply projecting forward anteromedially; margin of median portion shallowly convex to concave.

Clypeus usually with a distinct unpaired seta at the midpoint of the anterior margin; rarely the unpaired seta is masked by the development of a continuous row of stout setae along the anterior clypeal margin; in profile the unpaired median seta arises from the anteriormost point of the clypeal projection.

Frontal carinae restricted to the small frontal lobes.

Antennal scrobes absent.

Antennal fossae surrounded by fine, curved striolae or costulae.

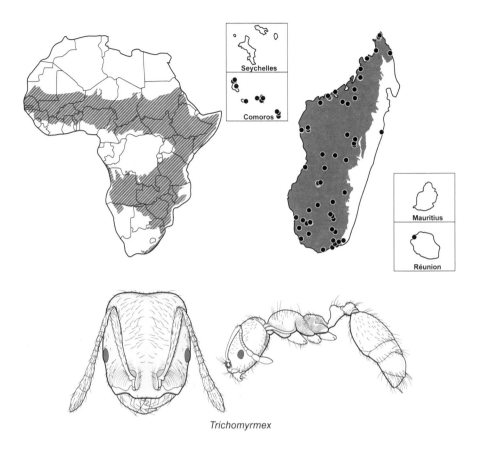

Trichomyrmex

Antenna with 12 segments; either gradually incrassate without a defined club or with a weakly to strongly demarcated club of 3–4 segments (usually 3).

Torulus with thet upper lobe visible in full-face view; maximum width of the torulus lobe is posterior to the point of maximum width of the frontal lobes.

Eyes present, relatively small, located in front of the midlength of the head capsule; eyes never reniform, never drawn out into a lobe or point anteriorly.

Promesonotal suture absent from the dorsum of the mesosoma, or a slightly impressed remnant may be discernible in largest workers. Metanotal groove present and impressed.

Propodeum unarmed, rounded to bluntly angular between the dorsum and declivity; the dorsum with transverse sculpture.

Propodeal spiracle approximately at the midlength of the sclerite; orifice of spiracle a vertical ellipse or short slit in *T. abyssinicus*, circular to subcircular in the remainder.

Metasternal process absent.

Tibial spurs: mesotibia 0 or 1; metatibia 0 or 1.

Petiole (A2) with the spiracle at or close to the midlength of the peduncle in *T. abyssinicus*, at or immediately in front of the anterior face of the node in others.

Abdominal segment 4 (first gastral) tergite overlaps the sternite on the ventral gaster; gastral shoulders present.

Sting present.

Main pilosity of dorsal head and body: simple, usually present and conspicuous on all surfaces but may be absent in places or rarely entirely absent.

AFROTROPICAL TRICHOMYRMEX

abyssinicus (Forel, 1894)

destructor (Jerdon, 1851)

   = *atomaria* (Gerstäcker, 1859)

   = *ominosa* (Gerstäcker, 1859)

emeryi (Mayr, 1895)

epinotalis Santschi, 1923

mayri (Forel, 1902)

   = *gracillimum karawajewi* (Wheeler, W.M., 1922)

oscaris (Forel, 1894)

   = *dispar* (Emery, 1895)

   = *destructor kalahariense* (Forel, 1910)

   = *solleri* (Forel, 1910)

   = *amblyops prossae* (Forel, 1916)

   = *amblyops bulawayense* (Forel, 1914) (homonym)

   = *destructor despecta* (Menozzi, 1931)

robustior (Forel, 1892)

MALAGASY TRICHOMYRMEX

destructor (Jerdon, 1851)

robustior (Forel, 1892)

VICINOPONE Bolton and Fisher, 2012   Dorylinae

1 species, Afrotropical   PLATE 35

DIAGNOSTIC REMARKS: Among the dorylines, *Vicinopone* has large eyes that are far forward on the head but has no ocelli. The antennae are 12-segmented, and the antennal sockets are fully exposed. The promesonotal suture is vestigial; the waist is of a single segment, which is elongate and barrel-shaped in profile, not marginate dorsolaterally. The propodeal spiracle is low on the side and situated behind the midlength of the sclerite, and propodeal lobes are present. The mesotibia lacks spurs, and the pretarsal claws have a preapical tooth.

DISTRIBUTION AND ECOLOGY: This monotypic genus is restricted to the Afrotropical region. It is entirely arboreal, with nests constructed in hollow twigs some distance above the ground. Although known from very few collections, they are widely scattered through forested Africa (Democratic Republic of Congo, Gabon, Ghana, and Tanzania), which implies that the genus is probably more common than it appears and is undercollected because of its purely arboreal life style. Queens are alate when virgin and nests are apparently polygynous, as 2 dealates were discovered in the nest of the type-series.

IDENTIFICATION OF SPECIES: Brown, 1975 (as part of *Simopone*); Bolton and Fisher, 2012.

DEFINITION OF THE GENUS (WORKER): Characters of subfamily DORYLINAE, plus:

Worker caste with considerable size variation.

Prementum not visible when mouthparts closed.

Palp formula 3,2. Maxillary palp very short: with mouthparts retracted the apex of

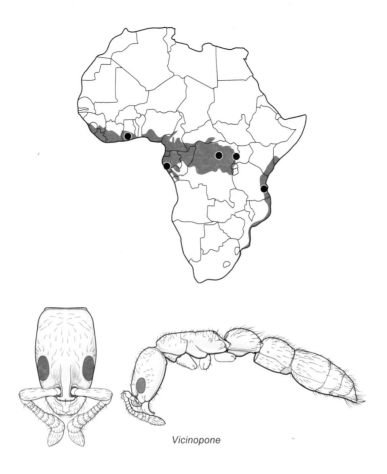

Vicinopone

the maxillary palp, when extended back on underside of head, does not reach the posterior margin of the buccal cavity.

Eyes present, large, in full-face view in front of the midlength of the head capsule; ocelli absent.

Frontal carinae present as a pair of raised ridges.

Antennal sockets fully exposed in full-face view.

Parafrontal ridges present but short and weak because of proximity of eye to front of head.

Antenna with 12 segments; apical antennomere large but subcylindrical, not swollen and bulbous.

Antennal scrobes absent.

Head capsule, in ventral or ventrolateral view, with a carina that extends down the posterolateral margin and onto the ventral surface, which it crosses to meet its opposite number at the ventral midline.

Promesonotal suture vestigial; sometimes a weak impression remains on the dorsum of the mesosoma.

Pronotal-mesopleural suture present on side of the mesosoma.

Metanotal groove vestigial; sometimes a weak impression remains.

Propodeal spiracle low on side and situated at or behind the midlength of the sclerite, not subtended by a longitudinal impression.

Propodeal lobes present.

Tibial spurs: mesotibia 0; metatibia 1, pectinate.

Metatibial gland absent.

Pretarsal claws of all legs with a single preapical tooth on the inner surface of each claw.

Waist of 1 segment (petiole = A2), not marginate laterally. Abdominal segment 3 large but subpostpetiolate, strongly separated from A4.

Petiole (A2) sessile.

Helcium projects from the anterior surface of A3 at about its midheight.

Prora of A3 merely a curved carina that separates the anterior face of the poststernite from the lateral and ventral surfaces.

Abdominal segment 4 with strongly developed presclerites.

Abdominal segments 5 and 6 without presclerites.

Pygidium (tergite of A7) large and distinct, flattened to concave dorsally, its apical margin evenly curved and equipped with a continuous row of minute denticles that are all approximately equal in size.

Sting present, large and fully functional.

AFROTROPICAL VICINOPONE

conciliatrix (Brown, 1975)

VITSIKA Bolton and Fisher, 2014   Myrmicinae

= *Myrmisaraka* Bolton and Fisher, 2014. syn. n. (see taxonomic comment below)   PLATE 35

16 species, Malagasy

DIAGNOSTIC REMARKS: Among the Malagasy myrmicines, *Vitsika* has 12-segmented antennae that usually terminate in an apical club of 3 segments, but uncommonly the club may be of 4 or 5 segments. The mandible has 6 or more teeth, and the lateral portions of the clypeus are not raised into a ridge or shield wall in front of the antennal sockets. Frontal carinae and antennal scrobes are usually distinct but in a couple of species are much reduced and the clypeus has a small notch at the midpoint of the anterior margin. The petiole node is high-domed to cuneate in profile, and the sting is simple and strongly developed.

DISTRIBUTION AND ECOLOGY: Restricted to Madagascar, the various species of this genus inhabit both tropical dry forest and rainforest, sometimes at great altitude. Nest sites are varied and include rotten logs, living or dead stems on plants, dead twigs on the ground, and rotten pockets on standing trees. One species appears to nest only in myrmecophytes. Foraging takes place in leaf litter and on vegetation. Queens may be alate or ergatoid; in some species both forms occur, sometimes with morphological intermediates also present.

TAXONOMIC COMMENT: *Vitsika* and *Myrmisaraka* were described in the same paper (Bolton and Fisher, 2014) and were separated on what appeared to be convincing morpho-

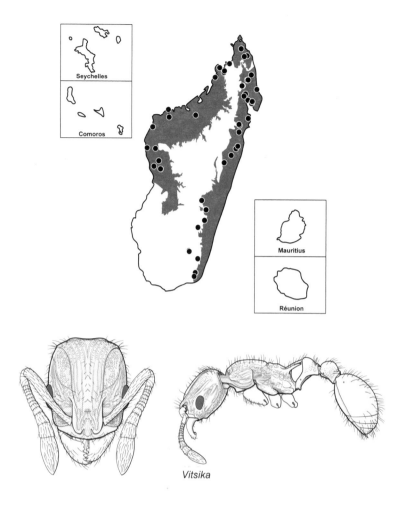

Vitsika

logical differences relating to the structure of the clypeus, the clypeal setae, the form of the frontal carinae and scrobes, and the profile shape of the mesosoma. However, soon after publication Philip S. Ward (pers. comm.) provided the authors with DNA evidence indicating that species ascribed to *Myrmisaraka* were nested within *Vitsika*, rendering *Vitsika* paraphyletic. The doublet *Vitsika+Myrmisaraka* had strong support as a monophyletic group. To rectify the situation, we propose the formal synonymy of the two, with *Vitsika* to be regarded as the senior name, as follows.

> *Vitsika* Bolton and Fisher, 2014: 68. Type-species: *Vitsika crebra* Bolton and
>   Fisher, 2014: 78, by original designation.
> = *Myrmisaraka* Bolton and Fisher, 2014: 33. Type-species: *Myrmisaraka producta*
>   Bolton and Fisher, 2014: 37, by original designation. Syn. n.
> The species originally described in *Myrmisaraka* are thus new combinations
>   in *Vitsika*, *V. brevis* (Bolton and Fisher) comb. n. and *V. producta* (Bolton and
>   Fisher) comb. n.

IDENTIFICATION OF SPECIES: Bolton and Fisher, 2014.

DEFINITION OF THE GENUS (WORKER): Characters of subfamily MYRMICINAE, plus:

Mandible triangular; masticatory margin with 6–11 teeth and longer than the basal margin.

Palp formula 5,3.

Stipes of the maxilla with a partial to strong transverse crest; when strong, the portion of stipes distal to the crest distinctly depressed and concave.

Clypeus posteriorly moderately broadly inserted between the frontal lobes; with longitudinal rugulae on median portion (no median carina present); 2 central rugulae may be enhanced to produce a weakly bicarinate appearance.

Clypeus with anterior margin entire or weakly notched; when entire, the margin with an unpaired median seta that arises just above a narrow anterior clypeal apron; when notched, usually with a closely approximated pair of short, fine setae that arise within the notch.

Clypeus with lateral portions not raised into a shielding wall or sharp ridge in front of the antennal sockets.

Frontal carinae usually present, divergent posteriorly, extending back almost to the posterior margin of the head; less commonly frontal carinae short, restricted to well-defined but narrow frontal lobes.

Antennal scrobes present above the eye, or absent.

Antenna with 12 segments, usually with a distinct 3-segmented apical club, but club 4- to weakly 5-segmented in 1 species.

Torulus with the upper lobe concealed by the frontal lobe in full-face view or its outer edge visible in full-face view.

Eyes present, located at or in front of the midlength of the side of the head capsule.

Head capsule without a median, longitudinal carina; occipital carina conspicuous.

Pronotal humeri weakly to distinctly angulate in dorsal view.

Promesonotum usually more or less evenly shallowly convex in profile, and the propodeal dorsum continuing the line of the promesonotum. Or, less commonly, the pronotum plus the anterior mesonotum swollen and distinctly convex in profile, with the dorsalmost point of the promesonotum on a considerably higher level than the propodeal dorsum.

Promesonotal suture absent; metanotal groove absent to distinct.

Propodeum bispinose; propodeal lobes small and rounded.

Propodeal spiracle at or behind the midlength of the sclerite.

Metasternal process absent; a pair of low carina arise anterior to the metasternal pit and diverge posteriorly, 1 on each side of the pit.

Tibial spurs: mesotibia 0 or 1; metatibia 0 or 1, simple; sometimes the spurs extremely reduced, hardly distinguishable from the setae at the tibial apices.

Petiole (A2) with a long anterior peduncle; the spiracle situated from slightly behind to distinctly in front of the midlength of the peduncle.

Petiole node narrow, high-domed to cuneate in profile.

Stridulitrum present on the pretergite of abdominal segment 4.

Abdominal tergite 4 (first gastral) does not broadly overlap the sternite on the ventral gaster; gastral shoulders absent.

Sting simple, strongly developed.

Main pilosity of dorsal head and body: simple, present on all dorsal surfaces of head and body. Scapes with elevated pubescence but without standing setae. Dorsal (outer) surfaces of mesotibiae and metatibiae with standing setae.

MALAGASY VITSIKA

acclivitas Bolton and Fisher, 2014	*miranda* Bolton and Fisher, 2014
astuta Bolton and Fisher, 2014	*obscura* Bolton and Fisher, 2014
brevis (Bolton and Fisher, 2014) comb. n.	*procera* Bolton and Fisher, 2014
breviscapa Bolton and Fisher, 2014	*producta* (Bolton and Fisher, 2014)
crebra Bolton and Fisher, 2014	comb. n.
disjuncta Bolton and Fisher, 2014	*suspicax* Bolton and Fisher, 2014
incisura Bolton and Fisher, 2014	*tenuis* Bolton and Fisher, 2014
labes Bolton and Fisher, 2014	*venustas* Bolton and Fisher, 2014
manifesta Bolton and Fisher, 2014	

VOLLENHOVIA Mayr, 1865 — Myrmicinae

2 species (+ 1 subspecies), Malagasy (Seychelles Islands) — PLATE 36

DIAGNOSTIC REMARKS: Among the myrmicines, *Vollenhovia* has 12-segmented antennae that terminate in a conspicuous apical club of 3 segments. The median portion of the clypeus is bicarinate, and antennal scrobes are absent. The propodeum is unarmed and rounded. The petiole (A2) is sessile anteriorly, or very nearly so, and the node is subtended by a large, laminate ventral process that is keel-like. The first gastral tergite (the tergite of A4) does not overlap the sternite on the ventral surface of the gaster.

DISTRIBUTION AND ECOLOGY: *Vollenhovia* currently contains 58 extant species, distributed throughout the Oriental and Malesian regions and on the islands of the Pacific Ocean. A few species exhibit tramping ability and have become widely distributed in the tropics and subtropics, although none has yet been recorded from the Afrotropical region. Two species have been found on the Seychelles, and although *V. piroskae* was originally described from there, it may represent an introduction from the usual range of the genus. *V. oblonga* is widespread in Indonesia, Malaysia, and the Philippines, and in the Seychelles, *V. oblonga alluaudi* could also represent an introduction. The species are found in forest leaf litter, with nests in tree stumps and rotten wood; occasional workers can also be found foraging on low vegetation.

IDENTIFICATION OF SPECIES: No review exists.

DEFINITION OF THE GENUS (WORKER): Characters of subfamily MYRMICINAE, plus:

Mandible triangular; masticatory margin with 5–8 teeth.

Palp formula 2,2, or 2,1.

Stipes of the maxilla without a transverse crest.

Clypeus posteriorly narrowly inserted between small, horizontal frontal lobes. Median

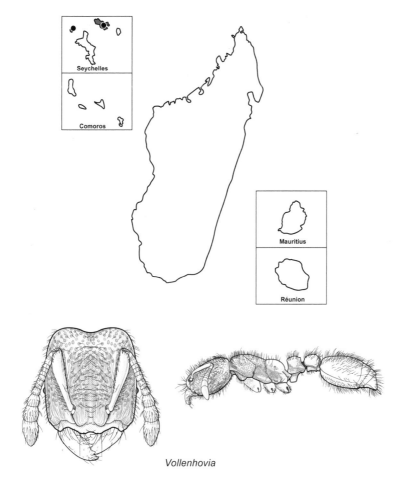

Vollenhovia

portion of the clypeus longitudinally bicarinate dorsally, the carinae arising between the frontal lobes.

Clypeus without an unpaired seta at the midpoint of the anterior margin; with a pair of setae present, 1 on each side of the midpoint.

Frontal carinae represented only by the frontal lobes, not extended posteriorly.

Antennal scrobes absent.

Antenna with 12 segments (11 in some extralimital species), with a conspicuous apical club of 3 segments.

Torulus with the upper lobe visible in full-face view.

Eyes present.

Promesonotal suture absent from the dorsum of the mesosoma; metanotal groove usually present but sometimes lacking.

Propodeum unarmed (rarely bidentate in some extralimital species).

Propodeal spiracle at about the midlength of the sclerite.

Metasternal process present or absent.

Tibial spurs: mesotibia 0; metatibia 0.

Petiole (A2) sessile or weakly pedunculate, with a pronounced ventral process.

Abdominal tergite 4 (first gastral) does not overlap the sternite on the ventral gaster; gastral shoulders absent.

Sting strongly developed.

Main pilosity of dorsal head and body: simple; generally conspicuous on dorsal surfaces but sometimes sparse.

MALAGASY VOLLENHOVIA

oblonga (Smith, F., 1860)
   oblonga alluaudi Emery, 1894
piroskae Forel, 1912

WASMANNIA Forel, 1893   Myrmicinae

1 species, Afrotropical, introduced   PLATE 36

DIAGNOSTIC REMARKS: Among the myrmicines, *Wasmannia* shows the following combination of characters: Antenna with 11 segments that terminate in an apical club of 2 segments. The clypeus is broadly inserted between the frontal lobes, and the lateral portions of the clypeus form a low ridge on each side, in front of the antennal sockets. The frontal carinae are long and extend almost to the posterior margin of the head; and antennal scrobes present.

DISTRIBUTION AND ECOLOGY: *Wasmannia* is a Neotropical genus that contains 10 extant species. One of these has been introduced elsewhere in the tropics and in the Afrotropical region has established itself in forested areas of Cameroun, Gabon, São Tomé and Principe, and perhaps other parts of the region. It has not yet been found in the Malagasy region. The single introduced species forages on trees.

IDENTIFICATION OF SPECIES: Longino and Fernández, 2007.

DEFINITION OF THE GENUS (WORKER): Characters of subfamily MYRMICINAE, plus:

Mandible short triangular; masticatory margin with 5 teeth.

Palp formula 3,2.

Stipes of the maxilla without a transverse crest.

Clypeus posteriorly broadly inserted between the frontal lobes. Anterior margin of median portion of the clypeus with a cuticular apron or flange that fits tightly over the closed mandibles, the apron at an angle to outline of the clypeus proper.

Clypeus with lateral portions forming a low ridge in front of the antennal sockets.

Clypeus with an unpaired seta at the midpoint of the anterior margin.

Frontal carinae extend to near the posterior margin of the head.

Antennal scrobes present, extending above the eyes.

Antenna with 11 segments, with an apical club of 2 segments.

Torulus with the upper lobe concealed by the frontal lobe in full-face view.

Eyes present, located in front of the midlength of the head capsule in full-face view.

Head capsule without a median, longitudinal carina.

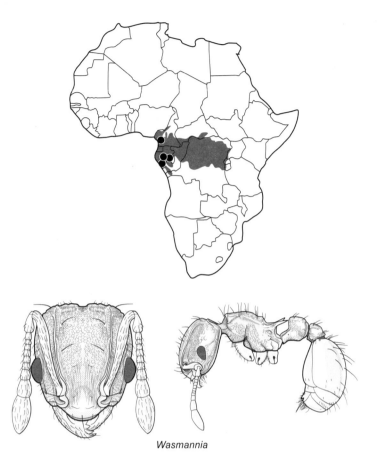

Wasmannia

Promesonotal suture absent from the dorsum of the mesosoma; pronotal humeri angulate.

Propodeum bispinose; propodeal lobes broadly rounded.

Propodeal spiracle relatively large, circular, located below propodeal spine and just behind midlength of the sclerite.

Metasternal process absent.

Tibial spurs: mesotibia 0; metatibia 0.

Abdominal segment 4 (first gastral) tergite overlaps the sternite on the ventral gaster; gastral shoulders weakly present.

Sting strongly developed.

Main pilosity of dorsal head and body: simple.

AFROTROPICAL WASMANNIA

auropunctata (Roger, 1863)
   = *atomum* (Santschi, 1914)

XYMMER Santschi, 1914   Amblyoponinae

1 species, Afrotropical   PLATE 36

DIAGNOSTIC REMARKS: Among the amblyoponines, *Xymmer* is characterized by the possession of apically pointed linear mandibles that do not close tightly against the clypeus, the absence of dentiform setae on the anterior clypeal margin, the presence of a short anterior peduncle on the petiole, the reduction of the petiolar sternite to a minute posterior sclerite, and the absence of spatulate setae on the head and body.

DISTRIBUTION AND ECOLOGY: This small genus is restricted to Madagascar and the Afrotropical region. There is only 1 described species, which is widespread but rare in the African rainforest zone from Ivory Coast to eastern Democratic Republic of Congo. Undescribed species are known from Madagascar. The single described species has been found in leaf litter, in rotten vegetation on the forest floor, and in abandoned termitaries.

IDENTIFICATION OF SPECIES: No review exists.

DEFINITION OF THE GENUS (WORKER): Characters of subfamily AMBLYOPONINAE, plus:

Mandibles linear, acute apically, crossing over near the apex when fully closed. Inner margin of the mandible with 7–8 teeth, of which teeth 3–6 are bidenticulate

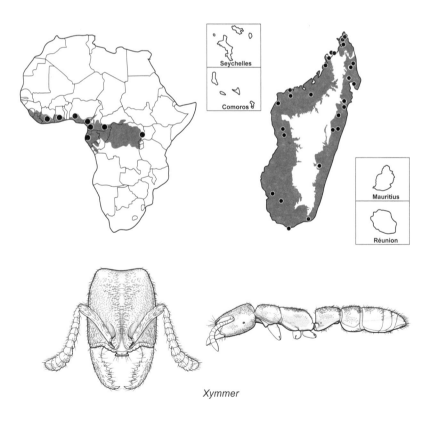

Xymmer

apically. When the mandibles are fully closed, there is a gap between the masticatory margins and between the basal margins of the mandibles and the clypeus.

Palp formula 3,2.

Clypeus narrow from front to back, so that antennal sockets are close to the anterior margin of the head.

Clypeus with a shallow, rectangular median lobe, the lobe angulate at each side; anterior margin of the lobe unarmed, without dentiform setae.

Frontal lobes present, separated.

Eyes absent.

Antenna with 12 segments, the funiculus gradually incrassate apically.

Metanotal groove absent.

Tibial spurs: mesotibia 1; metatibia 2, main spur pectinate.

Abdominal segment 2 (petiole) with a short anterior peduncle; in profile petiole with a steep anterior face and a long, weakly convex dorsum but without trace of a differentiated posterior face; petiole broadly attached to abdominal segment 3 (first gastral segment).

Petiole (A2) sternite reduced to a small medioventral sclerite.

Helcium with its dorsal surface arising very high on the anterior face of the first gastral segment (A3), so that in profile A3 above the helcium has no free anterior face.

Prora absent.

Abdominal segment 4 with distinctly differentiated presclerites.

AFROTROPICAL XYMMER

muticus (Santschi, 1914)

ZASPHINCTUS Mayr, 1866 Dorylinae

2 species, Afrotropical PLATE 36

DIAGNOSTIC REMARKS: Among the dorylines, *Zaspinctus* lacks a promesonotal suture. The waist consists of a single segment, the petiole (A2). Deep girdling constrictions are present between the second and third gastral segments (A4 and A5), and between the third and fourth gastral segments (A5 and A6), as well as between the first and second (A3 and A4), giving the gaster a distinctly irregular appearance. Only 1 other doryline genus shares this arrangement, *Aenictogiton*, but in that genus the promesonotal suture is conspicuous.

DISTRIBUTION AND ECOLOGY: Workers of several Afrotropical species are represented in collections, but named material remains restricted to the 2 male-based forms listed below, from Chad and Benin. Workers have been retrieved from litter and log mold samples and have been discovered in the soil, beneath surface objects.

TAXONOMIC COMMENT: *Zasphinctus* is currently recorded as a junior synonym of *Sphinctomyrmex*. It will be formally revived from synonymy by Marek Borowiec, in a genus-rank revision of Dorylinae that is in preparation.

IDENTIFICATION OF SPECIES: No review exists.

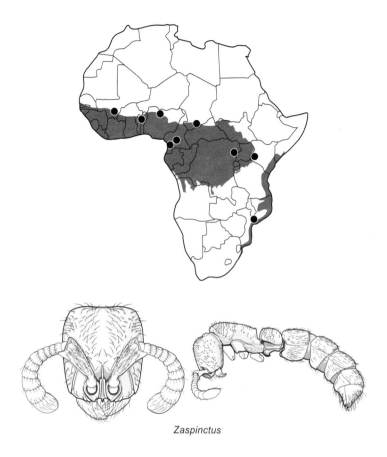

Zaspinctus

DEFINITION OF THE GENUS (WORKER): Characters of subfamily DORYLINAE, plus:

Prementum not visible when mouthparts closed.

Palp formula 3,3.

Eyes absent or present; when present, small and in front of the midlength of the head capsule in full-face view; ocelli absent.

Frontal carinae present as a pair of raised ridges.

Antennal sockets fully exposed.

Parafrontal ridges present.

Antenna with 12 segments; funiculus gradually incrassate apically.

Antennal scrobes absent.

Promesonotal suture absent on the dorsum of the mesosoma.

Pronotal-mesopleural suture absent on side of the mesosoma.

Metanotal groove absent.

Propodeal spiracle low on side and situated at or behind the midlength of the sclerite, not subtended by a longitudinal impression.

Propodeal lobes present.

Tibial spurs: mesotibia 1, pectinate; metatibia 1, pectinate.

Metatibial gland present, located distally on the ventral surface.

Pretarsal claws simple.

Waist of 1 segment (petiole = A2).

Petiole (A2) sessile; node not marginate dorsolaterally.

Helcium projects from the anterior surface of A3 at about its midheight.

Prora present, a simple flange or U-shaped cuticular ridge.

Abdominal segments 4, 5, and 6 with strongly developed presclerites; a distinct girdling constriction present on each of these segments.

Pygidium (tergite of A7) large and distinct, flattened to concave dorsally and armed laterally, posteriorly, or both, with a series or row of denticles or peg-like spiniform setae.

Sting present, large and fully functional.

AFROTROPICAL ZASPHINCTUS

chariensis (Santschi, 1915)

rufiventris (Santschi, 1915)

GLOSSARY OF MORPHOLOGICAL TERMS

The morphological terms are listed and described here in alphabetical order. Within each definition, cross-references to other defined terms are in *italics*. The *sclerites* of the *sting apparatus* and *male genitalia* are described because, although mostly or entirely internal in ants, they are ultimately derived from external appendages. The structure of the formicid *endoskeleton* is also outlined, as is the nomenclature of the wing *venation* and the *cells of the wings*. Omitted are the names of various forms of sculpture and of minor superficial features of the cuticle. These terms are of use in species-rank taxonomy but in higher taxa tend to be sporadic and not of universal application.

Abdomen The classical third *tagma* of the insect body. The abdomen in ants consists of 10 segments, of which the first 7 (A1–A7, from front to back) are visible in the female castes (workers and queens), while A8 is also exposed in males. Each of the *tergites* of segments A1–A8 bears a *spiracle*, which may be exposed or concealed. In the female castes, segments A8 and A9 are desclerotized, internal, and form parts of the *sting apparatus*, so that A7, because it is always the last visible segment, is usually referred to as being apical. In males, A8 is exposed but A9 is usually retracted and is the *gonosomite*, having the *male genitalia* attached to its posterior margin. Segment A10 is reduced in both sexes, at most a simple tergal arc of cuticle. In males, tergite 10 is sometimes fused to tergite 9 (= syntergite), and in many groups the *sclerite* bears a pair of pygostyles (= cerci) apically.

The terminology used to describe the ant abdomen may at first seem confusing. This is because 2 different terminology systems are in use and tend to be superimposed, but they are not strictly compatible.

1. A terminology based strictly on morphology simply numbers the visible abdominal segments from front to back. This system has the advantage of indicating homologous segments between different ant taxa, regardless of the specializations of individual segments or groups of segments in different groups of ants.

2. A more casual terminology based on observed subdivisions of the abdominal segments names various specialized segments or groups of segments. The advantage of this system is that the subdivisions are generally easily visible.

The first abdominal segment (A1) is the propodeum, represented only by its tergite (the sternite has been lost), which is immovably fused to the *thorax*. The body unit formed by the fusion of thorax and propodeum is termed the *mesosoma* (in some publications called the alitrunk or truncus, or uncommonly and inaccurately just the thorax).

The second abdominal segment (A2) is termed the *petiole* and is always specialized. It is usually reduced in size, always separated from the preceding propodeum by a complex narrow articulation, and is usually separated from subsequent abdominal segments by at least a constriction. In the vast majority of ants the petiole is distinctly isolated both anteriorly and posteriorly.

Abdominal segments 2 (petiole) to the apical are sometimes collectively called the *metasoma* (in contrast with the mesosoma, which consists of the thorax and propodeum). Thus the petiole (A2) may also be referred to as the first metasomal segment, A3 the second metasomal, and so on.

Abdominal segment 3 (A3) is termed the first gastral segment when it is full-sized and broadly articulated to the following segment (A4), but when reduced and distinctly isolated, it is commonly called the *postpetiole*. Abdominal segment 3 articulates with the preceding petiole by means of the *helcium*, which itself is formed from the reduced and specialized *presclerites* of A3, which fit within the posterior *foramen* of A2 (petiole). The anterior surface of the sternite of A3 may bear a cuticular *prora*, below the helcium.

The 1 or 2 isolated segments that follow the mesosoma may be called the *waist*. An older term, *pedicel*, should be abandoned, as it is used universally elsewhere in the Hymenoptera for the first funicular (second antennal) segment.

Abdominal segment 4 (A4) is the first gastral segment when the waist consists of petiole plus postpetiole, but A4 is the second gastral segment when the waist consists of the petiole alone. Abdominal segments 3 to the apex (when petiole [A2] alone is separated), or A4 to the apex (when petiole and postpetiole [A2 and A3] are separated), are collectively called the *gaster*, the apparent enlarged "abdomen" that comprises the terminal part of the body.

Each abdominal segment behind the first (*propodeum*) consists of a pair of sclerites (plates), a dorsal *tergite* (or *tergum*) and a ventral *sternite* (or *sternum*). These may all be similar, or some may be specialized by reduction, fusion, or subdivision into anterior (*presclerite*) and posterior (*postsclerite*) portions that are separated by a constriction (*cinctus*). Tergites and sternites may be referred to as abdominal or gastral, depending on whether an absolute count, or a count relative to the number of separated waist segments, is used. In workers and queens the last visible tergite, that of A7, is named the *pygidium*, and its corresponding sternite is the *hypopygium*. They have individual names because in some groups of ants the pygidium, the hypopygium, or both may exhibit a specialized morphology. In males the sternite of A9 is called the subgenital plate (= hypandrium, = hypopygium), as it shields the genital capsule ventrobasally.

Acidopore   The orifice of the formic acid-projecting system peculiar to, and diagnostic of, the female castes of subfamily Formicinae. The acidopore is formed entirely from the apex of the *hypopygium* (sternite of A7). Often it is plainly visible as a short nozzle, generally with a fringe of short *setae* (*coronula*) at its apex. However, in some genera there is no nozzle or setae and the acidopore takes the form of a semicircular to circular emargination or excavation in the apical hypopygial margin. In these taxa the posterior margin of the *pygidium* may overlap and conceal the acidopore when it is not in use, but the structure is revealed if the pygidium and hypopygium are separated.

Aculeus   See Sting apparatus.

Aedeagus   See Male genitalia.

Ala (pl. alae)   See Wings.

Alitrunk   See Mesosoma.

Anapleural sulcus   See Pleurite/pleuron.

Anepisternum   See Pleurite/pleuron.

Annulus (pl. annuli)   A simple, unsegmented ring of cuticle. For example, one of the funicular segments of the *antenna*, or the *torulus* (= annulus antennalis), around the antennal *foramen*.

Antenna (pl. antennae)   The antenna in ants is made up of a number of discrete segments, the antennomeres (sing. *antennomere*). The antennomeres consist of an elongated basal segment, the *scape* (= scapus), which is followed distally by 3–11 shorter segments in workers and queens, and 8–12 in males, and together constitute the *funiculus* (= flagellum), giving a total antennal segment count (= antennomere count) of 4–12 in females and 9–13 in males. The scape and funiculus meet at an angle so that in life the entire antenna appears bent (*geniculate*) between the two sections. The basal (first) funicular segment, the one that articulates with the apex of the scape, is sometimes called the *pedicel*. The scape articulates with the head in the antennal foramen (= antennal socket), a *foramen* located dorsally on the head, posterior to the *clypeus*. The funicular segments may all be simple (= *filiform*), or the segments may gradually enlarge toward the apex (= gradually incrassate), or a number toward the apex may be expanded into a distinctly differentiated *clava* (= club) that is usually of 2–3 segments but sometimes may be more. When a club is present, the antenna is said to be *clavate* or *claviform*.

    The antennal foramen itself is encircled by a narrow annular *sclerite*, the *torulus* (= antennal sclerite, = annulus antennalis), and the socket may be overhung and concealed by the *frontal lobe*, an anterior expansion of the *frontal carina*. At the base of the scape is a roughly ball-like *condylar bulb* (= articulatory bulb, = bulbus), which is the part of the scape that actually articulates within the socket. Just distal of the condylar bulb is a short constriction or neck, which may be straight or curved, distal of which the scape shaft proper commences.

Antennal fossa (pl. antennal fossae)   The antennal fossa is a depressed area in the cuticle of the dorsum of the *head*, present in some ant taxa, that surrounds and contains the *torulus* and antennal foramen, which consequently appear to be somewhat sunk into the surface of the head. In taxa with a fossa the *scape* is abruptly angled downward near its base, to allow the main part of the scape shaft a free antero-posterior sweep, despite the fact that its socket is on a lower level.

Antennal scrobe (= scrobe, = scrobis)   A longitudinal groove, impression, or excavation in the side of the *head*, which may extend above, below, or in front of the eye, which can accommodate at least the antennal scape but sometimes may accommodate the entire *antenna* when *scape* and *funiculus* are folded together. Scrobes vary in development from simple, broad, shallow grooves to extensive, deep trenches but are absent from the majority of ant genera.

Antennal socket/foramen/insertion   See Antenna.

Antennomere   See Antenna.

Anterior tentorial pits   A pair of *endophragmal pits* or impressions located anteriorly on the dorsal surface of the *head* capsule, at or very close to the posterior clypeal margin (*epistomal suture*) and usually close to the antennal sockets. The pits are small invaginations of the exoskeleton that indicate the points of attachment of the anterior arms of the internal skeleton (*tentorium*) of the head to the head capsule. The termination of the posterior arms of the tentorium are marked by a pair of *posterior tentorial pits* located close to the *occipital foramen*.

Apophyseal lines   Externally visible lines that mark the internal track of cuticular processes for muscle attachment.

Arolium (pl. arolia) A median, terminal lobe on the *pretarsus* (apical tarsomere) of any leg, between the pair of *pretarsal claws*. Arolia are uncommon in worker ants.

Axilla (pl. axillae) See Mesothorax.

Basal ring, basimere, basivolsella See Male genitalia.

Basitarsus (pl. basitarsi) See Tarsus.

Buccal cavity (= oral fossa) The anteroventral cavity in the *head* capsule, which contains the *labium* and *maxillae*. It is bounded anteriorly by the *labrum*, posteriorly and laterally by the ventral cuticle of the head (the *hypostoma* or the *genal bridge*). Within the buccal cavity the median appendage is the labium, which is flanked by a maxilla on each side of it.

Bulbus See Antenna.

Bulla (pl. bullae) See Pleurite/pleuron.

Calcar (= spur) See Tibal spur.

Calyx See Proventriculus.

Canthellus See Trulleum.

Cardo See Maxilla.

Cells of the wings Sections of the *venation* of the wings surround areas of wing membrane that are called cells. The cells are purely a function of the veins that form their boundaries, and the shape and number of cells diminishes with the contraction or loss of various veins or parts of veins. Cells that are entirely surrounded by vein sections are termed *closed cells*; those in which 1 or more boundaries are missing are termed *open cells*. In ants that show the most complete venation, there are 9 closed cells in the forewing (10 if the pterostigma is included), and 3 in the hindwing, together with a number of cells in each wing that are always open, as defined below.

Unfortunately, there is no single, agreed-upon system for naming wing cells, and several nomenclatures, some of them widely recognized but some idiosyncratic, are in use. The definitions outlined here indicate which sections of the venation form the boundaries of each cell—proximally, anteriorly, distally, and posteriorly—so that the identities of individual cells can be correlated across the various systems. The names given to the cells here are in common use but by no means universal.

Cells of the forewing (in alphabetical order):

Basal cell [BC] Bounded proximally by the wing base, anteriorly by Sc+R+Rs, distally by Rs·f1 and M·f1, posteriorly by M+Cu.

Costal cell [CC] Bounded proximally by the wing base, anteriorly by C, distally by Sc or the *pterostigma*, posteriorly by Sc+R+Rs and Sc+R.

First discal cell [DC1] Bounded proximally by M·f1, anteriorly by Rs+M, distally by M·f2 and 1m-cu, posteriorly by Cu·f1 and Cu·f2.

First subdiscal cell [SDC1] Bounded proximally by cu-a, anteriorly by Cu·f2, distally by Cu·f3 and Cu2, posteriorly by A·f2.

First submarginal cell [SMC1] Bounded proximally by Rs·f1, anteriorly by Sc+R and R·f1, distally by 2r-rs, posteriorly by Rs+M and Rs·f2-3. Note that in the very few taxa where 1r-rs is present, SMC1 is divided into proximal and distal portions by this cross-vein. In this uncommon circumstance, SMC1 (proximal) is bounded proximally by Rs·f1, anteriorly by Sc+R, distally by 1r-rs and posteriorly by Rs+M and Rs·f2. SMC1 (distal) is bounded proximally by 1r-rs, anteriorly by R·f1, distally by 2r-rs, and posteriorly by Rs·f3.

Marginal cell [MC]   Bounded proximally by 2r-rs, anteriorly by R·f2 and R·f3, distally by the junction of R·f3 and Rs·f5, or distally open when these veins fail to meet, posteriorly by Rs·f4-5. In ancestral homology the cell called MC here is strictly MC1. In some xyelids a second, more distal, marginal cell, MC2, is present. This is bounded proximally by Rs1 and distally by Rs2, but as no apocritan retains vein Rs2, MC2 is not represented.

Second discal cell [DC2]   Bounded proximally by 1m-cu and Cu·f3, anteriorly by M·f3-4, distally open, posteriorly by Cu1. In many aculeates there is an open third discal cell [DC3], but in ants this is never exhibited because cross-vein 2m-cu, which would form its proximal border, is universally absent. The absence of 2m-cu in ants means that the cell represented as DC2 is ancestrally DC2+3.

Second subdiscal cell [SDC2]   Bounded proximally by Cu2, anteriorly by Cu1, distally open, posteriorly by A·f3.

Second submarginal cell [SMC2]   Bounded proximally by Rs·f2 and M·f2, anteriorly by Rs·f3-4, distally by 2rs-m, posteriorly by M·f3.

Subbasal cell [SBC]   Bounded proximally by the wing base, anteriorly by M+Cu, distally by Cu·f1 and cu-a, or by cu-a alone when that vein is retracted, posteriorly by A·f1.

Third submarginal cell [SMC3]   Bounded proximally by 2rs-m, anteriorly by Rs·f5, distally open, posteriorly by M·f4. In many aculeates there is an open fourth submarginal cell [SMC4], but in ants this is never exhibited because cross-vein 3rs-m, which would form its proximal border, is universally absent. The absence of 3rs-m in ants means that the cell represented as SMC3 is ancestrally SMC3+4.

Cells of the hindwing (in alphabetical order):

Basal cell [BC]   Bounded proximally by the wing base, anteriorly by Sc+R+Rs, distally by Rs·f1, 1rs-m, M·f1 and M+Cu, posteriorly by M+Cu.

Costal cell [CC]   Bounded proximally by the wing base, anteriorly by C, distally by Sc+R, posteriorly by Sc+R+Rs.

Discal cell [DC]   Bounded proximally by M·f1, anteriorly by M·f2, distally open, posteriorly by Cu.

Marginal cell [MC]   Bounded proximally by Sc+R and Rs·f1, anteriorly by R, distally open, posteriorly by Rs·f2.

Subbasal cell [SBC]   Bounded proximally by the wing base, anteriorly by M+Cu, distally by cu-a, posteriorly by A·f1.

Subdiscal cell [SDC]   Bounded proximally by cu-a, anteriorly by M+Cu and Cu, distally open, posteriorly by A·f2.

Submarginal cell [SMC].   Bounded proximally by 1rs-m, anteriorly by Rs·f2, distally open, posteriorly by M·f2.

See also *Venation* and *Wings*.

Cephalon   See Head.

Cercus (pl. cerci)   See Abdomen.

Cervix   Strictly, the flexible intersegmental region between the *head* and the *prothorax*. It is usually shielded from above by a neck-like projection of the anterior *pronotum* and from below by closely approximated or fused anterior extensions of the *propleuron*. Sometimes the anterior portion of the pronotum, which covers and protects the true cervix, is also termed the cervix or the cervical portion of the pronotum.

Cinctus   See Girdling constriction.

Clava, clavate, claviform   See Antenna.

Clavus (= vannal lobe, = plical lobe), claval furrow (= vannal fold)   See Wings.

Claw   See Tarsus.

Clypeus (= epistoma)   Anterior *sclerite* of the dorsal *head*, bounded posteriorly by the *epistomal suture*, which is commonly referred to as the clypeal suture, or posterior clypeal margin. The median section of the epistomal suture, immediately anterior to the *frontal carinae* and antennal sockets, is sometimes called the *frontoclypeal suture*. The anterior clypeal margin usually forms the anterior margin of the head in full-face view (but a projection of the *labrum* may be anterior to the clypeus in some taxa). The body of the clypeus consists of a pair of lateral portions, or narrow bands of cuticle, on each side of a shield-like median portion. The median portion of the clypeus may be equipped with one or more longitudinal carinae or may be variously specialized in shape. Posteriorly, the median portion of the clypeus may end in front of the antennal sockets/frontal carinae or lobes, or may project backward between them. In some taxa the clypeus is extremely reduced and very narrow from front to back, bringing the antennal sockets very close to the anterior margin of the head.

Condylar bulb   See Antenna.

Coxa (pl. coxae)   The first, most basal segment of any leg; the leg segment that articulates within the coxal cavity (= coxal foramen) in the ventral thorax. The coxa of the prothoracic (fore) leg is often termed the *procoxa*, that of the mesothoracic (middle) leg the *mesocoxa*, and that of the metathoracic (hind) leg the *metacoxa*.

Cranium   See Head.

Cupula   See Male genitalia.

Cuspis   See Male genitalia.

Declivity   See Propodeum.

Digitus   See Male genitalia.

Distivolsella   See Male genitalia.

Endophragmal pit   A pit or pit-like impression in a *sclerite* that is an external indication of the point of attachment of part of the *endoskeleton*. Endophragmal pits that are universal in ants include the *anterior* and *posterior tentorial pits* of the *head*, and the *mesosternal* and *metasternal* pits of the ventral *mesosoma*. Some groups of ants also have endophragmal pits in other locations. For example, workers of the dorylomorph genera usually have a pit laterally on the mesosomal pleuron, either immediately posterior to the mesometapleural suture or even more posteriorly, closer to the *propodeal spiracle*.

Endoskeleton   As well as the extensive exoskeleton, worker ants have a small but significant endoskeleton that consists of sclerotized internal plates that serve as muscle attachments (the endoskeleton of alate forms remains to be investigated).

The endoskeleton of the head is the *tentorium*, formed from united anterior and posterior pairs of arms. In ants the fusion is complete and the tentorium takes the form of a pair of longitudinal struts attached to the head capsule anteriorly and posteriorly. In addition, each tentorial strut has an anterolateral side-branch that extends to the antennal socket. The only visible external indications of the tentorium are the *anterior tentorial pits* and *posterior tentorial pits*—endophragmal pits that mark the points at which the tentorial arms are attached to invaginations of the exoskeleton.

The *mesothorax* has a *mesendosternite*, apparently derived from the invaginated *sternum* of the segment. At maximum development the mesendosternite is a thin, longitudinal, roughly triangular, sclerite that extends the length of the segment and terminates posteriorly at the externally visible *mesosternal pit*. In most ant groups the dorsum of this endosclerite extends into a Y-shaped pair of apodemes that are inclined anteriorly and directed anterolaterally and dorsally, and there is usually a reinforcing cross-member between the apodemes, near their base. The apices of the apodemes are usually attached to the pleuron. In dorylomorphs the apodemes are inclined posteriorly, extend laterally from their point of origin, and are strongly attached to an invagination of the pleuron; the point of attachment is often visible externally as an *endophragmal pit*.

At the internal junction of the mesothorax and *metathorax*, just posterior to the *mesocoxal* cavities, is a raised ridge or low wall of transversely arched cuticle that appears to be derived from the invaginated, fused margins of the 2 segments right across their line of junction. Arising from the midpoint of this transverse cuticular wall, fused to its vertical midline anteriorly and extending posteriorly, is the *metendosternite*. At maximum development this sclerite is similar in shape to the mesendosternite and terminates posteriorly at the externally visible *metasternal pit*. In most ant groups the metendosternite has a pair of dorsal apodemes that may be very long, extending forward into the mesothoracic cavity above the mesendosternal apodemes, or they may fuse to the pleuron far in front of their point of origin. However, in the dorylomorphs the metendosternal apodemes are short and attached to the posterior surface of the mesendosternal apodemes.

Endosternite   See Endoskeleton.

Epimeral sclerite (= mesepimeral sclerite, = epimeral lobe)   In some groups of ant workers, and very commonly in alates, there is a small, sometimes detached, posterior lobe of the *mesopleuron* that covers or shields the orifice of the *metathoracic spiracle*, referred to as the epimeral sclerite (strictly the mesepimeral sclerite) or epimeral lobe. The name is derived from the ancestral morphology of the mesopleuron, where in more generalized forms the pleuron is divided by an oblique suture (pleural suture) into an anterior *mesepisternum* and a posterior *mesepimeron*. No trace of this division remains in ants. However, because of its position, this lobe is assumed, perhaps incorrectly, to have been derived from the ancestral mesepimeron.

Epinotum   An archaic name for the *propodeum*, used extensively in the past but only by myrmecologists. Propodeum is the recommended term, because it is universally used elsewhere in Hymenoptera morphology, and abandoning epinotum in favor of propodeum brings the terminology of ant morphology into line with the remainder of the Hymenoptera.

Epistoma, epistomal suture   See Clypeus.

Exocrine gland   A gland with secretory cells located below the cuticle and has pores, ducts, or 1 or more orifices that open to the surface.

Femur (pl. femora)   The third segment of any leg, counting from the basal coxal segment that articulates with the *thorax*. The femur is generally the longest leg segment and is separated from the *coxa* only by a small intermediate segment, the *trochanter*. The femur of the prothoracic (fore) leg is often termed the profemur, that of the mesothoracic (middle) leg the mesofemur, and that of the metathoracic (hind) leg the metafemur.

Fenestra (pl. fenestrae)   In general, a thin or translucent spot anywhere in the cuticle. In the wing *venation*, *"fenestra"* is the name applied to a translucent spot or apparent break in a vein that indicates the point at which a *flexion line* traverses the vein. In the most generalized ant wings, such fenestrae occur in Rs·f2, 2rs-m, cu-a, and Cu2 of the forewing, and in 1rs-m and cu-a of the hindwing.

Filiform (antenna)   With the antennal *funiculus* thread-like, the segments all of approximately the same width. The contrasting antennal form is *clavate* (club-like or clubbed), where the apical segments of the *antenna* are disproportionately enlarged.

Flagellum (= funiculus)   See Antenna.

Flexion line / fold line   See Wings.

Foramen (pl. foramina)   A natural opening or perforation in a *sclerite*. Usually an opening in one sclerite that accommodates the insertion of another. For example, the *occipital foramen* (= foramen magnum) is the posterior hole in the *head* capsule within which the articulation with the *prothorax* is accommodated. Similarly, the *coxal cavities* are the ventral foramina within which the coxae articulate with the thorax, and the *propodeal foramen* is the posterior orifice in the *mesosoma* within which the *petiole* (A2) articulates.

Frons   The area of the dorsal *head* posterior to the *frontoclypeal suture* above the antennal sockets and including the median *ocellus*. The posterior limit of the frons is vague in the many taxa where workers lack ocelli. Ancestrally, the anterior ocellus arises on the posteriormost part of the frons, while the pair of posterior ocelli arise on the *vertex*.

Frontal carinae (sing. frontal carina)   A pair of longitudinal cuticular ridges or flanges on the *head*, located dorsally behind the *clypeus* and between the antennal sockets. They are quite variable in length and strength of development among the various ant taxa, frequently being short and simple but sometimes extending back to the posterior margin of the head. In some groups the frontal carinae are vestigial or absent, but elsewhere they may be very strongly developed or form the dorsal margins of extensive *antennal scrobes*. In many groups the frontal carinae anteriorly, especially between the antennal sockets, are expanded into laterally projecting lobate extensions, the *frontal lobes*, which overhang and partially to entirely conceal the antennal sockets themselves. Frontal lobes may be the only expression of the frontal carinae in some groups. Sometimes the portion of the *torulus* closest to the cephalic midline is also raised and expanded into a small lobe that projects laterally. This may be visible below the frontal lobe or fused to the frontal lobe.

Frontal lobes   See Frontal carinae.

Frontal triangle (= supraclypeal area)   A small, usually triangular patch of cuticle located medio-dorsally on the *head*, immediately behind the *clypeus* and approximately between the antennal sockets or the anterior parts of the *frontal carinae*. Not apparent in many ant taxa and reduced or very narrow in some. In groups where the *frontal lobes* are medially very closely approximated, the frontal triangle may be compressed and longitudinal, and appears posterior to the frontal lobes as a narrow median *sclerite*.

Frontoclypeal suture   See Clypeus.

Funiculus (= flagellum)   See Antenna.

Furcula   See Sting apparatus.

Galea   See Maxilla.

Gaster A useful, convenient term for the swollen, apical portion of the body that forms the apparent *"abdomen."* Morphologically, the gaster consists of abdominal segments 3 to apex when the waist is of a single isolated segment (*petiole* = A2), but of abdominal segments 4 to apex when the waist is of 2 isolated segments (petiole plus *postpetiole* = A2 plus A3). In the second circumstance, the gaster, when of segments A4 to apex, may also be termed the *opisthogaster.*

Gena (pl. genae) Area of the side of the *head* bounded in front by the anterior margin of the head capsule and extending to the posterior margin of the head, below the eye when eyes are present. In ants the gena is expanded ventrally and forms the extensive ventral surface of the head, the *genal bridge.*

Genal bridge On the ventral surface of the *head*, the space between the *buccal cavity* anteriorly and the *occipital foramen* posteriorly is occupied by an extensive area of cuticle, divided by a midventral longitudinal line or groove. This surface is formed by the ventral expansion and midline fusion of the *genae* and is called the genal bridge. The extreme anterior margin of the genal bridge, which surrounds the buccal cavity, is the *hypostoma.*

Geniculate Bent like a knee joint. The term is usually used to describe the shape of the ant's *antenna* in life, where the *funiculus* is carried at a marked angle to the *scape.*

Girdling constriction (= cinctus) A constriction or sudden and marked narrowing of an abdominal segment, which usually extends around the entire circumference. For convenience it is usually stated in keys that girdling constrictions are present between two abdominal segments. This is not strictly correct as the constriction morphologically represents the separation between the *presclerites* and *postsclerites* of the more posterior segment. The greater part of the presclerites are usually inserted in the posterior end of the preceding segment and are concealed, leaving only the constriction and postsclerites visible externally.

Glossa (pl. glossae) See Labium.

Gonangulum, gonapophysis, gonocoxa, gonoplac, gonostylus See Sting apparatus.

Gonobase, gonocardo, gonocoxite, gonoforceps, gonolacinia, gonosomite, gonosquama, gonostipes, gonostylus See Male genitalia.

Guard setae (= guard hairs) A row or tuft of specialized *setae* that traverse and protect the orifice of the *metapleural gland*. These setae usually arise below the orifice of the gland and are directed upward across the orifice.

Gula Use of this term when referring to the ventral surface of the *head* capsule in ants is incorrect. Morphologically, the gula is a separated medioventral *sclerite* of the head that is bounded anteriorly by the *posterior tentorial pits*. In consequence, the posterior tentorial pits are distant from, and considerably anterior to, the *occipital foramen*. In ants the posterior tentorial pits are located adjacent to the occipital foramen; no gula sclerite is present.

Hamuli (sing. hamulus) See Wings.

Harpago See Male genitalia.

Head (= cephalon, = cranium, = prosoma) The classical first *tagma* of the insect body. The head capsule is the result of the fusion of 6 (or perhaps 7) embryonic segments. In adult ants these segments are completely fused and indistinguishable, but most retain appendages that are conspicuous. The appendages of cephalic segment 1 are fused and form the *labrum*; the appendages of segment 2 are the *antennae*; the appendages of segment 3 are embryonic only, unrepresented in adults; the appendages of segment 4 are the *mandibles*, of segment 5 the *maxillae*, and of segment 6 the *labium*. Some morphologists argue for the presence of a seventh embryonic segment, which would occur between 1 and 3 in the preceding list and bear the eyes as appendages.

Helcium   The very reduced and specialized *presclerites* of abdominal segment 3, which form a complex articulation within the posterior *foramen* of the *petiole* (A2). In general, the helcium is mostly or entirely concealed within the posterior foramen of the petiole but in some groups is partially visible.

Humerus (pl. humeri)   The anterolateral dorsal angle (shoulder) of the *pronotum*; frequently referred to as humeral angles.

Hypandrium   See Male genitalia.

Hypopygium   The *sternite* of abdominal segment 7 in workers and queens; the terminal visible gastral sternite in the female castes. In males the sternite of abdominal segment 9 is sometimes also referred to as the hypopygium, but *subgenital plate* (or hypandrium) is a better alternative term as it avoids the confusion caused by 2 nonhomologous *sclerites* having the same name.

Hypostoma   The narrow, U-shaped strip of cuticle immediately behind the *buccal cavity*, with posterior and lateral margins on the anteroventral surface of the *head*. In ants the hypostoma is often indistinct, sometimes indiscernible, and usually merely forms the anterior margin of the *genal bridge*, to which it is entirely fused. However, in some taxa a narrow suture remains detectable between the hypostoma and the genal bridge.

Hypostomal teeth   One or more pairs of triangular or rounded teeth that project forward from the *hypostoma*, toward or slightly into the *buccal cavity*.

Jugum (= jugal lobe, = anal lobe)   See Wings.

Katepisternum   See Pleurite/pleuron.

Labial palp (= labial palpus; pl. labial palpi)   A sensory palp, with a maximum of 4 segments, that arises anterolaterally on each side of the *prementum* sclerite of the *labium*. A count of the number of segments in a *maxillary palp* and a labial palp, in that order, is called the *palp formula*.

Labium   With the *head* in ventral view the labium is a longitudinal appendage situated medially within the *buccal cavity*; it is flanked on each side by a *maxilla*. The labium is formed from an ancestral pair of appendages, now indistinguishably fused together along the midline to form a single structure. The main *sclerite* of the labium is the *prementum*. Basal to the prementum, and attaching it to the head capsule, may be a small sclerite, the *postmentum*, but this area is frequently entirely membranous. The prementum bears a pair of *labial palps*, 1 on each side. At the distal end of the prementum is a lobe, the *glossa*, which may be simple or bilobed apically. The base of the glossa is sometimes flanked by a much smaller lobe on each side, the *paraglossa*, but in ants these are usually extremely reduced or absent. The glossae and paraglossae together are sometimes termed the *ligula*.

Labrum   Mouthpart *sclerite* that hinges on the anterior margin of the *clypeus* and usually folds back and down over the apices of the *maxillae* and *labium*, shielding them when the mouthparts are not in use. In most ants the labrum is a bilobed plate that is not visible in dorsal view, but in some taxa part of it projects forward beyond the anterior clypeal margin even when the mouthparts are retracted. Occasionally it is modified into 1 or more long, prominent labral lobes.

Lacinia   See Maxilla.

Lancet   See Sting apparatus.

Laterotergite   See Tergite/tergum.

Leg segments (= podites)   Each leg consists of a basal *coxa* that articulates with the *thorax*, followed in order by a small *trochanter*, a long and generally stout *femur*, a *tibia*, and a *tarsus*. The last consists of 5 small subsegments (= tarsal segments, = tarsomeres) and terminates apically in a pair of claws on the apical (*pretarsal*) segment; sometimes there is also a membranous lobe, the *arolium*, between the claws. The prefixes *pro-*, *meso-*, and *meta-*, applied to any of these terms, indicates that segment on the leg of a particular thoracic segment. For example, *procoxa* = the coxa of the fore (first, prothoracic) leg; *mesofemur* = the femur of the middle (second, mesothoracic) leg; *metatibia* = the tibia of the hind (third, metathoracic) leg.

Ligula   See Labium.

Malar area (= malar space)   The area of the *head* capsule between the base of the *mandible* and the anterior margin of the eye. Strictly a part of the *gena*, but sometimes its relative length is a useful concept in the taxonomy of some groups.

Male genitalia   The genitalia of male ants consist of 3 pairs of valves at the apex of the abdomen that are bounded basally by a ring-sclerite and shielded ventrobasally by the small sternite of A9, the *subgenital plate* (= hypopygium, = hypandrium). In some groups the genitalia may be completely retracted within the body, but in others they are permanently extruded, partially to almost entirely. Most sclerites of the male genitalia, and their subdivisions, have amassed a number of synonymous names. In the following discussion, the name given first is the one preferred here; alternative names follow, in parentheses.

   The basalmost sections of the stout outer pair of valves (the *parameres*, see below) are encircled by a cuticular *basal ring* (= gonobase, = gonocardo, = cupula, = lamina annularis). This genitalic complex of basal ring + parameres, sometimes called the phallobase, is attached to the apex of the ninth abdominal segment (= gonosomite).

   Of the 3 pairs of valves, the outermost valve on each side is the paramere (= gonosquama = gonoforceps), which is usually large and conspicuously sclerotized. In ants the paramere is often a single unit, not subdivided into proximal and distal portions, although in some groups the apical section may be narrowed, elongated, or otherwise modified. In some groups, however, the paramere is subdivided into proximal and distal portions by a transverse *sulcus*. In this instance the proximal section, closest to the basal ring, is the *basimere* (= gonostipes, = gonocoxite, = basiparamere), and the distal portion is the *telomere* (= gonostylus, = harpago). The median valve on each side is the *volsella*, which arises basally from the inner surface of the paramere. The volsella may be a simple lobe or may be split into a curved or hook-like dorsal lobe, the *digitus* (= gonolacinia), and a rounded lobe below it, the *cuspis* (= distivolsella), which together serve to clasp the female genitalia during copulation. Any part of the volsella that is proximal to its division into digitus and cuspis may be termed the *basivolsella* (= lamina volsellaris). The innermost pair of valves together constitute the *aedeagus* (= penis), the intromittent organ. Each aedeagal valve has been called a *penisvalva* (pl. penisvalvae), or *sagitta* (pl. sagittae). The latter is not recommended for ants as the term sagitta is extensively used among the apoids, where it always refers to a proximal lateral process on each aedeagal valve. The aedeagal valve itself may be divided into a basal apodeme, the *valvura*, and an apical plate, the *valviceps*, which usually has teeth or serrations on its ventral margin.

Mandalus See Trulleum.

Mandibles The appendages with which ants manipulate their environment. They are extremely variable in shape, size, and dentition, and are extremely important in ant taxonomy. Sometimes, but not always, the mandibles of the female castes are more strongly, or somewhat differently, developed than in conspecific males.

Margins: In full-face view, with the mandibles closed, the longitudinal margin or border of each mandibular blade that is closest to an anterior extension of the midline of the *head*, is the masticatory margin (= apical margin) and is usually armed with teeth. The base of this margin, close to the anterior margin of the *clypeus*, usually passes through a basal angle into a transverse or oblique basal margin. The 2 margins may meet through a broad or narrow curve or meet in an angle or tooth. When the mandibles are narrow or linear, the distinction between masticatory and basal margins may be lost by obliteration of the basal angle. The external margin (= lateral or outer margin) of each mandible forms its outer border in full-face view and may be straight, sinuate, or convex.

In many groups the dorsal base of the body of the mandible bears a number of specialized structures just distal to the mandibular articulation and proximal of the basal margin of the mandible, termed the *mandalus, canthellus,* and *trulleum*; see under the last named for a discussion of these.

Shape: In the vast majority of ants the mandibular margins form a triangular or subtriangular shape in full-face view but may be drawn out anteriorly while retaining the basic triangular shape and become elongate-triangular. In several discrete lineages the mandible has become linear; the blade is long and narrow, and the external and masticatory margins are approximately parallel or taper gradually to the apex; the whole blade may be straight or curved. Linear mandibles may evolve in 1 of 3 ways:

1. The base of the mandible narrows and the basal angle is obliterated, so that the basal and masticatory margins form a single margin.

2. The masticatory margin becomes elongated and the basal margin contracted.

3. The basal margin becomes elongated and the masticatory margin contracted.

Extremely curved mandibles, usually quite short and slender, and with few or no teeth on the masticatory margin, are termed *falcate*.

Dentition: The masticatory margin of each mandible is usually armed with a series of teeth or denticles (short or very reduced teeth), or a mixture of both, which generally extend the length of the masticatory margin. If teeth alone are present, or a combination of teeth and denticles, the mandible is dentate. If only tiny denticles occur on the margin, the mandible is denticulate; and if the margin lacks any form of armament, it is edentate. A natural gap in a row of teeth (as opposed to a site where teeth have been broken off or completely worn down) is a diastema (pl. diastemata), and an elongate mandible with an uninterrupted series of teeth may be described as serially dentate. Individual teeth are usually sharp and triangular in shape but may be rounded (crenulate), long, narrow, and spine-like (spiniform), or peg-like (paxilliform). Reduced teeth or denticles that occur between full-sized teeth are termed *intercalary*.

In general, the first, distalmost, or apical tooth—the one farthest away from the anterior clypeal margin—is the largest on the masticatory margin, although in some taxa, median or even basal teeth may be the largest. The tooth immediately preceding the apical is usually called the preapical tooth (= subapical tooth). Sometimes the term *preapical teeth* may be loosely applied to all the teeth that precede the apical. The tooth immediately distal of the basal tooth may be termed the prebasal (= subbasal) tooth.

In some genera, teeth or denticles may also be present on the basal margin of the mandible, but in most this margin is unarmed. Some taxa also have a basal lamella on the mandible. This is a thin, cuticular outgrowth from the margin, proximal to any teeth that may be present.

Maxilla (pl. maxillae) With the *head* in ventral view, the pair of maxillae are situated within the *buccal cavity*, one on each side of the central *labium*. The basal segment of the maxilla, which attaches the structure to the head capsule, is the *cardo*. Attached distally to the cardo is the main *sclerite* of the maxilla, the *stipes*, which toward its apex bears a *maxillary palp*. The inner surface of the stipes bears a lobe toward its distal end, the *lacinia*, the free margin of which is usually irregular, minutely dentiform, or even pectinate. At the distal apex of the stipes is a terminal hood or lobe, the *galea*, which often folds over the lacinia.

Maxillary palp (= maxillary palpus; pl. maxillary palpi) The segmented sensory palp on the *maxilla*, articulated to the *stipes*. The palp may have at most 6 segments, but these are variously reduced in number in different ant groups; only very rarely are the maxillary palps absent. A count of the number of segments in a maxillary palp and a *labial palp*, in that order, is called the *palp formula*.

Mayrian furrow See Mesothorax.

Mesendosternite See Endoskeleton.

Mesepimeron, mesepisternum See Epimeral sclerite.

Mesonotum See Tergite/tergum.

Mesopleuron See Pleurite/pleuron.

Mesoscutum, mesoscutellum See Mesothorax.

Mesosoma (= alitrunk, = truncus, = apparent "thorax") A convenience term for the second visible main section of an ant's body, following the *head*. Morphologically, the mesosoma consists of the 3 segments of the true thorax (*prothorax, mesothorax, metathorax*), to which is fused the *propodeum* (the *tergite* of AI), to form a single unit.

Mesosternal pit, mesosternal process See Metasternal pit/metasternal process.

Mesothoracic spiracle See Spiracle.

Mesothorax The second segment of the *thorax*, attached anteriorly to the *prothorax*, posteriorly to the *metathorax*, and bearing the mesothoracic (second, median) pair of legs and the first *spiracle*. In alate (winged) forms it also bears the *tegulae* and the anterior pair of *wings* (forewings).

The *tergite* of the mesothorax is the *mesonotum*. Anteriorly this is attached to the *pronotum*, either by a mobile suture (*promesonotal suture*) or frequently in the worker caste by fusion, in which instance the fused *nota* are termed the *promesonotum*. In alates the mesonotum is extensive and usually divided into a larger anterior *mesoscutum* (= scutum) and a smaller posterior *mesoscutellum* (= scutellum) by a transverse *scutoscutellar* suture. In alate queens and males the mesoscutum usually has a pair of narrow, incised lines, the *parapsidal grooves* (= parapsidal lines), which extend anteriorly from the scutoscutellar suture. In males the mesoscutum also frequently exhibits a pair of *notauli* (sing. notaulus), which in older publications are often called the Mayrian furrows or prescutal sutures. When present, each notaulus arises anterolaterally, near the anterior margin of the mesoscutum, and converges on its opposite number toward the midline posteriorly. In many groups the two notauli meet medially to form a V-shape, and in some the tail of the V-shape then extends posteriorly, so that the notauli are Y-shaped. On each side of the mesoscutum, covering the extreme base of the wing, is a small *sclerite*, the tegula. In alates, immediately posterior to the scutoscutellar suture, is the mesoscutellum. On each side, between mesoscutum and mesoscutellum, is a small, usually roughly triangular area, the *axilla* (pl. axillae), that extends down toward the base of the forewing. There is often a transverse depression, immediately posterior to the scutoscutellar suture, that links the axillae across the dorsum.

Posteriorly the mesonotum is ancestrally attached to the *metanotum* (tergite of the *metathorax*) by the *mesometanotal suture*, but in workers of some ant groups the mesonotum and metanotum are entirely fused. In many workers the metanotum is reduced to a transverse groove (*metanotal groove*), and in some the metanotum is entirely absent. In the last condition the posterior of the mesonotum is attached directly to the *propodeum*, and if a suture remains between them, it is the *notopropodeal suture*. The fused sclerite thus produced may be called the *notopropodeum*. In workers the mesonotum abuts the lateral *mesopleuron*, a long sclerite that extends down to the *mesocoxa*. Mesonotum and mesopleuron may be separated by a transverse *notopleural suture*, or the 2 sclerites may be fused together. In alates the articulation of the forewing occurs between the mesoscutum and the mesopleuron.

Immediately behind the posterodorsal corner of the mesopleuron there may be a small lobe, or small detached sclerite, the *epimeral sclerite* (or *epimeral lobe*), which is probably a detached section of the pleurite. When present, it conceals the orifice of the *metathoracic spiracle*, but this sclerite is absent in many groups (see *spiracle*).

The mesopleuron may be traversed by a horizontal *sulcus*, the anapleural sulcus, in which case the portion above the sulcus is the *anepisternum* (or *mesanepisternum*) and that below the sulcus is the *katepisternum* (or *mesokatepisternum*).

In profile view the upper portion of the mesopleuron abuts the lower portion of the side of the pronotum, from which it is separated by the *pronotal-mesopleural suture* (a lateral continuation of the *promesonotal suture*); below this is a long, free mesopleural margin against which the *procoxa* rests. Posteriorly, the mesopleuron is fused to the *metapleuron* by the oblique *mesometapleural suture*, though in workers of some groups these 2 sclerites are completely fused and the suture is obliterated.

The ventral surface of the mesothorax consists entirely of the pleurites, which have expanded across to the ventral midline and fused. The ancestral hymenopterous *sternite* of the mesothorax is internal and represented by the *mesendosternite*. The anterior margin of the ventral mesothorax often has a projecting median process that extends forward between the bases of the *procoxal cavities* and overlaps the posterior margin of the *prosternum*. On the ventral midline of the mesothorax, anterior to the *mesocoxal cavities*, is the *mesosternal pit*, an *endophragmal pit* that marks the attachment of the mesendosternite to the exoskeleton. This pit is sometimes accompanied by a paired, cuticular, *mesosternal process*. Immediately posterior to the mesocoxal cavities and the mesosternal pit is the arched suture that marks the junction of the mesothorax and the *metathorax*.

Metacoxal cavities The pair of foramina located posterolaterally in the ventral surface of the *metathorax*, within which the *coxae* (basal leg segments) of the metathoracic (hind, third) legs articulate. The metacoxal cavities are located on each side of the usually U-shaped or V-shaped *propodeal foramen* in which the base of the *petiole* (A2) articulates. In ventral view (with the coxae removed) the propodeal foramen may be entirely separated from the metacoxal cavity by an unbroken cuticular annulus that varies from a narrow bar to a broad band (metacoxal cavity closed); or the cuticular annulus may be interrupted by a linear break or suture that extends from the propodeal foramen to the metacoxal cavity (metacoxal annulus with a suture); or the annulus may have a wide gap so that the propodeal foramen is confluent with the metacoxal cavity (metacoxal cavity open).

Metanotal groove, metanotum   See Tergite/tergum.

Metapleural gland   An *exocrine gland*, common in female castes but very rare in males, whose orifice is on the *metapleuron*, usually situated at or near the posteroventral corner, above the level of the *metacoxa* and below the level of the *propodeal spiracle*. The swollen *bulla* of the metapleural gland is often more conspicuous than the gland's orifice and takes the form of a shallow blister or convex swelling on the metapleuron; the bulla sometimes extends almost to the propodeal spiracle. The orifice of the metapleural gland may be a simple pore or hole, or may be protected by cuticular flanges or other outgrowths, or by *guard setae* that arise below the orifice and extend across it. In a few groups of ants the metapleural gland has been lost in all female castes.

Metapleural lobe   See Propodeum.

Metapleuron (pl. metapleura)   See Pleurite/pleuron.

Metasoma   A collective term for abdominal segments 2 to apex, regardless of the absence or presence of abdominal constrictions that may occur between any of these segments.

Metasternal pit/metasternal process   The ventral surfaces of the *mesothorax* and *metathorax* each have an *endophragmal pit*, located on the midline anterior to the level of the *coxal cavities*. These pits mark the sites of attachment of the endoskeletal *mesendosternite* and *metendosternite* to the exoskeleton. In many groups of ants the pits are associated with a pair of cuticular projections, the mesosternal and metasternal processes. In most groups of ants the metasternal pit is distinctly anterior to the apex of the *propodeal foramen*, but in taxa where the foramen is extensive the pit may be extremely close to its apex.

Metathorax   The third segment of the *thorax*, attached anteriorly to the *mesothorax* and dorsally and posteriorly to the *propodeum* and bearing the metathoracic (third, hind) pair of legs and usually the second *spiracle*, though this may be lost in some groups. In alate (winged) forms it also bears the posterior (hind) pair of *wings*.

  The *tergite* of the metathorax is the *metanotum*. This *sclerite* is usually distinct in alates but extremely variably developed among workers of the various groups of ants. At its fullest development in workers the metanotum is a distinct transverse sclerite between the *mesonotum* and propodeum, separated from each by a transverse suture, the *mesometanotal suture* anteriorly and the *notopropodeal suture* posteriorly. In workers there is a very common morphoclinal reduction from this condition, where the metanotum becomes gradually shorter until it is represented only by a narrow transverse groove (*metanotal groove*), in which the true metanotum is represented only by the extreme base of the groove. In some groups even this groove is obliterated, so that the posterior margin of the mesonotum meets the anterior margin of the propodeum. This junction may be indicated by a feeble transverse impression (*notopropodeal suture*), or the sclerites may become fully fused, forming a *notopropodeum*. Conversely, the metanotum may remain present on the dorsum but become indistinguishably fused to the mesonotum, while retaining a strong suture between itself and the propodeum.

  Laterally, the *metapleuron* in workers forms an oblique sclerite that is usually roughly triangular. This is often separated from the *mesopleuron* by the *mesometapleural suture*, though in some the 2 sclerites are completely fused and no trace of a suture remains. Similarly, the dorsal junction of the metapleuron with the propodeum may be indicated by a suture, but again the 2 may be entirely fused to form a single sclerite, with no trace of a suture between them. In alates the metapleuron is usually more extensive. It is commonly divided by a short transverse *sulcus* into an upper *metanepisternum*, below the articulation of the hindwing, and a lower *metakatepisternum*. Close to the dorsal apex of the metapleuron is the *metathoracic spiracle*, which in alates is located between the mesopleuron and metapleuron and often shielded or concealed by the *epimeral sclerite*. Among workers, in groups where the metanotum forms a discrete dorsal sclerite, the spiracle is usually located dorsally or laterodorsally, on the side adjacent to the metanotum. Elsewhere, in workers where the metanotum is greatly reduced or absent, the spiracle is lower down on the side. It is sometimes open, sometimes concealed beneath the *epimeral sclerite*, and in some groups the spiracle has been lost. The posteroventral corner of the metapleuron, above the *metacoxa*, has the bulla and orifice of the metapleural gland.

The ventral surface of the metathorax consists entirely of the pleurites, which have expanded across to the ventral midline and fused. The ancestral hymenopterous sternite of the metathorax is internal and represented by the *metendosternite*. The ventral metathorax commences anteriorly in a curved suture immediately posterior to the *mesocoxal cavities* and *mesosternal pit*. On the ventral midline of the metathorax, anterior to the *metacoxal cavities*, is a *metasternal pit*, an *endophragmal pit* that marks the attachment of the metendosternite to the exoskeleton. This pit is sometimes accompanied by a paired, cuticular, *metasternal process*. The *metacoxal cavities* are located close to the posterior corners of the ventral surface, and the often extensive *propodeal foramen*, within which the *petiole* (A2) articulates, extends forward between them.

Metathoracic spiracle   See Spiracle.

Metatibial gland   A presumably *exocrine gland* that is located on the ventral surface, or more rarely the posterior surface, of the metatibia, usually just proximal of the metatibial spur.

Metendosternite   See Endoskeleton.

Mouthparts   The feeding appendages, located anteriorly and anteroventrally on the *prognathous* head capsule. The mouthparts consist of the *mandibles*, *maxillae*, and *labium*, of which the last 2 are located within the *buccal cavity*, on the underside of the head capsule.

Notaulus (pl. notauli)   See Mesothorax.

Notopropodeal suture   See Mesothorax and Metathorax.

Notum (pl. nota)   The name applied to any of the 3 ancestral thoracic tergites. Hence, *pronotum* is the notum of the *prothorax*, *mesonotum* of the *mesothorax*, and *metanotum* of the *metathorax*. In all worker ants each notum is a single *sclerite*, but in alate forms the mesonotum is usually subdivided. As the worker caste is derived ultimately from an alate female caste, the simple nota of the workers represent a secondary reversal to a more generalized condition.

Nuchal carina   A ridge situated posteriorly on the *head* capsule that separates the dorsal and lateral surfaces (*vertex* and *genae*) from the occipital surface.

Oblong plate   See Sting apparatus.

Occipital corners/margin (of head)   See Posterior corners/margin.

Occipital foramen (= foramen magnum)   The *foramen* located posteromedially in the *head* capsule, within which the membranous *cervix* articulates the head to the *prothorax*.

Occiput (= occipital surface)   The posterior surface of the *head* capsule, immediately above the *occipital foramen*. The occiput is usually vertical or nearly so above the foramen, and is separated from the *vertex* of the cephalic dorsum by the transverse posterior margin of the head capsule.

Ocellus (pl. ocelli)   A maximum of 3 simple, single-faceted eyes, which when present (absent in many worker ant taxa) are located in a triangle on the cephalic dorsum. Morphologically, the anterior (median) ocellus marks the posterior limit of the *frons*, and the posterior (lateral) pair are on the *vertex*.

Ommatidium (pl. ommatidia)   A single optical component (facet) of the compound eye.

Opisthogaster   See Gaster.

Palp Formula (PF)   A standardized way of indicating the number of segments in the maxillary and labial palps. The number of *maxillary palp* segments is given first, the number of *labial palp* segments second. Thus Palp Formula 6,4 indicates that the maxillary palp has 6 segments, the labial palp 4.

Paraglossa   See Labium.

Paramere   See Male genitalia.

Parapsidal groove   See Mesothorax.

Pedicel   An archaic term used in ants for the isolated body segments between *mesosoma* and *gaster*—namely, the *petiole* (A2) or petiole plus *postpetiole* (A2 plus A3). Use of the term is no longer recommended in this sense, as it is used elsewhere throughout the Hymenoptera as the name for the first *funicular* segment of the *antenna*.

Peduncle (of petiole = A2)   The relatively slender anterior section of the *petiole* that begins immediately posterior to the propodeal-petiolar articulation and extends back to the petiolar node or scale. It is quite variable in length and thickness, but when present in any form the petiole is termed *pedunculate*. When a peduncle is absent, so that the node or scale of the petiole immediately follows the articulation with the *propodeum*, the petiole is termed *sessile* (= apedunculate). If an extremely short or poorly defined peduncle occurs, the petiole is termed *subsessile*.

Penis, penisvalva   See Male genitalia.

Petiole (= abdominal segment 2 [A2], = first metasomal segment)   Morphologically, the second abdominal segment (A2), the segment that immediately follows the *mesosoma*. Anteriorly, it is always articulated within the *propodeal foramen* of the mesosoma. Generally the petiole takes the form of a node (nodiform) or of a scale (squamiform) of varying shape and size, but in some taxa it may be very reduced, represented only by a narrow subcylindrical segment that may be overhung from behind by the *gaster*. The petiole bears the second abdominal *spiracle* and usually consists of a distinct *tergite* and much smaller *sternite*. The tergite may have a differentiated *laterotergite*, low on each side and abutting the sternite. The sternite often has a specialized, depressed area anteroventrally, close behind the articulation, that is equipped with numerous short sensory hairs—the proprioceptor zone. In some groups the tergite and sternite of the petiole are immovably fused together (*tergosternal fusion*).

Phallobase   See Male genitalia.

Plectrum   See Stridulatory system.

Pleurite/pleuron (pl. pleura)   The lateral *sclerites* of the *thorax* proper, excluding the *propodeum*, which is morphologically the *tergite* of the first abdominal segment.
   The propleuron (pleuron of the prothorax) is relatively small in ants and is mostly or entirely overlapped and concealed by the extensive lateral part of the *pronotum* when viewed in profile but can always be seen clearly in ventral view (see *prothorax*). The mesopleuron (pleuron of the

mesothorax) is the largest pleurite. It may consist of a single sclerite that extends almost the entire height of the lateral mesothorax or may be divided by a transverse sulcus (the *anapleural sulcus*) into an upper *anepisternum* and a lower *katepisternum* (see *mesothorax*). The metapleuron (pleuron of the metathorax) is located posteriorly on the side of the *mesosoma*, mostly below the level of the propodeum in workers but more extensive in queens and males. The metapleuron bears, in the female castes of almost all ants, the *metapleural gland* (see *metathorax*). The ventral surfaces of the mesothorax and metathorax are formed by the ventral expansion of the pleurites and their fusion along the ventral midline; the true sternites of these segments are represented only by *endoskeletal* structures. The abdominal segments do not have pleurites, and each consists only of *tergite* (above) and *sternite* (below).

Podites   See Leg segments.

Posterior corners/margin (of head)   With the *head* in full-face view, the rounded to acute posterolateral angles, where the sides of the head curve into the posterior margin; the latter extends transversely between the corners. Earlier frequently referred to as occipital corners and occipital margin; however, these terms are not strictly accurate because the true occipital surface (*occiput*) is not involved.

Posterior tentorial pits   See Anterior tentorial pits.

Postmentum   See Labium.

Postpetiole (= abdominal segment 3 [A3])   A convenience term for what is morphologically the third abdominal segment when it is reduced in size and markedly separated from the *petiole* (A2) anterior to it and from the fourth abdominal segment (A4) posterior to it.

Postsclerite, poststernite, posttergite   See Presclerite.

Prementum   See Labium.

Presclerite   A distinctly differentiated anterior section of an abdominal *sclerite* (*tergite* or *sternite*) that is separated from the remaining posterior portion of the sclerite by a constriction, a ridge, or both.

In abdominal segments 3–7 (8 in males), it is usual for the posterior portion of each segment to overlap the anterior portion of the following segment. The overlapped area usually lacks sculpture and pilosity, but the absence of these features alone does not constitute a presclerite: there must be a constriction or ridge that separates the 2 zones. Presclerites derived from tergites are termed *pretergites*; those from sternites, *presternites*. The remainder of each sclerite, posterior to the constriction or ridge, is the *postsclerite, posttergite* dorsally and *poststernite* ventrally. A marked constriction that separates presclerites from postsclerites and extends around the entire circumference of a segment is a *girdling constriction* (= *cinctus*). The presclerites of abdominal segment 3 are reduced and form a specialized articulation within the posterior foramen of segment A2 (*petiole*), the *helcium*.

Prescutal suture   See Mesothorax.

Presternite, pretergite   See Presclerite.

Pretarsus, pretarsal claws (= ungues, sing. unguis)   See Tarsus.

Prognathous (head)   Among the ants the long axis of the *head* is horizontal or nearly horizontal, so that the head more or less continues the line of the long axis of the body, and the mouthparts, particularly the *mandibles*, are at the front of the head capsule. The ventral surface of the head has an extensive cuticular area (*genal bridge*) that widely separates the *buccal cavity* from the *occipital foramen*. This *prognathous* condition is in marked contrast to almost all the other families of Hymenoptera, where the long axis of the head is vertical or nearly vertical, the mandibles are ventral, and the buccal cavity and occipital foramen are closely approximated or even confluent—a condition termed *hypognathous*.

Because of the prognathous condition of the head, references to its orientation differ from what is usual in Hymenoptera. Apart from the mandibles being anterior, what is referred to as the frontal or anterior surface of the head elsewhere in the order is dorsal in ants.

Promesonotal suture   The transverse suture across the dorsal *mesosoma* that separates the *pronotum* from the *mesonotum*. In many groups of ants the promesonotal suture is fully developed, articulated, and flexible. The posterior margin of the pronotum slightly overlaps the anterior mesonotum and the 2 *sclerites* are linked by an intersegmental membrane so that they are capable of movement relative to each other. Elsewhere, and very commonly in workers, the suture is reduced from this condition. Initially in the sequence of reduction the suture is still present and distinct but inflexible, as the posterior pronotal margin has fused to the anterior margin of the mesonotum, and the intersegmental membrane has been lost. Beyond this fused condition, the suture shows a gradual morphoclinal reduction in size and degree of definition, eventually becoming nothing more than a faint line or weak impression across the dorsum, or often disappearing altogether. When fusion and obliteration of the suture is advanced, and there is little or no sign of separation of the 2 original sclerites, the resultant fusion sclerite is termed the *promesonotum*.

Promesonotum   See Promesonotal suture.

Promesothoracic suture   The promesothoracic suture consists of the transverse *promesonotal suture*, which links the *pronotum* and the *mesonotum* across the dorsal *mesosoma*, and the *pronotal-mesopleural suture*, which continues the promesonotal suture down the sides of the mesosoma and links the lower portion of the pronotum with the upper portion of the *mesopleuron*. The promesonotal suture is sometimes obliterated by fusion, but the pronotal-mesopleural suture is usually present; only uncommonly is it lost by complete fusion of the lower pronotum to the upper mesopleuron.

Pronotal-mesopleural suture   The approximately vertical suture on the side of the *mesosoma* that separates the lower portion of the lateral *pronotum* from the *mesopleuron*; it forms an extension of the *promesonotal suture* down the side of the mesosoma. Usually conspicuous, but in some groups the pronotal-mesopleural suture may be very reduced or obliterated, so that pronotum and mesopleuron are completely fused.

Pronotum   See Tergite/tergum.

Propodeal lobe (= metapleural lobe, = inferior propodeal plate)   See Propodeum.

Propodeal spiracle   See Propodeum.

Propodeum   Morphologically, the *tergite* of the first abdominal segment (the *sternite* of which is lost). It is immovably fused to the *thorax* and forms most of the posterior section of the *mesosoma* (= alitrunk). An older term for this *sclerite*, *epinotum*, should be abandoned.

The propodeal dorsum, which is sometimes referred to as the basal surface or base of the propodeum, is usually unspecialized but frequently terminates posteriorly in a pair of teeth or spines. The sloping posterior surface is the propodeal declivity and may bear a number of specializations. Most common of these is the development of a pair of *propodeal lobes* (= inferior propodeal plates). When present, these are situated at the base of the propodeal declivity, one on each side of the propodeal foramen, which is the posterior foramen of the mesosoma within which the *petiole* (A2) articulates. These lobes, which when present may vary considerably in shape and size, were frequently referred to as *metapleural lobes* in earlier publications, but this name should be abandoned as the lobes are morphologically part of the propodeum, not the metapleuron.

The side of the propodeum bears the *propodeal spiracle*, morphologically the first abdominal spiracle. Its shape, size, and location are variable and of considerable taxonomic value.

Prora (pl. prorae)   A cuticular process or prominence that projects forward from the anterior surface of abdominal *sternite* 3, below the *helcium*. Absent in many groups, when present it takes the form of a U-shaped ridge of cuticle, a tubercle of very variable size, or a distinct prow. In some specialized taxa the prora has become inserted between the tergal apices of the helcium and forms part of the articulation. In groups where abdominal segment 3 is very reduced in size, the prora may be called the anterior subpostpetiolar process.

Prosoma   See Head.

Prosternite   See Sternite/sternum.

Prothorax   The first of the 3 segments of the *thorax*, articulated anteriorly to the *head* by the membranous *cervix*, attached posteriorly to the *mesothorax*, and bearing the prothoracic (first, anterior) pair of legs. There is no *spiracle* on the prothorax; it has been lost in the Hymenoptera.

The *tergite* of the prothorax is the *pronotum*, always hypertrophied in the worker caste so that it is extensively present on the dorsal *mesosoma*. In alate queens and males the dorsal pronotum may be of a size similar to that seen in workers, but frequently its dorsal area is reduced, so that the pronotum may be represented only by a narrow anterior collar when seen in dorsal view or may be completely overhung by the anterior portion of the mesonotum (*mesoscutum*). Laterally, the pronotum extends down both sides of the segment in all castes and both sexes. It also extends for some distance medially on the ventral surface, behind the *procoxal cavities*, where it overlaps or fuses with the anterior margin of the ventral mesothorax. The posterior margin of the lateral portion of the pronotum usually covers and conceals the *mesonotal spiracle*, and in alate (winged) forms it extends posteriorly until it almost touches the *tegulae*. The *pleurites* of the prothorax are only partly visible in profile as they are largely concealed by the lateral parts of the pronotum (a condition termed *cryptopleury*), but the prothoracic pleurites are always conspicuous in ventral view. They are strongly attached or fused along the ventral midline but laterally are movably articulated to the pronotum, so that the 2 *propleurites* move as a single unit. Midventrally, between the procoxal cavities and posterior to the propleurites, is a small, usually shield-like sternite (*prosternum*), the posterior margin of which may be overlapped by an anteriorly projecting medioventral process of the mesothorax. The procoxal cavity is complex, being bounded anteriorly by the propleurite, medially by the prosternite, laterally by the pronotum, and posteriorly by the pronotum and anterior mesothorax. The pronotum articulates with the mesonotum at the *promesonotal suture* dorsally and with the mesopleuron at the *pronotal-mesopleural suture* laterally. These may be entirely flexible, with the sclerites linked by intersegmental membrane, but in the worker caste of many groups the pronotum may be completely fused to the mesonotum, producing a compound sclerite, the *promesonotum*. The promesopleural suture is usually conspicuous, only rarely are the pronotum and mesopleuron inseparably fused together. The prothorax does not have an *endoskeletal* sclerite.

Proventriculus   A muscular pump located in the intestine between the crop and the midgut. In all ants the proventriculus has a basal bulb, but in some the bulb is surmounted by a ring of

4 sclerotized *sepals*, collectively termed the *calyx*. Although an internal abdominal structure, the form of the proventriculus featured strongly in the early classifications of some subfamilies, so it is included here.

Psammophore A basket-like series of long, and usually stout, curved *setae* that arise from the ventral surfaces of the *head* and *mandibles* in some deserticolous ants, used for carrying sand grains. In some publications the setae of the psammophore are called ammochaete hairs.

Pterostigma (= stigma) A pigmented area that is usually present on the forewing of alate ants. When present, it is located immediately behind the leading edge of the wing, about half to two-thirds of the distance from the wing base to its apex and just distal of the hinge-like mechanism. It is bounded proximally and anteriorly by vein Sc, posteriorly and distally by vein R.

Pterothorax A term sometimes used to describe the form of the *thorax* in fully winged (alate) queens and males. In these alates the *notum* of the mesothorax (*mesonotum*) tends to be subdivided into an anterior *mesoscutum* (= scutum) and a posterior *mesoscutellum* (= scutellum), usually with a separately demarcated triangular area, the *axilla*, between them at each side. See the discussions under *mesothorax* and *metathorax*.

Pubescence Small to minute hair-like cuticular projections that are not socketed basally.

Pygidium The *tergite* of abdominal segment 7 in workers and queens; the terminal visible tergite of the *abdomen* in these castes.

Pygostyles See Abdomen.

Quadrate plate See Sting apparatus.

Radial flexion line See Wings.

Remigium See Wings.

Sagitta See Male genitalia.

Scape See Antenna.

Sclerite Functionally, a general term for any single plate of the exoskeleton (e.g., pronotal sclerite, abdominal sclerites); more specifically, an integumental plate in which the protein sclerotin has been deposited. In the case of ants, the latter applies to all parts of the exoskeleton.

Scrobe/scrobis See Antennal scrobe.

Scutoscutellar suture See Mesothorax.

Scutum, scutellum See Mesothorax.

Sepals See Proventriculus.

Sessile (petiole) See Peduncle.

Seta (pl. setae) Any stout, hair-like cuticular process that is socketed basally. Generally, as here, the terms *seta* and *hair* are interchangeable, but care must be taken to differentiate between setae and pubescence, as the latter may also sometimes be referred to as hairs.

Spiracle An orifice of the tracheal system by which gases enter and leave the body. Adult ants have 9 or 10 spiracles on each side of the body.

The spiracles of the *prothorax* have been lost, so the first spiracular opening occurs on the *mesothorax*. This *mesothoracic spiracle* is situated forward and quite high on the side of the segment and is usually concealed from view by a backward-projecting lobe of the *pronotum*; only rarely is its orifice open and clearly visible. The *metathoracic spiracle* may be dorsal (especially in those workers where the *metanotum* forms part of the dorsal *mesosoma*), lateral and open; lateral but concealed by a small, sometimes detached, lobe of the *mesopleuron* (the *epimeral sclerite*); or the metathoracic spiracle may be absent. Abdominal spiracles are always on the *tergite* of each segment. The *propodeal* (first abdominal) spiracle is usually the largest on the body. Behind this, on the *metasoma* (A2 to apex), spiracles are always visible on abdominal segments 2–4, but those on abdominal segments 5–7 are frequently overlapped and concealed by the posterior margin of the preceding tergite. A spiracle is also present on abdominal tergite 8, but in female castes this sclerite is always concealed; it is internal and forms part of the *sting apparatus* (the *spiracular plate*).

Spiracular plate See Sting apparatus.

Spur (= calcar) See Tibial spur.

Spur formula A simple statement of the number of *tibial spurs* that are present on the prothoracic, mesothoracic, and metathoracic legs, given in that order. Thus a spur formula of 1, 2, 2 indicates that the tibia of the prothoracic (fore) leg has 1 spur, and that of the mesothoracic (middle) and metathoracic (hind) legs each have 2 spurs.

Sternite/sternum (pl. sterna) The lower or ventral *sclerite* of a segment (the *tergite* is the upper sclerite on the thoracic segments and the *abdomen*; the *pleurites* are the lateral sclerites on the sides of the thorax). The sternite may be a simple, flat, or curved plate or may be specialized or subdivided on some segments. On the prothorax the sternite (*prosternum*) is small but visible in ventral view (see *prothorax*). The sternites of the mesothorax and metathorax are internal (*mesendosternite* and *metendosternite*, respectively), the ventral surfaces of these 2 segments being composed of extensions of the pleurites to the ventral midline, where they fuse (see *mesothorax* and *metathorax*). The sternite of the *propodeum* (A1) has been lost in the course of evolution, but those of the remaining visible abdominal segments are usually distinct, although the lateral margins of some may be difficult to discern because of fusion to the tergite (*tergosternal fusion*). The sternites of A8 and A9 are membranous in the female castes, internal, and associated with the *sting apparatus*. In males the sternites of A8 and A9 are visible, and that of A9 is generally called the *subgenital plate*. Abdominal sternites are usually simple but may be subdivided or otherwise specialized. The most common modification applies to abdominal sternites 3 and 4 (uncommonly also to A5 and A6), where distinct *presternites* (see under *presclerites*) may be differentiated.

Sting apparatus (= aculeus) Internal in the female castes (workers and queens) of the ant subfamilies in which it occurs, concealed within the apical visible segments of the *abdomen*. The sting apparatus is derived from parts of the ancestral exoskeleton of abdominal segments 8 and 9, subtended by a number of pairs of sclerites that are ultimately derived from the ancestral coxal homologues of those segments; 1 pair from segment A8 and 2 pairs from segment A9. Most sclerites of the sting apparatus have accumulated a number of synonymous names, as follows:

Gonangulum (= first valvifer, = triangular plate).
Gonapophysis 8 (= first gonapophysis, = first valvula, = lancet).
Gonapophysis 9 (= second gonapophysis, = second valvula, = stylet).
Gonocoxa 9 (= second valvifer, = second gonocoxa, = oblong plate).

Gonoplac (= third valvula, = gonostylus, = sting sheath).

Tergite A8 (= spiracular plate); frequently split into a pair of hemitergites, 1 on each side.

Tergite A9 (= quadrate plate); always split into a pair of hemitergites, 1 on each side.

Sternites A8 and A9 are membranous and usually inconspicuous.

The most ventral sclerites of the apparatus are the longitudinal pair of gonapophyses 8. These are attached to the base (anterior end) of the apparatus and curve posteriorly then upward, where they meet and lock with the pair of gonapophyses 9, the next sclerites dorsally. The latter are fused along their length and form the upper portion of the channel through which venom is transmitted; the lower portion of this channel consists of the pair of gonapophyses 8. Together, all these sclerites are sometimes termed the *terebra*. The bases of gonapophyses 9 arise directly from the bases of the next sclerites dorsally, gonocoxae 9, whose length is extended by the gonoplacs, which arise directly from the apex of each gonocoxa 9 and continue its line. Gonapophyses 8 are attached basally to a pair of roughly triangular gonangula, which also articulate with the bases of gonocoxae 9. The gonangula also articulate on each side with the bases of the hemitergites of A9, which extend posterodorsally. This entire system is covered from above by the arc of tergite A8, which bears the terminal abdominal spiracle. A final small sclerite, visible in some groups, is the *furcula*. This is a small, inverted Y-shaped sclerite at the extreme base of the sting apparatus; its arms are attached to the gonapophyses from which it is probably derived.

In 2 large subfamilies (Dolichoderinae and Formicinae), this complex venom-injecting apparatus has been abandoned in favor of repugnatorial glands or a formic acid projection system. However, dissection of these highly modified forms reveals some remnants of the original sting apparatus, here adapted to different, usually supportive, functions.

Sting sheath   See Sting apparatus.

Stipes (pl. stipites)   See Maxilla and also Male genitalia.

Stridulatory system   A sound-producing system present in the female castes of a number of ant subfamilies. The system consists of a *plectrum* (= stridulatory file), located on the posterior margin of abdominal segment 3 (usually, but not always, on the *tergite*) and an extremely finely grooved *stridulitrum* on the anterior portion of abdominal segment 4. Rapid to-and-fro movement of the plectrum along the stridulitrum produces a range of chirping or buzzing sounds.

Stridulitrum   See Stridulatory system.

Strigil   See Tibial spur.

Stylet   See Sting apparatus.

Subpetiolar process   A ventral cuticular projection of the *sternite* of the *petiole* (A2), either below the node or on its anterior *peduncle*; sometimes absent but when present quite variable in shape and size.

Subgenital plate   See Abdomen and Male genitalia.

Subsessile (petiole)   See Peduncle.

Sulcus (pl. sulci)   Strictly, an external groove or impression that corresponds to an internal ridge-like inflection of the cuticle and provides mechanical rigidity. The term is also used casually for any linear impression in the cuticle, without any obvious significance, and *sulcus* is often used interchangeably with *suture*.

Supraclypeal area   See Frontal triangle.

Suture   Strictly, a line of junction between two structural *sclerites*. The suture may be articulated, where the component sclerites are linked by flexible intersegmental membranes and retain the ability to move relative to each other, or may be fused together and immobile. The term *suture* is often used interchangeably with *sulcus*.

Tagma (pl. tagmata)   A fundamental unit of the body; an ancestral section of the body that is distinct from, or separated from, other body units in both form and function. In Insecta there are 3 ancestral tagmata: *head, thorax,* and *abdomen*; but in ants the second and third of these have become much modified by secondary evolutionary developments.

Tarsal claws   See Tarsus.

Tarsus (pl. tarsi)   The collective term for the 5 small apical subsegments (tarsomeres) of any leg. The first tarsal segment (first tarsomere, basal tarsomere) of each leg articulates with the *tibia* and is termed the *basitarsus*. The next 3 tarsomeres are not individually named but the fifth, apical (terminal) tarsomere is the *pretarsus* and bears a pair of *pretarsal claws* (= ungues, sing. unguis). The inner curvature of each claw may be a simple, smooth, concave surface, or may have 1 or more preapical teeth present, or the claw may be pectinate. Sometimes a membranous lobe, the *arolium*, is present between the claws. The tarsus of the prothoracic (fore) leg is often termed the *protarsus*, that of the mesothoracic (middle) leg the *mesotarsus*, and that of the metathoracic (hind) leg the *metatarsus*. The individually named tarsomeres may be referred to in a similar way—for instance, the basitarus of the prothoracic (fore), mesothoracic (middle), and metathoracic (hind) legs may be termed *probasitarsus, mesobasitarsus* and *metabasitarsus*, respectively. In some groups of ants the metabasitarsus bears a longitudinal groove that is probably the orifice of an *exocrine gland*.

Tegula (pl. tegulae)   See Mesothorax.

Telomere   See Male genitalia.

Tentorium   See Endoskeleton.

Tergite/tergum (pl. terga)   The upper *sclerite* of a segment (the *sternite* is the lower, the *pleurite* the lateral on the *thorax*). The tergite may be a simple flat or curved plate or may be specialized or subdivided on some segments. In terms of comparative morphology, each of the 3 ancestral dorsal plates of the thorax, 1 for each segment, is termed the *notum* (pl. nota). Thus the tergite of the prothorax is composed entirely of the *pronotum*. This sclerite is hypertrophied in worker ants and extends across the dorsum and down both sides of the segment, mostly or entirely concealing the *propleuron* (see *prothorax*). The *mesonotum*, tergite of the mesothorax, may be separated from the pronotum by the *promesonotal suture*, or the pronotum and mesonotum may be fused by obliteration of the suture in some workers, to form a single sclerite, the *promesonotum*. In alate forms the mesonotum is subdivided (see *mesothorax*). The *metanotum*, tergite of the metathorax, is usually present across the dorsum as a distinct sclerite in alates but is frequently reduced and sometimes entirely lost in workers. When the metanotum is extremely reduced, the mesonotum and *propodeum* are only separated by the *metanotal groove*—a transverse impression whose base represents the last vestige of the metanotum on the dorsum (see *metathorax*). The propodeum is the tergite of the first abdominal segment (A1). The remaining visible abdominal segments (A2–A7 in females, A2–A8 in males) have tergites that are usually simple but may be subdivided or otherwise specialized. The most common modification applies to abdominal tergites 3 and 4 (uncommonly also to A5 and A6), where distinct *pretergites* and *posttergites* (see under *presclerites*) may be differentiated. In general, the abdominal tergites are free and attached to their respective sternites by flexible intersegmental membranes, but in some groups there is *tergosternal fusion* in segments A2 (petiole), A3, and A4. The *petiole* (A2) in some groups has a small lower section of the tergite split off from the main part of the sclerite by a distinct suture on each side, where they flank the sternite. These are called *laterotergites*, and in some taxa these sections of the tergite are mobile with respect to both the remainder of the tergite and also the sternite.

Tergosternal fusion   A condition of the *abdomen* where the *tergite* and *sternite* of a single segment fuse together. This may occur in some or all of segments A2 (*petiole*), A3, and A4; tergosternal fusion never occurs posterior to A4. The absence or presence of tergosternal fusion varies throughout the Formicidae, but in individual subfamilies it tends to be fairly consistent for most of the segments. For example, in workers tergosternal fusion is distributed as follows through the various subfamilies (u = unfused, p = partially fused, f = fused).

	A2	A3	A4		A2	A3	A4
Myrmeciinae	u	u	u	Leptanillinae	f	f	u
Pseudomyrmecinae	u	u	u	Martialinae	f	f	u
Aneuretinae	f	u	u	Ponerinae	u	f	f
Dolichoderinae	f	u/p	u	Heteroponerinae	u	f	
Formicinae	f	u/p	u	Ectatomminae	u/f	f	f
Myrmicinae	f	u/f	u	Proceratiinae	u/f	f	f
Amblyoponinae	u/p	u/p/f	u/f	Agroecomyrmecinae	f	u/f	f
Apomyrminae	p	f	u	Paraponerinae	f	f	f
Dorylinae	u/f	f	u				

Thorax   The classical second *tagma* of the insect body. In ants and other Hymenoptera, the apparent thorax consists of the usual 3 leg-bearing body segments of the true thorax (*prothorax, mesothorax, metathorax*), to which the tergite of the first abdominal segment (the *propodeum*) is immovably fused. This modification means that the combined "true thorax + propodeum" cannot strictly be called the thorax, as it is not homologous with the term as used otherwise throughout the Insecta. Several names have been utilized in the recent past for "true thorax + propodeum," of which 3—mesosoma, alitrunk, and truncus—have been frequent. All 3 names are somewhat misleading as far as the ants are concerned, but all are improvements over "thorax," which is morphologically inaccurate. Currently the term *mesosoma* has gained ascendency and is the name recommended in this guide.

Tibia (pl. tibiae)   The fourth segment of any leg, counting from the basal segment (*coxa*) that articulates with the *thorax*. At its apex the tibia frequently bears 1 or 2 *tibial spurs*. The tibia of the prothoracic (fore) leg is often termed the *protibia*, that of the mesothoracic (middle) leg the *mesotibia*, and that of the metathoracic (hind) leg the *metatibia*.

Tibial spur (= calcar)   Located at the apex of each *tibia*, the tibial spur is 1 or 2 basally socketed spurs. The forelegs (prothoracic legs) have a single pectinate tibial spur that is modified as part of a specialized antennal cleaning device, the *strigil*. The mesothoracic (middle) and metathoracic (hind) tibia, also referred to as *mesotibia* and *metatibia*, may each have 2, 1, or no spurs present. When present the mesotibial and metatibial spurs may be pectinate, barbed, or simple cuticular spikes. If 2 spurs are present on a tibia, it is usual for 1 to be larger than the other; in such instances the larger spur is often pectinate, while the smaller spur is simple. A simple count of the number of tibial spurs on each of the 3 legs, from front to back, is the *spur formula*.

Torulus (pl. toruli) (= antennal sclerite, = annulus antennalis)   The small annular *sclerite* that surrounds the *antennal socket* (*antennal foramen*). The torulus may be horizontal, or the part closest to the midline of the *head* may be elevated in some taxa to such an extent that the antennal socket is almost vertical. The upper arc of the torulus may be indistinguishably fused to the inner wall of the *frontal carina*, may remain discrete, or may even project laterally as a small torular lobe below the frontal carina. In some groups where the frontal carina is very slender, the torulus projects laterally beyond the outer margin of the frontal carina and becomes visible in full-face view.

Triangular plate   See Sting apparatus.

Trochanter   The second segment, counting from the base, of any leg; the small segment between the *coxa* and *femur*. In all recent ants the trochanter is a single segment but represents the result of fusion of an ancestral pair of small segments. The trochanter of the prothoracic (fore) leg is often termed the *protrochanter*, that of the mesothoracic (middle) leg the *mesotrochanter*, and that of the metathoracic (hind) leg the *metatrochanter*.

Trulleum   A basin-shaped depression near the dorsal base of the *mandible*, close to its articulation. It is bounded distally by the *basal margin* of the mandibular blade and, in those groups where it occurs, is visible just in front of the anterior clypeal margin when the mandibles are open. The trulleum in many groups is closed along its inner (medial) border by a raised ridge of cuticle, the *canthellus*. The canthellus may extend to, and even fuse with, the basal margin of the mandible (canthellus closed) or may fail to reach the basal margin (canthellus open). Proximal to both these structures, in the cuticle of the extreme dorsal base of the mandible, is a small, unsclerotized impression of variable shape, the *mandalus*. It contrasts strongly with the fully sclerotized surrounding cuticle of the mandibular base.

Unguis   See Tarsus.

Valviceps, valvura   See Male genitalia.

Valvifer, valvula   See Sting apparatus.

Venation (= neuration)   The configuration of the veins in the wings of ants and other aculeates is very much modified from the ancestral insect pattern. Within the Formicidae the venation is complex and extremely variable. There is a finely stepped series of reductions from a formicid ancestral common pattern, still represented in many subfamilies in which the venation is most complete. In general, the venation of ants is very similar to that seen elsewhere in the vespoid and apoid lineages, but in ants the cross-veins 3rs-m and 2m-cu are always absent, and there is never a *fenestra* in M·f3.

A number of conventions are useful in understanding venation and vein nomenclature:

1. The leading edge of a wing when fully open, as if in flight, is its anterior margin; the trailing edge, in the same circumstances, is the posterior margin.

2. The abbreviations of the names of the main longitudinal veins that run from the wing base (proximal) toward the apex (distal) are written with an initial capital letter—for example, M (Media), Cu (Cubitus). These abbreviations of the names are usually reduced from their forms as expressed in the ancestral insect wing. This is purely for convenience and brevity; the main vein homologues, as they appear in the ancestral insect wing, are noted below.

3. The abbreviations of names of the secondary veins (cross-veins) that extend between the main veins are written in lowercase throughout—for example, 2r-rs, cu-a—always with the most anterior vein noted first.

4. The plus sign (+) is used to indicate sections of veins that are indistinguishably fused together longitudinally—for example, Rs+M. In fused veins, the one that was ancestrally anterior, closest to the leading edge, is named first.

5. The ampersand (&) can be used to indicate sections of separate veins that are fused end to end so that they appear as a single, continuous vein. For example, the so-called stigmal vein, a hook-shaped vein that arises from the pterostigma in some groups, is actually composed of the second radial-radial sector cross-vein (2r-rs) fused to the fourth and fifth free abscissae of the radial sector vein (Rs); thus the stigmal vein is properly 2r-rs&Rs·f4-5. Note that 2r-rs is written first because it is nearest to the anterior margin of the wing.

6. Sections of main veins that occur between cross-veins, or between fused sections of main veins, are termed *free abscissae* and are represented ·f after the abbreviation of a main vein, then followed by a number; for example, M·f3 is the third free abscissa of the median vein (M). This convention is useful to indicate where sections of veins have been lost or are free from fusion with other veins.

The veins, taken in order of occurrence commencing with the anterior margin (leading edge), are listed in the most complete venation pattern seen in ants.

Forewing longitudinal veins:

Costal vein (= Costa) [C]   Undivided, never fused with any other vein, and sometimes absent.

Subcostal vein (= Subcosta) [Sc]   In the proximal part of its length Sc is fused with R and Rs (Sc+R+Rs), which is almost always the thickest vein section in the wing. Beyond the point of divergence of Rs, Sc remains fused to R (Sc+R) until close to the base of the *pterostigma* where they divide, after which the apical portion of Sc extends along the anterior margin of the pterostigma and R along its posterior margin. In ancestral homology the vein called Sc here is correctly Sc2, because in some Hymenoptera a more proximal Sc1 occurs that is absent in ants.

Radial vein (= Radius) [R]   In its basal section fused to Sc and Rs (Sc+R+Rs), then continuing as Sc+R after the divergence of Rs. At the pterostigma Sc separates from R, which thereafter shows 3 free abscissae (R·f1–R·f3), of which R·f1 and R·f2 form the posterior margin of the *pterostigma*, and R·f3 extends along the leading edge of the wing distal of the pterostigma and termination of Sc. In ancestral homology the vein called R here is correctly R1, because in many insect groups R branches into several veins toward its apex, R1, R2, R3, etc., of which R1 is the only remnant in Hymenoptera.

Radial sector [Rs]   Proximally fused with Sc and R (Sc+R+Rs). Beyond its point of divergence, Rs has a maximum of 5 free abscissae (Rs·f1–Rs·f5) and always has a section fused with M (Rs+M), which is present between Rs·f1 and Rs·f2. The latter, in more generalized forms, has a distinct *fenestra* through which the *radial flexion line* passes. Rs·f5 may curve anteriorly to the leading edge, where it meets R·f3, but often ends before reaching the margin. In its distant ancestry, Rs probably originated as a vein independent of R, but in all Hymenoptera, and almost all insects, R and Rs are fused basally. In ancestral homology the vein called Rs here is correctly Rs1. In a very few Hymenoptera, Rs branches in its distal half into Rs1 and a more posterior Rs2.

Median vein (= Media) [M]   Basally fused with Cu (M+Cu), then with a free abscissa (M·f1) before its fused section with Rs (Rs+M). Distal of this, M has a maximum of 3 further free abscissae (M·f2-4). In ancestral homology the vein called M here is correctly MP (media posterior), the posterior member of 2 original median veins. The anterior median vein (MA, media anterior) has been lost in Hymenoptera and most other insects.

Cubital vein (= Cubitus) [Cu]   Basally fused with M (M+Cu), then with a maximum of 3 free
   abscissae (Cu·f1-3) before branching, near the posterior wing margin, into Cu1 and Cu2.
   Cu2 sometimes anastomoses with the anal vein, but when free it has a fenestra through
   which the *claval furrow* passes. In ancestral homology the vein called Cu here is correctly
   CuA (cubitus anterior), and its terminal branches are CuA1 and CuA2. This is because
   it is the anterior member of 2 original cubital veins, the posterior of which (CuP, cubitus
   posterior) has been lost in Hymenoptera.

Anal vein [A]   A single unbranched, unfused vein that is closest to the posterior margin of the
   wing, with 3 free abscissae (A·f1-3) at maximum. In ancestral homology the vein called A
   here is most probably A1. In some groups of Hymenoptera 2, 3, or rarely 4 anal veins may
   be present (A1, A2, etc.).

Forewing cross-veins (a maximum of 5):

First radial-radial sector cross-vein [1r-rs]   Extremely rare in ants; when present it extends from
   R, close to the proximal end of the *pterostigma*, to Rs, where it marks the junction of Rs·f2 and
   Rs·f3. Because 1r-rs is only rarely present, this section of Rs is most often seen as Rs·f2-3.

Second radial-radial sector cross-vein [2r-rs]   Extends from R at the base of the pterostigma
   (where it marks the junction of R·f1 and R·f2) to Rs (where it marks the junction of Rs·f3
   and Rs·f4). In the most generalized ant venations 2r-rs is proximal of 2rs-m.

Second radial sector-median cross-vein [2rs-m]   Extends from the junction of Rs·f4 and Rs·f5
   to the junction of M·f3 and M·f4. In more generalized forms, 2rs-m has a distinct *fenestra*
   through which the *radial flexion line* passes. In the most generalized ant venations, 2rs-m
   is distal of 2r-rs. The cross-vein 1rs-m is not present in any Hymenoptera, having been
   obliterated by the proximal end of the fusion of Rs with M (Rs+M). A third radial sector-
   median cross-vein, 3rs-m, is never developed in Formicidae but very common in other
   aculeates.

First median-cubital cross-vein [1m-cu]   Extends from the junction of M·f2 and M·f3 to the
   junction of Cu·f2 and Cu·f3. A second median-cubital cross-vein, 2m-cu, is always absent in
   Formicidae, though common in apoids and vespoids.

Cubital-anal cross-vein [cu-a]   In forms with the most generalized venation, cu-a extends from
   the junction of Cu·f1 and Cu·f2 to the junction of A·f1 and A·f2. Extremely commonly, cu-a
   is retracted toward the wing base and arises from M+Cu. There is usually a *fenestra* in cu-a
   through which the *claval furrow* passes.

Hindwing veins   These are basically the same as in the forewing but somewhat simplified.
   For instance, the hindwing has no pterostigma and no section Rs+M, the main veins have
   fewer abscissae, the apical abscissa of Cu is simple, and there are no r-rs or m-cu cross-
   veins. Cross-veins 1rs-m and cu-a have fenestrae, through which the radial flexion line and
   claval furrow respectively pass. In a considerable number of ant taxa M·f2 has disappeared
   from the hindwing, so the apparently single vein that extends between Rs·f2 and Cu is
   actually 1rs-m&M·f1.

See also *Wings* and *Cells of the wings*.
   The minimal forewing venation observed in ants consists merely of a very faint remnant of
Sc+R+Rs.

Vertex   The portion of the cephalic dorsum that lies immediately in front of the *occiput*. In those
groups where *ocelli* are absent, the area can be only vaguely defined, but in those that possess
ocelli the vertex is the area from immediately behind the anterior ocellus to the occiput containing

the posterior pair of ocelli. Ancestrally, the anterior ocellus is on, and marks the posterior limit of, the *frons*.

Volsella See Male genitalia.

Waist An informal collective term for the 1 or 2 isolated and reduced abdominal segments that occur between the *mesosoma* and *gaster*. When only the *petiole* (A2) is isolated, the waist is said to be 1-segmented, but in those taxa where the *postpetiole* (A3) is also separated, the waist is said to be 2-segmented.

Wings (alae, sing. ala) Alate ants, winged queens and males, have 2 pairs of wings, as do all winged Hymenoptera. The forewings are largest and articulate with the *mesothorax*, where their extreme base is shielded by a *tegula* on each side. The hindwings are smaller and articulate with the *metathorax*. The forewing and hindwing on each side are held together by a series of small hooks, the *hamuli* (sing. hamulus), that arise from the leading edge of the hindwing and engage folds on the posterior margin of the forewing. This ensures that the forewing and hindwing on each side beat as a single unit in flight. In ants the hamuli usually arise only from the hindwing vein R, but in a few groups another patch of hamuli is present more basally, on vein C.

About two-thirds of the way along the anterior forewing there is usually a pigmented patch, surrounded by veins, the *pterostigma* (= stigma), that is only very rarely absent. The main membranous area of each wing is the *remigium*, which contains the entire area between the leading edge and the *claval furrow* (= vannal fold), immediately anterior to vein A. Posterior to the claval furrow, on both forewing and hindwing, is a membranous area termed the *clavus* (= claval lobe, = vannal lobe, = plical lobe), which usually terminates in a slight *claval notch* (= preaxillary excision, = vannal notch) in the posterior wing margin. In some taxa, proximal of the clavus on the hindwing is the *jugal lobe* (= jugum, = anal lobe).

Within the membrane of the wings there are usually also 2 *flexion lines* or *fold lines* that allow the wing to flex during active flight. In the forewing the *radial flexion line* runs parallel and immediately posterior to veins Sc+R+Rs and Sc+R. Toward its distal end the flexion line curves anteriorly toward the base of the *pterostigma*, where it is contiguous with the hinge-like mechanism, a weakened area at the pterostigmal base that allows the outer part of the forewing to deform during flight. From there the radial flexion line extends for a short distance posteriorly, then curves distally and extends outward toward the wing margin. In many groups the track of the radial flexion line is marked by a distinct *fenestra* in vein section Rs·f2 and in cross-vein 2rs-m. Just anterior to vein A, the most posterior longitudinal vein, is the *claval furrow* (= vannal fold). This passes through a fenestra in cross-vein cu-a and usually also a fenestra in vein Cu2.

The same flexion lines occur in the hindwing, but the radial flexion line is simpler. It runs longitudinally, posterior to Sc+R+Rs, often through a fenestra in cross-vein 1rs-m, then out toward the wing margin posterior to Rs·f2. The claval furrow runs just anterior to A, with a fenestra in cross-vein cu-a. In those taxa where the hindwing retains a *jugal lobe*, the lobe is separated from the clavus by the *jugal fold*, which in ants is often represented by a cleft in the membrane. In addition to these features, each wing usually bears a conspicuous series of longitudinal and transverse veins, collectively termed the *venation*, together with a series of cells, which are areas of membrane enclosed by particular veins or sections of veins. Both of these are discussed separately here in *Venation* and *Cells of the wings*.

REFERENCES

Most references apply to relatively recent (post-1960) taxonomic studies that provide keys for the identification of Afrotropical and Malagasy species within the genera defined in this guide. A number of older key references are cited, either when no recent improvements have been attempted or to illustrate how the taxonomy has expanded and changed, but these should be consulted with care as they frequently contain subspecific and infrasubspecific (unavailable) names. In addition, a number of other references are included that contain historical information relating to the development of taxonomy and classification in the regions under consideration here or relating to improvements in the phylogeny.

Alpert, G.D. 2007. A review of the ant genus *Metapone* Forel from Madagascar. *Memoirs of the American Entomological Institute* 80: 8–18.

Arnold, G. 1915. A monograph of the Formicidae of South Africa. Part 1. (Ponerinae; Dorylinae). *Annals of the South African Museum* 14: 1–159.

Arnold, G. 1916. A monograph of the Formicidae of South Africa. Part 2. (Ponerinae; Dorylinae). *Annals of the South African Museum* 14: 159–270.

Arnold, G. 1917. A monograph of the Formicidae of South Africa. Part 3. (Myrmicinae). *Annals of the South African Museum* 14: 271–402.

Arnold, G. 1920. A monograph of the Formicidae of South Africa. Part 4. (Myrmicinae). *Annals of the South African Museum* 14: 403–578.

Arnold, G. 1922. A monograph of the Formicidae of South Africa. Part 5. (Myrmicinae). *Annals of the South African Museum* 14: 579–674.

Arnold, G. 1924. A monograph of the Formicidae of South Africa. Part 6. (Camponotinae). *Annals of the South African Museum* 14: 675–766.

Arnold, G. 1926. A monograph of the Formicidae of South Africa. Appendix. *Annals of the South African Museum* 23: 191–295.

Arnold, G. 1951. The genus *Hagensia* Forel. *Journal of the Entomological Society of Southern Africa* 14: 53–56.

Arnold, G. 1952. New species of African Hymenoptera. No. 10. *Occasional Papers of the National Museum of Southern Rhodesia* 2 (no. 17): 460–493.

Baroni Urbani, C. 1977. Materiali per una revisione della sottofamiglia Leptanillinae Emery. *Entomologica Basiliensia* 2: 427–488.

Baroni Urbani, C., Bolton, B., and Ward, P.S. 1992. The internal phylogeny of ants. *Systematic Entomology* 17: 301–329.

Baroni Urbani, C., and De Andrade, M.L. 2003. The ant genus *Proceratium* in the extant and fossil record. Museo Regionale di Scienze Naturali–Torino *Monographie* 36: 492 pp.

Bernard, F. 1953. La réserve naturelle intégrale du Mt Nimba. 11. Hyménoptères Formicidae. *Mémoires de l'Institut Français d'Afrique Noire* 19 (1952): 165–270.

Blaimer, B.B. 2010. Taxonomy and natural history of the *Crematogaster (Decacrema)*-group in Madagascar. *Zootaxa* 2714: 1–39.

Blaimer, B.B. 2012a. Untangling complex morphological variation: Taxonomic revision of the subgenus *Crematogaster (Oxygyne)* in Madagascar, with insight into the evolution and biogeography of this enigmatic ant clade. *Systematic Entomology* 37: 240–260.

Blaimer, B.B. 2012b. Taxonomy and species-groups of the subgenus *Crematogaster (Orthocrema)* in the Malagasy region. *ZooKeys* 199: 23–70.

Blaimer, B.B. 2012c. A subgeneric revision of *Crematogaster* and discussion of regional species-groups. *Zootaxa* 3482: 47–67.

Blaimer, B.B., and Fisher, B.L. 2013a. How much variation can one ant species hold? Species delimitation in the *Crematogaster kelleri*-group in Madagascar. *PloS ONE* 8 (7): 31 pp. e68082. doi:10.1371/journal.pone.0068082.

Blaimer, B.B., and Fisher, B.L. 2013b. Taxonomy of the *Crematogaster degeeri*-species-assemblage in the Malagasy region. *European Journal of Taxonomy* 51: 1–64.

Bolton, B. 1972. Two new species of the ant genus *Epitritus* from Ghana, with a key to the world species. *Entomologist's Monthly Magazine* 107 (1971): 205–208.

Bolton, B. 1973a. The ant genus *Polyrhachis* F. Smith in the Ethiopian region. *Bulletin of the British Museum (Natural History)* (Entomology) 28: 283–369.

Bolton, B. 1973b. A remarkable new arboreal ant genus from West Africa. *Entomologist's Monthly Magazine* 108 (1972): 234–237.

Bolton, B. 1974a. A revision of the Palaeotropical arboreal ant genus *Cataulacus* F. Smith. *Bulletin of the British Museum (Natural History)* (Entomology) 30: 1–105.

Bolton, B. 1974b. A revision of the ponerine ant genus *Plectroctena* F. Smith. *Bulletin of the British Museum (Natural History)* (Entomology) 30: 309–338.

Bolton, B. 1975a. A revision of the ant genus *Leptogenys* Roger in the Ethiopian region, with a review of the Malagasy species. *Bulletin of the British Museum (Natural History)* (Entomology) 31: 235–305.

Bolton, B. 1975b. A revision of the African ponerine ant genus *Psalidomyrmex* André. *Bulletin of the British Museum (Natural History)* (Entomology) 32: 1–16.

Bolton, B. 1976. The ant tribe Tetramoriini: Constituent genera, review of smaller genera and revision of *Triglyphothrix* Forel. *Bulletin of the British Museum (Natural History)* (Entomology) 34: 281–379.

Bolton, B. 1979. The ant tribe Tetramoriini: The genus *Tetramorium* Mayr in the Malagasy region and in the New World. *Bulletin of the British Museum (Natural History)* (Entomology) 38: 129–181.

Bolton, B. 1980. The ant tribe Tetramoriini: The genus *Tetramorium* Mayr in the Ethiopian zoogeographical region. *Bulletin of the British Museum (Natural History)* (Entomology) 40: 193–384.

Bolton, B. 1981a. A revision of the ant genera *Meranoplus* F. Smith, *Dicroaspis* Emery, and *Calyptomyrmex* Emery in the Ethiopian zoogeographical region. *Bulletin of the British Museum (Natural History)* (Entomology) 42: 43–81.

Bolton, B. 1981b. A revision of six minor genera of Myrmicinae in the Ethiopian zoogeographical region. *Bulletin of the British Museum (Natural History)* (Entomology) 43: 245–307.

Bolton, B. 1982. Afrotropical species of the myrmicine ant genera *Cardiocondyla, Leptothorax, Melissotarsus, Messor,* and *Cataulacus. Bulletin of the British Museum (Natural History)* (Entomology) 45: 307–370.

Bolton, B. 1983. The Afrotropical dacetine ants. *Bulletin of the British Museum (Natural History)* (Entomology) 46: 267–416.

Bolton, B. 1985. The ant genus *Triglyphothrix* Forel a synonym of *Tetramorium* Mayr. *Journal of Natural History* 19: 243–248.

Bolton, B. 1986. A taxonomic and biological review of the tetramoriine ant genus *Rhoptromyrmex. Systematic Entomology* 11: 1–17.

Bolton, B. 1987. A review of the *Solenopsis* genus-group and revision of Afrotropical *Monomorium* Mayr. *Bulletin of the British Museum (Natural History)* (Entomology) 54: 263–452.

Bolton, B. 1990a. The higher classification of the ant subfamily Leptanillinae. *Systematic Entomology* 15: 267–282.

Bolton, B. 1990b. Army ants reassessed: The phylogeny and classification of the doryline section. *Journal of Natural History* 24: 1339–1364.

Bolton, B. 1994. *Identification Guide to the Ant Genera of the World.* Cambridge, Mass. 222 pp.

Bolton, B. 1995. *A New General Catalogue of the Ants of the World.* Cambridge, Mass. 504 pp.

Bolton, B. 2000. The ant tribe Dacetini. *Memoirs of the American Entomological Institute* 65: 1028 pp.

Bolton, B. 2003. Synopsis and classification of Formicidae. *Memoirs of the American Entomological Institute* 71: 370 pp.

Bolton, B. 2007. Taxonomy of the dolichoderine ant genus *Technomyrmex* Mayr based on the worker caste. *Contributions of the American Entomological Institute* 35 (1): 1–150.

Bolton, B., and Belshaw, R. 1993. Taxonomy and biology of the supposedly lestobiotic ant genus *Paedalgus. Systematic Entomology* 18: 181–189.

Bolton, B., and Brown, W.L., Jr. 2002. *Loboponera* gen. n. and a review of the Afrotropical *Plectroctena* genus group. *Bulletin of the Natural History Museum (Entomology Series)* 71: 1–18.

Bolton, B., and Fisher, B.L. 2008a. The Afrotropical ponerine ant genus *Asphinctopone* Santschi. *Zootaxa* 1827: 53–61.

Bolton, B., and Fisher, B.L. 2008b. The Afrotropical ponerine ant genus *Phrynoponera* Wheeler. *Zootaxa* 1892: 35–52.

Bolton, B. and Fisher, B.L. 2008c. Afrotropical ants of the ponerine genera *Centromyrmex* Mayr, *Promyopias* Santschi gen. rev., and *Feroponera* gen. n., with a revised key to genera of African Ponerinae. *Zootaxa* 1929: 1–37.

Bolton, B., and Fisher, B.L. 2011. Taxonomy of Afrotropical and West Palaearctic ants of the ponerine genus *Hypoponera* Santschi. *Zootaxa* 2843: 1–118.

Bolton, B., and Fisher, B.L. 2012. Taxonomy of the cerapachyine ant genera *Simopone* Forel, *Vicinopone* gen. n., and *Tanipone* gen. n. *Zootaxa* 3283: 1–101.

Bolton, B., and Fisher, B.L. 2014. The Madagascan endemic myrmicine ants related to *Eutetramorium*: Taxonomy of the genera *Eutetramorium* Emery, *Malagidris* nom. n., *Myrmisaraka* gen. n., *Royidris* gen. n., and *Vitsika* gen. n. *Zootaxa* 3791: 1–99.

Bolton, B.,, and Marsh, A.C. 1989. The Afrotropical thermophilic ant genus *Ocymyrmex. Journal of Natural History* 23: 1267–1308.

Boudinot, B.E., and Fisher, B.L. 2013. A taxonomic revision of the *Meranoplus* F. Smith of Madagascar with keys to species and diagnosis of the males. *Zootaxa* 3635: 301–339.

Brady, S.G. 2003. Evolution of the army ant syndrome: The origin and long-term evolutionary stasis of a complex of behavioral and reproductive adaptations. *Proceedings of the National Academy of Sciences USA* 100: 6575–6579.

Brady, S.G., and Ward, P.S. 2005. Morphological phylogeny of army ants and other dorylomorphs. *Systematic Entomology* 30: 592–618.

Brady, S.G., Fisher, B.L., Schultz, T.R., and Ward, P.S. 2014. The rise of army ants and their relatives: Diversification of specialized predatory doryline ants. *BMC Evolutionary Biology* 14 (93): 14 pp.

Brady, S.G., Schultz, T.R., Fisher, B.L., and Ward, P.S. 2006. Evaluating alternative hypotheses for the early evolution and diversification of ants. *Proceedings of the National Academy of Sciences USA* 103: 18172–18177.

Brown, W.L., Jr. 1952. Revision of the ant genus *Serrastruma*. *Bulletin of the Museum of Comparative Zoology at Harvard College* 107: 67–86.

Brown, W.L., Jr. 1953. Revisionary studies in the ant tribe Dacetini. *American Midland Naturalist* 50: 1–137.

Brown, W.L., Jr. 1954. The ant genus *Strumigenys* Fred. Smith in the Ethiopian and Malagasy regions. *Bulletin of the Museum of Comparative Zoology at Harvard College* 112: 3–34.

Brown, W.L., Jr. 1958. Contributions toward a reclassification of the Formicidae. 2. Tribe Ectatommini. *Bulletin of the Museum of Comparative Zoology at Harvard College* 118: 175–362.

Brown, W.L., Jr. 1960. Contributions toward a reclassification of the Formicidae. 3. Tribe Amblyoponini. *Bulletin of the Museum of Comparative Zoology at Harvard College* 122: 145–230.

Brown, W.L., Jr. 1964. Genus *Rhoptromyrmex*, revision of, and key to species. *Pilot Register of Zoology* cards 11–19.

Brown, W.L., Jr. 1974a. *Concoctio* genus nov.; *C. concenta* species nov. *Pilot Register of Zoology* cards 29–30.

Brown, W.L., Jr. 1974b. *Dolioponera* genus nov.; *D. fustigera* species nov. *Pilot Register of Zoology* cards 31–32.

Brown, W.L., Jr. 1975. Contributions toward a reclassification of the Formicidae. 5. Ponerinae, tribes Platythyreini, Cerapachyini, Cylindromyrmecini, Acanthostichini, and Aenictogitini. *Search Agriculture* 5. Entomology (Ithaca) 15: 1–115.

Brown, W.L., Jr. 1976. Contributions toward a reclassification of the Formicidae. Part 6. Ponerinae, tribe Ponerini, subtribe Odontomachiti. Section A. Introduction, subtribal characters, genus *Odontomachus*. *Studia Entomologica* (n.s.) 19: 67–171.

Brown, W.L., Jr. 1978. Contributions toward a reclassification of the Formicidae. Part 6. Ponerinae, tribe Ponerini, subtribe Odontomachiti. Section B. Genus *Anochetus* and bibliography. *Studia Entomologica* (n.s.) 20: 549–652.

Brown, W.L., Jr., Gotwald, W.H., and Lévieux, J. 1971. A new genus of ponerine ants from West Africa, with ecological notes. *Psyche* 77 (1970): 259–275.

Dalla Torre, C.G. de. 1893. *Catalogus Hymenopterorum, hucusque descriptorum systematicus et synonymicus* 7: 289 pp. Lipsiae.

Dlussky, G.M., and Fedoseeva, E.B. 1988. Proiskhozhdenie i rannie etapy evolyutsii murav'ev. Pp. 70–144. In Ponomarenko, A.G. *Melovoi Biotsenoticheskii Krizis i Evolyutsiya Nasekomykh.* Moskva. 228 pp.

Eguchi, K., Bui, T.V., General, D.M., and Alpert, G.D. 2009. Revision of the ant genus *Anillomyrma* Emery, 1913. *Myrmecological News* 13: 31–36.

Emery, C. 1895. Die Gattung *Dorylus* Fab. und die systematische Eintheilung der Formiciden. *Zoologische Jahrbücher: Abtheilung für Systematik, Geographie, und Biologie der Thiere* 8: 685–778.

Emery, C. 1902. Note mirmecologiche. *Rendiconto delle Sessioni della R. Accademia delle Scienze dell'Istituto di Bologna* (n.s.) 6 (1901): 22–34.

Emery, C. 1910. In Wytsman, P. *Genera Insectorum*. Hymenoptera, Fam. Formicidae, subfam. Dorylinae. Fasc. 102. Brussells. 34 pp.

Emery, C. 1911. In Wytsman, P. *Genera Insectorum*. Hymenoptera, Fam. Formicidae, subfam. Ponerinae. Fasc. 118. Brussells. 124 pp.

Emery, C. 1913. In Wytsman, P. *Genera Insectorum*. Hymenoptera, Fam. Formicidae, subfam. Dolichoderinae. Fasc. 137 (1912). Brussells. 50 pp.

Emery, C. 1921. In Wytsman, P. *Genera Insectorum*. Hymenoptera, Fam. Formicidae, subfam. Myrmicinae. Fasc. 174A: 1–94. Brussells.

Emery, C. 1922. In Wytsman, P. *Genera Insectorum*. Hymenoptera, Fam. Formicidae, subfam. Myrmicinae. Fasc. 174B: 95–206. Brussells.

Emery, C. 1924. In Wytsman, P. *Genera Insectorum*. Hymenoptera, Fam. Formicidae, subfam. Myrmicinae. Fasc. 174C (1922): 207–397.

Emery, C. 1925. In Wytsman, P. *Genera Insectorum*. Hymenoptera, Fam. Formicidae, subfam. Formicinae. Fasc. 183. Brussells. 302 pp.

Fernández, F. 2003. Revision of the myrmicinae ants of the *Adelomyrmex* genus-group. *Zootaxa* 361: 1–52.

Fischer, G., and Fisher, B.L. 2013. A revision of *Pheidole* Westwood in the islands of the southwest Indian Ocean and designation of a neotype for the invasive *Pheidole megacephala*. *Zootaxa* 3683: 301–356.

Fischer, G., Azorsa, F., and Fisher, B.L. 2014. The ant genus *Carebara* Westwood: Synonymisation of *Pheidologeton* Mayr under *Carebara*, establishment and revision of the *C. polita* species group. *ZooKeys* 438: 57–112.

Fischer, G., Hita Garcia, F., and Peters, M.K. 2012. Taxonomy of the genus *Pheidole* Westwood in the Afrotropical zoogeographical region: Definition of species groups and systematic revision of the *Pheidole pulchella* group. *Zootaxa* 3232: 1–43.

Fisher, B.L. 1998. Ant diversity patterns along an elevational gradient in the Réserve Spéciale d'Anjanaharibe-Sud and on the western Masoala Peninsula. *Fieldiana: Zoology* 90: 39–67.

Fisher, B.L. 2000. The Malagasy fauna of *Strumigenys*. Pp. 612–696. In Bolton, B. The ant tribe Dacetini. *Memoirs of the American Entomological Institute* 65: 1028 pp.

Fisher, B.L. 2006. *Boloponera vicans* gen. n. and sp. n. and two new species of the *Plectroctena* genus group. *Myrmecologische Nachrichten* 8: 111–118.

Fisher, B.L. 2009. Two new dolichoderine ant genera from Madagascar: *Aptinoma* gen. n. and *Ravavy* gen. n. *Zootaxa* 2118: 37–52.

Fisher, B.L., and Smith, M.A. 2008. A revision of Malagasy species of *Anochetus* Mayr and *Odontomachus* Latreille. *PloS ONE* 3 (5): 23 pp. e1787. doi:10.1371/ journal.pone.0001787.

Forel, A. 1891. In Grandidier, A. *Histoire Physique, Naturelle et Politique de Madagascar* 20. Histoire naturelle des Hyménoptères. 2 (fascicule 28). Les Formicides. Paris. Pp. 1–231.

Forel, A. 1892. In Grandidier, A. *Histoire Physique, Naturelle et Politique de Madagascar* 20. Histoire naturelle des Hyménoptères. 2 (supplément au 28 fascicule). Les Formicides. Paris. Pp. 229–280.

Gauld, I. D., and Bolton, B., editors. 1988. *The Hymenoptera*. Oxford University Press + British Museum (Natural History). Second edition, 1996. 332 pp.

Gotwald, W.H., Jr. 1982. Army Ants. Pp. 157–254. In Hermann, H.R., ed. *Social Insects* 4. New York. 385 pp.

Goulet, H., and Huber, J.T. 1993. *Hymenoptera of the world: An identification guide to families*. Centre for Land and Biological Resources Research, Ottawa, Canada. 668 pp.

Hawkes, P.G. 2010. A new species of *Asphinctopone* from Tanzania. *Zootaxa* 2480: 27–36.

Heterick, B.E. 2006. A revision of the Malagasy ants belonging to genus *Monomorium* Mayr, 1855. *Proceedings of the California Academy of Sciences* 57: 69–202.

Hita Garcia, F., and Fischer, G. 2014. Additions to the taxonomy of the Afrotropical *Tetramorium weitzeckeri* species complex, with the description of a new species from Kenya. *European Journal of Taxonomy* 90: 1–16.

Hita Garcia, F., Fischer, G., and Peters, M.K. 2010. Taxonomy of the *Tetramorium weitzeckeri* species group in the Afrotropical zoogeographical region. *Zootaxa* 2704: 1–90.

Hita Garcia, F., and Fisher, B.L. 2011. The ant genus *Tetramorium* Mayr in the Malagasy region—introduction, definition of species groups, and revision of the *T. bicarinatum, T. obesum, T. sericeiventre,* and *T. tosii* species groups. *Zootaxa* 3039: 1–72.

Hita Garcia, F., and Fisher, B.L. 2012a. The ant genus *Tetramorium* Mayr in the Malagasy region—taxonomy of the *T. bessonii, T. bonibony, T. dysalum, T. marginatum, T. tsingy,* and *T. weitzeckeri* species groups. *Zootaxa* 3365: 1–123.

Hita Garcia, F., and Fisher, B.L. 2012b. The ant genus *Tetramorium* Mayr in the Malagasy region—taxonomic revision of the *T. kelleri* and *T. tortuosum* species groups. *Zootaxa* 3592: 1–85.

Hita Garcia, F., and Fisher, B.L. 2013. The *Tetramorium tortuosum* group revisited—taxonomic revision of the Afrotropical *T. capillosum* species complex. *ZooKeys* 299: 77–99.

Hita Garcia, F., and Fisher, B.L. 2014a. The ant genus *Tetramorium* Mayr in the Afrotropical region: Synonymisation of *Decamorium* Forel under *Tetramorium*, and taxonomic revision of the *T. decem* species group. *ZooKeys* 411: 67–103.

Hita Garcia, F., and Fisher, B.L. 2014b. The hyper-diverse ant genus *Tetramorium* Mayr in the Malagasy region—taxonomic revision of the *T. naganum, T. plesiarum, T. schaufussii,* and *T. severini* species groups. *ZooKeys* 413: 1–170.

Hita Garcia, F., and Fisher, B.L. 2014c. Taxonomic revision of the cryptic ant genus *Probolomyrmex* Mayr in Madagascar. *Deutsche Entomologische Zeitschrift* 61: 65–76.

Hita Garcia, F., and Fisher, B.L. 2015. Taxonomy of the hyper-diverse ant genus *Tetramorium* Mayr in the Malagasy region-first record of the *T. setigerum* species group and additions to the Malagasy species groups with an updated illustrated identification key. *ZooKeys* 512: 121–153.

Hita Garcia, F., Hawkes, P.G., and Alpert, G.D. 2014. Taxonomy of the ant genus *Proceratium* Roger in the Afrotropical region with a revision of the *P. arnoldi* clade and description of four new species. *ZooKeys* 447: 47–86.

Ketterl, J., Verhaagh, M., and Dietz, B.H. 2004. *Eurhopalothrix depressa* sp. n. from southern Brazil, with a key to the Neotropical taxa of the genus. *Studies in Neotropical Fauna and Environment* 39: 45–48.

LaPolla, J.S. 2004a. *Acropyga* of the world. *Contributions of the American Entomological Institute* 33 (3): 1–130.

LaPolla, J.S. 2004b. Taxonomic review of the ant genus *Pseudolasius* in the Afrotropical region. *Journal of the New York Entomological Society* 112: 97–105.

LaPolla, J.S., and Fisher, B.L. 2005. A remarkable new species of *Acropyga* from Gabon, with a key to the Afrotropical species. *Proceedings of the California Academy of Sciences* (4) 56: 601–605.

LaPolla, J.S., and Fisher, B.L. 2014a. Two new *Paraparatrechina* species from the Seychelles, with notes on the hypogaeic *weissi* species-group. *ZooKeys* 414: 139–155.

LaPolla, J.S., and Fisher, B.L. 2014b. Then there were five: A reexamination of the ant genus *Paratrechina*. *ZooKeys* 422: 35–48.

LaPolla, J.S., Brady, S.G., and Shattuck, S.O. 2010. Phylogeny and taxonomy of the *Prenolepis* genus-group of ants. *Systematic Entomology* 35: 118–131.

LaPolla, J.S., Cheng, C.H., and Fisher, B.L. 2010. Taxonomic revision of the ant genus *Parapara-trechina* in the Afrotropical and Malagasy regions. *Zootaxa* 2387: 1–27.

LaPolla, J.S., Hawkes, P.G., and Fisher, B.L. 2011. Monograph of *Nylanderia* of the world, part 1: *Nylanderia* in the Afrotropics. *Zootaxa* 3011: 10–36.

LaPolla, J.S., Hawkes, P.G., and Fisher, J.N. 2013. Taxonomic review of the ant genus *Paratrechina*, with a description of a new species from Africa. *Journal of Hymenoptera Research* 35: 71–82.

Longino, J.T. 2012. A review of the ant genus *Adelomyrmex* Emery, 1897, in Central America. *Zootaxa* 3456: 1–35.

Longino, J.T. 2013. A review of the Central American and Caribbean species of the ant genus *Eurhopalothrix* Brown and Kempf, 1961, with a key to New World species. *Zootaxa* 3693: 101–151.

Longino, J.T., and Fernández, F. 2007. Taxonomic review of the genus *Wasmannia*. *Memoirs of the American Entomological Institute* 80: 271–289.

Mayr, G. 1863. Formicidarum index synonymicus. *Verhandlungen der k.k. Zoologisch-Botanischen Gesellschaft in Wien* 13: 385–460.

Mayr, G. 1865. *Reise der Österreichischen Fregatte Novara um die Erde in den Jahren 1857, 1858, 1859, unter den befehlen des Commodore B. von Wüllerstorf-Urbair*. Zoologischer Theil. Formicidae. Vienna. 119 pp.

Mayr, G. 1872. Formicidae Borneenses collectae a J. Doria et O. Beccari in territorio Sarawak annis 1865-1867. *Annali del Museo Civico di Storia Naturale di Genova* 2: 133–155.

Mbanyana, N., and Robertson, H.G. 2008. Review of the ant genus *Nesomyrmex* in southern Africa. *African Natural History* 4: 35–55.

Menozzi, C. 1929. Revisione delle formiche del genere *Mystrium* Roger. *Zoologischer Anzeiger* 82: 518–536.

Moreau, C.S. 2009. Inferring ant evolution in the age of molecular data. *Myrmecological News* 12: 201–210.

Moreau, C.S., and Bell, C.D. 2013. Testing the museum versus cradle tropical biological diversity hypotheses: Phylogeny, diversification, and ancestral biogeographic range evolution of the ants. *Evolution* 67. 18 pp.

Moreau, C.S., Bell, C.D., Vila, R., Archibald, S.B., and Pierce, N.E. 2006. Phylogeny of the ants: Diversification in the age of angiosperms. *Science* 312: 101–104.

Ouellette, G.D., Fisher, B.L., and Girman, D.J. 2006. Molecular sytematics of basal subfamilies of ants using 28S rRNA. *Molecular Phylogenetics and Evolution* 40: 359–369.

Overson, R., and Fisher, B.L. 2015. Taxonomic revision of the genus *Prionopelta* in the Malagasy region. *ZooKeys* 507: 115–150.

Prebus, M. 2015. Palearctic elements in the old world tropics: A taxonomic revision of the ant genus *Temnothorax* Mayr for the Afrotropical biogeographical region. *ZooKeys* 483: 23–57.

Prins, A.J. 1982. Review of *Anoplolepis* with reference to male genitalia, and notes on *Acropyga*. *Annals of the South African Museum* 89: 215–247.

Prins, A.J. 1983. A new ant genus from southern Africa. *Annals of the South African Museum* 94: 1–11.

Raignier, A., and van Boven, J.K.A. 1955. Etude taxonomique, biologique et biométrique des *Dorylus* du sous-genre *Anomma*. Annales du Musée Royal du Congo Belge, Tervuren (n.s. in quarto). *Sciences Zoologiques* 2: 1–359.

Rakotonirina, J.C., and Fisher, B.L. 2013a. Revision of the *Pachycondyla wasmannii*-group from the Malagasy region. *Zootaxa* 3609: 101–141.

Rakotonirina, J.C., and Fisher, B.L. 2013b. Revision of the *Pachycondyla sikorae* species-group in Madagascar. *Zootaxa* 3683: 447–485.

Rakotonirina, J.C., and Fisher, B.L. 2014. Revision of the Malagasy ponerine ants of the genus *Leptogenys* Roger. *Zootaxa* 3836: 1–163.

Rigato, F. 2002. Three new Afrotropical *Cardiocondyla* Emery, with a revised key to the workers. *Bollettino della Società Entomologica Italiana* 134: 167–173.

Robertson, H.G. 2002. Revision of the ant genus *Streblognathus*. *Zootaxa* 97: 1–16.

Robertson, H.G., and Zachariades, C. 1997. Revision of the *Camponotus fulvopilosus* (De Geer) species-group. *African Entomology* 5: 1–18.

Roger, J. 1863. Verzeichniss der Formiciden-Gattungen und Arten. *Berliner Entomologische Zeitschrift* 7 (Beilage): 1–65.

Santschi, F. 1913a. Clé analytique des fourmis africaines du genre *Strumigenys* Sm. *Bulletin de la Société Entomologique de France* 1913: 257–259.

Santschi, F. 1913b. Clé dichotomique des *Oligomyrmex* africains. *Bulletin de la Société Entomologique de France* 1913: 459–460.

Santschi, F. 1914. Formicides de l'Afrique occidentale et australe du voyage de Mr. le Professeur F. Silvestri. *Bollettino del Laboratorio di Zoologia generale e agraria della R. Scuola superiore d'Agricoltura in Portici* 8: 309–385.

Santschi, F. 1919. Nouvelles fourmis du Congo Belge du Musée du Congo Belge, à Tervuren. *Revue Zoologique Africaine* 7: 79–91.

Santschi, F. 1921. Quelques nouveaux formicides africains. *Annales de la Société Entomologique de Belgique* 61: 113–122.

Santschi, F. 1924. Descriptions de nouveaux formicides africains et notes diverses. 2. *Revue Zoologique Africaine* 12: 195–224.

Santschi, F. 1925. Révision des *Myrmicaria* d'Afrique. *Annales de la Société Entomologique de Belgique* 64 (1924): 133–176.

Santschi, F. 1929. Etude sur les *Cataglyphis*. *Revue Suisse de Zoologie* 36: 25–70.

Santschi, F. 1930. Formicides de l'Angola. Résultats de la Mission scientifique suisse en Angola (1928-1929). *Revue Suisse de Zoologie* 37: 53–81.

Santschi, F. 1936. Étude sur les fourmis du genre *Monomorium* Mayr. *Bulletin de la Société des Sciences Naturelles du Maroc* 16: 32–64.

Santschi, F. 1939. Contribution au sous-genre *Alaopone*. *Revue Suisse de Zoologie* 46: 143–154.

Saux, C., Fisher, B.L., and Spicer, G.S. 2004. Dracula ant phylogeny as inferred by nuclear 28S rDNA sequences and implications for ant systematics. *Molecular Phylogenetics and Evolution* 33: 457–468.

Schmidt, C.A. 2013. Molecular phylogenetics of ponerine ants. *Zootaxa* 3647: 201–250.

Schmidt, C.A., and Shattuck, S.O. 2014. The higher classification of the ant subfamily Ponerinae, with a review of ponerine ecology and behavior. *Zootaxa* 3817: 1–242.

Seifert, B. 2003. The ant genus *Cardiocondyla*—a taxonomic revision of the *C. elegans*, *C. bulgarica*, *C. batesii*, *C. nuda*, *C. shuckardi*, *C. stambuloffii*, *C. wroughtonii*, *C. emeryi*, and *C. minutior* species groups. *Annalen des Naturhistorischen Museums in Wien* 104 B: 203–338.

Sharaf, M.R., and Aldawood, A.S. 2013. First occurrence of the *Monomorium hildebrandti* group in the Arabian Peninsula, with description of a new species *M. kondratieffi* sp. n. *Proceedings of the Entomological Society of Washington* 115: 75–84.

Shattuck, S.O. 1991. Revision of the dolichoderine ant genus *Axinidris*. *Systematic Entomology* 16: 105–120.

Shattuck, S.O. 1992. Generic revision of the ant subfamily Dolichoderinae. *Sociobiogy* 21: 1–181.

Snelling, R.R. 1979. *Aphomomyrmex* and a related new genus of arboreal African ants. *Contributions in Science* 316: 3–8.

Snelling, R.R. 1992. Two unusual new myrmicine ants from Cameroon. *Psyche* 99: 95–101.

Snelling, R.R. 2007. A review of the arboreal Afrotropical ant genus *Axinidris*. *Memoirs of the American Entomological Institute* 80: 551–579.

Snelling, R.R., and Longino, J.T. 1992. Revisionary notes on the fungus-growing ants of the genus *Cyphomyrmex*, *rimosus*-group. Pp. 479–494. In Quintero, D., and Aiello, A., editors. *Insects of Panama and Mesoamerica: Selected studies*. 692 pp.

Taylor, R.W. 1965. A monographic revision of the rare tropicopolitan ant genus *Probolomyrmex* Mayr. *Transactions of the Royal Entomological Society of London* 117: 345–365.

Taylor, R.W. 1967. A monographic revision of the ant genus *Ponera* Latreille. *Pacific Insects Monograph* 13: 1–112.

Taylor, R.W. 1990. New Asian ants of the tribe Basicerotini, with an on-line computer interactive key to the twenty-six known Indo-Australian species. *Invertebrate Taxonomy* 4: 397–425.

Terron, G. 1974. Découverte au Cameroun de deux espèces nouvelles du genre *Prionopelta* Mayr. *Annales de la Faculté des Sciences du Cameroun* 17: 105–119.

Wang, M. 2003. A monographic revision of the ant genus *Pristomyrmex*. *Bulletin of the Museum of Comparative Zoology* 157: 383–542.

Ward, P.S. 1990. The ant subfamily Pseudomyrmecinae: Generic revision and relationship to other formicids. *Systematic Entomology* 15: 449–489.

Ward, P.S. 2006. The ant genus *Tetraponera* in the Afrotropical region: Synopsis of species groups and revision of the *T. ambigua*-group. *Myrmecologische Nachrichten* 8: 119–130.

Ward, P.S. 2007. Phylogeny, classification, and species-level taxonomy of ants. *Zootaxa* 1668: 549–563.

Ward, P.S. 2009. The ant genus *Tetraponera* in the Afrotropical region: The *T. grandidieri* group. *Journal of Hymenoptera Research* 18: 285–304.

Ward, P.S., Brady, S.G., Fisher, B.L., and Schultz, T.R. 2010. Phylogeny and biogeography of Dolichoderinae ants: Effects of data partitioning and relict taxa on historical inference. *Systematic Biology* 59: 342–362.

Ward, P.S., Brady, S.G., Fisher, B.L., and Schultz, T.R. 2015. The evolution of myrmicine ants: Phylogeny and biogeography of a hyperdiverse ant clade. *Systematic Entomology* 40: 61–81.

Ward, P.S., and Downie, D.A. 2005. The ant subfamily Pseudomyrmecinae: Phylogeny and evolution of big-eyed arboreal ants. *Systematic Entomology* 30: 310–335.

Weber, N.A. 1950. The African species of the genus *Oligomyrmex* Mayr. *American Museum Novitates* 1442: 1–19.

Weber, N.A. 1952. Studies on African Myrmicinae, I. *American Museum Novitates* 1548: 1–32.

Wheeler, W.M. 1911. A list of the type species of the genera and subgenera of Formicidae. *Annals of the New York Academy of Sciences* 21: 157–175.

Wheeler, W.M. 1922. The ants of the Belgian Congo. *Bulletin of the American Museum of Natural History* 45: 1–1139.

Wild, A.L. 2007. Taxonomic revision of the ant genus *Linepithema*. *University of California Publications in Entomology* 126: 1–151.

Yoshimura, M., and Fisher, B.L. 2012. A revision of the Malagasy endemic genus *Adetomyrma*. *Zootaxa* 3341: 1–31.

Yoshimura, M., and Fisher, B.L. 2014. A revision of the ant genus *Mystrium* in the Malagasy region with description of six new species and remarks on *Amblyopone* and *Stigmatomma*. *ZooKeys* 394: 1–99.

INDEX

Names in the genus-group that are currently valid are in *lowercase italic*. Names in the genus-group that are currently junior synonyms, or that are misidentifications for the Afrotropical and Malagasy regions, are in lowercase roman. In **bold** is the first page of the main taxon account; (key) indicates the page on which the taxon is identified in the key; (fig.) indicates the page for head and profile illustration and map of each genus; and "pl#" indicates the plate number for color images of each genus.